Microstructure and Properties of Catalysts

MATERIALS RESEARCH SOCIETY SYMPOSIUM PROCEEDINGS   VOLUME 111

Microstructure and Properties of Catalysts

Symposium held November 30-December 3, 1987, Boston, Massachusetts, U.S.A.

EDITORS:

M. M. J. Treacy
Exxon Research and Engineering Company, Annandale, New Jersey, U.S.A.

J.M. Thomas
The Royal Institution, London, United Kingdom

J. M. White
The Univerity of Texas at Austin, Austin, Texas, U.S.A.

MRS MATERIALS RESEARCH SOCIETY
Pittsburgh, Pennsylvania

ENGI
.085-305-55

This work was supported by the Air Force Office of Scientific Research, Air Force Systems Command, USAF, under Grant Number AFOSR 87-0345-F.

Published by:

Materials Research Society
9800 McKnight Road, Suite 327
Pittsburgh, Pennsylvania 15237
telephone (412) 367-3003

Library of Congress Cataloging in Publication Data

Microstructure and properties of catalysts : symposium held November 30-December 3, 1987, Boston Massachusetts, U.S.A. / editors, M.M.J. Treacy, J.M. Thomas, J.M. White.

    p.   cm. — (Materials Research Society symposium proceedings, ISSN 0272-9172 ; v. 111)
    Bibliography: p.
    Includes indexes.
    ISBN 0-931837-79-0
    1. Catalysts—Congresses.   I. Treacy, M.M.J. (Michael M.J.)   II. Thomas, J.M. (John Meurig)   III. White, John M., 1938-   .   IV. Series: Materials Research Society symposia proceedings : v. 111.
QD505.M53   1988   88-10026
541.3'95—dc19   CIP
ISBN 0-931837-67-7

Manufactured in the United States of America

Manufactured by Publishers Choice Book Mfg. Co.
Mars, Pennsylvania 16046

Contents

QD 505
M53
1988
ENGI

*Invited Paper

v

*Invited Paper

*Invited Paper

*Invited Paper

Preface

This proceedings contains most of the papers which were presented at the symposium on "Microstructure and Properties of Catalysts" held in Boston, Massachusetts, November 30 – December 3, 1987, and sponsored by the Materials Research Society. The symposium provided a forum for materials scientists, synthesis chemists, physical chemists, chemical engineers, physicists and theorists, to compare and discuss the latest results which establish clear relationships between structure and catalytic properties. We have resisted the temptation to group papers under conventional subject headings, since, by the very nature of the symposium, papers are interdisciplinary. Rather than impose an editorial bias, we have kept the papers in the same order they were presented at the conference. Thus, rather than by sections, these proceedings are organized by mood; beginning with the minutiae of small particle characterization, passing on to zeolites, oxides and pillared clays, and aspects of their synthesis, and then moving on towards the expansive details of catalytic processes. Reading the contributions in these areas will quickly reveal that these, too, are inter–related. Subjects such as high resolution electron microscopy, which permeate the field of materials science, are to be found sprinkled more or less uniformly throughout this volume.

It has been a professional pleasure to edit these proceedings. We would like to acknowledge all our contributors for the quality of their work and for the care and pride taken in preparing their camera–ready manuscripts; and also the staff of the Materials Research Society for their untiring help with organizing the symposium and assembling this proceedings. Finally, we would like to extend our thanks to Margie Hillpot for her secretarial services both before and after the conference.

M. M. J. Treacy
J. M. Thomas
J. M. White

MATERIALS RESEARCH SOCIETY SYMPOSIUM PROCEEDINGS

ISSN 0272 - 9172

MATERIALS RESEARCH SOCIETY SYMPOSIUM PROCEEDINGS

MATERIALS RESEARCH SOCIETY CONFERENCE PROCEEDINGS

Tungsten and Other Refractory Metals for VLSI Applications, R. S. Blewer, 1986; ISSN: 0886-7860; ISBN: 0-931837-32-4

Tungsten and Other Refractory Metals for VLSI Applications II, E.K. Broadbent, 1987; ISSN: 0886-7860; ISBN: 0-931837-66-9

Ternary and Multinary Compounds, S. Deb, A. Zunger, 1987; ISBN:0-931837-57-x

Tungsten and Other Refractory Metals for VLSI Applications III, Victor A. Wells, 1988, ISSN 0886-7860; ISBN 0-931837-84-7

Atomic and Molecular Processing of Electronic and Ceramic Materials: Preparation, Characterization and Properties, Ilhan A. Aksay, Gary L. McVay, Thomas G. Stoebe, 1988, ISBN 0-931837-85-5

THE ANALYSIS OF VAPOR-DEPOSITED Pd PARTICLES
SUPPORTED ON THIN OXIDES

Helmut Poppa,* F. Rumpf,** R. D. Moorhead,*** and C. Henry**†
*IBM Research Division, Almaden Research Center, 650 Harry Road, San Jose, California 95120-6099
**Department of Chemical Engineering, Stanford University, Stanford, California 94350
***NASA Ames Research Center, Moffett Field, California 94035

ABSTRACT

Problems of analyzing by TEM and TPD (Temperature Programmed Desorption) the structure and micromorphology of small Pd particles vapor deposited in UHV onto clean oxide supports are discussed. Particle changes induced during extended exposures to high intensity electron irradiation of a number of electron transparent support materials such as Al_2O_3, SiO_2, TiO_2, MgO, and mica are examined. Different damage mechanisms are evaluated and experimental means of reducing the damage are explored. The possibility of extracting particle morphology information from a detailed analysis of CO thermal desorption spectra is also investigated. Evidence suggests that it may be possible to obtain micromorphology information down to very small particle sizes from TPD measurements if the effects of intrinsic particle morphology can be separated from the influence of diffusion of support species.

†Permanent address: CRMC2-CNRS, Campus Luminy, 13288 Marseille, Cedex 9, France.

INTRODUCTION

The surface and electronic properties of small metal particles, prepared by controlled nucleation and growth from the vapor phase onto planar (thin film) oxide supports under UHV conditions, can be studied by surface analytical techniques such as AES, XPS, SIMS, TPD, and IRAS (Infrared Absorption Spectroscopy). Their "bulk" properties, as for instance, their crystallographic microstructre, size, and shape are accessible through TEM/TED measurements, and integrated experimental approaches [1,2] make it possible, in principle, to study these properties on one and the same particulate sample. An integrated experimental approach permits the correlation of particle properties from different types of measurements which is usually essential for the meaningful interpretation of size and support effects.

In this report, only TEM and TPD results of evaporated particles supported on several catalysis relevant oxide substrates such as a(amorphous) $- Al_2O_3$, a-SiO_2, a-TiO_2, $(1\bar{1}02)$-Al_2O_3, (111)-MgO, and muscovite mica will be discussed. It will be demonstrated, in particular, that the potential for introducing specimen artifacts during high-resolution TEM under presently used standard imaging conditions is often overlooked, and we will also discuss the advantages and problems associated with the interpretation of particle morphology data obtained from the temperature programmed desorption of CO.

RESULTS AND DISCUSSION

(1) High Resolution TEM

The Pd particles to be analyzed by TEM were evaporated in UHV onto thin film supports. Extremely thin amorphous supports were prepared by ex-situ reactive sputter deposition of alumina, titania, and silica on ultrathin carbon support films on TEM specimen grids and were cleaned and conditioned in-situ by an rf oxygen discharge before Pd deposition. Alternatively, UHV-cleaved mica, epitaxial deposits of MgO on mica, and pre-thinned and oxygen discharge cleaned sapphire discs were used as single-crystal supports. The overall goal of these support preparation procedures being the production of very thin, stoichiometric, and hopefully stable supports the TEM background structures of which would cause minimal interference during the high resolution imaging of the particles [3]. The particulate TEM specimens were removed from the UHV preparation system and transferred to a 200 kV instrument for routine high-resolution imaging. In order to assess the effects of different support materials upon the imaging results, each Pd/oxide specimen was systematically exposed to differing electron beam imaging intensities for increasing times of exposure, and the same specimen area was monitored. The following beam exposure conditions were used:

(a)   Low-beam current density exposure at $i = 0.1$ A/cm^2 at 3 min intervals for up to 15 min. (This current density corresponds approximately to an image brightness at 100,000× magnification necessary for a 2 s photographic exposure on Kodak Electron Image film);

(b)   High-beam current density exposure at $i = 1$ A/cm^2 again for up to 15 min total exposure (corresponding approximately to an image brightness at 300,000× magnification necessitating a photographic exposure of 2 s). This i is 5 to 10 times lower than the current densities often used routinely in today's medium-voltage/very-high resolution TEMs.

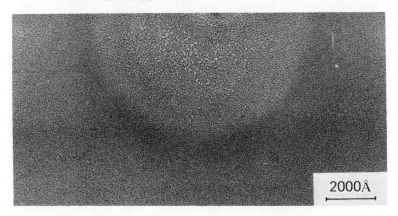

2000Å

Figure 1. Low magnification of e-beam exposed area of a Pd/a-SiO$_2$ sample showing marked beam effect.

Figure 1 is a low magnification micrograph of the most beam sensitive sample studied: Pd/SiO$_2$. After 15 min exposure at high i, it is very obvious that the specimen area exposed to the beam has been severely affected. The details of this change are similar in all oxides examined; only the severity of the beam induced change is different. In a-Al$_2$O$_3$ for instance, together with TiO$_2$ one of the more stable substrate materials investigated here, the changes caused by high i exposure are shown in Fig. 2 in which the sample specimen area is shown after 3 min increments of exposure. It is obvious that in spite of the amorphous phase contrast background structure from the support Pd particles of varying size can be clearly distinguished, although little more than a general particle outline is seen. The fact that Pd(111) lattice planes are clearly resolved in some of the Pd areas attests to the intrinsically good resolution in these micrographs. The most striking result of increasing beam damage is, however, easy to recognize: The Pd deposits spread out on the support surface, and in doing

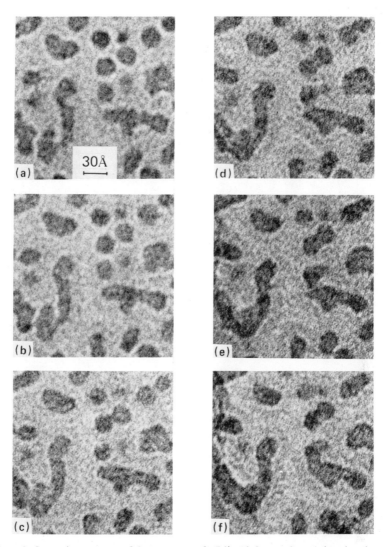

Figure 2. Increasing exposure of the same area of a Pd/a-Al$_2$O$_3$ sample to an imaging electron beam of approximately 1 A/cm^2 current density. The time between successive micrographs is 3 min. Particles are spreading by beam-induced interaction with the support.

so, coalesce with neighboring Pd particles. The high i exposure series of Fig. 3 shows, in principle, the same beam influence except the support material is TiO$_2$ in this case. However, when the current density is reduced to 0.1 A/cm^2, as in Fig. 4, no corresponding major particle change can be detected during long beam exposures. Of the single crystal support materials tested, mica in Fig. 5 shows particularly severe radiation damage effects concerning Pd particle shapes (extensive spreading) although the bulk of the mica support was not yet amorphized, as seen from persistence of lattice plane images even after long exposures. This hints at the

4

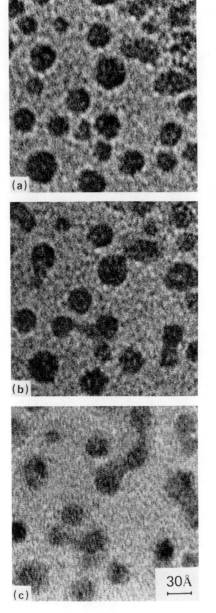

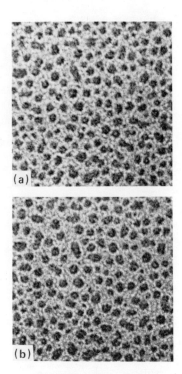

Figure 4. Lower magnification exposure of a Pd/a-TiO$_2$ sample area (a) to a low current density (0.1 A/cm^2) electron beam for 9 min with negligible beam effect.

Figure 3. Increasing exposure of a Pd/a-TiO$_2$ sample to an imaging electron beam of about 1 A/cm^2 current density. The exposure time between (a) and (b) is 6 min and the time between (b) and (c) is 14 min. Particle spreading/coalescence is seen.

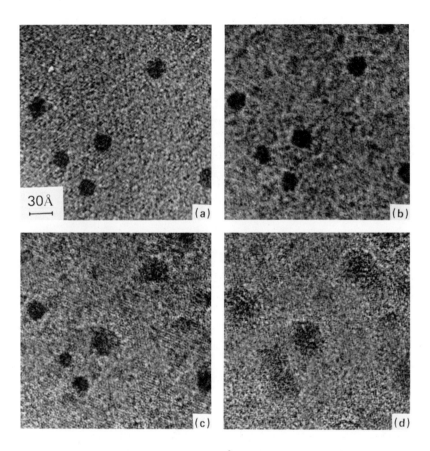

Figure 5. High-current density exposure (1 A/cm^2) exposure of a Pd/mica sample; the exposure time between successive micrographs is 6, 9, 12 min, respectively. Severe particle spreading and interaction with the support is obvious.

nature of the particle/support interaction as a surface effect. However, even in this case of an obviously strong Pd/support interactions, low i exposures would not lead to obvious particle shape changes. Sapphire was expected to be one of the most stable support materials. Instead, Fig. 6 indicates also extensive particle shape changes during high i imaging which, in this case, was accompanied by severely interfering cavity formation in the sapphire bulk (see Fig. 6b).

What is the origin of these electron beam induced changes of Pd particles supported on the various oxides? Changes that are negligible at imaging current densities of about 0.1 A/cm^2 and that are extensive at 1 A/cm^2 for exposure times of up to 15 min for the 200 kV electrons used here must be due to beam-induced changes of the interfacial/surface free energies involved: e_s = support/residual gas surface energy, e_{Pd} = Pd/residual gas surface energy, and e_i = support/Pd interfacial energy. The experimentally observed spreading (wetting) of the Pd particles is a consequence of either a decrease of e_i and/or e_{Pd}, or an

6

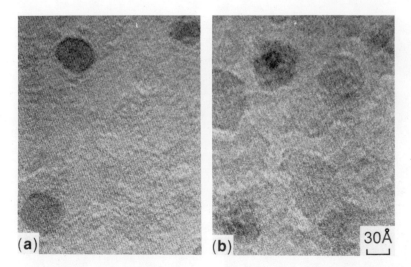

Figure 6. High-current density (1 A/cm^2) exposure of a Pd/($1\bar{1}02$)-sapphire sample (a) for 12 min (b). In addition to severe Pd/support interactions, the single crystal support film is also seen to be heavily damaged in (b).

increase of e_s, or the interrelated change of all these surface energies causing an overall decrease of the wetting angle α according to

$$e_{Pd} \cos \alpha + e_i = e_s. \tag{1}$$

Because of the standard TEM vacuum environment prevailing in the present studies, the possibility of (i) severe oxidation of the Pd surfaces by beam-dissociated residual water vapor, and (ii) beam-induced compositional changes of the compound support interfaces must certainly be considered. In principle [4,5], primary compositional changes can be produced by electron/electron interactions (bond breaking "radiolysis," which decreases with increasing beam energy E_o), or by electron/nucleus interactions (knock-on damage, which increases with E_o). Since the secondary compositional changes are the result of the diffusion of primary damage products, the mechanism for which can be of thermal or nonthermal (i.e., knock-on driven) nature, the effect of sample temperature during irradiation should provide important clues as to the damage mechanism during TEM. The TEM exposure studies of Pd/a-Al$_2$O$_3$ were, therefore, duplicated at liquid nitrogen temperatures [6]. The results were similar to the room temperature (RT) exposures indicating that a nonthermal diffusion mechanism must be dominating. Low-T TEM would also be expected to increase the condensation of residual H$_2$O causing increased Pd oxidation, i.e., spreading of particles. Additional evidence against a residual gas-induced beam damage mechanism was obtained by e-beam exposure experiments in the Stanford UHV in-situ TEM where residual gas pressures of about 6×10^{-10} Torr [7] exist at the site of the specimen. The rate of beam-induced specimen change was again not affected substantially (although it must be noted that the TEM specimen was not heated intentionally before or during TEM to desorb adsorbates possibly stemming from the specimen transfer in air). It must be concluded, therefore, that beam damage of reactive metals on oxide supports under high-resolution TEM conditions at 100-200 kV is inescapable, is a serious problem at beam current densities appreciably larger than about 0.1 A/cm^2, and is expected to become an even more serious problem with increasing primary beam energies.

These results of high-resolution, high-beam current density TEM studies pose the general question of what kind of useful microstructural information can be obtained with

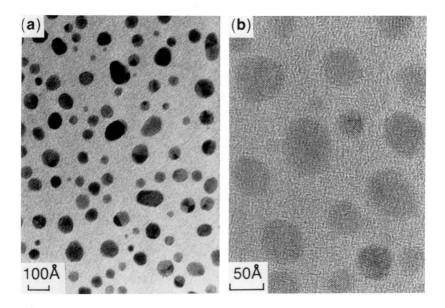

Figure 7. Pd particles on (1$\bar{1}$02)-sapphire analyzed by lower-voltage (75 kV) high-contrast TEM (a) compared to high-voltage (1000 kV [29]) high-resolution low-contrast TEM (b).

various TEM techniques and with different instruments. The overall microstructure, crystallinity, and projected shape of supported metal particles can be measured easily with medium resolution, lower beam energy instruments (75-120 kV) down to about 5 nm particle size with relatively high image contrast (Fig. 7a). It is also possible to achieve a line resolution of approximately 0.2 nm with these instruments so that imaging some support and particle lattice planes becomes feasible, which can add information on the particle crystallinity and orientation with respect to the substrate. Three-dimensional shapes can also be studied to a limited degree in particles larger than about 5 nm by weak-beam dark-field techniques [8], and at the present time, efforts of profile imaging particles deposited on support edges parallel to the imaging electron beam [9] are underway in several laboratories. However, the often most needed information about particulate deposits — the size distribution and the average distance of the particles — can be rather easily obtained and is the essential information in connection with many important applications. For the shape analysis of particles between 1 and 5 nm, higher point resolutions than are usually routinely available in commercial TEMs are a necessity in order to resolve edge and corner structures in this size region. Although better lattice images are a by-product of such advanced medium-voltage instruments (200-500 kV), these usually do not add vital information about three-dimensional particle morphology; on the contrary, the overall particle contrast is sharply reduced (Fig. 7b). The most direct way of detecting details of a supported particles third dimension is, of course, "profile TEM" [9] but the preparational possibilities for such metal deposits are still quite limited and beam-induced particle changes must be guarded against just as much as in projectional "planar-view TEM." To what extent AFM (Atomic Force Microscopy) with its enormous z-sensitivity will become a viable tool in this respect, only the future will tell.

8

(2) Thermal Desorption Spectroscopy (TDS)

Considering the TEM results, it seems imperative to search for other, less intrusive methods for investigating the micromorphology of supported metal particles such as TDS, PAX (Photoelectron Spectroscopy of Adsorbed Xenon [10]), and IRAS (Infrared Adsorption Spectroscopy [11]). TDS or TPD (Temperature Programmed Desorption) is one of the least demanding surface analysis techniques in terms of additional instrumentation, but it is not trivial in its application to metal particles supported on planar insulating substrates. This is mainly because of the experimental difficulties encountered when attempting to uniformly and controllably heat such a support system [12,13]. Establishing a reliable temperature scale applicable to the actual particle surfaces can be a demanding project [14]. Since the desorption energy E_{des} of such probe gases as CO depend on the orientation of a single-crystal metal surface and on the coordination of special surface adsorption sites, such as corner and edge sites, it should, in principle, be possible to distinguish by TDS between particle shapes that are defined by differing crystallographic facets. As an example, Fig. 8 shows several

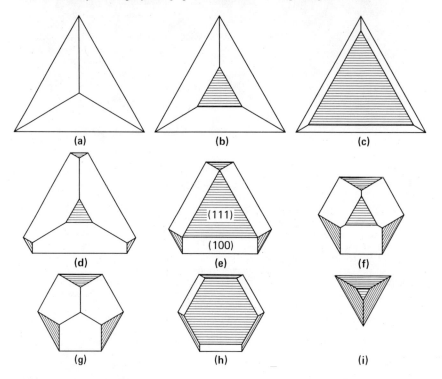

Figure 8. Varying schematic shapes of a {111}-oriented f.c.c. particle as a function of the proportion of {100} and {111} facets exposed.

morphology modifications of a {111}-oriented particle exposing {111} and {100} low surface energy facets, although it is an open question at present whether and in what size range such facetted morphologies might actually exist [15-17]. TPD spectra of small supported metal particles (Pd, Ni, Ru, Pt) to date have mainly been evaluated in terms of overall TPD peak shape and amplitude changes as a function of particle size and thermal processing [18-22].

Recently [23], a more detailed analysis of the spectra of small Pd particles grown on the same substrate but under different conditions has lead to the tentative interpretation of the CO peaks from about 3 nm particles in terms of desorption from individual crystal facets. The corresponding desorption energies from {111} and {100} facets correlated reasonably well with the respective values from extended single-crystal surfaces. However, more systematic and detailed studies under improved experimental conditions and as a function of particle size are needed to establish in what size range and following what thermal processing conditions (see later) such a simple picture can apply. In order to unravel subtle changes in the shape of desorption spectra, it would be helpful to deconvolute the observed TPD peaks. This is particularly difficult for the case of CO/Pd because of the existence of various adsorption states on a single facet. (On Pd{111} there are, for instance, the β_1 and β_2 states of doubly and triply coordinated CO [24]). We have, therefore, empirically deconvoluted the saturated "equilibrium" spectrum of Fig. 9 which was obtained from about 3 nm Pd particles grown on hydroxylated mica [23] by subtracting a normalized {111} desorption spectrum. The latter was derived from the spectrum of predominantly {111} surfaces exposing Pd particles grown

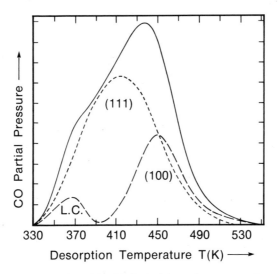

Figure 9. Empirical deconvolution of the CO thermal desorption spectrum from a Pd deposit (about 3 nm particle size) on hydroxylated mica into contributions from {111}, {100} facets and low-coordination (L.C.) adsorption sites.

on dehydroxylated mica [25]. As a result of this subtraction, a low coordination {L.C.} peak and a {100} peak − so designated because of its good fit in peak temperature with desorption from bulk {100} Pd − are leftover, indicating that these particles exhibit an appreciable proportion of {100} facets. This result is corroborated by the CO coverage dependence of this {100} peak in contrast to the different coverage dependence of the {111} −β_2 peak and by combined AES/TEM results [26] suggesting more three-dimensional particle shapes than in the case of the flat epitaxial {111} particles. The L.C. peak is believed to be due to corner and edge adsorption sites because it always increases strongly with particle size in all of our studies and in the calculations of such sites as a function of size [27].

The stability of vapor-deposited metal particles during thermal cycling is yet another complex problem involving effects of particle size, shape, and metal/support material combinations. Although the Pd particles are intentionally grown at temperatures higher than the maximum TPD temperature experienced later, desorption peak changes are always observed upon successive TPD cycles. The support materials tested include mica, α-Al$_3$O$_3$, a-Al$_2$O$_3$, and vapor grown MgO. Figure 10 demonstrates the difference of the desorption spectra for the first desorption of CO from Pd particles grown at 600 K on mica and on MgO{111}. Although these desorption "peaks" are quite similar in overall shape and position,

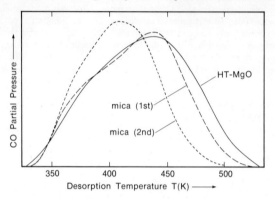

Figure 10. Two successive thermal desorptions of CO from a particulate Pd deposit on mica and one desorption trace from similar size (3 nm) Pd particles on {111}-MgO (grown epitaxially at elevated temperature − HT − on mica).

their behavior during subsequent successive desorption is very different. While on (dehydroxylated) mica, the second desorption (also shown in Fig. 10) practically represents the final shape of the spectrum, which under these experimental conditions and for particles of this size and shape, is maintained for a large number of additional desorption, this is not the case of Pd/HT-MgO. Here a continuous change in peak shape and position is found, which strongly depends on particle size [28]. Whether this behavior is due to morphological particle changes or whether support interactions also play an important role is not clear yet. However, a definite deactivation of the particle surface − strongly size dependent again − was found for Pd grown on room temperature deposited MgO [28]. This has been attributed to the deactivating effect of support species that diffuse onto (or underneath) the Pd particle surfaces and adds some support to the view that in all these particle/support systems the diffusion of support species has to be considered to various extents.

CONCLUSIONS

The detailed analysis of the structure and micromorphology of small reactive metal particles (Pd) supported on planar refractive oxide supports is fraught with difficulties when conducting high-resolution TEM and TPD measurements. For HR-TEM studies, it is imperative to consider the deleterious effects of high current density imaging electrons. Both dissociative and knock-on radiation damage are caused and the consequences of this damage are increasingly apparent for dosages (of 200 kV electrons) larger than about 300 A s cm^{-2}. (This corresponds to an exposure time of 5 min at an image intensity of 100,000× magnification which would necessitate a photographic exposure time of about 2 s for standard Electron Image plate material). The beam-induced damage increases with beam energy and cannot be reduced substantially by either microscopy at low temperature (LN)$_2$ or by TEM in an UHV environment because knock-on promoted diffusion of support species seems largely responsible for the observed spreading of the particles.

Details of TPD data interpretation represent the main difficulty at the present state of the method when using only weakly interacting probe gases for adsorption. Evidence exists that desorption from different particle facets with nearly bulk single-crystal desorption energies (for CO) can be used to characterize particle morphologies down to sizes of about 2 nm. However, some particle restructuring during thermal cycling (in the presence or absence of CO) is observed. The restructuring is emphasized for very small particles, in particular, as is the influence of specific support materials (mica or HT-, RT-MgO). More work is needed to distinguish between genuine restructuring of clean particles and the effects of diffusing support species. Systematic studies with even smaller vapor-deposited metal particles under varying deposition conditions and on differently oriented substrates should be fruitful.

ACKNOWLEDGMENTS

This work was partially supported through NASA Grant NCC2-394. Thanks for collaboration are due to Dr. Klaus Heinemann of Eloret Institute and to Dr. J. Mardingly of Lockheed Research Laboratories in Sunnyvale, California.

LITERATURE

1. H. Poppa, Vacuum 34,1081 (1984).
2. H. Poppa, Ultramicroscopy 11, 105 (1983).
3. H. Poppa, R.D. Moorhead, K. Heinemann, Thin Solid Films 128, 251 (1985).
4. L.W. Hobbs, Ultramicroscopy 23, 339 (1987).
5. R.F. Egerton, Ultramicroscopy 23, 305 (1987).
6. Low-T TEM provided by Dr. J. Mardingly, Lockheed Research Laboratories, Sunnyvale, California.
7. K. Heinemann, H. Poppa, J. Vac. Sci. Techn. A4, 127 (1986).
8. M.J. Yacaman, K. Heinemann, H. Poppa, CRC Critical Rev. Sol. State Mat. 10, 243 (1983).
9. A.K. Datye, MRS National Symposium, Boston, December 1987; L.D. Schmidt, MRS National Symposium, Boston, December 1987.
10. J. Hulse, J. Kuppers, K. Wandelt, G. Ertl, Surf. Sci. 6, 453 (1980).
11. Y. Takasu, R. Unwin, B. Tesche, A.M. Bradshaw, M. Grunze, Surf. Sci. 77, 219 (1978).
12. M. Thomas, J.T. Dickinson, H. Poppa, G.M. Pound, J. Vac. Sci. Techn. 15, 568 (1978).
13. D.L. Doering, H. Poppa, J.T. Dickinson, J. Vac. Sci. Techn. 17, 198 (1980).
14. R. Koch, R. Browning, to be published.
15. M. Drechsler, Surface Mobilities on Solid Materials, edited by V. Thien Binh, NATO Advanced Study Institute (Plenum Press, New York/London 1981), p. 405.
16. J.C. Heyraud, J.J. Metois, J. Cryst. Growth 50, 571 (1980).
17. T. Wang, C. Lee, L.D. Schmidt, Surf. Sci. 163, 181 (1985).
18. C. Park, W.G. Durrer, H. Poppa, J.T. Dickinson, J. Catalysis 95, 361 (1985).
19. D.L. Doering, H. Poppa, J.T. Dickinson, J. Catalysis 73, 104 (1982).
20. W.G. Durrer, H. Poppa, J.T. Dickinson, C. Park, J. Vac. Sci. Techn. A 3, 1545 (1985).
21. E. Gillet, S. Channakhone, V. Matolin, J. Catal. 97, 437 (1986).
22. E.I. Altman, R.J. Gorte, Surf. Sci. 172, 71 (1986).
23. R. Koch, H. Poppa, JVST A (5), 1845 (1987).
24. M.P. Kiskinova, G.M. Bliznakov, Surf. Sci. 123, 61 (1982).
25. R. Koch, H. Poppa, Thin Solid Films 151, 365 (1987).
26. R. Koch, to be published.
27. R. Van Hardeveld, F. Hartog, Surf. Sci. 15, 189 (1969).
28. C.R. Henry, H. Poppa, Proc. AVS National Symposium, Anaheim/California, Novevember 1987, to be published.
29. Micrograph acquired at the National Facility for Electron Microscopy, Univeristy of California, Berkeley, California.

COLLOIDAL ROUTES TO CATALYSTS: Pt-Au

PAUL A. SERMON[*], K. KERYOU[*], J.M. THOMAS[**] AND G.R. MILLWARD[***]
*Department of Chemistry, Brunel University, Uxbridge, Middlesex UB8 3PH, UK
**Davy Faraday Research Laboratories, The Royal Institution of Great Britain, 21 Albemarle Street, London W1X 4BS, UK
***Department of Physical Chemistry, University of Cambridge, Lensfield Road, Cambridge CB2 1EP, UK

ABSTRACT

A colloidal route to particles of $Pt_{100-x}Au_x$ of high dispersion is reported; their activities and selectivities in hydrocarbon conversion are consistent with alloy catalysts prepared by other routes.

INTRODUCTION

Since the time of Faraday [1] and Graham [2] it has been known [3] that colloidal particles of group VIII (and IB metals) could be prepared by reduction of their aquated metal ions in low concentration. Elegant work by Turkevich and others [4] has optimised reduction conditions for the preparation of metal particles of a narrow size distribution, i.e. mono-dispersed sols. The ultimate size of such particles is decided by a balance between an increase in the free energy (arising from entropy loss associated with conversion of free ions into sol particles), a decrease in free energy arising from reduction of the metal salt and a free energy term arising from changes in the interfacial area of the sol particles presented to the solution [3,5]. Monometallic particles of group VIII (or IB) metals so produced are quite small and sometimes show unusual crystallographic packing when examined [6], e.g. five-fold symmetry.

It was appreciated many years ago [7] that such small metal particles would show intriguing catalytic properties. However, early analyses in aqueous dispersion [8] were probably diffusion-limited and the reactions investigated may have caused sol destabilisation. There is thus advantage in adsorbing the colloidal metal particles onto a support [9] before probing their adsorptive and catalytic properties.

Recent years have seen a movement away from monometallic to multimetallic heterogeneous catalysts; alloys are therefore of increasing catalytic importance [10], and are traditionally prepared by methods of co-impregnation, co-adsorption or co-precipitation. One early study has considered the potential of colloidal routes to bimetallic metal particles of value in catalysis [11]; here we have explored [12] sol routes to the preparation of bimetallic particles as they could have advantages over the traditional approaches. Such production of uniform bimetallic catalytic particles of colloidal size has not been otherwise reported. We chose $Pt_{100-x}Au_x$ ($0 \leqslant x \leqslant 100$) because (i) of the wealth of catalytic knowledge on the respective monometallic sols [4], (ii) as x increases in bulk alloys of Pt and Au, the activity of hydrocarbon isomerization and hydrogenolysis decreases but activity for dehydrogenation increases [13], and (iii) of the limited bulk miscibility of these metals.

EXPERIMENTAL

Materials

Hydrated H_2PtCl_6 and $HAuCl_4$ were obtained from Johnson Matthey and were of Specpure purity.

Sol Preparation

$Pt_{100-x}Au_x$ ($0 \leqslant x \leqslant 100$) bimetallic sols were prepared by reducing mixed aqueous solutions to H_2PtCl_6 and $HAuCl_4$ (<30 mg.dm^{-3} of each) with a 1% trisodium citrate solution at 373K in a manner similar to that used for monometallic sols [4,12,14] and were purified by ion-exchange. These were adsorbed onto graphite (Fluka; 99.9% purity).

Catalyst Testing

The activity and selectivity of graphite-supported sol particles in cyclohexene conversion and n-butane hydrogenation- dehydrogenation-isomerization were determined in flow micro-reactors as described previously [15].

Characterisation

The average particle sizes of the Pt_xAu_{100-x} sol particles in aqueous dispersion and after supporting on graphite were found by electron microscopy. High resolution electron microscopy was also used to examine the internal and surface structure of these sols; the Jeol 200CX electron microscope employed having been fitted with an improved pole-piece and stage [16] with which a point-to-point resolution of 0.196 nm was achievable.

The total composition of the alloys was determined in two ways: globally by means of atomic absorption (AA) analysis of the digested solid and locally (2 nm diameter spot) from the intensities of the L_α X-ray emission peaks in a scanning transmission electron microscope (STEM).

Micro-Auger analysis was kindly carried out by Jeol and was used to indicate surface Pt:Au ratios which could be compared with bulk values to indicate any surface enrichment.

RESULTS

Characterisation

Several noteworthy features emerged for $Pt_{100-x}Au_x$ particles. First, the individual colloidal particles possess a high degree of crystalline order (see Fig. 1), with well-defined (111) and (200) planes present in sol particles of all compositions, but with no preferential particle morphologies defined by γ_{hkl} or the presence of the graphite [17]. Second, the size distribution of the sols at each composition is narrow (2.3 ±0.9 nm, 4.8 ±1.8 nm, 12.3 ±1.4 nm, 14.3 ±4.0 nm, and 24.7 ±6.4 nm for x = 0, 10, 50, 90 and 100, respectively) and the average size increases progressively with Au content. This is unexpected since an earlier report [11] indicated that the average size increased marginally from 3.2 to 6.0 nm as x increased from 0 to 75. Average radii computed from nucleation theory [5] assuming attainment of equilibrium are much smaller than measured average radii of all compositions (see Fig. 2). Third, no significant segregation of one metal with respect to the other was noted, but then Pt and Au lattice parameters are too close to allow distinction. Nevertheless, in sols prepared with the nominal composition corresponding to x = 50 (i.e. $Pt_{50}-Au_{50}$) STEM analysis of nine individual particles of diameter 12.0 ±0.1 nm yielded an average composition corresponding to an x value of 47 ±3. Particles in Pt-rich sols were somewhat less homogeneous than Au-rich ones. AA analysis revealed total Pt:Au ratios very close to those expected (i.e. for sols with x = 10, 25 and 50, Pt:Au ratios were found to be 9.0, 2.7 and 1.0, respectively). However, micro-Auger analysis of sol particles from aqueous suspension with x = 10, 25 and 50 revealed no surface Au when first prepared, but only after sputtering.

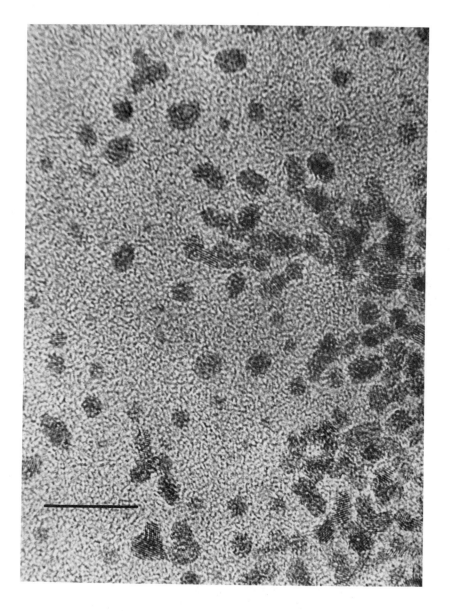

Fig. 1a Typical high resolution micrograph of colloidal metallic
particles of nominal composition $Pt_{100-x}Au_x$ with x = 0.
Scale line = 10 nm.

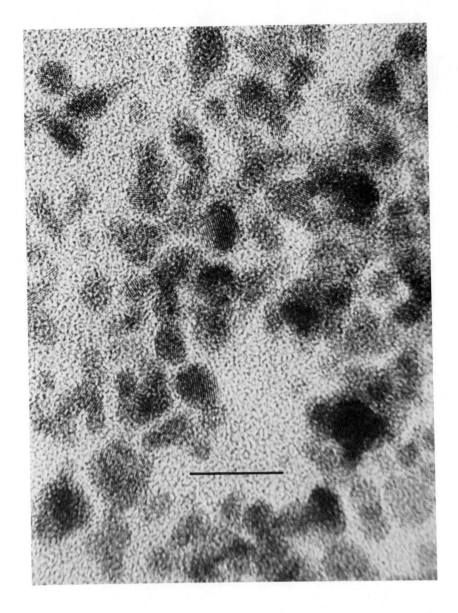

Fig. 1b Typical high resolution micrograph of colloidal metallic
particles of nominal composition $Pt_{100-x}Au_x$ with x = 10.
Scale line = 10 nm.

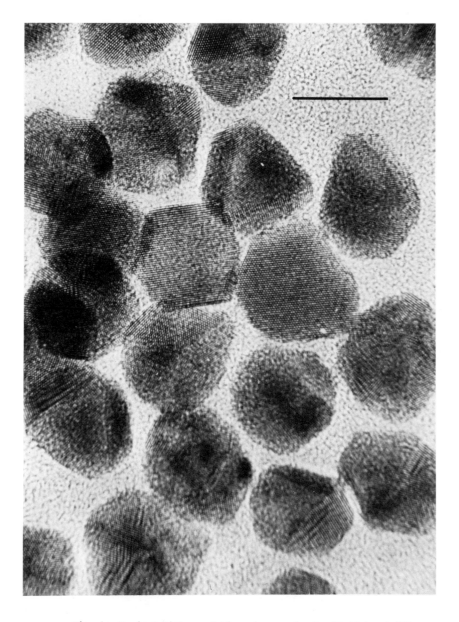

Fig. 1c Typical high resolution micrograph of colloidal metallic
particles of nominal composition $Pt_{100-x}Au_x$ with x = 50.
Scale line = 10 nm.

18

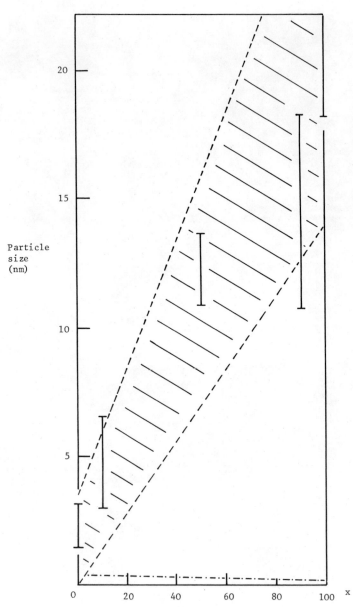

Fig. 2 Particle sizes of $Pt_{100-x}Au_x$ colloidal samples as a function of x, and compared with those predicted by nucleation theory (·—··—··—·).

Hence the sol particles first prepared appear to have Au-rich centres and Pt-rich exteriors; this egg-like structure is to be expected since $AuCl_4^-$ reduces faster at 373K than $PtCl_6^{2-}$ using tri-sodium citrate. However, Auger analysis after catalytic pretreatment did confirm particle homogeneity.

Catalytic Properties

Figure 3 shows that the isothermal activities of these supported sols towards the hydrogenolysis and isomerization of n-butane decrease steadily as their gold content (and x) increases and dehydrogenation passes through a maximum. In broad terms these catalytic results are consistent with those for bulk Pt-Au [13] and Ni-Cu [18] alloys. As expected, addition of Au disrupts local Pt-Pt coordination thereby making the structure-sensitive hydrogenolysis [19] less favourable. However, structure-insensitive dehydrogenation, which is sustained on very small ensembles, is seen to be facilitated here.

Consider n-butane conversion a little more fully; conditions used and results are given in Table 1. The total conversion of n-butane decreases as the Au content of $Pt_{100-x}Au_x$ particles on graphite increases at 633K, and the selectivity to hydrogenolysis within this total conversion decreases (as already indicated in Fig. 3). Furthermore if S_1, S_2 and S_3 are the selectivities with which CH_4, C_2H_6 amd C_3H_8 are produced within hydrogenolysis, then Table 1 shows that selectivity to CH_4 and C_3H_8 increased markedly with Au addition.

Table 1   Activities and Selectivities in n-butane hydrogenolysis*(H), isomerisation (I) and dehydrogenation (DH)

$Pt_{100-x}Au_x$ samples present as 1-2% metal on graphite	T(K)	% conver- sions	Selectivities			Hydrogenolysis Selectivity		
			H	I	DH	S_1	S_2	S_3
x = 0	603	0.48	0.13	0.83	0.05	0.59	0.95	0.51
	613	0.83	0.16	0.79	0.05	0.57	0.96	0.50
	623	1.17	0.16	0.72	0.12	0.57	0.95	0.51
	633	1.50	0.18	0.70	0.12	0.58	0.93	0.52
x = 9	613	0.53	0.13	0.65	0.23	0.59	0.86	0.56
	633	0.77	0.11	0.54	0.35	0.61	0.84	0.57
	643	0.82	0.09	0.47	0.44	0.61	0.84	0.57
	653	0.73	0.10	0.54	0.36	0.62	0.84	0.57
	663	1.05	0.06	0.36	0.58	0.62	0.83	0.57
	673	1.22	0.04	0.29	0.67	0.62	0.83	0.57
x = 25	613	0.29	0.006	0.57	0.43	0.70	0.00	1.10
	633	0.41	0.004	0.39	0.60	0.70	0.00	1.10
	653	0.60	0.004	0.29	0.73	0.73	0.00	1.09
	673	0.95	0.003	0.17	0.83	0.97	0.20	0.87

*measured at 4.2 kPa C_4H_{10}, 12.7 kPa N_2 and 84.4 kPa H_2.

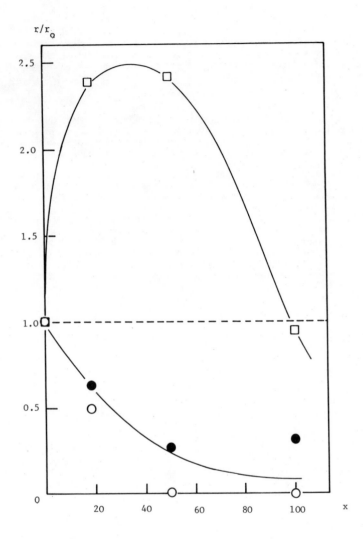

Fig. 3 Rates [5] of dehydrogenation (□), isomerization (●), and hydrogenolysis (O) of n-butane over graphite-supported sol-derived $Pt_{100-x}Au_x$ particles relative to those exhibited when x=0 (r_o).

Now the relative probabilities of central and terminal C-C bond breakage in hydrogenolysis [15] may be characteristic of either different types of butane adsorption site

$$
C_4H_{10} \quad
\begin{array}{l}
\nearrow \\
\\
\searrow
\end{array}
\quad
\begin{array}{l}
\overset{(1-F)}{C_4H_{10} \longrightarrow} \quad C_1 + C_3 \longrightarrow \\
\\
\underset{(F)}{C_4H_{10} \longrightarrow} \quad 2C_2 \longrightarrow
\end{array}
$$

or different electronic-geometric properties induced on Pt ensembles by Au ligand atoms

$$
C_4H_{10} \longrightarrow C_4H_{10}
\begin{array}{l}
\overset{(1-F)}{\nearrow} \quad C_1 + C_3 \longrightarrow \\
\\
\underset{(F)}{\searrow} \quad 2C_2 \longrightarrow
\end{array}
$$

As x increases so the probability F decreases dramatically. Bearing in mind mobility of surface atoms it may be difficult to differentiate the above effects of Au; nevertheless, both activities and selectivities measured do indicate that the bimetallic sol particles prepared have properties expected of alloy surfaces produced by other methods. This is promising.

Interestingly, the activity of the graphite-supported Pt sol in cyclohexene hydrogenation was greater than EUROPT-1, but the reverse was true in butane hydrogenolysis, despite similar Pt particle sizes. Whether this is a reflection of the different local Pt particle morphologies caused by these two preparative routes or the two supports (i.e. graphite and silica) resulting in some local preferential orientation of the Pt particles [17] is uncertain.

CONCLUSIONS

It is encouraging that these bimetallic particles of colloidal size show catalytic properties that can be related to those of the bulk alloys while at the same time offering considerable scope for fine-tuning of their specific performance.

This method of mono-dispersed alloy preparation is only limited by our ability to reduce the component aquated (or solvated) ions simultaneously to produce stable colloidal particles; it is probably applicable to many alloys of catalytic interest and where catalytic probes may quantify the extent of alloying [20]. Further work is in progress.

REFERENCES

1. M. Faraday, Philos. Trans. Roy. Soc. London 147, 145 (1857).
2. T.H. Graham, ibid 151, 183 (1861).
3. D.H. Everett, Chem. Britain 377 (1980).
4. J. Turkevich, P.C. Stevenson and J. Hillier, Discuss. Faraday Soc. 11, 55 (1951); D.N. Furlong, A. Launikonis, W.H.F. Sasse and J.V. Sanders, J. Chem. Soc. Faraday Trans. I 80, 571 (1984); J. Turkevich, R.S. Miner and L. Babenkova, J. Phys. Chem. 90, 4765 (1986).
5. E.K. Rideal, Discuss. Faraday Soc. 11, 9 (1951).
6. D.G. Duff, A.C. Curtis, P.P. Edwards, D.A. Jefferson, B.F.G. Johnson, A.I. Kirkland and D.E. Logan, Angew. Chem. 26, 678 (1987).

7. E.K. Rideal, J. Amer. Chem. Soc. 42, 748 (1920); M.A.G. Rocasolano, Compt. Rend. 1502 (1920); 234 (1921); C. Ernst, Z. Physik. Chem. 37, 448 (1901).
8. G. Senter, Proc. Roy. Soc. 74, 577 (1905); H.J.S. Sand, Proc. Roy. Soc. 74, 356 (1905).
9. J. Turkevich and G. Kim, Science 169, 873 (1970).
10. S. Naito and M. Tanimoto, J. Chem. Soc. Chem. Commun. 363 (1987).
11. R.S. Miner, S. Namba and J. Turkevich, in Proc. 7th Intl. Congr. Catalysis, (1981), Tokyo (T. Seiyama and K. Tanabe, Eds.), p.160.
12 P.A. Sermon, J.M. Thomas, K. Keryou and G.R. Millward, Angew. Chem. 26, 918 (1987).
13. R.C. Yates and G.A. Somorjai, J. Catal. 103, 208 (1987); J.K.A. Clark, I. Manninger and T. Baird, ibid 54, 230 (1978); A.O. Cinneide and F.G. Gault, ibid 38, 311 (1975); J.R.H. van Schaik, P.R. Dessing and V. Ponec, ibid 38, 272 (1975); J.W.A. Sachtler, M.A. van Hove, J.P. Biberian and G.A. Somorjai, Phys. Rev. Lett. 45, 1601 (1980); W.M.H. Sachtler and G.A. Somorjai, J. Catal. 89, 35 (1984).
14. B.V. Enustun and J. Turkevich, J. Amer. Chem. Soc. 85, 3317 (1963).
15. G.C. Bond and Xu Yide, J. Chem. Soc. Faraday Trans. I 80, 969 (1984); P.A. Sermon, G. Georgiades, M.S.W. Vong, M.A. Martin Luengo and P.N. Reyes, Proc. Roy. Soc. 410A, 353 (1987).
16. D.A. Jefferson, J.M. Thomas, G.R. Millward, K. Tsuno, A. Harriman and R.D. Brydson, Nature (London) 323, 428 (1986); J.L. Tirado, J.M. Thomas, D.A. Jefferson, G.R. Millward and S.W. Charles, J. Chem. Soc. Chem. Commun. 365 (1987).
17. N.L. Wu and J. Phillips, Surf. Sci. 184, 463 (1987); P.J.F. Harris, Surf. Sci. 185, L465 (1987).
18. J.H. Sinfelt, J.L. Carter and D.J.C. Hayes, J. Catal. 24, 283 (1972); J.H. Sinfelt, Adv. Catal. 23, 147 (1974).
19 J.H. Sinfelt, Bimetallic Catalysts, Wiley, New York, 1985.
20 S.M. Augustine and W.M.H. Sachtler, J. Catal. 106, 417 (1987).

ACKNOWLEDGEMENTS

The authors thank the University of Gezira (Sudan) and the British Council for support of KK and BP for support of GRM and the SERC for equipment provision.

STRUCTURE OF SUPPORTED METAL CATALYSTS DERIVED FROM MOLECULAR BIMETALLIC CLUSTERS

M. J. Kelley*, A. S. Fung, M. R. McDevitt#, P. A. Tooley+ and
B. C. Gates
Center for Catalytic Science and Technology, Department of
Chemical Engineering, University of Delaware, Newark DE 19716
*and Engineering Technology Laboratory, Building 304,
Experimental Station, E. I. du Pont de Nemours & Co., Inc.,
Wilmington, DE 19898.
now at Department of Chemistry, Drexel University,
Philadelphia, PA 19104
+ now at Phillips Petroleum Company, Bartlesville, OK.

ABSTRACT

Re-containing molecular bimetallic clusters were used as precursors
for gamma alumina supported catalysts. Characterization by XPS and XAS
after hydrogen reduction showed Re valence states in the order Re
monometallic > Re-Os bimetallic > Re-Pt bimetallic > zero. These results
suggest that industrial bimetallic catalysts are not best understood in
terms of alloys, but rather as multifunctional, with each element
contributing its own activity.

Twenty years' operating experience since the original patent /1/ has
established firmly that rhenium addition to Pt,Cl:Al$_2$O$_3$ naphtha reforming
catalysts is beneficial. The most visible effect is extended life through
increased carbon tolerance, even though more or less comparable weight
gains due to coke deposition take place /2,3/.
There is still a lively debate as to how this comes about /27/. One
line of thinking holds that Pt-Re catalysts are multifunctional: Re
contributes activity toward removal of coke precursors. In effect, Re
protects the rest of the catalyst, whose activity otherwise remains
substantially unaltered. The strongest evidence is that mixed beds of
separately supported Pt and Re monometallics perform very much like the
bimetallic in actual reforming /4,5/ as well as certain model reactions
/6/. Evidence also comes from characterization studies showing that after
calcination and reduction comparable to industrial practice, Re is not
metallic /7,8/. This Re surface species acts catalytically to render coke
precursors innocuous /4/.
Another point of view is that Re alloys with the Pt and combines with
sulfur to subdivide the Pt surface into such small ensembles that coke
deposition is greatly inhibited compared to the main reforming reactions
/9/. By alloy we mean that both Pt and Re are zero-valent and connected to
each other by metallic bonds. Because the meaning of "metal" becomes more
complex for the very small entities envisioned here, some researchers
prefer terms such as "bimetallic clusters". Experimental evidence includes
model reaction studies showing that the performance of the separately
supported monometallics is not equivalent to that of co-impregnated
bimetallic /10, 11/ and EXAFS studies showing that Pt and Re are near
neighbors at the atomic scale /12/.
This is not merely an academic dispute. Carbon deposition is a
pervasive problem in industrial catalysis, shortening catalyst life and
excluding otherwise desirable operating conditions. A clear understanding
of how Re's beneficial impact comes about might well point to strategies
for catalyst materials science that could be applied to other systems.

Part of this conflict arises from the experimental problems posed by the catalyst itself. Typical industrial preparation schemes apply the metals as salts from aqueous solution, followed by drying, calcination, reduction and sulfiding; the chloride content is raised to a level sufficient to provide the necessary acid sites. The strong interaction of Re with Al_2O_3 is well documented /13/ so that it is easy to see how a conventionally prepared catalyst with less than perfect co-dispersion could include Re metal, Re ions and Pt-Re alloy particles of different composition. Consequently the study of supported Pt-Re alloys is made more difficult, since the presence of these other species and their activity are hard to exclude.

One strategy to circumvent this problem is to use what may be viewed as pre-formed alloy particles: bimetallic molecular clusters having ligands that do not leave potentially active species such as chloride ions. Moreover, the metals are formally zero-valent and bonded to each other as well as to the ligands. The clusters used here are listed in Table 1 and their syntheses are described in the references. For convenience in reading, we refer to them henceforth in short-hand fashion in terms of their metal nuclearity; cluster means the intact molecular entity. "Re-2 cluster" means $Re_2(CO)_{10}$; Re-2 means a species derived from an Re-2 cluster.

MATERIALS AND PREPARATION

We chose the Re_2Pt cluster as a catalyst precursor because it allows us to address directly the central issue, even though it departs from the 1:1 metal ratio more typical of industrial catalysts. We included $ReOs_3$ as a second system because Os's potency as a hydrogenation catalyst is well recognized. We reasoned that if a key to alloy formation is a more reducing environment brought about by atomic hydrogen supplied to Re by the partner metal, the effect should be clearly visible here. The monometallics were chosen to represent separately the environment of each partner in the bimetallic.

The purity of the synthesized or purchased clusters was verified by IR spectroscopy in all cases. We then deposited them onto pre-dried Degussa "C" gamma alumina (200 m^2/gm, particle size 50 micrometers and below) from freshly distilled hexanes or tetrahydrofuran. Overnight evacuation removed the solvent. The first study phase consisted of forming these materials into self-supporting wafers and determining their response to pretreatment by in-situ transmission IR, using a heated, evacuable cell. Details will be reported in a future publication. These results led us to conditions that give rise to well-defined states for detailed spectroscopic studies.

SPECTROSCOPIC STUDIES

For the XPS experiments, the catalyst powder was pressed into aluminum holders, which were then mounted onto the sample probe inside the glove box. The transfer to the surface analysis machine was carried out under the protective nitrogen blanket retained by a plastic bag attached within the glove box. After evacuation for 30 min at 10^{-5} torr in the pre-chamber, we supplied either He or H_2 as required by the treatment sequence and raised the temperature using a heater in the specimen rod. After cooling in H_2 and subsequent evacuating, the specimen was shifted into the analysis chamber and evacuated for about 30 min, reaching a base pressure less than $5x10^{-9}$ torr. The PHI Model 550 was operated with an Al anode (10 kV, 60 ma, 1486.6 eV x-rays) and sufficient scans taken to obtain adequate signal to noise ratio. We examined the Re-1 complex (only) at 77 K to minimize its degradation. Reference material binding energies were adjusted to ClS = 284.6 eV. The catalysts were referenced to Al 2P at

Table 1: Materials Investigated

Designation	Cluster	Preparation	Loading (a.)
-	H_2OsCl_6	(b.)	2.0
Os-3	$Os_3(CO)_{12}$	(b.)	2.0
Re-1	$[Re(CO)_3OH]_4$	/28/	(c.)
Re-2	$Re_2(CO)_{10}$	(b.)	1.6
Re-3	$H_3Re_3(CO)_{12}$	/14/	1.4
ReOs-3	$H_3ReOs_3(CO)_{13}$	/15/	2.6
Re2Pt	$Re_2Pt(CO)_{12}$	/16/	2.4
Pt	$Pt(CH_3)_2COD$	/21/	(d.)

a. Weight percent

b. Purchased from Strem Chemicals, Inc. and used without further purification

c. Used in bulk form as a reference material for XPS

d. Used at 0.88 % loading as a physical mixture with equal parts of Re-3 loaded at 1.4 %

Table 2: Binding Energies (eV) by XPS

	Re $4f_{7/2}$	Os $4f_{7/2}$	C 1s	Re/Al 2s
Re foil	40.0	-	(284.6)	-
$[Re(OH)(CO)_3]_4$	40.2	-	(284.6)	-
ReO_2	42.5	-	(284.6)	-
Re_2O_7	46.4	-	(284.6)	-
Re-Pt Alloy (60/40)/22/	40.5	-	(284.6)	-
Re-Pt:Al_2O_3 /23/	42.5	-	(284.6)	0.14
Re-Pt:Al_2O_3 /8/	42.2	-	-	-
(Precursor- all supported on alumina)				
$H_3Re_3(CO)_{12}$**	42.4, 47.0	-	284.4	-
$H_3Re_3(CO)_{12}$*	46.7	-	284.6	0.16
$H_3ReOs_3(CO)_{13}$**	47.1	49.9	284.6	-
$H_3ReOs_3(CO)_{13}$*	41.6	50.6	284.6	-
$Re_2(CO)_{10}$*	40.3	-	284.3	5.21
$Re_2Pt(CO)_{12}$*	40.5	-	284.3	0.57
$H_3Re_3(CO)_{12}/Pt(Me)_2COD$*	47.7	-	284.4	0.17

*H_2 pre-heat, H_2 reduction at 400°C, 2 hours

**He pre-heat, H_2 reduction at 400°C, 4 hours

74.5 eV and the resulting C1S values indicate the adequacy of the charge
referencing. Binding energy for Re and Os always refers to the 4f7/2 peak.
Table 2 summarizes the results.

Most of the x-ray absorption spectra (XAS) were obtained at Station
C-2 at CHESS (Cornell High Energy Synchrotron Source) and the rest on line
X-11-A at NSLS (National Synchrotron Light Source at the Brookhaven
National Laboratory). We also re-analyzed previously reported /7,8/ data.
All spectra were obtained in transmission from pressed wafers sized to give
unity absorption increase at the Re L3 edge. The sample cells were the
same ones used in the previous work /7/, having an internal gas volume that
can be heated or cooled and allowing the x-ray beam to enter and exit
through beryllium windows. All data were collected at 77 K, as measured by
a thermocouple attached to the sample support ring.

For the present we focus on edge peak resonance, the so-called "white
line", at the Re L3 edge. The white line represents photon absorption by
transitions from the core level underlying the edge (2p 3/2 for the L-3) to
unfilled bound states as permitted by dipole selection rules (chiefly 5d
here). Its intensity is an indicator of the unfilled density of states,
increasing from right to left across the periodic table /17/. Differences
in L-3 edge white line intensity among compounds of the same element have
been correlated quantitatively with the number of unoccupied d-states /18/.
As the detailed understanding of the white line continues to grow, it
becomes increasingly clear that reality is more complex /19,20/.
Nonetheless, the white line proves a useful "fingerprint" for comparing
states among Pt-Re catalysts /8,12/.

RESULTS AND DISCUSSION

XPS gave 50.6 eV for the Os binding energy after H_2 reduction of Os-3,
the same value as reported by others /29,30/ for bulk Os metal. We see no
binding energy shifts that might be attributed to particle size effects,
even though transmission electron microscopy (not reported here) showed no
discernable particles. A previous study in our laboratory /24/ of a
catalyst prepared by 400°C, H_2 reduction of 0.34% Os from Os-3 on the same
alumina found by TEM no particles larger than 1 nm and by XPS an Os binding
energy of 50.9 eV.

The Re-2 and Re-3 monometallics gave different results. After direct
H_2 treatment, the binding energy of reduced Re-2 is nearly that of the Re
foil. The intensity ratio of the Re 4f7/2 to the Al 2s is about ten times
that of other materials (see Table 2). Repeated experiments gave
essentially the same result as did substituting tetrahydrofuran for hexanes
in the deposition. Quantitation of the XPS peak intensities gives surface
Re contents for Re-3 suggestive of large metal particles. We conclude that
the initial cluster physisorption results in migration to the support
particle exterior during the preparation. The lack of reaction with the
support is not entirely surprising, since Re-2 is more stable than Re-3.
For example, only the latter can react with the support by deprotonation on
basic surface sites /31/. Therefore Re-2's value as a starting material for
highly dispersed catalysts is problematical. Applying the same reduction
to Re-3 led to the same binding energy as Re_2O_7 and a low Re/Al peak
intensity ratio, consistent with the strong interaction reported by many
workers. The conversion of the initially zerovalent metal in the cluster
to a surface species having an oxidation state near +7 shows that the
alumina surface is a potent oxidizing agent.

The results from the bimetallics carry on this theme (Fig. 1). When
$ReOs_3$ is thermally decarbonylated under He, the Re binding energy after
reduction is near that of Re-3, in contrast to a value less than that of
ReO_2 when H_2 is used instead. Thus in the absence of a reducing
environment, alumina is able to oxidize Re to a species quite resistant to
subsequent reduction. Moreover, Os in both bimetallic catalysts was

reduced to metal and would therefore be expected to supply atomic hydrogen to the Re entities, which would be in its immediate atomic neighborhood. The environment thus created was still not sufficiently reducing to convert Re to a zerovalent species.

The binding energy of Re in reduced industrial /8/ or laboratory-made conventional /23/ Pt-Re bimetallics is equivalent to that in ReO_2. This can be viewed as evidence for strong interaction with chloride ions or with the support. The value for the bimetallic cluster after reduction is far lower (Fig. 2), equal to that found for metal and the Re-1 complex. Re binding energies for these two species are sufficiently close (Table 2) that we cannot distinguish by XPS which is present on the catalyst surface. Comparing decarbonylation in He or H_2 prior to H_2 reduction (the same treatments applied to $ReOs_3$ above) shows perhaps a better defined peak for metal in the latter case. Nonetheless, the difference from equivalently treated $ReOs_3$ is clear: Re is in a lower valent state after reduction when Pt is in its atomic neighborhood rather than Os. Others have found that Pt is more effective than Os for promoting H_2 spillover on WO_3 /25/ and SnO_2 /26/. Re's low valence state in the cluster-derived materials constrasts with the industrial and laboratory-made conventional catalysts, which show only the higher valent species.

Pt and Re-3 clusters were put on separate support particles and physically mixed, and then subjected to the same reduction as the bimetallic cluster. The Re binding energy was high, showing no promotion of reduction by the Pt. This points to the importance of the relative locations of Pt and Re on a microscopic scale. For conventionally made catalysts, the relative positions of the two elements on the support surface will be affected by both the original deposition and by regeneration practices, emphasizing the critical nature of these catalyst preparation steps.

The x-ray absorption studies were limited to Pt-Re materials because the extended absorption fine structures overlap too extensively for Re-Os, even though the white lines could still be studied. Fig. 3 shows that the white lines of the industrial catalyst and Re-3 are nearly identical to ReO_2 after reduction and differ greatly from Re metal, whereas that of reduced Re_2Pt lies in between (Fig. 4). The XPS results discussed above also suggested a yet-to-be-identified intermediate species. Thus the while line results provide a basis of discrimination not available from the XPS alone.

CONCLUSIONS

The conclusions to which these findings lead are:

o   There is no evidence for formation of a Pt-Re alloy in any of the materials after reduction. The best interpretation of the results is that Re is present as a low-valent ionic species, similar to Re 4^+ for the industrial material and similar to Re^+ for the bimetallic cluster.

o   Atomic-scale proximity of Re to a metal which can dissociate hydrogen facilitates its reduction. Pt is more effective than Os in this respect.

o   The oxidation state of Re after reduction reflects the history prior to reduction: decarbonylation of metal cluster precursors in the absence of H_2 leads to a more oxidized species.

We now return to the original questions: what is the state of Re and what is its mechanism of action in real reforming catalysts? The partner metal strongly influenced Re's state only when the two metals were in

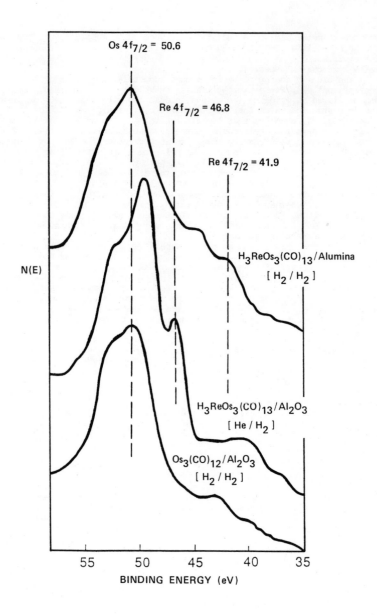

Figure 1: X-ray Photoelectron Spectra of Re-Os Materials

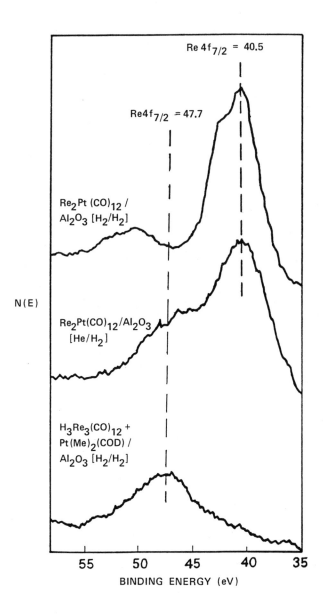

Figure 2: X-ray Photoelectron Spectra of Re-Pt Materials

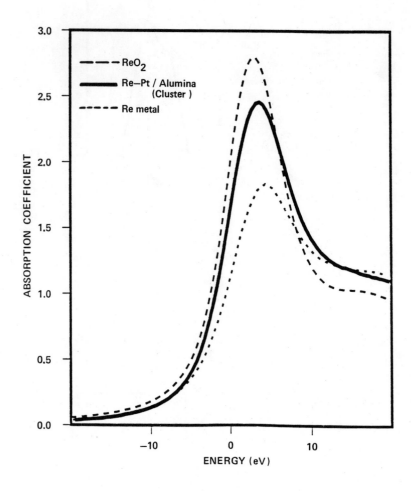

Figure 3: Comparison of Re L₃ absorption edges (10534 eV) for reduced cluster-derived catalyst and bulk reference materials.

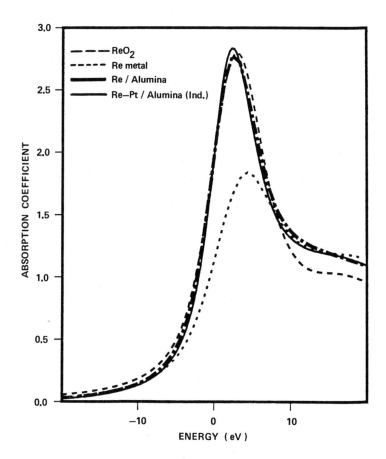

Figure 4: Comparison of Re L₃ absorption edges (10534 eV) for reduced industrial bimetallic and monometallic catalysts and bulk reference materials.

closest proximity: when the bimetallic cluster precursor was used. While these materials may exhibit interesting catalytic properties in their own right, they do not represent what is present in the industrial naphtha-reforming catalysts. Removing the Re only a few tens of micrometers from the Pt by making a physical mixture of the separately supported clusters led always to an Re ionic species strongly bound to the alumina surface. We therefore infer that multifunctionality is the best representation of how Re improves the performance of naphtha reforming catalysts /27/. This does not exclude that Re CAN be in Pt's atomic neighborhood or indeed that catalysts can be made to Pt-Re alloys. However, evidence is lacking that alloy formation or atomic-scale proximity is needed to accomplish the benefits observed in naphtha reforming.

ACKNOWLEDGEMENTS

We are grateful to the National Science Foundation for support of University of Delaware activities under grant CBT-8317140 and providing x-ray absorption facilities at the Cornell High Energy Synchrotron Source (CHESS). We also are indebted to the National Synchrotron Light Source supported by the Department of Energy under grant DE-AC02-76CH00016. We appreciate the help of the staff of NSLS beamline X-11 and the Department of Energy Division of Materials Science for the support of its development under grant DE-AS05-80ER40742. Brian Strohmeier, John Vohs and Rama Hegde assisted in the XPS and Larry Dodd provided some of the computer codes used for data manipulation. Finally, we owe a special debt to Prof. D. C. Koningsberger of the Eindhoven University of Technology.

REFEFENCES

1. H.E.Klucksdahl; U.S. Patent 3,415,737 (1968).

2. J.H.Sinfelt; "Bimetallic Catalysts", Wiley/EXXON (1983).

3. J.L.Carter, G.B.McVicker, W.Weissman, W.S.Kmak and J.H.Sinfelt; Appl.Cat.3, 327 (1982).

4. R.J.Bertolacini and R.J.Pellet; in "Catalyst Deactivation", Elsevier, 72 (1980).

5. G.W.Roberts and K.E.Hastings; Ger.Offen. 26 27 822 (1977).

6. R.W.Coughlin, K.Kawakami and A.Hasan; J.Catal.88, 150 (1984).

7. D.R.Short, S.M.Khalid, J.R.Katzer and M.J.Kelley; J.Catal.72, 288 (1981).

8. J.H.Onuferko, D.R.Short and M.J.Kelley; Appl.Surf.Sci.19, 227 (1984).

9. P.Biloen, J.N.Helle, H.Verbeek, F.M.Dautzenberg and W.M.H.Sachtler; J.Catal.63, 119 (1980).

10. M.A.Pacheco and E.E.Petersen; J.Catal.96, 499 (1985).

11. V.K.Shum, J.B.Butt and W.M.H.Sachtler; J.Catal.96, 371 (1985).

12. G.Meitzner, G.H.Via, J.H.Sinfelt and F.W.Lytle; ACS Petrol. Div.Prepr.32, 733 (1987).

13. H.C.Yao and M.Shelef; J.Catal.44, 392 (1975).

14. M.A.Andrews, S.W.Kirtley and H.D.Kaesz; Inorg.Synth.17, 66 (1977).

15. J.Knight and M.J.Mays; J.Chem.Soc.- Dalton Trans., 1022 (1972).

16. M.A.Urbancic, S.R.Wilson and J.R.Shapley; Inorg.Chem.23, 2954 (1984).

17. F.W.Lytle; J.Catal.43, 376 (1976).

18. F.W.Lytle, P.S.P.Wei, R.B.Greegor, G.H.Via and J.H.Sinfelt;
    J.Chem.Phys.70, 4849 (1979).

19. J.A.Horsely; J.Chem.Phys.76, 1451 (1982).

20. e.g., P.Lagarde, D.Raoux and J.Petiau; "EXAFS and Near Edge Structure
    IV", J.Physique Coll.C8 (1986).

21. H.C.Clark and L.E.Manzer; J.Organomet.Chem.59, 411 (1973).

22. M.Alnot, A.Cassuto, R.Ducros, J.J.Ehrhardt and B.Weber; Surf.Sci.114,
    L48 (1982).

23. P.S.Kirlin, B.R.Strohmeier and B.C.Gates; J.Catal.98, 308 (1986).

24. H.Knozinger, Y.Zhao, B.Tesche, R.Barth, R.Epstein, B.C.Gates and
    J.P.Scott; Farad.Disc.Chem.Soc.72, 53, 1981.

25. A.J.von Hegedus, T.Millner, J.Neugebauer and K.Sasv'ari;
    Z.Anorg.Allg.Chem.281, 64 (1955).

26. R.Frety, H.Charcosset and Y.Trambouze; Ind.Chim.Belge 38, 501 (1973).

27. M.J.Kelley and D.B.Dadyburjor; in "Catalyst Deactivation",
    E.E.Petersen and A.T.Bell eds., Marcel Dekker, 125 (1987).

28. M.Herberhold, G.Suss, J.Ellerman and H.Gabelein; Chem.Ber.111,
    2931 (1978).

29. B.Folkesson; Acta Chem.Scand.27, 297 (1973).

30. A.Berndtsson, R.Nyholm, H.Martensson, R.Nilsson and J.Hedman;
    Phys.Stat.Sol.B 93, K103 (1979)

31. P.S.Kirlin, F.A.DeThomas, J.W.Bailey, K.Moller, H.S.Gold, C.Dybowski
    and B.C.Gates; Surf.Sci.175, L707 (1986).

STRUCTURE AND REACTIVITY OF RH IN THE SMSI STATE

EHRICH J. BRAUNSCHWEIG, A. DAVID LOGAN, and ABHAYA K. DATYE
Department of Chemical & Nuclear Engineering, University of New Mexico,
Albuquerque, NM 87131, U.S.A.

ABSTRACT

Hydrogenolysis of n-butane and CO hydrogenation were used to characterize the behavior of TiO_2 supported Rh in the SMSI state. The Rh was
supported on a model support consisting of nonporous spherical particles of
TiO_2. The simple geometry of the support permits the metal crystallites to
be imaged 'edge-on' in a transmission electron microscope. This allows
detailed examination of the metal surface and the metal-oxide interface as a
function of pretreatment. After high temperature reduction in H_2, there is
no significant change in the morphology of the Rh crystallites. However,
the presence of amorphous overlayers 0.2-0.4 nm thick is clearly evident on
the Rh surface. These overlayers can only be partially removed after oxidation at 473 K. The presence of these overlayers is well correlated with the
drop in hydrogen chemisorption and altered reactivity. While the butane
hydrogenolysis activity of Rh is considerably suppressed in the SMSI state,
the CO hydrogenation activity is increased by a factor of 3. The H_2 uptake
and the butane hydrogenolysis activity can be only partially restored after
oxidation at 473 K. Treatment at 773 K in oxygen is necessary for restoring
the pre-SMSI behavior.

INTRODUCTION

The behavior of noble metals such as Pt and Rh supported on TiO_2 has
been studied extensively since the original observation that high temperature reduction (HTR, H_2 @ 773 K) leads to a marked suppression in the
chemisorption uptake of H_2 [1]. In recent years, there is increasing
evidence that this drop in chemisorption is caused, at least in part, by the
encapsulation of the metal by TiO_x overlayers [2-4]. The mechanisms for the
removal of these oxide overlayers and the extent to which the properties of
the metal can be restored are not yet well understood.
In previous work, Baker et al. [5] observed changes in the morphology
of TiO_2 supported Pt after HTR. They used model catalysts consisting of
thin planar oxide films where the metal was deposited by evaporation. In
the vicinity of the Pt crystallites, the TiO_2 was observed to transform to
Ti_4O_7 after HTR; such a transformation was not seen in the vicinity of Ag
crystallites after similar treatment [6]. Thus, the presence of a metal
capable of dissociating H_2 appears to be essential to the creation of the
SMSI state. On supported Rh, however, Singh et al. [7] could not detect the
presence of reduced Ti_4O_7 after HTR, neither was any pronounced flattening
of the metal crystallites observed. A problem in the use of TEM for the
study of small particles in heterogeneous catalysts is that generally only
the projected outline of the metal crystallite is observed through the oxide
support. This precludes observation of the surfaces of the metal and the
three-dimensional shapes of the metal crystallites. Using the contrast from
the metal crystallite to infer three dimensional shape is misleading, as
shown by Treacy and Howie [8]. To circumvent some of these limitations, we
have used nonporous oxide particles of simple geometric shape as model
support [9]. This permits observation of the metal crystallites in a direction parallel to the oxide surface (edge-on) and normal to the oxide surface
(projected outline). The objective of this work is to study the morphology

and structure of metal crystallites in the SMSI state using TEM, and to
relate this information to the catalytic behavior of the metal.

EXPERIMENTAL

The model catalyst support consisted of sub-micron, spherical particles
of TiO_2. The primary particles are single crystals and range in size from
25 - 250 nm. X-ray diffraction shows a distribution of anatase and rutile
phases with the peak heights being 70:30 anatase:rutile. The BET surface
area of the powder is 10-15 m^2/gm. The TiO_2 powder was prepared by striking
an AC discharge between two high purity Ti electrodes in a controlled atmosphere of O_2 and Ar. The catalyst was prepared by non-aqueous impregnation
using Rh(III)2,4-pentanedionate (Ru(acac)$_3$) as the precursor. The catalyst
was air dried at 383 K before reducing in H_2. The temperature was gradually
raised to 423 K, maintained for 5 hrs, and then raised to 473 K to complete
the reduction. This is necessary to prevent sublimation of the precursor
prior to decomposition. The catalysts were examined in a JEOL JEM-2000 FX
electron microscope which has a point resolution of 0.3 nm. The catalyst
powder was supported on holey carbon films supported on Cu grids. No solvents were used during sample preparation to avoid any artifacts due to
contamination. Reactivity measurements were performed in a U-tube quartz
microreactor with the catalyst held in place with quartz wool plugs. The
flow reactor was equipped with mass flow controllers for metering H_2, He,
CO, n-butane and oxygen to the reactor. This allowed measurement of the CO
hydrogenation and butane hydrogenolysis activity successively on the same
catalyst. Chemisorption measurements were performed using an all-glass
static volumetric apparatus with gas uptake being monitored by a Baratron
capacitance manometer.

RESULTS

Figure 1a shows an electron micrograph of the catalyst after low temperature reduction (LTR, H_2 @ 473 K). Small Rh crystallites having an
average diameter of 5 nm are visible in the micrograph. The edge-on views
show that the Rh surface is clean and free of any amorphous material.
Examination of the same sample at a higher resolution than 0.3 nm confirmed
that despite exposure of the sample to air during sample preparation, there
was no detectable contamination or oxide overlayer on the metal surface
[10]. This catalyst sample was then treated in hydrogen at 773 K (HTR) and
examined in the TEM. The same area of the sample as imaged in fig. 1a is
now seen in fig. 1b after HTR. An amorphous overlayer approximately 0.2-0.4
nm is visible on the Rh surface. A similar overlayer is also visible on the
surfaces of the TiO_2 oxide support. Examination at 400 kV reveals no lattice fringes in this overlayer, suggesting that it is amorphous [10]. The
Rh crystallites do not seem to change shape significantly. At most there is
a slight tendency to becoming more spherical. Similarly, the TiO_2 support
is also unaffected structurally and we were unable to detect the presence of
Ti_4O_7 on the surface. Fig. 1c shows an electron micrograph of the same
region of the sample after oxidation at 473 K overnight and LTR. While the
oxide overlayers appear to be reduced in thickness, the Rh metal surfaces
still appear irregular.
The presence of the amorphous overlayers is correlated very well with
the H_2 chemisorption ability of the Rh. Table I shows the chemisorption
uptake on this catalyst after various pretreatments. The chemisorption
measurements were performed on a larger batch of the same catalyst imaged in
figs. 1a-c. It is seen that oxidation at 473 K overnight is necessary to
restore 80% of the original H_2 uptake of the catalyst. The data in Table I

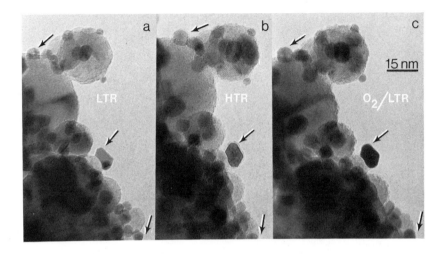

Figure 1. Micrographs of the Rh/TiO$_2$ catalyst after various pretreatments

suggests that oxidation at 773 K may be necessary to restore completely the H$_2$ uptake of the catalyst. At this temperature, the Rh metal crystallites get completely oxidized to Rh$_2$O$_3$ [11]. Due to the 90% volume expansion upon oxidation, there is considerable restructuring of the Rh crystallites and hence, during LTR, it is conceivable that the amorphous oxide overlayers may phase segregate and separate from the Rh metal. However, in our work, we have been unable to detect small TiO$_2$ on the catalyst after this treatment.

Figure 2 shows the activity of Rh in the hydrogenolysis of n-butane as a function of pretreatment. The turnover frequency (TOF) is based on the chemisorption uptake of the fresh catalyst and is not corrected for any variation in the uptake due to pretreatment. The reaction mixture consisted of 20:1 H$_2$:n-butane with no diluent. The total pressure was kept at 101 kPa. The activity shown is after a 10 min. reaction period. These catalysts do not show any appreciable variation in reaction rate if the reaction mixture is allowed to flow continuously over the catalyst. Except in the SMSI state, all measurements were made at 473 K. Due to the low reactivity in the SMSI state, the value reported in fig. 2 has been extrapolated from higher temperature measurements using an Arrhenius plot. The activation energy was 35 kcal/mol. Figure 3 shows the product selectivity in butane hydrogenolysis also as a function of pretreatment. The CO hydrogenation activity under similar conditions is shown in figure 4. These measurements were made in the same reactor system and so reflect differing reactivities of the same catalyst. We observed that the butane hydrogenolysis reactivity of the catalyst was lower if performed after the CO hydrogenation. Hence, in all of the experiments reported in figures 2-4, the butane hydrogenolysis was performed before the CO hydrogenation. The CO hydrogenation activity was not affected by whether butane hydrogenolysis had been performed on the catalyst earlier.

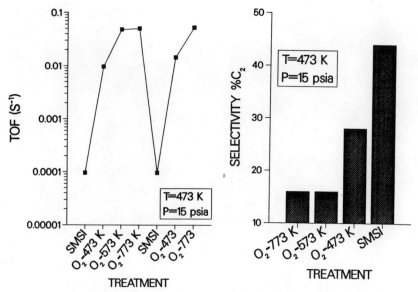

Figure 2. Turnover Frequency of Rh/TiO$_2$ as a function of treatment for n-Butane Hydrogenolysis.

Figure 3. Product Selectivity as a function of treatment for n-Butane Hydrogenolysis.

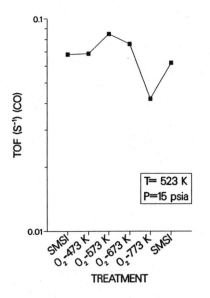

CHEMISORPTION UPTAKE

TREATMENT	UPTAKE (μmole/gram)
LTR	17.5
HTR	2.9
O$_2$ @ 473 K 2 hours/LTR	6.1
O$_2$ @ 473 K Overnight/LTR	14.7
O$_2$ @ 573 K Overnight/LTR	15.5
O$_2$ @ 673 K Overnight/LTR	15.5
O$_2$ @ 773 K Overnight/LTR	18.0

Table I. Hydrogen Chemisorption tptake for Cataylst Rh/TiO$_2$ after various treatments.

Figure 4. CO Turnover Frequency of Catalyst Rh/TiO$_2$ as a function of treatment for CO Hydrogenation.

DISCUSSION

The observations in figures 1a-c show that there are no significant morphological changes in the Rh crystallites as a result of HTR in H_2. The absence of underlying support in edge-on views makes it possible to study the surfaces of small metal crystallites. While 0.2-0.4 nm amorphous films can be detected on the surfaces of the small Rh crystallites, we are unable to determine the chemical composition of these overlayers. Spectroscopic evidence from previous work suggests that these overlayers must be sub-oxides of Ti that migrate onto the metal surface after HTR [12]. The mechanism for the removal of these overlayers upon oxidation also cannot be specified at this time. It is possible that stoichiometric TiO_2 does not wet the metal and tends to recrystallize on the support. It is also clear that these oxide overlayers must be sufficiently porous for the gas molecules to diffuse through during the reaction step. The edge-on view does not provide any indication of the coverage of the metal and it is possible that the films are patchy and allow access to part of the metal.

The reactivity trends reported in figs. 2-4 are consistent with those reported in earlier work [13,14]. The presence of the amorphous overlayers causes significant drop in the chemisorptive capacity and the activity in butane hydrogenolysis. Besides a drop in the catalyst activity, a significant alteration in the product selectivity was observed in the SMSI state in this study. Also remarkable is the almost complete reversibility in the chemisorption uptake and the reactivity after high temperature oxidation. Surprisingly, while the butane hydrogenolysis varies by over 3 orders of magnitude, the CO hydrogenation activity is hardly affected at all. The drop in the butane hydrogenolysis activity can be related to the structure sensitivity of this reaction. The presence of the oxide overlayers dramatically reduces the number of sites available for this reaction. The altered selectivity suggests that the Rh sites that remain have a different surface structure. For instance, on single crystal Ir [15], it has been shown that the (110) surface has a much higher ethane reactivity in the butane hydrogenolysis compared with the (111) surface. Similar comparisons are not available for Rh surfaces and hence understanding of these selectivity variations will have to await further work. The increase in CO hydrogenation activity which occurs concurrently with a sharp drop in the H_2 chemisorption uptake suggests that the sites that are accessible to the CO/H_2 mixture are considerably more reactive than those present on the initial Rh surface. This lends support to the hypothesis of Burch and Flambard [16] that the CO hydrogenation may occur on the periphery of the oxide overlayers where special sites are created at the Rh/TiO_x interface.

CONCLUSIONS

This work has shown that the use of model supported catalysts permits the correlation of catalyst microstructure to reactivity. By using oxide particles of simple geometric shape, we are able to study the surfaces of small metal crystallites. These model catalysts have sufficiently high metal surface areas so that reactivity studies can be performed in a conventional flow reactor. We have observed the presence of sub-nm oxide overlayers on the surfaces of Rh crystallites subjected to HTR. These oxide overlayers cause the butane hydrogenolysis activity to drop by three orders of magnitude while the CO hydrogenation activity goes up by 2-3 times, based on the initial H_2 chemisorption uptakes. These activity changes are reversible upon oxidation and LTR, however we have found that oxidation at 773 K may be necessary to completely reverse the effects of these oxide overlayers. This happens to be the temperature when bulk oxidation of the Rh is also complete. Thus, while mild oxidation of the oxide overlayer helps to reduce the coverage, it is not sufficient for complete removal which occurs only when the bulk of the metal gets oxidized.

ACKNOWLEDGEMENTS

Financial support for this work from NSF via grant CBT-8707693 is gratefully acknowledged. Electron microscopy was performed at the Electron Microbeam Analysis Facility located within the Department of Geology and the Institute for Meteoritics at the University of New Mexico.

REFERENCES

1. S. J. Tauster, S. C. Fung, and R. L. Garten, J. Am. Chem. Soc. 100, 170 (1978).
2. J. Santos, J. Phillips, and J.A. Dumesic, J. Catal. 81, 147 (1983).
3. D. N. Belton, Y.-M. Sun, and J. M. White, J. Am. Soc. 106, 3059 (1984).
4. H. B. Sadeghi and V. E. Henrich, J. Catal. 87, 279 (1984).
5. R. T. Baker, E. B. Prestridge, and R. L. Garten, J. Catal. 56, 390 (1979).
6. R. T. Baker, E. B. Prestridge, and L. L. Murrell, J. Catal. 79, 348 (1983).
7. A. K. Singh, N. K. Pande, and A. T. Bell, J. Catal. 94, 422 (1985).
8. M. M. J. Treacy and A. J. Howie, J. Catal. 63, 265 (1980).
9. A. K. Datye and A. D. Logan, Proc. 44th meeting of the Electron Microsc. Soc. Amer., pg 772, ed. G. W. Bailey, San Francisco Press, 1986.
10. A. D. Logan, E. J. Braunschweig, A. K. Datye, and D. J. Smith, submitted to Langmuir.
11. S. Chakraborti, A.K. Datye and N.J. Long, J. Catal., 108, in press.
12. R. A. Demmin, C. S. Ko, and R. J. Gorte, Strong Metal Support Interactions, ACS Symposium Series 298, eds. R. T. K. Baker, S. J. Tauster, and J. A. Dumesic., p 48 (1986).
13. M. A. Vannice and R. L. Garten, J. Catal. 56, 236 (1979).
14. D.E. Resasco and G.L. Haller, J. Catal. 82, 279 (1983).
15. J. R. Engstrom, D. W. Goodman, and W. H. Weinberg, J. Am. Chem. Soc. 108, 4653 (1986).
16. R. Burch and A. R. Flambard, J. Catal. 78, 389 (1982).

CARBON FORMATION ON SUPPORTED METAL CATALYSTS

R.T.K. BAKER
Chemical Engineering Department, Auburn University, Auburn, AL 36849-3501

Abstract

The potential for carbon formation exists in any system in which hydrocarbons undergo thermal decomposition. It is well known that certain metals can increase the carbon yield by catalyzing the growth of both filamentous and graphitic types of deposit. The highest catalytic activity for carbon deposition is exhibited by the ferromagnetic metals and in particular, iron.

We have used a combination of controlled atmosphere and high resolution electron microscopy techniques to study the formation of the various types of carbon on metal surfaces. In this paper the emphasis is placed on the fundamental aspects surrounding the growth of filamentous carbon. The qualitative and quantitative data obtained from these studies have enable us to develop a mechanism for the growth of filamentous carbon and also provided insights into methods of inhibiting the growth of this form of carbon. Continuous observation of many filamentous carbon growth sequences has shown that both the addition of a second metal to the catalyst and also the strength of the metal-support interaction can have a profound effect on the mode by which carbon filaments grow.

Results and Discussion

The formation of filamentous carbon was reviewed by Baker and Harris [1] and it is well established that the ferromagnetic metals, iron, cobalt and nickel are the most active catalysts for carbon filament growth from hydrocarbons. Figure 1 is a transmission electron micrograph showing the formation of filamentous carbon on cobalt particles supported on graphite after reaction in ethylene at 750°C.

In the past the major focus in this area has been to develop methods for inhibiting the growth of filamentous carbon on metal surfaces since the deposit can lead to plugging of tubes and deactivation of catalyst particles [2-6]. Recent work by Tibbetts [7] has spurred intense interest concerning the structural application of this material to fiber-reinforced composites.

Based on the accumulated data from controlled atmosphere and high resolution electron microscopy studies, a model was developed to account for the formation of filamentous carbon from the metal catalyzed decomposition of certain hydrocarbons [8,9]. The key step in the model is diffusion of carbon through the metal particle from the hotter leading surface on which hydrocarbon decomposition occurs to the cooler trailing faces at which carbon is deposited from solution. Growth ceases when the leading face is covered with carbon built up as a consequence of overall rate control by the carbon diffusion process. During the growth process the metal catalyst is frequently carried away from the support surface and

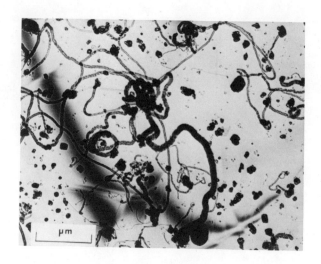

FIGURE 1. Transimssion electron micrograph of filamentous carbon produced by the decomposition of ethylene on Co particles supported on Graphite.

TABLE I

Comparison of Measured Activation Energies for
Filament growth with those for Carbon Diffusion
through the Corresponding Metal Catalysts

CATALYST	ACTIVATION ENERGY FOR	
	CATALYZED FILAMENT GROWTH ($kcal.mole^{-1}$)	DIFFUSION OF CARBON ($kcal.mole^{-1}$)
Nickel	34.7[3]	33.0 - 34.8[13]
α-Iron	16.1[9]	10.5 - 16.5[14]
γ-Iron	33.9[12]	33.3 - 37.4[15]
Nickel-Iron	33.6[4]	34.0[16]
Cobalt	33.0[9]	34.7[17]
Vanadium	27.6[11]	27.8[18]
Molybdenum	38.8[11]	41.0[19]
Chromium	27.1[9]	26.6[20]

remains at the tip of the filament. This characteristic is indicative of a relatively weak metal/support interaction. In other cases the strength of the interaction between the metal and the support is sufficient to prevent catalyst particles from being lifted from the support by the conventional mode of filament growth. As a consequence, the only available pathway for removal of excess dissolved carbon from the particle is via extrusion mode [10,11]. During this type of filament growth, the catalyst particle remains attached to the surface of the support. Justification for the claim that diffusion of carbon through the catalyst particle is the rate determining step is based on the remarkable agreement between the measured activation energies for filament growth and those for diffusion of carbon through the respective metals, Table I.

A comparison of the kinetic parameters obtained from a series of experiments performed by CAEM [4] designed to investigate the inhibiting effect of various additives on carbon filament formation is presented in Table II.

TABLE II

Effect of Various Additives on the Growth of Filamentous Carbon
from Ni-Fe/C_2H_2 [4]

ADDITIVE	ONSET TEMPERATURE ($^\circ$C)	RATE OF FILAMENT GROWTH AT 850°C (nm.sec^{-1})	APPARENT ACTIVATION ENERGY (kcal.mole^{-1})
Virgin	480	413	33.6
Alumina	650	428	32.9
Titania	635	220	31.5
Tungsten Oxide	700	12.6	32.7
Tantalum Oxide	680	34.7	30.5
Molybdenum Oxide	670	6.6	30.9
Boron Oxide	475	204	31.1
Silica	620	2.1	--

Inspection of the data shows that the rate of formation of filamentous carbon from the Ni-Fe/C_2H_2 system varies quite markedly with the different additives. Perhaps the most intriguing result is, that with the exception of SiO_2, none of these additives appear to alter the activation energy for the filament growth process, i.e., they do not affect the rate determining step, the diffusion of carbon through the catalyst particles.

It is possible that once the oxides are incorporated within the catalyst their role is that of a component for reducing carbon solubility in the particle. Available bulk metallurgical information supports this hypothesis for addition of Mo, Si and W to the C-Fe binary system; as little as 1% W addition to Fe reduces C solubility by 50%. Unfortunately, comparable data is not available for Ta and B, however, it is expected that these materials should exert a similar qualitative effect. Moreover, there is no reason to suppose that the additives would behave any differently with the C-Fe-Ni ternary system.

From a consideration of the proposed model for carbon filament formation one can readily predict the consequences of a reduction in carbon solubility in the catalyst on the subsequent filament growth characteristics. From a structural standpoint one would expect to see a reduction in the density of the inner core region and the resultant build-up of excess carbon at the particle surface causing premature deactivation of the catalyst.

Furthermore, at any given temperature the rate of diffusion of those carbon atoms which did dissolve would also be reduced since the driving force for diffusion, the temperature gradient created in the particle, would not be a maximum, lacking the full contribution from the heat of solution of carbon in the alloy. All these features were observed experimentally, varying in degree with the different additives.

From the results of this investigation it is possible to establish a classification of the roles of various additives on inhibition of filamentous carbon growth.

(i) Some materials provide a physical barrier towards hydrocarbon adsorption and decomposition on the metal surface, but spall at elevated temperatures, e.g. Al_2O_3 and TiO_2.

(ii) Other additives reduce carbon solubility in the metal catalyst particles, but have no effect on carbon diffusion through the particle, e.g. MoO_3, WO_3 and Ta_2O_5, and finally,

(iii) There are cases where an additive can reduce both the solubility of carbon and its diffusion through the catalyst particle, e.g. SiO_2.

Although the proposed model gives a plausible explanation for the growth of "whisker-like" filaments, and has been the basis for the development of methods of inhibiting filamentous carbon formation on metal surfaces, it does have a number of deficiencies:

(a) It fails to account for the reported formation of filaments produced from the methane decomposition, an endothermic process.

(b) The explanation of the formation of the graphitic skin of the filaments is inadequate.

(c) The model does not take into account the chemical nature of either the surface or bulk of the catalyst particles, and

(d) The model does not predict the deviations which occur in filament growth characteristics (branching and multidirectional growth) when additives are introduced into the catalyst particles.

Over the past few years there has been considerable debate over two aspects of this mechanism, the driving force for carbon diffusion and the nature of the catalytic species. Rostrup-Nielsen and Trim [21] argued that the driving force for the diffusion of carbon through the metal is probably not due to the existence of a temperature gradient, but rather to a carbon concentration gradient. In contrast, rather elegant experiments performed by Yang and Yang [22] using the gold decoration technique to probe the growth of carbon deposits from the interaction of nickel with various hydrocarbons, seem to support the notion that a temperature gradient is the driving force for carbon diffusion. However, the major disagreement is concerned with the chemical state of the catalyst, where some workers claim

that the active entities are carbides [23-25] and others favor the metallic
state [12,26].

A variation in these investigations has been to examine the manner by
which additives to the ferro-magnetic metals modify the growth process and
produce the more intricate filament conformations shown schematically in
Figure 2.

**INFLUENCE OF METAL ADDITIVES TO THE
CATALYST PARTICLE
ON THE FILAMENT GROWTH CHARACTERISTICS**

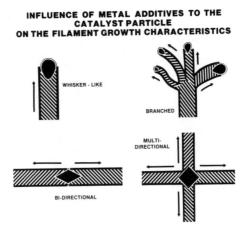

FIGURE 2. Schematic representation of different types of growth observed in
carbon filaments.

Nishiyama and Tamai [27,28] studied carbon formation from the
decomposition of benzene by various copper-nickel alloys over the range 580
to 900°C. They found that filamentous carbon predominates at the lower
temperatures region and that alloys containing 40 to 80 % nickel had much
higher catalytic activity for the formation of filamentous than did pure
nickel. Bernardo and coworkers [29,30] have used a variety of techniques to
study the formation of carbon filaments from the copper-nickel catalyzed
decomposition of methane. In addition to following the kinetics of the
carbon deposition process these workers examined the morphological
characteristics of the filaments as a function of the catalyst composition.
Conventional type of filaments were produced from catalyst particles
containing less than 80 at% copper. At higher copper concentrations a new
type of structure was created, where at least six carbon filaments
originated from a single catalyst particle. Similar multi-directional
carbon deposits were also observed by Rostrup-Nielsen [31] in filaments
formed on nickel catalysts containing a high sulphur coverage. The
symmetrical arrangement of these multi-growth filaments strongly suggests

that carbon filaments development occurs from specific crystallographic planes of the catalyst particle (possibly a single crystal), as shown by Audier and coworkers [32,33]. Although bi-directional growth has been reported for several binary catalyst particles, e.g. Fe-Sn [43] and Fe-Ni [4], multi-directional growth from a single catalyst particle is a rare phenomenon. An example of the bi-directional type of filament growth is presented in Figure 3, which is a scanning micrograph of an iron-nickel specimens which had been reacted in ethylene at 800°C. The location of the catalyst particles within the filaments is indicated.

FIGURE 3. Carbon filaments produced by the decomposition of ethylene   at 800°C on iron-nickel catalyst. (The catalyst particle is shown by the arrow).

In the cases of extrusion, branched and multi-directional filaments the effects of carbon concentration gradients, temperature gradients and active faces for hydrocarbon decomposition versus carbon precipitation become more complex. As a consequence these growth modes can no longer be treated by a simple one dimensional growth/diffusion model involving temperature gradients or carbon concentration gradients as these concepts become difficult to justify. It appears that a significant aspect of the mechanism for these growth structures must revolve around the surface modification of catalyst particles as a result of an interaction with the support or from the effects of adding a second component to the catalyst particle. It is clear there is a need for further mechanistic studies in order to increase the understanding of carbon filament growth.

References

1. R.T.K. Baker and P.S. Harris in "Chemistry and Physics of Carbon", Vol 14 (P.L. Walker, Jr. and P.A. Thrower, eds.) Dekker, New York, 1978, p.83.

2. L.F. Albright and R.T.K. Baker, eds., "Coke Formation on Metal Surfaces", ACS Symposium Series, 202. American Chemical Society, Washington, D.C., 1982.

3. A.Dyer, ed., "Gas Chemistry in Nuclear Reactors and Large Industrial Plants", Heyden, London 1980.

4. R.T.K. Baker and J.J. Chludzinski, J. Catal. 64, 464 (1980).

5. M.J. Bennett and J.B. Price, J. Mater. Sci 16, 170 (1981).

6. G.W. Horsley and J.A. Cairns, Appl. Surf. Sci. 18, 273 (1984).

7. G.G. Tibbetts, Appl. Phys. Letters 42, 66 (1983).

8. R.T.K. Baker, M.A. Barber, P.S. Harris, F.S. Feates and R.J. Waite, J. Catal. 26, 51 (1972).

9. R.T.K. Baker, P.S. Harris, R.B. Thomas and R.J. Waite, J. Catal. 30, 86 (1973).

10. R.T.K. Baker and R.J. Waite, J. Catal. 37, 101 (1975).

11. R.T.K. Baker, J.J. Chludzinski, N.S. Dudash and A.J. Simoens, Carbon 21, 463 (1983).

12. R.T.K. Baker, J.J. Chludzinski and C.F.R. Lund, Carbon 25, 295 (1987).

13. S. Diamond and C. Wert, Trans. AIME 239, 705 (1967).

14. D.W. Morgan and J.A.Kitchener, Trans. Farad. Soc. 50 51 (1964).

15. C. Wells, W. Baz and R.T. Mehl, Trans AIME 188, 553 (1950), R.P. Smith, trans AIME 230, 476 (1964).

16. R.P. Smith, Trans. AIME 236, 1224 (1966).

17. J.J. Koveenskaj, Phys. Metlas Metallage (USSR) 16, 107 (1963).

18. F.A.Schmidt and J.C. Warner, J. Less-Common Metals 26, 325 (1972).

19. P.S. Rudman, Trans. AIME 239, 1949 (1967).

20. E.V. Barison, P.L. Gruzin and S.V. Zemskii, Zashch. Pokrytiya, Metal 2, 104 (1968).

21. J.R. Rostrup-Nielsen and D.L. Trimm, J. Catal. 48, 155 (1977).

22. R.T. Yang and K.L. Yang, J. Catal. 93, 182 (1985).

23. A. Sacco, Jr., P. Thacker, T.N. Chang and A.T.S. Chiang, J. Catal. 85, 224 (1984).

24. A.J.H.M. Kock, P.K. Bokx, E. Boellard, W. Klop and J.W. Geus, J. Catal. 96, 468 (1985).

25. J.R. Bradley, Y.-L. Chen and H.W. Sturner, Carbon 23, 715 (1985).

26. C.A. Bernardo and L.S. Lobo, J. Catal. 37, 267 (1975).

27. Y. Nishiyama and Y. Tamai, J. Catal. 33, 98 (1974).

28. Y. Nishiyama and Y. Tamai, J. Catal.. 45, 1 (1976).

29. C.A. Bernardo, I. Alstrup and J.R. Rostrup-Nielsen, J. Catal. 96, 517 (1985).

30. M.J. Tavares, C.A. Bernardo, I. Alstrup, and J.R. Rostrup-Nielsen, J. Catal. 100, 545 (1986).

31. J.R. Rostrup-Nielsen, J. Catal. 85, 31 (1984).

32. M. Audier, M. Coulon and A. Oberlin, Carbon 18, 73 (1980).

33. M. Audier, A. Oberlin and M. Coulon, J. Cryst. Growth 55, 549 (1981).

34. R.T.K. Baker, P.S. Harris, and S. Terry, Nature 253, 37 (1975).

Acknowledgements

The author would like to thank Drs. B.J. Tatarchuk and N.M. Rodriguez (Chemical Engineering Dept., Auburn University) for stimulating discussions.

HIGH TEMPERATURE LIMIT FOR THE GROWTH OF CARBON FILAMENTS ON CATALYTIC IRON PARTICLES

GARY G. TIBBETTS AND ERIC J. RODDA
Physics Department, General Motors Research Laboratories, Warren, Michigan 48090-9055

ABSTRACT

Carbon filaments of macroscopic length may be produced by heating transition metal catalyst particles in a hydrocarbon-hydrogen mixture. Because the nanometer-sized catalyst particles contain relatively many surface atoms, they melt at a temperature considerably below that of bulk material; this depression of the melting point is difficult both to calculate and measure. This work addresses the fundamental question of whether the catalyst particles are in the liquid or solid state during filament growth.

We present measurements of the average number of filaments grown per unit area of substrate as a function of temperature which show that the the activity of the catalyst particles drops precipitously above 1030°C and nearly disappears above 1120°C. This compares with the melting point of the Fe-C eutectic of 1154°C. TEM observations show that the average diameter of an active catalyst particle is 15±7 nm. Calculations based on the model of Ross and Andres [1] for the melting temperature of small clusters of atoms predict a drop in activity above 1050°C similar to our observations.

INTRODUCTION

The study of the growth of carbon filaments on transition metal catalyst particles is a venerable one [2]. In over three decades of mostly qualitative work on this subject, there has been much speculation and observation on the growth mechanism of these interesting structures, but insufficient quantitative measurement. The success of the VLS theory of Wagner and Ellis [3] and its remarkable applicability to the growth of a broad variety of whiskers has convinced some researchers that the catalytic particles are in the liquid state during filament growth.

In the following paper we will show that the decrease in the number of fibers which may be grown at temperatures over 1030°C may be attributable to the melting of the iron particles comprising the catalyst at these high temperatures.

EXPERIMENT

The iron catalyst particles used in this work were obtained from the Ferrofluidics Corp., USA. Our transmission electron microscopy showed that the particles averaged 12 nm in diameter with a standard deviation of 5 nm. These particles were deposited on mullite substrates. The four semi-cylindrical substrates were placed in a furnace as shown in Figure 1 and heated in a flow of He to the growth temperature. Upon attaining this temperature, they were exposed to a 10% by volume mixture of 0.9999 methane in 0.99999 hydrogen for four minutes. The total flow rate was 500 cm^3/min. After a He purge, the filaments were heated to 1120°C where they were thickened with no further lengthening in a 29% mixture of methane in hydrogen. The thickening procedure was identical for all samples so that the number of fibers growing per unit area could be obtained from the total mass, average diameter, and average length of fibers produced.

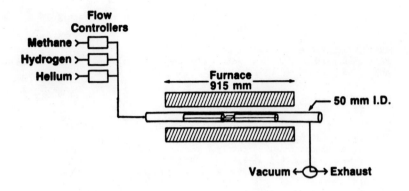

Figure 1. Apparatus. The tubular growth reactor is of 50 mm I.D. mullite. Both furnace temperature and gaseous flow rate are computer controlled.

Calculation of Particle Melting Points

A truly effective theory of melting points for these catalyst particles is a very difficult problem requiring knowledge of the surface structure and surface free energy of small particles saturated or supersaturated with carbon. However, the classic theory of Pawlow [4] as modified by Ross and Andres [1] can be shown to give good agreement with experiment for such diverse systems as Au and Ar. This theory attributes the premature melting of small particles to their excess surface energy in the solid phase, which may be diminished by melting at temperatures below the bulk melting point T_o. The melting point T of a particle containing i atoms is given by:

$$T/T_o = 1 - A\left(\frac{3}{i}\right)^{1/3} + \left[\frac{T_3}{T_o} - 1 + A\right]\left(\frac{3}{i}\right) \tag{1}$$

where

$$A = \frac{1}{(12\pi)}1/3 \frac{\left(\sigma_s v_s^{2/3} - \sigma_\ell v_\ell^{2/3}\right)}{|\Delta h|} \tag{2}$$

Here v_s and v_ℓ are the molar volumes in the solid and liquid phases, respectively, and $|\Delta h|$ is the molar heat of fusion. This theory utilizes a trimer binding energy T_3 [5] to circumvent the failure of the non-atomic Gibbs-Thompson thermodynamic description of very small particles. Furthermore, the theory requires the surface free energies σ_s and σ_ℓ of solid and molten bulk material, which are available for iron [6], but not for iron saturated with carbon. Accordingly, we use the parameters for iron, except that the Fe-C eutectic temperature of 1154°C is used for To instead of the melting point of pure iron. Figure 2 shows the relative melting point T of iron particles as a function of diameter from the above theory of Ross and Andres compared to the bulk value T_o.

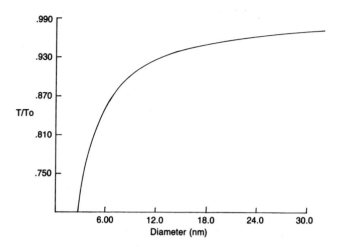

Figure 2. Melting temperature T of Fe particles of diameter D calculated from the theory of Ross and Andres relative to the bulk melting point T_o.

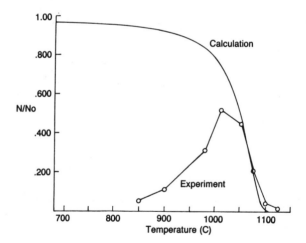

Figure 3. The calculation shows the reduction in the relative number N/N_o of solid catalyst particles with increasing temperature assuming the initial size distribution shown in Fig. 2 and using the theory of Ross and Andres. The experimental values are 1/1000 the number of fibers per cm^2 grown at temperature T.

RESULTS AND DISCUSSION

From the above specification of the normal distribution of catalytic particle diameters, it is easy to calculate the fraction f of catalytic particles of diameter larger than a specified diameter D. A fraction 1-f of the particles will have a diameter below D and will melt before they can grow a filament at temperature T. This result is displayed in Figure 3 where it is labeled "calculation". Note that the fraction of effective catalytic particles has dropped to 1/2 at 1040°C and is nearly 0 at 1100°C.

Our measurements for the number of filaments grown per cm^2 are also plotted in Figure 3. This plot is normalized to give good agreement to the calculated number of solid particles N/N_0. The number of filaments rises sharply above 800°C as the carbon potential increases. However, the dramatic fall in filament production between 1030 and 1120°C is very similar to that exhibited by the calculation. The similarity between theory and experiment is compelling reason to believe that the loss in activity by iron catalyst particles above 1050°C is due to their melting.

CONCLUSIONS

1. The number of carbon fibers which can be grown per unit area of substrate begins to diminish over 1030°C and becomes nearly negligible over 1120°C.
2. Calculations show a substantial fraction of our catalyst particles melting above 1030°C. When the distribution of catalyst particle diameters is properly accounted for, a predicted yield as a function of temperature may be calculated which gives good agreement with measurements of the number of fibers growing per cm^2.
3. The upper temperature limit for the growth of long carbon filaments by iron catalyst particles is thus consistent with the deactivation of the catalyst by melting.

ACKNOWLEDGMENTS

The authors would like to thank J. G. Gay, B. M. Clemens and T. W. Capehart for helpful discussions.

REFERENCES

1. J. Ross and R. P. Andres, Surface Science 106, 11 (1981).
2. R. T. K. Baker, and P. S. Harris, Chemistry and Physics of Carbon, edited by P. L. Walker and P. A. Thrower (Dekker, New York, 1978), Vol. 14, p. 83.
3. R. S. Wagner and W. C. Ellis, Trans. AIME 233, 1053, (1965).
4. P. Pawlow, Z. Physik. Chem. 65, 545 (1909).
5. M. D. Morse, Chem. Rev. 86, 1049 (1986).
6. H. Wawra, Z. Metall. 66, 396 (1975).

ENTROPY AND MELTING IN SMALL PARTICLES

P.M.AJAYAN and   L.D.MARKS
Department of Materials Science and Engineering
Northwestern University, Evanston, IL-60208.

ABSTRACT

We describe theoretical results for part of the Gibbs free
energy surface as a function of shape and temperature for small
particles. We predict the existence of true 'quasi-melting' at
temperatures much below the conventional melting points and also
phase   transitions   between   different stable forms as the
temperature is varied. A phase diagram for small particles is
constructed for the first time, showing the stability ranges for
different particle shapes as a function of size and temperature.

INTRODUCTION

Recent observations of small fcc metal particles in a high
resolution electron microscope, under intense fluxes, have shown
that they exist in a variety of multiply twinned particle shapes
(MTP's) and transform between these structures at a rate of about
10 Hz[1-4]. This dynamic fluid-like behaviour implies that at
sizes less than 100Å the particles undergo a sort of 'quasi-
melting' [2,5], existing in a hybrid state of solid and liquid.
The   understanding of   such processes is important both
fundamentally, since small particles act as a bridge between
atoms and solid state continua, and in industry for heterogeneous
catalysts. Monte   Carlo   simulations   and   pair-potential
calculations by various authors [6-9] have indicated anomalous
properties for atom clusters containing up to a few hundred atoms
but for sizes greater than 1-10nm such analyses become too
tedious and impractical and hence continuum models assume
importance.
In our earlier analysis [10,11] we proposed a model to
calculate the static potential energy surface for a range of
particle shapes for different values of surface anisotropies. We
considered a range of assymetric Dh particle shapes characterised
by b, the fractional distance of an off-centric disclination
from the geometric center. The calculation was done using earlier
continuum models [12,13] with the help of modified Wulff
construction for the surface energy and inhomogeneous isotropic
elasticity for strain energy. In this paper we include
vibrational and surface entropy contributions in these particles
and construct the Gibbs free energy surfaces. For the first time
a phase diagram is constructed showing the different surface
morphologies as functions of temperature and particle radius. The
results seem to be consistent with earlier experimental
observations [14,15].

ENTROPY

The evaluation of the entropy enables us to understand the
behaviour of small particles at different temperatures and hence
we move from a previous static model to a more realistic dynamic
one. The main part of entropy arises from the changes in the

vibrational spectrum due to defects (in this case an eccentric disclination) and the anisotropy in the surface energy. The entropy of mixing is the same for all MTP shapes since our model presumes a degeneracy of five for the disclination position in all geometries, except for a single crystal and a bicrystal and is hence not important.

Vibrational entropy

To calculate this we follow the perturbation approach used by Huntington et al [16], who considered contributions due to the elastic distortion in the core of the defect, the temperature dependence of the elastic strain energy outside the core, and the effect of the nett dilatation on the normal mode frequencies. For a disclination core the convergence of the strain energy integral is much faster than for a dislocation core and the strains in the core are unreasonable only for large particles [13]. Hence we assume that the distortion of the core of the disclination is small and neglect the contribution of entropy from this region.

The entropy contribution from the stored elastic energy W comes from the temperature dependence of the elastic constants. This gives

$$dS = -(dW/dT) = -(dW/d\mu) \; (d\mu/dT). \qquad (1)$$

Here μ is the shear modulus and T the temperature. Now from the disclination solutions we had found earlier [11] that the strain energy density is:

$$W = E_D^2 V(1-b^2)^2 \; \mu \; / \; 4(1-\nu) \qquad (2)$$

where $E_D = 0.0205$, ν =Poissons ratio and V=Volume of the particle. Hence

$$dS = E_D^2 V(1-b^2)^2 \; (d\mu/dT) \; / \; 4(1-\nu). \qquad (3)$$

The value of $d\mu/dT$ used is $-0.38 \; 10^8$ Pascal degK as in [16].

The Stress and Strain components due to the disclination were calculated as before [11]. The nett dilatation was evaluated by numerically integrating (trapezoidal rule was used for integration) the strain solutions from the center to the outer boundary of the particle. The volume changes were positive suggesting expansion, though negligible (<0.1%) and were largest around b = +/-0.6.

The entropy contribution resulting from the volume change was found from the Gruneisen relation since the dilatation induces a size dependent image force which acts to alter the phonon spectrum. The change in the frequency due to dV, in atomic volumes, is

$$d \; / \; = - \beta \; dV/N \qquad (4)$$

where N is the total number of atoms affected by the elastic field and β is the Gruneisen constant. This gives rise to an entropy contribution

$$dS = -3 \; \beta \; dV \; k. \qquad (5)$$

where k is the Boltzmann constant and value of Gruneisen constant is taken as 1.96, a common value for fcc metals.

Surface entropy

It has been shown that the anisotropy of surface free energy is temperature dependent [17] and with clean surfaces this should logically stem from the entropy contribution to the surface free energy. For a small particle this could be quite substantial since a major portion of the total energy comes from the surface. We can write the free energy from the surface as $G = H - TS$, and we can write H and S as

$$H = E_w \, \gamma_H^{111} \, V^{2/3} \text{ and } S = E_w \, \gamma_S^{111} \, V^{2/3}$$

$$\text{where } \gamma_S^{111} = -(d\gamma_H^{111}/dT) \qquad (6)$$

The value of H was evaluated as a function of b in our earlier papers [10,11] where E_w is a dimensionless energy parameter, the value of which depends only on the shape and not the volume. The approximate value of specific surface entropy, $d\gamma_H^{111}/dT$, taken from Kirchner et al [19] was -1.2 ergs/cm^2 degK and the surface entropy contribution dS was calculated for different values of b.

RESULTS

The total entropy contribution is the sum of the three terms calculated above which are then added to the enthalpy found earlier [10]. Normalized values of G are plotted against b for four temperatures in figure 1; at 0^{0}K, room temperature (300^{0}K), near epitaxial temperature for fcc metals (600^{0}K), and a much higher temperature (1000^{0}K).

The Boltzmann distribution function can be written as:

$$P = \exp \{-E_b/KT\} / \sum_b \exp \{-E_b/KT\} \qquad (7)$$

The denominator was integrated numerically. The values of P are plotted against b in figure.2 for two different particle radii.

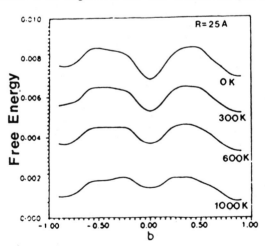

Figure 1. Shows the Gibbs free energy surfaces as a function of temperature and particle shape.

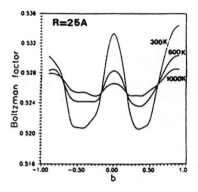 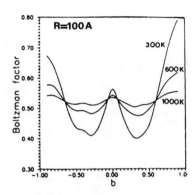

Figure 2: Boltzmann distribution curves for particle shapes at three different temperatures and two particle sizes.

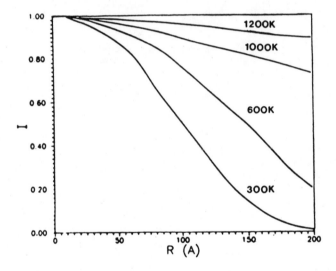

Figure 3. Shows the index of quasi-melting as a function of particle size.

Then we define an 'index of quasi-melting', I, as the ratio of the minimum to maximum value of P occuring for a particular particle radius and temperature and plot this value against particle size in figure.3. As the value of I approaches 1 the minima on the free energy surfaces turn to saddle points and disappear, making them essentially flat. This indicates the

transition from stable morphologies to metastable shapes and
finally to randomly fluctuating configurations or quasi-melting.
From the above results, for an empirical value of I=0.98 we plot
the quasi-melting temperatures against particle radius in
figure.4. With the data for the conventional melting in small
particles [18] together with the stability ranges for the Dh and
Ic [12,13] we convert figure.4 into a phase diagram for small
particles.

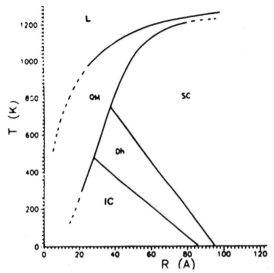

Figure 4. The phase diagram for small particles. L, QM, SC, Ic
and Dh denote Liquid, Quasi-Molten, Single crystal, Icosohedral
and Decahedral structures respectively. Dotted lines are
extrapolated.

DISCUSSION

    From the above results we have demonstrated the true quasi-
melting behaviour of small particles. Below a certain particle
radius and much below the melting temperatures the particle
structure floats in a shallow probability well of varying
geometric shapes, showing little selection tendency. In other
words as the temperature is raised, aided with entropy the
particle moves from anisotropy to isotropy, from the realm of
rigidity to the realm of quasi-shapes, exhibiting the phenomenon
of quasi-melting. Note that temperature is not the only mechanism
of inducing quasi-melting; the results of the in-situ microscopy
work [1-4] indicate that an electron beam can raise the effective
temperature high enough. We should mention that preliminary work
(A.K.Petford, private communication) has indicated that heat
alone will induce quasi-melting.
    Another important result that we obtained from the free
energy curves is that around 600K, a typical epitaxial
temperature for most fcc metals, the most stable form of the
particle undergoes a phase transition for a small particle: ie
compared to a symmetrical decahedral particle at room temperature
the stable form becomes the single crystal at 600K. It was also

noticed that as the particle size increases the phase transition temperature goes up which means that for larger sizes, once formed, the MTP remains stable till higher temperatures. It would be interesting to relate the phase transitions our curves predict to the true epitaxial nature of particles for instance one could say epitaxy could be much more stable in regions where the single crystal phase is contiguous with the molten. Though the statistical occupancy of the different local minima on the potential energy surface relate directly to the Boltzmann factor that we calculated, the value of the index of quasi-melting used to extrapolate the melting temperatures are quite arbitrary though seems reasonable. It would also be interesting to see the index of quasi-melting in a reciprocal relation to the surface anisotropy ratio, α, which in a strong faceting model for the MTP's is the ratio of the energy of (100) facet to that of (111) facet. Thus the value of I=0.98 that we have taken corresponds to a value of 1.02 for α, whereas we started with an anisotropy value of 1.1. Since the value of $\alpha =1$ corresponds to perfect isotropy, we have reduced anisotropy to an extent that is suggestive of quasi-melting.

ACKNOWLEDGEMENTS

This work would have been impossible without major contributions at earlier stages of the models by Professors John Dundurs and Archie Howie and was funded by the NSF on grant number DMR 8514779.

REFERENCES

1. S.Iijima and T.Ichihashi, Japanese J. App. Phy., 24, L125(1985).
2. S.Iijima and T.Ichihashi, Phys.Rev.Letters, 56, 616(1986).
3. S.Iijima, J.Electron Microscopy, 34, 249(1985).
4. J.O.Bovin, R.Wallenberg and D.J Smith, Nature, 317, 47(1985).
5. L.D.Marks, P.M.Ajayan and J.Dundurs, Ultramicroscopy, 20, 78(1986).
6. M.R.Hoare and P.Pal, J.Crystal Growth, 17, 77(1972).
7. J.Farges, M.F.Deferaudy, B.Raoult and J.Torchet, Surf.Sci, 106, 95(1981).
8. J.J.Burton, J.Chem.Phys., 52, 345(1970).
9. R.Stephen Berry, Julius Jellinek and Grigory Natanson, Physical Rev., 30, 919(1984).
10. P.M.Ajayan, L.D.Marks and J.Dundurs,Mat.Res.Soc.Symp.Proc. 82, 469(1987).
11. J.Dundurs, P.M.Ajayan and L.D.Marks, Phil.Mag, in press.
12. L.D.Marks, Phil.Mag, 49, 81(1984).
13. A.Howie and L.D.Marks, Phil.Mag, 49, 95(1984).
14. S.Ino, J.Phys.Soc.Jap., 21, 346(1966).
15. S.Ino and T.Ogawa, J.Phys.Soc.Jap., 22, 1369(1967).
16. H.B.Huntington, G.A.Shirn and E.S. Wajda, Phys.Rev., 99, 1085(1955).
17. H.O.K.Kirchner and G.A.Chadwick, Phil.Mag., 22, 447(1970).
18. J.P.Borel, Surf.Sci., 106, 1(1981).

PERSPECTIVE ON THE USE OF GAS ADSORPTION FOR
PARTICLE-SHAPE CONTROL IN SUPPORTED METAL CATALYSIS

A-C. Shi, K. K. Fung, J. F. Welch, M. Wortis[1], R. I. Masel[*] Departments of Chemical Engineering, Physics, and Materials Research Lab, University of Illinois, Urbana Il. 61801

Abstract

This paper explores the possibility of using gas adsorption for particle-shape control in supported metal catalysis. Calculations are done to identify the kinds of shape changes which are possible due to treatment in an appropriate adgas. Experiments are done to try to verify the concepts. The results show that gas adsorption can be used for particle shape control in supported metal catalysis.

Introduction

Traditional supported metal catalysts consist of small metal particles on an inert support. While a typical supported metal catalyst is fairly inhomogeneous, recent work with single crystal catalysts has shown that one can get more active and selective catalysts, if one can control the distribution of the sites on the catalyst surface appropriately. There are a number of examples where the reactivity of one face of a metal is several orders of magnitude larger than that of another face[1]. The data on selectivity is less clear, but there are a few examples where the selectivity of one face of a metal is an order of magnitude larger than the selectivity of another face of the metal. Thus, it seems that, if one could find a way to control the particle shapes in a supported metal catalyst as to produce an enhanced concentration of especially active or selective faces, one should be able to obtain a significant improvement in the catalyst's performance.

Unfortunately, at present, particle-shape control in supported metal catalysis is largely a black art. There have been several observations of particle shape changes during various treatments such as those in references [2-6]. However, no systematics are available. Thus, it is not obvious how to control particle shapes in a supported metal catalyst.

The objective of the work reported here is to see whether the particle shapes in a supported metal catalyst could be controlled by treating the catalyst in an appropriate adgas. The idea is to heat the catalyst in an adgas at high temperature, wait for the particle shapes to equilibrate, and then rapidly quench the catalyst in order to freeze in the desired shape. One would then clean off the adsorbate to activate the catalyst.

Fundamentally, equilibrium particle shapes are determined by the variation in the surface free energy with orientation. If there are no variations in the surface free energy with orientation, as in a liquid, then at equilibrium all of the particles in the catalyst would be spherical. With a metal, certain crystal faces have a lower free energy that others. Facets with low surface energy tend to grow at the expense of other higher energy faces[7]. As a result, theoretically, metal particles can be faceted rather than spherical. However, note that, at equilibrium, all of the particles in the supported metal catalyst should have the same shape[8]. Hence, one could use an equilibration process to obtain a catalyst with a fixed distribution of surface orientations.

At present, there have only been a few detailed measurements of the equilibrium

---

[*] Correspondence Should be Addressed to this Author
[1] Present Address: Physics Dept, Simon Fraser Univ., Burnaby B.C. V5A 1S6 Canada

crystal shape of clean transition metal particles[3, 9-12]. One of the surprises in the data so far is that at equilibrium, clean metal particles are usually almost spherical. Nearly spherical particles arise theoretically only when there are small variations in the surface free energy with orientation. The best available evidence is that the variation in the surface free energy with orientation for a typical transition metal particle is only about 1-3 kcal/mole.

To put the 1-3 kcal/mole variation in the surface energy in perspective, note that the heat of adsorption of O_2 on platinum can vary with orientation by as much as 20 kcal/mole[1]. Thus, when one adsorbs a gas, one makes a big perturbation in the surface free energy. That perturbation can have a significant effect on the particle shape.

Phenomological Calculations

We have done simple phenomological calculations to see what types of shape changes one would expect due to the adsorption of an adgas. Metois' equilibrium crystal shape [9] was used to estimate the free energy of a clean platinum particle as a function of crystal face. Data on the variation of the heat of adsorption as a function of crystal face was then used to estimate how the free energy would change in a gaseous environment. Herring's method was then used to estimate a crystal shape[7]. For simplicity, it was assumed that the adsorption followed a Langmuir adsorption isotherm, and that the heat of adsorption was constant.

Figures 1a, b, c show the results of the calculations. Figure 1a shows the shape predicted for a clean crystal. The free energies calculations were fit to Metois's data, so it is not surprising that the particle shape calculated for a clean platinum particle looks much like that observed by Metois. However it is interesting to examine the results in 1b and c. Figure 1c, shows the crystal shape predicted for a crystal heated in hydrogen. According to McCabe and Schmidt[12], the initial heat of adsorption of hydrogen is about 10 kcal/mole higher on Pt(100) than on Pt(111) and about 5 kcal/mole higher than on step surfaces. As a result, when a platinum catalyst is exposed to 1 atm of hydrogen at temperatures above 800 K, where atoms are mobile, (100) facets should grow at the expense of step surfaces and (111). The equilibrium shape works out to be nearly a cube. A similar effect also occurs in oxygen. The heat of adsorption of oxygen is about 20 kcal/mole higher on Pt(100) than on Pt(111)[1]. Thus, (100) facets should grow when the sample is exposed to an oxygen atmosphere; the equilibrium shape is again almost a cube. In contrast, however, the heat of adsorption of N_2 varies little with crystal face[13]. As a result, no particular face should grow preferentially when a platinum catalyst is exposed to nitrogen. According to the calculations, the crystal shape should be nearly a sphere.

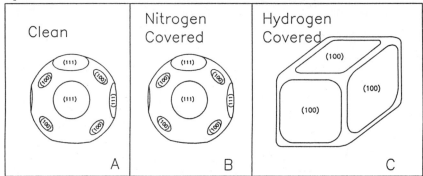

Figure 1. The equilibrium crystal shape calculated for clean and adsorbate covered platinum particles.

Experiments

Experiments were done to see if the kinds of particle-shape changes expected theoretically do indeed occur experimentally. The experiments involved evaporating about 100 Å of silicon onto a sodium chloride substrate, removing the silicon from the substrate, oxidizing the silicon to silica, evaporating platinum onto the silica, treating the resultant sample in hydrogen, nitrogen, or oxygen, then examining the shapes of the platinum particles in the catalyst with an electron microscope. In previous work Wang, Lee, and Schmidt[2] measured the changes in the particle shapes which occur when a platinum particle is exposed to hydrogen and nitrogen. They found results similar to those predicted above. We have repeated their experiment and found that we get square particles in a H_2 atmosphere and rounded particles in an N_2 atmosphere, as shown in figure 2. Notice the close correspondence between the computational results in figure 1 and the experimental results in figure 2. While there are some details missing, the general particle shapes expected theoretically are the shapes one observes experimentally, at least in the data in figure 2.

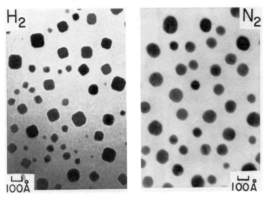

Figure 2 The Particle shapes observed after evaporating platinum onto a smooth silica substrate then annealing for 6 hours in hydrogen or nitrogen at 900 K.

Figure 3 The particle shapes observed after evaporating platinum onto a rough silica substrate then annealing in hydrogen for 6 hours at 900 K. Similar results were also seen after 9, 24 and 48 hours. Note that the scale is different in figures 2 and 3.

Surprisingly, however, one does not obtain the same shapes all of the time. The data in figure 2 were taken by depositing platinum particles on a silica substrate which had been stretched flat. When a topographically rough silica substrate is used, the results are quite different. Figure 3 shows a good example of that. The data in figure 3 were taken by depositing platinum onto a silica support which had some rough regions and then treating the particle in H_2 as above. In this particular experiment, the particle morphology was found to vary over the surface of the catalyst. The particles on the flat regions of the support were similar to those in figure 2a. However, on the rough regions of the support, the shapes changed. Figure 3 shows the transition region, where the particles roughen, and then lose their distinct shape. Notice the sharp transition in the particle morphology that occurs as the substrate roughness changes.

At present it is not clear why the particle shape is affected by the morphology of the support. Winterbottom[8] showed that if one deposits a metal particle on a flat substrate, the particle-shape will be truncated by the substrate, but the general particle shape will not change. There is some speculation in the literature that if one deposits a particle on a substrate which is topographically rough on a length scale comparable to the particle size[14], the particle shapes will change, but the evidence is not very clear. The results in figure 3 however show that particle shapes do change as the characteristics of the substrate change. The phemonological model seems to fit the data on the flat substrates reasonable well. However, more work needs to be done to understand what happens on the topographically rough substrates.

Lattice Gas Model.

Still, the results above show that, if one chooses the conditions and substrates appropriately, one can use gas adsorption to change particle shapes in the way expected from an analysis of the variations in the heat of adsorption with crystal face. Therefore, one can use an analysis of the possible variations in the heat of adsorption with crystal face to answer the question, 'what kinds of shape changes are possible due to gas adsorption, and how does one obtain them?'

Our recent calculations[15] have answered that question in a fairly general way for ordered overlayers of atomic adsorbates. Our approach was to assume that the adsorbate-surface interaction could be written as a pairwise sum of first-nearest-neighbor and second-nearest-neighbor forces of arbitrary form and strength. We then calculated how the free energy of the each face of the metal would vary as the strength of the interaction changed. The Wulff construction was then used to calculate the particle-shape.

Figure 4 shows all of the generically different kinds of surface shapes possible for the adsorption of an atomic adsorbate with first and second nearest neighbor pairwise interactions on an FCC substrate at 0 °K in the absence of surface reconstructions. Surprisingly, the calculations show that only 13 generic types of particle shapes are possible with an atomic adsorbate with pairwise interactions (the relative sizes of the faces can be varied arbitrarily). If one chooses an adsorbate which stabilizes flat surfaces, (100) or (111) facets grow, while, if one uses an adsorbate which stabilizes steps of different orientations, (110), (210) or (311) facets grow. However, according to the calculations one cannot get other kinds of facets to grow in the presence of an atomic adsorbate with pairwise interactions. Hence, the only particle shapes that one can get are those with (100), (111), (110), (210) and (311) facets. Simple geometry shows that only 13 combinations are possible.

Physically, what is happening is that with an atomic adsorbate, one can find an interaction to stabilize steps or terraces, but not both. The (111), and (100) faces are flat, so they can be stabilized with an interaction which just stabilizes terraces, while the (110), (210) and (311) faces are completely stepped, so they can be stabilized with an interaction that just stabilizes steps. However, all of the other faces in the stereographic triangle have a step, terrace structure, so one would need a more complicated interaction to stabilize them.

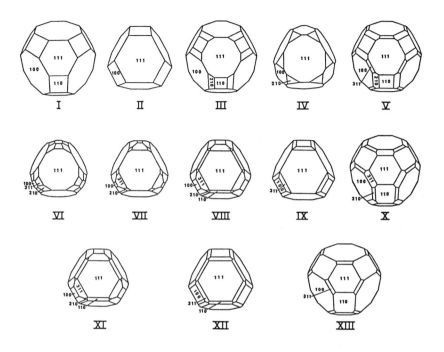

Figure 4. The generic crystal shapes which can be stabilized on an FCC surface at 0 °K due to the adsorption of an atomic adsorbate with first nearest neighbor and second nearest neighbor interactions.

Consider, for the moment, what one would have to do to stabilize (410) facets. The (410) faces has a step terrace structure, with (100) terraces, and (110) steps. One could imagine trying to form a (410) surface by starting with a (100) facet and adding an adsorbate which stabilizes (100)x(110) steps. Notice, however, that if one added more adsorbate, one would get more steps, and eventually one would produce a (210) surface. One could of course, imagine trying to produce a (410) facet, by carefully controlling the adsorbate concentration. However, according to the calculations, if one produced a (410) facet in that way, it would not be stable; one could always lower the free energy of the system, by changing the particle into a hill and valley structure, with (100) and (210) facets. As a result, one could not stabilize (410) facets with an adsorbate which merely stabilizes (100)x(110) steps. Instead, one would need an interaction which stabilizes both steps and terraces. According to the calculations, that is not possible with an atomic adsorbate with first- and second-nearest-neighbor interactions only.

A similar argument holds for all the other faces in the stereographic triangle except (100), (111), (110), (210) and (311). As a result, according to the calculations the only faces one can stabilize due to adsorption of an atomic adsorbate, with first nearest neighbor and second nearest neighbor pairwise interactions are (100), (111), (110), (210), and (311). It is interesting to note, therefore, that in all the experiments which have been done so far, the only faces which have been observed to be stabilized in the presence of an atomic adsorbate are (111), (100), (110) and (210).

Of course, the situation is entirely different with molecular adsorbates. According to our analysis, if the molecule were big enough, the steric hindrances would be sufficient to keep the steps apart. One might also tailor the molecule so that one part of the molecule stabilizes the steps, and another part of the molecule stabilizes the terraces. At present, we have not completed our calculations on the effects of molecular adsorbates on the equilibrium crystal shape. However, it appears so far that we can stabilize any given face, if we choose the molecule/surface interaction appropriately.

Conclusion

The results here show that particle-shape control in supported metal catalysis is feasible. The experimental results show that the particle shapes change when the catalyst is treated in an appropriate adgas. The shapes observed experimentally show close correspondence to the shapes expected from an analysis of the variations in the heat of adsorption with crystallographic orientation, provided the substrates are flat. Calculations so far indicate that (111), (100), (110), (210) and (311) facets can be stabilized in the presence of an atomic adsorbate, while other faces can be stabilized in the presence of a molecular adsorbates. Clearly, there is much more to be done. However, it appears so far that gas adsorption can be used for particle-shape control in supported metal catalysis.

Acknowledgement

This work was supported by the National Science Foundation under grant DMR 86-12860, and by grants from the General Motors Corporation, Amoco, Shell and Chevron USA. This work made use of the University of Illinois Center for the Microanalysis of Materials which is supported as a national facility by the National Science Foundation.

Literature Cited

1. See for example, M. Asscher, J. Carrazza, M. M. Khan, K. B. Lewis, G. A. Somorjai, J. Catalysis, 98(1986) 277., R. I. Masel, Catalysis Rev. 28(1986) 335.
2. T. Wang, C. Lee, L. D. Schmidt, Surface Sci. 163(1985), 181.
3. M. Drechsler, in Vu. Thien Binh ed. Surface Mobilities on Solid Materials, Plenum, NY (1983).
4. M. J. Yacaman, A. Gomez, Appl. Surface Sci., 19 (1984)348. Appl. Catalysis 13 (1984) 1.
5. E. Ruckenstein, I. Sushumna, J. Catalysis, 97(1986)1. S. H. Lee, E. Ruckenstein, J. Catalysis, 107(1987) 23.
6. W. M. Robertson, P. G. Shewmon, J. Chem. Phys. 39(1963) 2330.
7. G. Wulff, Z. Krist. 34(1901) 449. C. Herring, Phys Rev 82(1951) 87.
8. W. L. Winterbottom, Acta Met. 15(1967), 303.
9. J. C. Heyraud, J. J. Metois, Acta Met. 28 (1980) 1789.
10. T. Barsotti, J. M. Bermond, M. Drechsler, Surface Sci. 146(1984) 467.
11. S. K. Menon, P. L. Martin, Ultramicroscopy 20(1986) 93.
12. R. McCabe L. D. Schmidt, Proc. 7th Inter. Conf. Solid Surfaces (1977) 1201.
13. K. Schwaha, E. Bechtold, Surface Sci. 66(1977) 45.
14. R. K. P. Zia, J. Taylor, unpublished work
15. A-C. Shi, Phys. Rev. B, to appear

MÖSSBAUER AND EXAFS STUDY OF ^{57}Fe LABELLED Pt/SiO$_2$ CATALYSTS

AECL-9677

J.A. Sawicki*, J.H. Rolston*, S.R. Julian**, T. Tyliszczak*** and
K.D. McCrimmon*
* Chalk River Nuclear Laboratories, Chalk River, Ontario K0J 1J0, Canada
** Department of Physics, University of Toronto, Toronto, Ontario M5S 1A7,
Canada
*** Materials Science Center, McMaster University, Hamilton, Ontario
L8S 4M1, Canada

ABSTRACT

    Specimens of highly dispersed Pt/Silicalite catalyst, decorated with
^{57}Fe by impregnation in aqueous ferric nitrate solution, have been studied
after decoration, and various pretreatments including exposure to carbon
monoxide, by in situ Mössbauer spectroscopy. The form of Pt was followed
by EXAFS analysis near the Pt L_{III} edge. The microstructural studies
were combined with CO chemisorption and infrared absorption measurements.
It is observed that iron in Pt clusters is very reactive even at ambient
temperatures and that the interaction of Fe with CO is strongly accelerated
by platinum.

INTRODUCTION

    The catalytic performance of silica-supported Pt-Fe catalysts for
hydrocarbon reduction and CO oxidation is largely determined by the
electronic state of iron in bimetallic clusters [1-3], since iron, as the
most electropositive metal among group VIII transition elements, acquires
electrons from platinum and forms very strong bonds with carbon and oxygen
atoms. Due to a fairly weak interaction between iron and silica, the
dispersion and other final properties of such catalysts depend very much on
fabrication procedures, and especially are affected by acidity during
impregnation [1,4].

    Mössbauer spectroscopy of ^{57}Fe was used earlier for characterization
of Pt-Fe/silica catalysts with rather large contents of iron (1-30 wt%)
[4-7]). In the present paper, we report on the studies of the interaction
between carbon monoxide and Pt-Fe/Silicalite catalysts with iron contents
below ~ 0.1 wt% (molar ratio Fe/Pt ~ 1/40).

EXPERIMENTS AND RESULTS

    The initial Pt catalyst was obtained by impregnation of crystalline
silica (Silicalite S-115) with Pt solution, resulting in a loading of
~ 10 wt%Pt, metal area of 14.3 m^2/g (from CO adsorption), average
particle size ~ 3 nm and H/Pt ratio of ~ 0.6. The 250 mg batches
were then impregnated with iron concentrations up to ~ 0.1 wt% Fe,
using acidic (pH~ 1) ferric nitrate solution highly enriched in ^{57}Fe
(~ 95%). After drying at 90 - 150°C and calcination, the decorated
catalysts were examined by X-ray diffraction, infrared absorption and
measurements of metal area and hydrogen-deuterium exchange rates.

    Mössbauer spectroscopy, EXAFS and infrared absorption measurements
were carried out under well controlled conditions, in an all-glass cell
with provisions for in-situ thermal treatments at temperatures up to 300°C
in vacuum, O$_2$, H$_2$ and CO (Fig. 1). Analyzed specimens were pressed into 2
cm^2 discs with the thicknesses optimized for infrared (~ 10 mg/cm^2),

Mössbauer and EXAFS ($\sim$ 50 mg/cm^2) measurements.

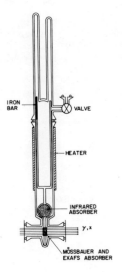

IRON
BAR

VALVE

HEATER

INFRARED
ABSORBER

γ, x

MÖSSBAUER AND
EXAFS ABSORBER

Figure 1

Microreactor cell used for in
situ infrared spectroscopy,
Mössbauer spectroscopy and EXAFS
measurements in transmission
geometries.

Carbon monoxide uptake was analyzed by Fourier transform infrared
absorption (FTIR), using a Bomem spectrophotometer. Characteristic spectra
are shown in Fig. 2. On exposure of the as-calcined specimen to carbon
monoxide at ambient temperature (spectra 2-5), one observes a gradual
growth of the absorption band at 2075 cm^{-1} (for CO adsorbed on Pt), whereas
the 2120 cm^{-1} band (for CO adsorbed on PtO) stays the same size. Long
exposures to CO resulted in $\sim$ 80% reduction of the catalysts (spectrum
5) as compared to a sample reduced in H$_2$ (spectrum 6). Iron decoration did
not change the IR absorption spectra in any significant way. Measurement
of the H-D exchange rate in the vapour phase indicated however that 0.1 wt%
Fe-10 wt% Pt catalyst was a factor of 3 less active than the initial Pt
catalyst.

Mössbauer spectra were measured using a 30 mCi ^{57}CoRh source. Fig. 3
shows the patterns observed and Table 1 gives the results of least square
fits to the data points.

When interpreting Mössbauer spectra of Pt-Fe catalysts one must be
aware of the coincidence between the isomer shifts of ferric (Fe^{3+}) oxides
and dilute Fe in Pt alloys [8-10], since in both cases the IS $\sim$ 0.35
mm/s, despite obvious differences in the electron structure of iron in
these two cases. The large positive isomer shift for Fe in Pt is due to
the fact that platinum donates a significant fraction of its valence
electrons to electropositive iron.

The comments about the Mössbauer spectra showin in Fig. 3 are as
follows:

(i) Decoration of Pt/Silicalite catalysts with iron ions produces highly
dispersed superparamagnetic particles of ferric hydroxide (Fe(OH)$_3$) and/or
oxide (α-Fe$_2$O$_3$), which are represented by an asymmetric quadrupole doublet
with an isomer shift IS=0.33-0.38 mm/s and QS=0.77-0.89 mm/s (attributed

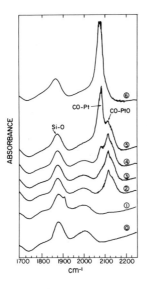

Figure 2. FTIR absorption
spectra of Fe-decorated
Pt/Silicalite catalyst measured
at room temperature for as
calcined specimens: (0) virgin
Silicalite;(1) Pt-Fe/Silicalite
calcined in O_2 at 180°C,
measured in vacuo; (2- 5)
measured in 13 kPa CO as a
function of exposure time; (2)
0.5 h; (3) 1.5 h; (4) 4 h and
(5) 45 h. Adsorption of CO on
catalyst fully reduced in H_2 is
represented by curve (6).

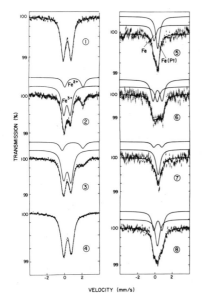

Figure 3. Mössbauer spectra of
10 wt% Pt-0.1 wt%
Fe/Silicalite catalysts taken
at 20°C for the following
specimens: (0) virgin
Silicalite; (1) calcined at
500°C in air; (2) exposed to 13
kPa CO at 20°C; (3) exposed to
CO at 200°C; (4) re-opened to
air; (5) calcined at 300°C and
reduced in H_2 at 200°C; (6)
kept in vacuo for 30 days; (7)
exposed to CO at 20°C for 1 d;
(8) exposed to air.

Table 1. The summary of the results. IS - isomer shift with respect to metallic iron, QS - quadrupole splitting and W - linewidth (FWHM). The last two columns give conclusions about the chemical form of Fe and Pt drawn from the Mössbauer and EXAFS data, respectively.

Specimen	IS mm/s	QS mm/s	W mm/s	Form of Fe	Form of Pt
1. calcined; kept in air	0.33(1)	0.92(1)	0.62(1)	Fe^{3+}	PtO_2
2. calcined; exposed to CO at 20°C	0.32(2) 1.13(2)	0.76(2) 2.09(4)	0.58(2)	Fe^{3+} 27%Fe^{2+}	PtO_2
3. calcined; exposed to CO at 100°C	0.36 0.96	0.83 2.36	0.66(1)	Fe^{3+} 20%Fe^{2+}	PtO_2
4. re-exposed to air	0.35	0.90	0.67(1)	$Fe3^+$	PtO_2
5. reduced in H_2 at 200°C; kept in H_2	-0.08(2) 0.35(1)	– –	0.54(3)	Fe met 67%Fe(Pt)	Pt met
6. reduced in H_2 at 200°C; kept in vacuo	0.34(2) 0.30(3)	1.04(4) –	0.64(4)	Fe^{3+} 27%Fe(Pt)	Pt met
7. exposed to CO at 20°C	0.31(4) 0.35(2)	0.96(9) –	0.58(7)	Fe^{3+} 71%Fe(Pt)	Pt met
8. exposed to air at 20°C	0.35(1) 0.29(1)	0.95(2)	0.58(2)	Fe^{3+} 43%Fe(Pt)	Pt met +PtO_2

to Fe oxide clusters on SiO_2 [5]). Drying of the specimen at mild temperatures up to 150°C and calcination at temperatures up to 300°C did not cause any significant change in the spectra. After the sample was heated at 500°C in air, however, the splitting increased markedly, up to QS=1.08 mm/s, which can be attributed to formation of Pt-Fe oxide clusters [5] (Fig. 3-1). A lack of ferromagnetic splitting at 77 K is consistent with high dispersions characteristic of depositions at low pH [4].

(ii) The spectra measured during exposure of, as-deposited or calcined specimens, to CO at 20°C indicated reduction of about 30% of Fe^{3+} to Fe^{2+}, which is represented by an additional quadrupole doublet (IS=1.1 mm/s, QS=2.1 mm/s) (Fig. 3-2). Appearance of Fe^{2+} can be attributed to chemisorption of CO groups, most probably only on those Fe atoms which are in intimate contact with Pt atoms which act as reduction promoters. The fraction of Fe^{2+} species decreases with the exposure temperature to about 20% at 100°C (Fig. 3-3), and rapidly drops to zero on re-exposure to air (Fig. 3-4). The low amount of Fe^{2+} can be attributed to the fact that CO easily decomposes at the 100 surface of Fe [12].

(iii) An in-situ measurement, made after prolonged reduction of the sample at 500°C, which consisted of three evacuation and H_2 exposure cycles, indicated complete reduction of iron into metallic form. The spectrum (Fig. 3-5) shows a superposition of two single lines, representing small particles of metallic γ-iron (IS=-0.08 mm/s) and Fe diluted in Pt (IS=0.35 mm/s), respectively. After a long stay (20 d) in low vacuum (~ 1 Pa), the iron oxidized back to largely Fe^{3+} (Fig. 3-6). On exposure of reduced catalyst to 13 kPa CO, the spectrum (Fig. 3-7) has reverted to that of metallic-like iron, which is attributed to chemisorption of carbon monoxide groups and re-reduction to metallic form. Brief (1 day) exposure to air resulted in partial (~ 60%) conversion of Fe to Fe^{3+} (Fig. 3-8).

Supplementary data about the microstructure of the catalysts were obtained from EXAFS analysis near the Pt L_{III} edge [13-14]. EXAFS measurements were performed at Cornell High Energy Synchrotron Source (CHESS) facility using a Si(220) double crystal monochromator and a 1x10 mm^2 X-ray beam exit slit. Representative data from EXAFS analysis are shown in Fig. 4. Experimental data were treated in a standard way using a least square fitting to the inverse Fourier filtered K-weighted data as well as by a procedure for parameter determination from phase shift and amplitude functions. Experimental data of $PtO_2 \cdot H_2O$ and metallic platinum were used as models.

(i) Visual inspection of the spectra allows a clear distinction between the metallic and oxidized platinum clusters. The very close similarity between the results for the Pt catalyst and $PtO_2 \cdot H_2O$ is to be noted, whereas the differences between metallic platinum and platinum in reduced catalyst are pronounced, apparently due to lowering of the coordination number in the latter case [13,14]. The predominant form of platinum, as determined from EXAFS measurements, is given in Table 1.

(ii) From standard numerical analysis of the data more structural details could be extracted. In particular, from the radial distribution for the calcined catalyst one finds only one oxygen shell, and no metal bonds are detected. In the case of reduced and CO blanketed catalyst one can see mainly Pt atoms at a distance of 0.277 nm (same as for Pt metal) with somewhat smaller coordination number (92% of Pt metal). One can also see a small contribution (~ 8%) probably from carbon neighbors (surface contribution). On exposure of the sample to air for 1 day one observes two types of near neighbors: Pt atoms (distance of 0.276(3) nm and mean coordination number 0.57(15), both similar to Pt metal) and oxygen atoms (distance 0.181(9) nm and mean coordination number 0.32(10), characteristic of $PtO_2 \cdot H_2O$).

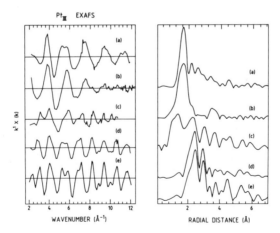

Pt_{III} EXAFS

WAVENUMBER (Å⁻¹)

RADIAL DISTANCE (Å)

Figure 4. K-weighted Pt L_{III} edge EXAFS curves (left) and their Fourier transforms (right) obtained at 20°C for $PtO_2 \cdot H_2O$ (a); as deposited (calcined) Pt-Fe catalyst (b); Pt-Fe catalyst reduced in H_2 and measured in air (c) and in CO (d); 3 μm platinum foil (e).

CONCLUSIONS

Combined Mössbauer and EXAFS studies have been carried out for Fe-decorated Pt/Silicalite catalysts, wherein the chemisorption of CO was followed in situ by infrared absorption measurements. The Mössbauer spectroscopy analysis indicated the ability of iron in Pt clusters to react with carbon monoxide. In the presence of CO, in the calcined catalyst about 30% of Fe^{3+} atoms are reduced to the ferrous Fe^{2+} state at room temperature, and this fraction drops to about 20% in the 100-200°C range. In pre-reduced and passivated catalyst, iron can be rapidly and almost completely re-reduced in CO even at room temperature, but on exposure to air quickly re-oxidizes to the ferric Fe^{3+} state. EXAFS analysis performed on the same specimen has shown that in this case platinum clusters remain largely metallic. Thus, platinum strongly promotes the reduction of iron not only in hydrogen but also in carbon monoxide.

Acknowledgements

The authors would like to thank Dr. E.A. Symons for discussions and helpful comments on the manuscript of this paper.

References

1. L. Guczi, Cat. Rev. Sci. Eng. 23, 329 (1981).

2. H. Tøpsoe, B.S. Clausen and S. Mørup, Hyperfine Interactions 27, 231 (1986).

3. C.H. Bartholomew and M. Boudart, J. Catal. 29, 278 (1973).

4. M.A. Vannice and R.L. Garten, J. Mol. Catal. 1, 201 (1975/6).

5. I. Dezsi, D.L. Nagy, M. Eszterle and L. Guczi, React. Kinet. Catal. Lett. 8, 301 (1978); J. Phys. (Paris) 40, C2-76 (1979).

6. J.W. Niemantsverdriet, J.A.C. van Kaam, C.F.J. Flipse, and A.M. van der Kraan, J. Catal. 96, 58 (1985).

7. J.W. Niemantsverdriet, J. van Grondelle and A.M. van der Kraan, Hyperfine Interactions 28, 867 (1986).

8. D. Palaith, C.W. Kimball, R.S. Preston and J. Crangle, Phys. Rev. 178, 795 (1969).

9. B.D. Sawicka and J.A. Sawicki, J. Phys. 40, C2-576 (1979).

10. R.L. Garten and J.H. Sinfelt, J. Catal. 62, 127 (1980).

11. J.H. Sinfelt and G.H. Via, J. Catal. 56, 1 (1979).

12. O.L.J. Gijzeman, T.J. Vink, O.P. van Pruissen and J.W. Gens, in press.

13. J.H. Sinfelt, G.H. Via and F.W. Lytle, J. Chem. Phys. 76, 2779 (1982); Cat. Rev. Sci. Eng. 26, 1 (1984).

14. F.W. Lytle, R.B. Greegor, E.C. Marques, D.R. Sundstrom, G.H. Via and J.H. Sinfelt, J. Catal. 95, 546 (1985).

ENERGETICS OF ADSORBED DIMERS VIA
SELF-CONSISTENT SCATTERING THEORY

Peter J. Feibelman
Sandia National Laboratories, P. O. Box 5800, Albuquerque, NM 87185

ABSTRACT

A new approach to the surface electronic structure problem, based on a self-consistent scattering theory of point defects, permits 1st-principles calculations for an isolated cluster of adatoms and defects on an otherwise perfect infinite crystalline surface. A first numerical application of the method explains important observations concerning the interaction and diffusion of adatom dimers on Field Ion Microscope tips. Further studies will shed light on questions such as: What impurity species migrate to what kind of surface defects? What is the contribution of substrate atom positional relaxation to adsorption and diffusion barrier energies?

INTRODUCTION

This article reviews the development of a new formulation of the surface electronic structure problem which is particularly well-suited to the calculation of the energy hypersurface underlying fundamental surface chemical phenomena [1-3]. Such phenomena, including energy transfer from an impinging molecule to a surface, dissociation, adspecies vibration, surface diffusion, recombination and desorption, generally involve low partial coverages of reactants and low symmetries. Therefore the usual battery of techniques available to the surface physicist, founded on translational periodicity and growing rapidly in cost as symmetry is lowered, do not appear to be the techniques of choice. At the same time the quantum chemistry approach to surface electronic structure, i.e., modeling the substrate as a small cluster of atoms, is suspect because few if any of the atoms that are to represent the surface actually has the coordination of a surface atom. Since bond-order bond-length correlations imply that the strength of a surface atom - adatom bond depends on the surface atom's coordination, convergence of cluster calculations with respect to cluster size is a major issue, which however is generally too expensive to study.

Mat. Res. Soc. Symp. Proc. Vol. 111. ©1988 Materials Research Society

First principles scattering theory avoids the need for high symmetry and high coverages as well as the artificial effects of cluster termination. The tradeoff is the necessity of developing a major new computer code. In what follows, I explain the basic ideas of the scattering theory of surface electronic structure [1,2], and review first results [3] of numerical work based on the method - they provide a basis for understanding what have seemed to be surprising though general results for separation and diffusion barrier energies in Field Ion Microscope (FIM) observations of adsorbed metal dimers [4].

POINT-DEFECT PROBLEMS VIA SCATTERING THEORY

The size of any electronic structure calculation is determined by the number of inequivalent atoms included. Thus a cluster calculation grows with the number of symmetry inequivalent constituent atoms, while a periodicity-based calculation grows with the number of atoms in the unit cell. From this viewpoint it may seem absurd to consider studying a surface point defect, for which the number of inequivalent atoms is formally infinite. However it is not absurd because of the phenomenon of screening, according to which the 1-electron potential "heals" to its perfect system value at microscopic distances from a "defect" (which might be an adatom or a vacancy or a cluster of either or both of these). Because of this rapid healing, a surface point defect can be treated as a localized scatterer of 2-dimensional Bloch waves, and one can take advantage of the fact that the size of a scattering theory problem is determined by the size of the scatterer.

This idea is not new. It was introduced in the 1950's by Koster and Slater (KS) [5] to study point defects in bulk solids. More recently, using large, fast computers, a self-consistent formulation of the KS method has been applied to a variety of problems involving the spectroscopy and energetics of defects and impurities in both bulk semiconductors and metals [6]. The work reviewed here represents a first adaptation of the self-consistent KS method to surface problems. It shows that the scattering theory approach is viable in the surface context, and can lead to important new insight into surface chemical phenomena.

Since it is the localization effected by screening that makes point defect problems feasible, I formulate the problem of a surface defect in a basis of spatially localized orbitals. In general, this choice of basis leads to an eigenvalue problem of the form,

$$\Sigma_j \ (\ H_{ij} - E^{(n)} \ S_{ij} \ ) \ c^{(n)}_j = 0 \tag{1}$$

where $c^{(n)}_j$ is the coefficient of orbital j in the nth eigenvector, whose eigenvalue is $E^{(n)}$, H_{ij} is the 1-electron Hamiltonian matrix, and S_{ij} is the overlap matrix in the (usual) case that the basis orbitals are not orthogonal. For a point defect in an otherwise perfect crystal, however, H_{ij} and S_{ij} are of dimension $\infty \times \infty$. Therefore Eq. 1 cannot be solved directly. The scattering theory method involves reformulating Eq. 1 as two finite and soluble problems.

The key idea is to write H_{ij} (and similarly S_{ij}) as,

$$H_{ij} = H^{(0)}_{ij} + H^{(1)}_{ij} , \tag{2}$$

where $H^{(0)}_{ij}$ represents the properties of the clean substrate, and separately, the isolated adatoms, while $H^{(1)}_{ij}$ has only a finite number of non-zero matrix elements, and represents the interaction of the adatoms among themselves and with the substrate. Specifically, $H^{(0)}_{ij}$ is defined to be the block diagonal matrix,

$$H^{(0)}_{ij} = \left(\begin{array}{c|c} H^{(a)}_{ij} & 0 \\ \hline 0 & H^{(cl)}_{ij} \end{array} \right) , \tag{3}$$

with $H^{(cl)}_{ij}$ equal to the Hamiltonian of the clean substrate and $H^{(a)}_{ij}$ a diagonal matrix, whose diagonal matrix elements represent the energy levels of isolated adatoms. The eigenvalue problem,

$$\Sigma_j \ (\ H^{(0)}_{ij} - E^{(0)(n)} \ S^{(0)}_{ij} \ ) \ c^{(0)(n)}_j = 0 \tag{4}$$

is easy to solve, because it comprises disjoint problems for the clean substrate, which is translationally periodic, and the adatoms.

To put Eq. 1 into a scattering theory form, a Green's function matrix $G^{(0)}_{ij}(Z)$ is defined by,

$$\Sigma_j \ (\ Z \ S^{(0)}_{ij} - H^{(0)}_{ij}) \ G^{(0)}_{jk}(Z) = \delta_{ik} \tag{5}$$

where Z is a complex number with the dimensions of energy and δ_{ik} is a Kronecker delta. Using Eq. 3, it is easy to prove that Eq. 1 can be rewritten in the form,

$$c_i^{(n)} = c_i^{(0)(n)} + \Sigma_{jk} \, G_{ij}^{(0)} (E^{(0)(n)} + i\delta)(H_{jk}^{(1)} - E^{(0)(n)} \, S_{jk}^{(1)}) \, c_k^{(n)} \quad . \tag{6}$$

Eq. 6 is a scattering theory equation for the c's, analogous to the Lippmann-Schwinger equation of elementary quantum mechanics. Since $H_{jk}^{(1)}$ and $S_{jk}^{(1)}$ are finite matrices, because the adatom induced potential and the basis orbitals are localized, Eq. 6 represents a <u>finite</u> and therefore <u>soluble</u>, set of simultaneous linear equations. Once Eq. 6 has been solved, a sum of products of the c's yields the electron density near the scatterer. Thus local density theory [7] can be used to determine the local 1-electron potential and corresponding value of $H_{jk}^{(1)}$, and one can iterate until an input guess for $H_{jk}^{(1)}$ produces the same $H_{jk}^{(1)}$ as output, to within a desired tolerance. In this way the isolated adsorbate problem can be solved self-consistently, and an adsorption energy hypersurface can be computed.

In fact, rather than solving Eq. 6, one generally works with an equivalent equation for the Green's function matrix, $G_{ij}(Z)$, of the full Hamiltonian. This avoids the need for performing summations over many eigenstates n, and permits the convenience of working with Z "off the energy shell," i.e. in the complex plane. $G_{ij}(Z)$ is defined by

$$\Sigma_j \, (\, Z \, S_{ij} - H_{ij}) \, G_{jk}(Z) = \delta_{ik} \quad . \tag{7}$$

Eqs. 7 and 4, lead to the "Dyson equation,"

$$G_{il}(Z) = G_{il}^{(0)}(Z) + \Sigma_{jk} \, G_{ij}^{(0)}(Z)(H_{jk}^{(1)} - Z \, S_{jk}^{(1)}) \, G_{kl}(Z) \quad . \tag{8}$$

while Eqs. 7 and 1 enable us to prove that $G_{ij}(Z)$ has the spectral representation,

$$G_{jk}(Z) = \Sigma_n \, c_j^{(n)} \, c_k^{(n)*} / (Z - E^{(n)}) \quad . \tag{9}$$

Eq. 8 is the scattering theory problem; similarly to Eq. 6, it is a finite set of simultaneous linear equations corresponding to indices i and l such that $(H_{il}^{(1)} - Z \, S_{il}^{(1)})$ is non-zero. The formal expression of Eq. 9 provides the means for carrying out iterations to self-consistency and for the evaluation of output quantities such as the total energy. According to Cauchy's theorem,

$$(2\pi i)^{-1} \oint_C dz\ G_{jk}(Z) = \Sigma_n\ c_j^{(n)}\ c_k^{(n)*} \ , \qquad (10)$$

if C is a contour that surrounds the $E^{(n)}$ corresponding to the occupied states. The right-hand side of Eq. 10 is the system density matrix, from which it is straightforward to compute quantities such as the electron number density, the total energy, etc.

ENERGETICS OF A SELF-ADSORBED Al DIMER ON Al(001)

The first application of the scattering theory method to a problem of scientific interest [3] yielded an interpretation of long-standing, surprising results concerning the energetics of transition metal dimers adsorbed on Field Ion Microscope tips. The first surprising result is that the energy to separate the atoms of a dimer while leaving them both adsorbed is generally no more than 1/6 the energy of a bond in a comparable solid. The second is that the barrier to dimer diffusion is often smaller than for a monomer of the same species. The scattering theory calculations, performed for an Al dimer adsorbed on a rigid 2-layer Al(001) film, gave results that were qualitatively similar. The energy to separate the Al dimer was found to be only 0.07eV in comparison to the cohesion per bond of 0.557meV in bulk Al. The diffusion barrier for the Al dimer (with one atom hopping at a time) was found to be 0.66eV, while the barrier for a single Al was 0.80eV.

The key to understanding these results is in the bondlength changes that occur when two Al's form a dimer on the Al(001) surface. The calcula-ions indicate that a single ad-Al resides in a 4-fold hollow site at an equilibrium height of 3.30 bohr above the outer layer. When a dimer is formed each atom moves 0.09 bohr toward its partner, and at the same time higher above the surface by 0.30 bohr. One can understand why the Al's move higher in terms of a simple bond-order - bond-length argument: Each Al only has three valence electrons. In forming a dimer each Al must provide an electron for the dimer bond. This electron is removed from the Al surface bond, weakening and lengthening it. This discussion leads to

following interpretation of the FIM separation and diffusion energy
results:

The energy of a bond in a solid is defined to be the cohesive energy
divided by the number of bonds. Since the cohesive energy is calculated by
removing an atom from the solid to infinity, it corresponds to the energy
difference between electrons in bonding orbitals and in non-bonding orbitals. But when two atoms are separated on a surface, the electrons have
the freedom to rehybridize. Thus when the inter-adatom bond is ruptured,
the electrons need not go into non-bonding orbitals. Rather they will
rebond to the surface. Thus the dimer separation energy involves the
difference between two bonding situations and is much smaller than the
cohesion per bond in a comparable solid.

The reduction in diffusion barrier for a dimer relative to a monomer
can be understood via the same argument. Each adatom only has a finite
number of valence electrons. If it forms a bond to another adatom, it does
so at the cost of weakening its bond to the surface. Since the weakened
adatom-surface bond is longer, the atoms of a dimer reside higher above the
surface than a monomer where the potential seen by the adatoms is less
corrugated. This means that the diffusion barrier is lower. In addition
to this effect, one also finds that when the first atom of a dimer begins
to surmount a barrier, its partner will find it energetically profitable to
rebond to the surface, allowing the dimer bond to lengthen. Thus the partner can recoup energy in rebonding to the surface, that the first atom is
giving up in surmounting a bridge. This reduces the total energy needed by
the pair to achieve a hop by one of its members.

Although the scattering theory calculations discussed here were
performed for an Al-dimer on Al(001), it seems likely that the same
concepts will be valid for transition metal dimers on transition metal
substrates. Calculations for this case are in progress.

FURTHER CALCULATIONS USING THE SCATTERING THEORY METHOD

It is easy to envision further calculations of great interest, once
the scattering method has been fully implemented. One project now underway
concerns the question: What atoms migrate on surfaces to what kind of

defects? Defects are thought to be chemically active, as well as attractive to catalytic "poisons" because they are sites of unsaturated valence. These ideas imply that low coverages of poisons might derive their effectiveness in inhibiting reactions from the fact that they migrate to and block the small number of sites at which surface reactivity is highest. But it is not clear that <u>all</u> poison species will migrate to <u>any</u> sort of defect. To see what is at issue, consider the case of coadsorbed sulfur and Al atoms on an Al(001) surface in contrast to the Al dimer just discussed. The ad-Al may be thought of as a defect to which the S will or will not migrate. In general, in forming a bond to the surface, any adatom reduces the power of its surface atom neighbors to bind to other atoms. Thus the reason that an Al dimer forms is that the direct interaction of the two Al adatoms overcomes the weakening of their bonds to the surface. However a S atom is smaller than an Al: the ad-S radius is found to be 1.19Å compared to the Al metallic radius of 1.43Å. Thus a bond between adsorbed S and Al atoms can only form if one of them is rather far from a 4-fold hollow site. This means that the price of increasing the interaction of the adsorbed S and Al atoms is reducing the attraction of at least one them to the surface. This atomic size effect might well be expected to yield a net <u>repulsion</u> between adsorbed S and Al atoms; this idea is supported by preliminary calculations indicating a repulsion of about 0.25eV between an S and an Al adsorbed in neighboring hollows.

Another project in progress concerns the contributions to surface energetics of substrate atom positional relaxation. Since the formation of an adatom-substrate bond might involve the weakening of the bonds between nearby surface atoms, one might expect a substrate to pucker out in the neighborhood of an adsorbate. The energy associated with this sort of lattice relaxation will be comparable to thermal energies, and therefore should be considered in a realistic calculation of, e.g. a dimer diffusion barrier or separation energy. (Recall that the Al dimer calculations reported above were performed for a <u>rigid</u> Al substrate.) The first step in the calculation is to show that the surface atoms are in elastic equilibrium when there are no adatoms. I have found that this is indeed the case for Al(001), with a realistic outer layer separation (i.e., close to ideal [8]), but only if a very good localized orbital basis is used in the scattering calculations [9]. The next step is to show that the local surface force constants of the clean film can be evaluated. Finally one will be ready to study contributions of substrate elasticity to adatom energetics.

REFERENCES

1. A.R. Williams, P.J. Feibelman and N.D. Lang, Phys. Rev. B35, 5433 (1982).

2. P.J. Feibelman, Phys. Rev. B35, 2626 (1987).

3. P.J. Feibelman, Phys. Rev. Lett. 58, 2766 (1987).

4. An extensive list of references is given in Ref. 3.

5. G.F. Koster and J.C. Slater, Phys. Rev. 95, 1167 (1954).

6. J. Bernholc, N.O. Lipari and S.T. Pantelides, Phys. Rev. B21, 3545 (1980); G.A. Baraff and M. Schlüter, ibid. 19, 4965 (1979); P. Braspenning, R. Zeller, A. Lodder and P.H. Dederichs, Phys. Rev. B29, 703 (1984).

7. The Theory of the Inhomogeneous Electron, S. Lundqvist and N.H. March, eds. (Plenum, New York, 1983).

8. M. van Hove, S.Y. Tong and N. Stoner, Surf. Sci. 54, 259 (1976); Groupe d'Etudes de Surface, ibid. 62, 567 (1977); N. Masud, R. Baudoing, D. Aberdam an dC. Gaubert, ibid. 133, 580 (1983); P.E. Viljoen, B.J. Wessels, G.L.P. Berning and J.P. Roux, J. Vac. Sci. and Tech. 20, 204 (1982).

9. P.J. Feibelman, to be published.

NMR AND QUANTUM EFFECTS IN METAL PARTICLES

W.P.HALPERIN* AND INSUK YU**
*Northwestern University, Department of Physics and Astronomy and
Materials Research Laboratory, Evanston, Illinois 60208
**Seoul National University, Department of Physics, Seoul 151,
Korea

1. INTRODUCTION

The considerable interest in metallic clusters spans a wide
range. In this paper we discuss quantum effects and we suggest
that there may be important interrelationships between such
behavior and heterogeneous catalysis. Of course, the physical and
chemical behavior of the metal particle are the key to
understanding principles of catalytic activity. The quantum
effects are found at progressively lower temperatures where the
metallic properties change. In both cases a fundamental question
is posed, "Do these particles exhibit metallic behavior?" One of
the common features of research in these two areas is the important
role played by nuclear magnetic resonance(NMR) techniques since the
signatures of a metal and an insulator are so dramatically
different. The issue of metallic or non-metallic behavior is
therefore best addressed with this experimental approach and we
will review the current situation by comparing a number of results
on quantum size effects deduced from NMR experiments.

2. QUANTUM SIZE EFFECTS

2.1 Theory of the Quantum Size Effect in Simple Metals

2.1.1 Single Particle Behavior

It is obvious that the electronic structure of an atom is
rather different from that of the metal from which it can be
formed. The atom has a finite set of discrete energy levels. These
are filled starting with the lowest proceeding to an uppermost
occupied level determined by the number of available electrons,
which is to say the atomic valence. In contrast the bulk metal has
an essentially continuous set of levels in ranges of energy
referred to as energy bands. These also can be viewed as being
filled from the bottom of the band up to the Fermi energy.
However, an intermediate situation holds for the metallic cluster.
In this case there are a set of levels, much more closely spaced
than for an atom but discrete nonetheless. At low temperatures, $k_B T$
$\ll E_F$ these are systematically filled up to the Fermi energy, E_F.
The details of the energy level spectrum for a specific particle
are fixed by the shape of the particle and conditions at the
surface. The average spacing, δ, between levels is determined by

the particle size varying slightly from one element to another[1].
This spacing is approximately 1 °K for 100 Å diameter and increases
inversely with the volume,

$$\frac{\delta}{k_B} = \frac{A}{V} \; 10^{-18} \; cm^3 \; K$$

(1)

where A is a dimensionless constant of order one (for example 0.74,
Li; 0.97, Mg; 0.68, Al; 0.93, Cu; 1.44, Ag; 0.84, Sn; 0.12, Pt;
1.26, Au; and 0.56, Pb). Since there is a gap of order δ in the
electronic excitation spectrum at the Fermi energy, the metal
cluster will appear to become an insulator[2] with decreasing
temperature, $\delta \ll k_B T$. This *quantum size effect* was first
recognized by Fröhlich in 1937 and has since recieved considerable
attention in the experimental and theoretical literature. From Eq.
1 one can see that a 20 Å particle of silver will have an average
level spacing at room temperature! To date the only clear evidence
for the quantum size effect appears in measurements of the magnetic
susceptibility, mostly from NMR experiments.

Greenwood, Brout and Krumhansl[3] showed that the NMR Knight
shift would be a simple way to measure this phenomenon since the
Knight shift is proportional to the electronic susceptibility. The
quantum size effect on the susceptibility is most simply stated for
simple free electron-like metals. Then the particle will have a
number of doubly degenerate levels equal to one half of the number
of valence electrons. If the particle shape is irregular then this
may be the only degeneracy that can be expected, otherwise for a
highly symmetric cluster the average level spacing could be much
larger. A particle with an odd number of valence electrons will
have a highest occupied level containing just one electron which
can be either spin up or down in small applied magnetic fields.
Such a particle behaves magnetically as a localized spin with
magnetic moment of one Bohr magneton, μ_B. An electronic spin flip
can be accomplished rather easily in this case, the magnetic
susceptibility being given by,

$$\chi_{odd} = \mu_B^2 \; / \; k_B T$$

(2)

For an even electron numbered particle the highest occupied level
is doubly occupied. The measurement of magnetic susceptibility
requires flipping a spin. This can only happen if one of the two
electrons is promoted to the next highest energy level
approximately Δ_i ($\approx \delta$) above the Fermi energy. Since the level
spectrum, and hence Δ_i, can vary from one particle to another
depending on particle size and shape we give it an index i to label
the particle. The spin flip transition becomes more improbable
where $\Delta_i \approx \delta \gg k_B T$ such that at low temperatures the

susceptibility will decrease to zero exponentially,

$$\chi_{even} = \frac{8\,\mu_B^2}{k_B T}\; e^{\,-\Delta_i/k_B T}$$

(3)

This cross over to nonmetallic behavior associated with the quantum size effect can be determined by direct measurements of the electronic magnetic susceptibility, Kimura *et al* [4], or through NMR measurements of the Knight shift.

In a metal the Larmor frequency, ν, of the nucleus appears shifted relative to that of an insulating reference compound, ν_0, containing the same element,

$$\nu = \nu_o\,(1 + K)$$

(4)

where the Knight shift K is proportional to the electronic susceptibility.

$$K \propto \chi$$

(5)

For metals that are not well approximated by free electron considerations, such as the transition metals, there are other possible contributions to the Knight shift[1] which we will ignore. Knight shifts at room temperature vary from a record −3.05 % for Pd to a negligble −0.0025 % for Be.

With the results of Eq 2,3,4, and 5 it is easily seen that the NMR frequency for particles with an odd number of electrons will be shifted more and more as the temperature decreases. For even numbered particles the shift will go to zero exponentially with decreasing temeprature.

2.1.2 Ensembles of Particles

One serious amendment to the above discussion requires averaging over all particles in an ensemble. This requires averaging over those particles having the same volume but different shapes or surface conditions that determine the details of the electron level spectrum; these are particles with the same δ. Finally one must average over the size distribution in the sample; particles with different δ. It was recognized by Kubo[5] and subsequent workers (reviewed in[1]) that the electron level distribution functions depend on the symmetry of the electron Hamiltonian, different for the case of weak and strong spin-orbit interaction, and weak or strong magnetic fields, or combinations thereof. The electron spectrum near the Fermi energy for a particle can be specified by a probability distribution function for the nearest neighbor energy level spacing Δ near the Fermi energy, $P(\Delta,\delta)$. The magnetic susceptibility is then found by averaging $\chi(\Delta)$, taken from Eq. 2 or 3, over this distribution,

$$\chi = \int P(\Delta)\,\chi(\Delta)\,d\Delta$$

(6)

Generally one need only consider the form for the distribution functions $P(\Delta,\delta)$ at small Δ, corresponding to the low temperature limit where,

$$P_a(\Delta,\delta) \propto (\Delta/\delta)^a \qquad \Delta/\delta \to 0$$

(7)

The index a depends on the symmetry of the electron Hamiltonian (a = 1, 2, or 4):

a	distribution	magnetic field	spin-orbit interaction
1	orthogonal	small small	small large (even particles)
2	unitary	large	large
4	symplectic	small	large (odd particles)

We can summarize the relevant changes that are introduced by generalizing the discussion to include an ensemble of particles of common size, fixed by δ, but with a distribution of surface conditions and particle shapes:

a) The electronic susceptibility, and hence the Knight shift, for even electron particles decreases with the temperature as a power law T^a rather than exponentially as indicated in Eq.3. Qualitatively we can write the Knight shift for even particles in a form relative to the Knight shift in bulk material, K_p,

$$\frac{K}{K_p} = C_a \left(\frac{k_B T}{\delta}\right)^a$$

(8)

where C_a is a dimensionless constant: $C_1 = 3.82$, $C_4 = 1.01 \times 10^3$ [1].

b) The electronic susceptibility of the odd particles at a given temperature will vary from particle to particle depending on its size, more specifically depending on δ/k_BT. This dependence produces a broad distribution of Knight shifts rendering the odd particle part of the NMR spectrum difficult to observe. Consequently it is reasonable to assume that NMR is sensitive only to the even electron particles.

2.1.3   The Quantum Size Effect and the Spin-orbit Interaction

Of the many elements which may seem to provide a suitable compromise between considerations of ease in fabrication and facility for NMR measurement, most have a significant spin-orbit interaction. Then spin is not a good quantum number and the electron levels for the simple metal, doubly degenerate in spin, are replaced with Kramers doublets having more or less spin up and spin down character depending on the strength of the spin-orbit interaction. For the purposes of the discussion here it suffices to state the result of a perturbation theory treatment[6] of the spin-orbit interaction for the even particles, the contribution to the NMR line from the odd particles being difficult to observe as was noted above.

$$\frac{K}{K_p} = \frac{hv_F (\Delta g_\infty)^2}{3 \, \delta \, d} + C_a \left(\frac{k_BT}{\delta}\right)^a$$

(9)

Eq. 9 indicates that the Knight shift at zero temperature should be nonzero. This result is valid only in the limit of small spin-orbit interaction and therefore should be interpreted with caution. The strength of the spin-orbit interaction is most easily expressed in terms of the electron spin resonance g-shift, Δg_∞.

The Fermi velocity is v_F. Since $\delta \propto d^{-3}$ we note that the zero temperature Knight shift is expected to be proportional to d^2. For very large spin-orbit interaction Eq. 9 does not hold and the Knight shift and electronic susceptibility should remain equal to that of the bulk metal as given below in Eq. 10.

2.1.4 Large Magnetic Fields

On application of a large magnetic field or for very large spin-orbit interaction it can be shown that the even and odd particle magnetic susceptibilities revert back to their bulk values, χ_p.

$$\chi_{even} = \chi_{odd} = \chi_P$$

(10)

84

However, as the magnetic field, H, is increased from zero, the even particle suceptibility at T ≈ 0 increases in a manner direct; determined by the near-neighbor level distribution function[1], $P(\Delta,\delta)$. This is easily measured by the Knight shift experiment from which it is possible, in principle, to make a direct determination of the appropriate energy level statistics for a given system.

$$\frac{\chi_{even}}{\chi_P} = \frac{K}{K_P} = R(x_o) + \frac{(\pi k_B T/\delta)^2}{6} \left.\frac{d^2 R}{dx^2}\right|_{x=x_o} + \cdots \qquad T\to 0$$

(11)

where $x_o = 2\pi\mu_B H/\delta$ and the two-level correlation function, $R(x)$, approaches the near-neighbor level distribution function, $P(x)$, for small x. This may be the most direct way, if not the only way, by which energy level statistics can be inferred from experiment. The several possibilities for $R(x)$ that have been considered for these statistics, a= 1,2, and 4, are shown in Fig. 1.

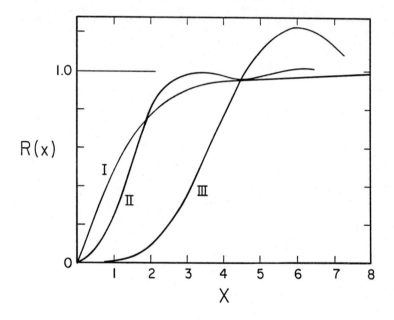

Figure 1. The behavior of the Knight shift at low temperatures is given by the two-level correlation function for different statistical distributions (orthogonal, I; unitary, II, and symplectic, III) as a function of applied magnetic field x $=2\mu_B\pi H/\delta$.

2.3 Experimental Evidence for the Quantum Size Effect

Knight shift measurements as a function of temperature and particle size have been reported for Cu, Sn, and Pt. Both Cu and Sn are cases of metals with modestly large spin-orbit interaction while Pt presumably has a very large spin-orbit interaction. NMR results for other elements are less complete and have been reviewed elsewhere[1]. In Fig. 2 the measurements of Yee and Knight[7] are shown for a range of particle sizes from 25 Å to 110 Å. These data appear to have a linear temperature dependence with a non-zero intercept at T=0, consistent with the notion that Cu has a significant spin-orbit interaction. The linear dependence on temperature corresponds to the prediction of the othogonal ensemble, a=1. Using the results of Eq. 9 we infer for the 25 Å data set that δ/k_B = 100 K. This result is qualitatively consistent with the prediction δ/k_B = 176 K of Eq. 1 assuming spherical particles.

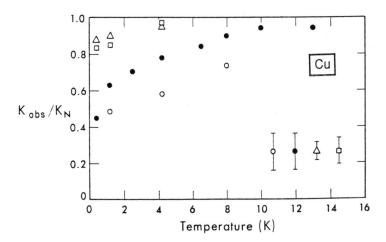

Figure 2. The temperature dependence of the Knight shifts of copper particles reported by Yee and Knight for average sizes 25 Å(o), 40 Å (●), 100 Å (Δ) and 110 Å (□). The measured Knight shifts are normalized to the Knight shift of the bulk metal labeled K_N.

The zero temperature intercept of the Knight shift from Yee and Knight[7] and from Kobayashi[8] are displayed in Fig. 3. These data appear to be consistent with each other and show that the zero temperature Knight shift decreases with decreasing particle size again qualitatively consistent with the theoretically expected behavior given by Eq. 9; however the prediction, $K(0)/K_p \propto d^2$ is not precisely born out, shown in the figure as a light solid line. Still, one finds that the 25 Å data of Yee and Knight extrapolated to zero temeprature combined with experimental values[1] for Δg_∞ =3.1×10^{-2} and Eq. 9 indicate that $\delta/k_B = 119$ K. This is quite close to the average energy level spacing for these particles inferred independently from the other data ($\delta/k_B = 100$ K and $\delta/k_B = 176$ K.)

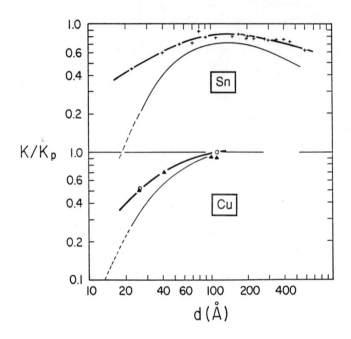

Figure 3. The residual Knight shifts at T=0 are shown for a collection of measurements on Cu and Sn as a function of particle size. The light solid lines are theoretically expected size dependences for weak spin-orbit interaction. The heavy lines are guides to the eye.

For the case of Sn particles a composite of data is displayed in Fig. 3 from Fukagawa et al, Nomura et al, Kobayashi et al, and Wright[9]. Since tin is a superconducting metal it is necessary to separate the effects of reduced size on the superconductivity from corresponding effects on the electron level spectrum. Again from a qualitative viewpoint this can be performed by focusing first on the large particles. Here the effects on superconducting behavior dominate. The superconducting energy gap that opens up at the Fermi energy produces a reduction in the Knight shift that can be accounted for at zero temperature by substituting for δ the superconducting energy gap Δ_s. This is approximately size independent for such large particles. From Eq. 9 we see that the zero temperature Knight shift should increase with decreasing particle size, $K(0)/K_p \propto d^{-1}$. Although the observed increase in Knight shift in Fig. 3 is not as strong as predicted by this argument, shown by the light solid line, an increase is indeed seen with reduced particle size. At smaller sizes, d<100 Å, where superconductivity is quenched, the effect of size on the electron level spectrum is apparent since the Knight shift decreases with decreasing particle size. Here also the dependence on size is not as dramatic ($\propto d^2$) as would be expected from Eq. 9, although the qualitative features do appear to be quite convincing.

Finally, Knight shift measurements on Pt particles from Yu et al[10] show no deviation from bulk behavior for diameters as small as 33 Å at 4.2 K. Although we do not have a quantitative determination of the spin-orbit interaction in platinum it can be expected to be very large owing to its high atomic number. Here also it appears that there is consistency with the predictions of the quantum size effect theory for the case of large spin-orbit interaction, Eq. 10.

3. SUMMARY

We may summarize the experimental situation as being qualitatively consistent with the predictions of the quantum size effect theory. The reduction of the Knight shift on decreasing the temperature signals cross-over to non-metallic behavior and this appears to be confirmed for both Cu and Sn, and can be expected to be the case for other elements under similar conditions of particle size and reduced temperature. The weakest aspect is that the level statistics can not be unambigously inferred from the data that are presently available. In this regard it may be possible to obtain additional information from Knight shift measurements as a function of magnetic field strength.

In a domain more closely allied to heterogeneous catalysts and their properties, magnetic resonance behavior of the surface atoms of platinum clusters on η-alumina has been clearly identified by Slichter and colleagues[11] at temperatures below 77 K. Generally, it was found that the Knight shift is close to zero or is very substantially reduced for the surface atoms and that the spin-lattice relaxation time is two to three orders of magnitude longer compared to the atoms in the metallic interior of the particle. This suggests that the surface atoms have non-metallic behavior below 77 K. At high temperatures, T>290 K, they have measured the transferred hyperfine interaction to the carbon

nucleus of an adsorbed CO molecule which appears to be in a
(weakly) metallic state. At high temperatures then one can infer
that the surface of the platinum cluster is (weakly) metallic. How
this evolves with reduced temperature is not yet known. It is
likely that there is evidence in these experiments for a discrete
nature of the electron level spectrum for the surface atoms. The
interrelationships between the two subjects of quantum effects and
surface effects are a fascinating area for continuing NMR research.

Acknowledgement is made for support from the National Science
Foundation through grants, DMR 8520280, INT-8505560, and
DMR-8620275.

4. REFERENCES

1. W.P. Halperin, Rev. Mod. Phys. 58, 533 (1986).

2. H. Fröhlich, Physica(Utrecht) 4, 406 (1937).

3. D.A. Greenwood, R. Brout, and J.A. Krumhansl, Bull. Am. Phys.
   Soc. 5, 297 (1960).

4. K. Kimura, S. Bandow, and S. Sako, Surf. Sci. 156, 883 (1985).

5. R. Kubo, J. Phys. Soc. Jpn. 17, 975 (1962).

6. J. Sone, J. Phys. Soc. Jpn. 42, 1457 (1977).

7. P. Yee and W.D. Knight, Phys. Rev. B 11, 3261 (1975).

8. S. Kobayashi, J. Phys. (Paris) Colloq. 38, C2-121 (1977).

9. Y. Fukugawa, S. Kobayashi, and W. Sasaki, J. Phys. Soc. Jpn.
   51, 1095 (1982); K. Nomura, S. Kobayashi, and W. Sasaki, J.
   Phys. Soc. Jpn. 48, 37 (1980);S. Kobayashi, in Proceedings of
   the 13th Inter. Conf. on Low Temp. Phys., Univ. Colorado,
   edited by K.D. Timmerhaus, W.J. O'Sullivan, and E.F. Hammel
   (Plenum, New York) 4, 315 (1974);F. Wright Jr., Phys. Rev. 163,
   420 (1967).

10. I. Yu, A.A.V. Gibson, E.R. Hunt, and W.P. Halperin, Phys. Rev.
    Lett. 44, 348 (1980); I. Yu and W.P. Halperin, J. Low Temp.
    Phys. 45, 189 (1981).

11. S.L. Rudaz, J.-P. Ansermet,P.-K. Wang, C.P. Slichter, and J. H.
    Sinfelt, Phys. Rev. Lett. 54, 71 (1985); C.D. Makowka, C.P.
    Slichter, and J.H. Sinfelt, Phys. Rev. B 31, 5663 (1985).

SECONDARY CATION EFFECTS ON SODIUM AND POTASSIUM ZEOLITE SYNTHESES AT Si/Al$_2$ = 9:

PART 1- PHASE DEVELOPMENT IN THE SODIUM SYSTEM

DAVID E. W. VAUGHAN
Exxon Research and Engineering Company,
Route 22 East, Annandale, N.J., U.S.A., 08801

ABSTRACT

The influence of Group 1A elements, and various alkyl ammonium cations, on a metastable faujasite synthesis at a silica/ alumina ratio of 9 are reported. Co-crystallization of faujasite, gmelinite and gismondine (P) occurs in reactions with Li$^+$, K$^+$ and Rb$^+$, but the trapping of Cs$^+$ in the FAU sodalite cage (CSZ-1/ CSZ-3) inhibits subsequent reactions to GIS, and blocks the formation of GME. Morphological effects of these cations include platelet formation and the promotion of small crystals. TMA addition promotes mazzite analogues; and DEDM stabilizes FAU, and at 140°C., induces the crystallization of ECR-1. The derivation of all these phases from a restricted number of secondary building units is discussed.

INTRODUCTION

One of the recurring questions in zeolite synthesis is the relative importance of cations in directing crystallization to specific products, compared to the influence of the framework–forming polymer present in the initial gel or reaction medium. It is well known that different sources of silica and alumina (volcanic glass, gels, meta-kaolin, minerals etc.), in otherwise identical composition reaction systems, may completely change the reactant products, and there is much published work on the influence of cations and cation combinations in directing reactions to specific products[1]. Recent advances in applying various spectroscopy techniques to zeolite synthesis problems [2,3,4,5,6], is beginning to reveal important details of the range of polymers present in crystallizing zeolite gels and the component silicates and other raw materials, indicating possible leads to synthesis mechanisms. This paper examines the influence of cation combinations on the reaction products from a highly reactive and metastable gel system, and seeks to rationalize the evolution of different phases on the basis of common secondary building units (SBU) present in the products, proposing possible crystallization mechanisms or pathways to explain the multiple product combinations.

It is well established that a waterglass derived gel of composition:

$$3Na_2O: Al_2O_3: 9SiO_2: 140H_2O$$

Mat. Res. Soc. Symp. Proc. Vol. 111. ©1988 Materials Research Society

yields high purity faujasite (FAU) in seeded systems [7,8], but does not readily crystallize in the unseeded mode. Minor agitation during the seeded crystallization induces the development of an otherwise latent phase, gmelinite (GME), apparently promoted by collision breeding of GME nuclei. The high level of instability of the gel, caused in part by the low OH/Si ratio, makes it particularly sensitive to minor perturbations, and therefore an appropriate composition in which to study cation structure direction. It is less well known that, in the analogous unseeded potassium system, this stoichiometry yields high purity Linde L (LTL)[9], but the synthesis is insensitive to seeding in the mode used for the sodium system.

This synthesis program was designed to evaluate the perturbations in these reactions when a small fraction of of the primary base metal (Na or K) is substituted by other Group 1 elements (Li, Na, K, Rb, Cs), tetra-alkylammonium cations such as tetra-methyl(TMA), tetra-ethyl(TEA), tetra-propyl(TPA), tetra-butyl(TBA) or bis-(dihydroxyethyl) dimethyl-ammonium(DEDM), and various alkyl-amines. As with most zeolite synthesis reactions we observe the progressive crystallization of successively more stable phases. It is therefore imperative to monitor the crystallization as a function of time intervals that allow the recognition of all the crystalline products. The roles of large cations as templates and reaction components have been reviewed by several authors [1,10,11,12], and on that basis, and this writer's experiences, they can be summarized as follows:

Cation Role	Effect
o Large basic cation	None
o Structure directing template	Novel structure
o Nucleus stabilizer	Novel structure
o Surface poison	Morphology; reaction rate
o Solution or SBU complex	Si/Al; reaction rate

In some experiments the addition of a large template molecule has no influence whatever; the molecule is excluded from the zeolite structure, having no effect on structure, morphology or composition. This mode is observed in some Y-FAU syntheses with TMA. In other systems the template causes a new zeolite to form which occludes the template: typical of this mode is TMA-mazzite (MAZ) - identified with the synthetic zeolites Omega and ZSM-4 [13,14]. Alternatively, there may be no change in structure, but the Si/Al ratio may change, as is observed in the TMA modified syntheses of Linde A (LTA) which yield TMA occluding ZK-4 [15]. Some systems require only sufficient template to induce the nucleation of a particular phase, for that phase to dominate the subsequent crystallization. This is observed in some offretite (OFF) and MAZ syntheses [16]. There are numerous cases of morphological changes induced by trace amounts of impurities or secondary cations. Selective absorption of some molecules on particular crystal faces will slow or arrest growth on those faces, causing major morphology changes, and these have been reported for numerous diverse chemical systems. Alternatively the non-selective absorption on all faces will arrest growth on the whole crystal with the propagation of very small crystals, a phenomenon that has been reported elsewhere for FAU [17]. Some of these observations are reported in the experiments to follow, particularly for FAU.

Recent studies of anion influences on a FAU synthesis composition having much higher Si/Al and OH/Si than studied here, focus on the stable phases rather than the first, meta-stable, materials to crystallize [18]. In the experiments to follow based on the 9/1 Si/Al$_2$ formulation, anions such as Cl$^-$, Br$^-$,NO$_3^-$, and SO$_4^=$ at the levels investigated here, have no major effect on the FAU products.

EXPERIMENTAL

The experiments described in this paper were typically 0.02 to 0.03 molar in the basic stoichiometry:

$$(3-x)N_2O : xM_2O : Al_2O_3 : 9SiO_2 : 140H_2O$$

where N was either Na or K, M was Li, Na(when N=K), K(when N=Na), Rb, Cs, TMA, TEA, TPA, TBA or DEDM. x was varied from zero to 0.5. Only the case with N=Na is reported in this paper, and the M=K experiments will be reported in subsequent papers. To achieve such a low "effective base" level (OH/Si) using commercial silicates requires the partial neutralization of the high base associated with the silicate. This can be done either by acid neutralization, or addition of part of the Al as an acidic salt, such as sulfate, chloride etc.. The standard procedure involved making a basic aluminate by dissolving Al$_2$O$_3$.3H$_2$O (Alcoa-C31) in a hot concentrated basic hydroxide solution, then adding the cooled aluminate to the diluted sodium or potassium silicate (PQ "N" or Ksil-6 silicates) with vigorous mixing in a blender, followed by addition of the remaining Al as a solution of the acidic aluminum salt (in these cases Al$_2$(SO$_4$)$_3$.18H$_2$O), and finally homogenizing the gel for about three minutes. If seeded, as was done with N=Na, the seeds were added to the silicate prior to the aluminate addition, and had a composition with Si/Al=13.3 [7,8]. Seeding was at 5%wt. based on Al - ie. 5% of the Al originated from the seed source. This level induces crystallization in less than one day, obviating the need for a cold age nucleation step, but does not have a significant influence on crystal size in the base composition. These seeds do not effectively nucleate the potassium system, and analogous potassium composition gels, or mixed sodium- potassium gels, do not yield nuclei [18].

The typical experimental procedure for a seeded triplet experiment (about 300g or 220ml.), using the base composition with N=Na, and x=0, was as follows (~0.075 molar):

6.4g NaOH were dissolved in 15g H$_2$O; 8.1g C-31alumina were added, and the solution refluxed until the alumina was dissolved, at which time the aluminate solution was cooled to room temperature, and its weight adjusted to the base amount. 154.1g PQ Corp. N silicate (28.7wt.% SiO$_2$; 8.9wt.% Na$_2$O;) were added with mixing in a blender together with 40g H$_2$O and 28.5g seeds, then the aluminate solution was added with continued mixing, followed by 20.4g. alum in 27.5g. H$_2$O. After homogenizing the gel was divided between three reaction vessels- a 125ml. Teflon bottle for 100°C. experiments, and two Parr acid digestion autoclaves or Hoke 75 ml. test tube pressure cylinders for experiments at 140°C.. These were heated in forced air ovens for the duration of the experiments.

Samples were taken at intervals over three to five days, depending on the reaction rates reflected in the degree of crystallinity of the initial samples, and analysed by x-ray diffraction using a Siemens D500 diffractometer and CuK_α radiation. Selected samples were evaluated by scanning electron microscopy (SEM) to demonstrate morphology and crystal size effects. As powder x-ray analysis of partly crystalline and multi-phase materials has a detection limit in the range of 3-5%, SEM was a useful tool for observing low levels of secondary phases. Chemical analyses were made using Jarrel-Ash or Leeman Labs. ICPES. Initial and final products are reported in the tables. Intermediate timed samples usually comprised mixtures of the reported phases. Known phase compositions will be reported using the codes recommended by the International Zeolite Association Structure Commission [20].

RESULTS OF EXPERIMENTS WITH SODIUM AS THE PRIMARY CATION

Group 1A Cations

Numerous experiments have been carried out by several workers to evaluate the influence of Group 1 metals on faujasite syntheses, and these are largely confirmed in the seeded experiments summarised in Table 1, carried out at 100°C. Figure 2-2 is an example of the base case FAU product for crystallization of the pure sodium gel. In the absence of agitation, faujasite is fully crystallized in about eighteen hours, and is progressively replaced by zeolite P (GIS), which is the stable product. Even minor agitation induces the co-crystallization of GME together with FAU, both zeolites progressively reacting to pure P (GIS) within 36 to 48 hours. Presumably both phases nucleate in this gel, but the FAU has a faster growth rate than GME. (Many FAU products that are apparently pure on the basis of x-ray diffraction analysis, often show low levels of GME in SEM micrographs.) However, GME undergoes collision breeding of nuclei in agitated systems, whereas FAU does not, resulting in the development of higher net levels of growth on the more numerous GME nuclei. In highly agitated gels of this composition GME may initially develop as a nearly pure product, reacting to GIS as a function of time. However, the GME product is a highly mis-stacked structure, as shown by its broad x-ray diffraction peaks (Figure 3-2) and low n-hexane capacity.

TABLE 1
GROUP 1A CATION EFFECTS ON THE SODIUM SYSTEM
(3-x) Na_2O: x M_2O: Al_2O_3: 9 SiO_2: 140 H_2O.

M	Initial Product	Final Product
Li	FAU	GIS(P)
Na	FAU	GIS(P)
K	platelet FAU	MER
Rb	platelet FAU	MER
Cs	CSZ-1/CSZ-3	POL

All of the initial products in these multiple cation crystallizations are modifications of faujasite, demonstrating several of the cation effects described in the introduction. The Na+ base case experiment crystallizes a product comprising largely 0.5 to 1μ regular, distorted or intergrown octahedra, as shown in Figure 1. The addition of Li+ produced very small crystallite materials (<0.1μ). Previous work in high lithium containing systems reported the formation of ZSM-2 and 3 [21] - FAU like materials having mis-stacked layers[22], recognized as rotated [111] sheets of sodalite cages in FAU. The ZSM-2/3 structure has apparent hexagonal symmetry, (Figure 2-1) but the 'c' axis cannot be fixed. At the compositions studied here, no such structural variations were seen in the x-ray diffraction spectra.

FIGURE 1: Scanning electron micrographs of Na-FAU (left), showing regular and distorted intergrown octahedra; and the platelet Na, Cs-FAU designated CSZ-1 (right). Markers are 1μ.

Cs+ shows a more distinctive effect on the morphology, yielding thin platelet crystals. Initial low levels of Cs+ (x up to about 0.15) are trapped in the sodalite cages of FAU (Cs+ cannot be readily exchanged into or out of FAU), showing little effect on the x-ray diffraction patterns, but a large proportion of platy crystals are present. Cs+ is included in the structure during synthesis, and higher Si/Al ratios are also observed, giving a novel FAU modification designated CSZ-3 [23]. Higher levels of Cs+ induce the growth of very thin platelet, hexagonally shaped, crystals having aspect ratios generally greater than about 10, and a higher Si/Al ratio than the base FAU. The x-ray diffraction patterns have some similarity with ZSM-2/3, but the spectra also show strong x-ray absorption by Cs+, and apparent hexagonal symmetry [24]. Recent high resolution electron microscopy analyses of CSZ-1 and CSZ-3 have shown that both derivatives have the FAU connectivity, confirming that CSZ-3 has conventional cubic symmetry, but that CSZ-1 has a rhombohedral distortion [25]. Modelling of this rhombohedral form produces an x-ray diffraction pattern essentially identical to the experimental pattern. This rhombohedral distortion is carried through to the dealuminated form, the ^{29}Si MASNMR spectrum for which confirms the presence of four Si sites [26], compared to one in conventional FAU. A characteristic of CSZ-1 crystals observed edge on in the electron microscope, is that they all have at least one twin plane through the center of the crystals. Detailed modelling by Treacy et al. [27] of the strain field in the whole crystal caused by this twin, shows that it is sufficient to explain the rhombohedral distortion. Previous proposals interpreted CSZ-1 as Breck-6 or a related modification [28,29]. However, the more recent work reports no evidence for any Breck-6 component in CSZ-1 or CSZ-3 [25].

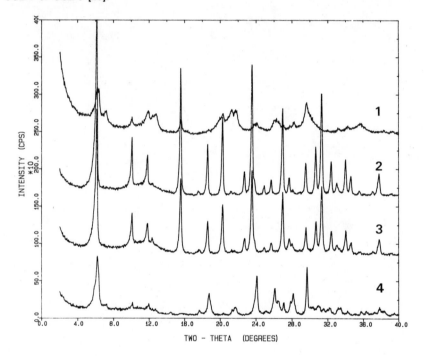

Figure 2: X-ray diffraction patterns of (1) ZSM-2 [21]; (2) Na FAU(Y) base material; (3) Na,K FAU platelet material; (4) Na,Cs FAU(CSZ-1).

K$^+$ and Rb$^+$ induce the formation of some squashed octahedra that have the appearance of thick plates, but the product is otherwise similar to the base product. Higher levels of K$^+$ yield higher proportions of plates, to the point where all the material is of that morphology at the point where x=0.5 [18].

Remarkably small amounts of M (eg. x=0.1) have a pronounced morphological influence on all these products, but the samples usually comprise mixed morphology crystals - some crystals are clearly not changed by the addition of M. This seems to be a selective surface poisoning effect, such that the proportion of plates is proportional to the M^+ concentration in the gel, at least at low concentrations. The x-ray diffraction patterns of K, Na-platelet FAU (Figure 2-3) is identical to that for the Na FAU standard (Figure 2-2).

All of the above phases are metastable in the reaction medium, and as a function of time are replaced by the apparently thermodynamically stable products - all well known analogues of minerals. These are the synthetic P analogue of gismondine (GIS); the Barrer M or Linde W [30] analogues of merlinoite (MER); and pollucite (ANA). The last is unusual, since in non-Cs syntheses analcite is characteristic of higher temperature experiments. Cs has a remarkable effect in nucleating pollucite, and at even very low levels drives the crystallization to the ANA polymorph. In all these experiments the first product was well crystallized after less than 24 hours, and was significantly replaced by the more stable component after 48 hours.

Identical experiments run at 140°C. dramatically compressed the crystallization sequence, so that after only about twelve hours the stable phase was the predominant, or sole, component. Several experiments run at 160°C. yielded ANA products for all cations within 24 hours.

Tetra-alkylammonium Cations

The influence of tetra-alkylammonium cations are summarized in Table 2 for experiments made at 100°C. and 140°C. More extensive substitution of M with these cations has been reported elsewhere [31].

TABLE 2
ALKYL-AMMONIUM CATION EFFECTS ON THE SODIUM SYSTEM
(3-x) Na_2O: x M_2O: Al_2O_3: 9 SiO_2: 140 H_2O.

T°C.	M	Initial Product	FinalProduct
100°	$(CH_3)_4N$	FAU	MAZ
140°	..	MAZ	MAZ
100°	$(CH_3)_2(C_2H_4OH)_2$	FAU(ECR-4)	FAU(ECR-4)
140°	..	ECR-1	ECR-1/ANA
100°	$(C_2H_5)_4N$	FAU	GIS
140°	..	FAU/GME	GIS/ANA
100°	$(C_3H_7)_4N$	FAU/GME	GIS
140°	..	FAU/GME	GIS/ANA
100°	$(C_4H_9)_4N$	FAU/GME	GIS
140°	..	FAU/GME	ANA

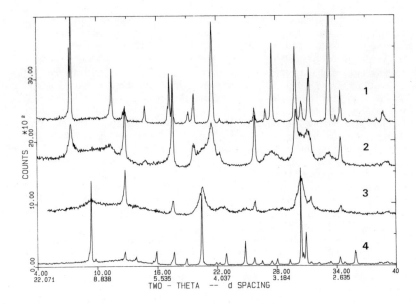

FIGURE 3: The GME materials synthesized in these reactions is highly intergrown in a manner reported by Kokotailo and Ciric for some mineral GME [32]. Chabazite may be the primary intergrowth, as indicated by the x-ray diffraction patterns for: (1) Nova Scotia GME mineral; (2) Synthetic GME from these experiments; (3) Synthetic Na G "near CHA"; (4) Nova Scotia CHA mineral.

Although faujasite is clearly a major product in the presence of most alkylammonium cations at 100°C in this composition range, only in the case of DEDM is the template trapped in the FAU structure. For the other organic cations there is no evidence for their inclusion in the structure from thermo- gravimetric analyses, chemical analyses or cation/ aluminum balance. DEDM is located in the FAU supercage, as shown by the TGA low temperature weight loss in Figure 4, and this inclusion complex stabilizes the FAU in its reaction solution so that it does not readily degrade to the denser GIS product characteristic of the pure Na compositions, or other templated reactions.

The higher Si/Al ratio DEDM-FAU has been designated ECR-4 [33]. TEA, TPA and TBA alkylammonium cations are not occluded in the crystalline products, but the latter two do seem to promote the co-crystallization of GME (repeat experiments show that occasionally GME does not form, and FAU/GME weight ratios may vary). These alkyl ammonium cations greatly influence crystallite size, and in some cases products were less than 0.1μ in diameter, resulting in peak broadening in their x-ray diffraction patterns. However, it has been reported that high seeding levels promote the formation of very small FAU crystallites [17], and it is possible that alkylammonium surfactant properties make seeding more effective in the templated gels, and lower seeding in these compositions may yield larger crystals. It is well known that TMA promotes the crystallization of analogues of MAZ (ZSM-4 [13] and Linde Ω [14]), but FAU preceded it as a transient initial product at these low template fractions, probably indicating faster nucleation for FAU. MAZ is the stable product at 100° and 140°C., and as the TMA is trapped in the gmelinite cages of MAZ, the occluded template may be the stabilising factor in inhibiting sequential reaction of MAZ to GIS. In the cases of TEA, TPA and TBA, not occluded in any products, the final products are all GIS.

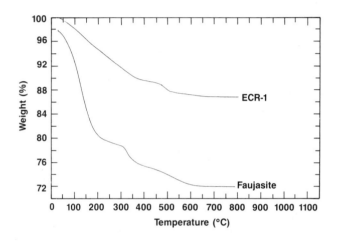

FIGURE 4: Thermogravimetric analyses of DEDM containing FAU (ECR-4) with the template in the supercage, and for ECR-1 with the DEDM in the 12-ring channel.

Reactions at 140°C. are similar to those at 100°C., except that ANA replaces GIS as the stable product, and DEDM templates a new zeolite, ECR-1, in place of FAU [32]. ECR-1 is composed of alternating sheets of MAZ and mordenite (MOR) [33], and constitutes a "boundary phase" at the overlap of the crystallization fields of the two constituents. At higher Si/Al ratios than investigated here, MOR is the dominant and stable zeolite. ECR-1 is unusual, in that it is the first example of a combination of the essential elements of entirely dissimilar structures, which hitherto have not been recognized as possible coexistent phases. In the presence of even trace amounts of TMA, MAZ replaces ECR-1, in a manner indicative of nucleation control by TMA. ANA commonly co-crystallizes with ECR-1 at this composition, and in some cases the latter seems to be the secondary growth, indicated by a frequent tendency to grow on the surface of large ANA crystals, possibly indicating a seeding influence of ANA. Seeding of several ECR-1 compositions with ANA did not induce the growth of ECR-1.

CORRELATION AND DISCUSSION OF RESULTS

The first crystalline phases to form in the Group 1A modified Na+ reactions are all various modifications of FAU, with the most pronounced effect being to give platelet modifications. The proportionality of the degree of plate formation with level of M component at low M levels, may reflect surface adsorption modifying growth rates. However, if an intermediate building unit comprised sheets of double six ring connected sodalite cages, equivalent to 111 planes in FAU, then one can explain the various mis-stacked, twinned and platelet modifications as secondary cation induced growth modifications. The lingering presence of GME in all but the Li+ and Cs+ modified syntheses would indicate that the main SBU are either 4-ring or 4-, 6-ring clusters. Both FAU and GME can be built from the (43,6) cluster, by connecting two clusters through a mirror plane (GME), or by connecting the same after a 60° relative rotation (FAU). This is also the common building unit for LTA and SOD, which also co-crystallize with FAU(X) from lower Si/Al ratio sodium aluminosilicate gels. However, GME may also be constructed from double crankshaft chains of 4-rings, which are common structural units of GIS and MER which co-crystallize with, and progressively replace, FAU and GME across the range of compositions studied in the Na+ experiments.

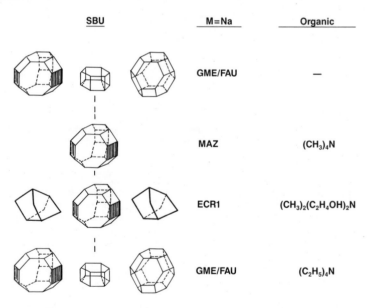

	SBU		M=Na	Organic
			GME/FAU	–
			MAZ	(CH₃)₄N
			ECR1	(CH₃)₂(C₂H₄OH)₂N
			GME/FAU	(C₂H₅)₄N

FIGURE 5: Building polyhedra characteristic of the several zeolites crystallizing in the composition range reported in the Tables.

The zeolites in these experiments that do not fall into the above groups are MAZ and ECR-1. The last is unusual in that it represents a boundary phase comprising components of two zeolites that have not been recognized as co-crystallizing, and require the existence 5-ring SBU in addition to 4-, 6-ring SBU. MAZ can be built from 4-ring chains having a different configuration to the double crankshaft, but this can be readily obtained by minor reorganization of the (43,6)-ring SBU.

CONCLUSIONS

Wide variation in secondary cations yield several zeolites that can be readily derived from common SBUs. Kokotailo has summarised several of these relationships[34]. The present descriptions reinforce the role of the gel in determining the SBU, and the cations in the ways that the SBUs are directed to the formation of larger structural units. These latter seem most likely to have sheet structures, in a form similar to those found in clays. The most pronounced effects seen in these experiments are in the various modifications of the FAU structure, which show compositional, morphological and stacking influences of changes in the nature and level of secondary cation M. Subsequent papers will further explore this system, examining the influences of secondary cations when N=K.

ACKNOWLEDGMENTS

Much of the earlier work on FAU modifications was done with collaborators E.W.Albers, M.G.Barrett and G.C.Edwards. The present work reflects the able synthesis support of K.G.Strohmaier, and SEM analyses by S.B.Rice and M.Brons.

REFERENCES

1. R.M.Barrer, "Hydrothermal Chemistry of Zeolites", Academic Press (London), (1982).
2. G.Engelhardt and D.Michel, "High Resolution Solid State NMR of Silicates and Zeolites", J.Wiley (London),Ch.3, 75 (1987).
3. P.K.Dutta and D.C.Shieh, J. Phys. Chem., 90, 2331 (1986).
4. A.V.McCormick, A.T.Bell and C.J.Radke, Proc. 7th. Intl. Zeolite Conf., Ed. Y.Murakami, A.Iijima and J.W.Ward, Kodansha/Elsevier, 247(1986).
5. E.J.J.Groenen, A.G.T.G.Kortbeek, M.Mackay and O.Sudmeijer, Zeolites, 5, 403 (1986).
6. P.Bodart, J.B.Nagy, Z.Gabelica and E.G.Derouane, J.Chim.Phys.-Chim.Biol., 83 , 777 (1986).
7. P.K.Maher. E.W.Albers and C.V.McDaniel, U.S. Patent3,671,191 (1972).
8. D.E.W.Vaughan, G.C.Edwards and M.G.Barrett, U.S. Patent 4,340,573 (1982).
9. D.E.W.Vaughan, U.S. Patent 4,657,749 (1986).
10. D.W.Breck, "Zeolite Molecular Sieves", J.Wiley (New York), (1974).
11. L.D.Rollmann, "Zeolites: Science and Technology", Ed. F.R.Ribeiro, A.E. Rodrigues, L.D.Rollmann and C.Naccache, NATO ASI Ser. E, No.80, 109 (1984).
12. B.M.Lok, T.R,Cannan and C.A.Messina, Zeolites,3, 282 (1983).

13. J.Ciric, U.S. Patent 3,923,639 (1975).
14. E.M.Flanigen and E.R.Kellberg, U.S. Patent 4,241,036 (1980); British Patent, 1,178,186 (1970).
15. G.T.Kerr and G. Kokotailo, J.Amer.Chem. Soc., 83, 4675 (1961).
16. E.W.Albers and D.E.W.Vaughan, U.S. Patent 3,947,482 (1976).
17. E.W.Albers, G.C.Edwards and D.E.W.Vaughan, U.S. Patent 3,755,538 (1973).
18. G.C.Edwards, D.E.W.Vaughan and E.W.Albers, U.S.Patent 4,175,059 (1979).
19. N.Dewaele, P.Bodart, Z.Gabelica and J.B.Nagy, "Zeolites: Synthesis, Structure, Technology and Applications.". Ed. B. Drzaj, S.Hecevar, and S.Pejovnik, Elsevier Press (Amsterdam), 119, (1985).
20. W.M.Meier and D.H.Olson, "Atlas of Zeolite Structures", 2nd. Edition, Butterworth Press (London), (1987).
21. J.Ciric, U.S. Patent 3,411,874;3,415,736 (1968).
22. G.T.Kokotailo and J.Ciric, "Molecular Sieve Zeolites - 1", Ed. E.M.Flanigen and L.B.Sand, Amer. Chem. Soc. Adv. Chem. Ser. 101, 109 (1971).
23. D.E.W.Vaughan and M.G.Barrett, U.S. Patent 4,333,859 (1982).
24. M.G.Barrett and D.E.W.Vaughan, U.S. Patent 4,309,313 (1982).
25. M.M.J.Treacy, J.M.Newsam, R.A.Beyerlein, M.E.Leonowicz and D.E.W.Vaughan, J. Chem. Soc. Chem. Commun., 1211 (1986).
26. R.A.Beyerlein, J.M.Newsam, M.T.Melchior and H.Malone, J.Phys.Chem., accepted for publ., (1987).
27. M.M.J.Treacy, J.M.Newsam, D.E.W.Vaughan, R.A.Beyerlein, S.B.Rice and C.B.de Gruyter, Mater. Res. Soc. Symp. Proc. - this volume.
28. J.M.Thomas, M.Audier, and J.Klinowski, J. Chem. Soc. Chem. Commun., 1221, (1981).
29 G.R.Millward, J.M.Thomas, S.Ramdas and M.T.Barlow, Proc. 6th. Intl. Zeolite Conf., Ed. D.H.Olson and A.Bisio, Butterworth, London,793, (1984).
30. J.D.Sherman, "Molecular Sieves-II", Amer.Chem.Soc. Symp.Ser. 40, Ed. J.R.Katzer, 30 (1977).
31. D.E.W.Vaughan and K.G.Strohmaler, Proc. 7th. Intl. Zeolite Conf., Ed. Y.Murakami, A.Iijima and J.W.Ward, Kodansha/ Elsevier (Tokyo),207 (1987).
32. G.T.Kokotailo and S.L.Lawton, Nature ,203, 621, (1964)
33. D.E.W.Vaughan, U.S. Patent 4,714,601 (1987).
34. M.E.Leonowicz and D.E.W.Vaughan, Nature, 329, 819 (1987).
35. G.Kokotailo, Proc. 6th. Intl. Zeolite Conf., Ed. D.H.Olson and A.Bisio, Butterworth, London,760, (1984).

FORMATION OF ZEOLITES: A MOLECULAR DESCRIPTION

PRABIR K. DUTTA*, M. PURI AND D.C. SHIEH
Department of Chemistry
The Ohio State University
Columbus, Ohio   43210

ABSTRACT

Raman spectroscopy of the solution, solid and gel phases present during crystallization of zeolite X was investigated. The vibrational data indicate that an amorphous aluminosilicate solid, composed primarily of four membered aluminosilicate rings is in contact with monomeric silicate ions during the prenucleation stages of the zeolite formation. No intermediate building blocks specific to zeolite X could be discerned from the vibrational spectra. The influence of a series of monovalent cations on the crystallization process was also examined, and a model of zeolite formation has been proposed.

INTRODUCTION

Crystallographers estimate that only a few percent of the possible zeolite structures have been synthesized to date [1]. The lack of knowledge, at a molecular level, of the chemistry of zeolite synthesis has hampered the ability of researchers to create new zeolitic structures. However, the determination of the chemical transformation that occurs during zeolite synthesis is a very difficult experimental problem [2].

We have developed Raman spectroscopy as a tool to monitor the structural changes during zeolite synthesis and have applied it to study the formation of zeolites A, Y and ZSM-5 [3-5]. This paper examines the formation of zeolite X. Spectroscopic studies of the liquid and solid phases were carried out. A direct spectroscopic study of the gel like solid phase present during synthesis is reported for the first time. The role monovalent cations play in altering the routes of zeolite crystallization is probed and a model for zeolite synthesis is proposed.

EXPERIMENTAL SECTION

Chemicals - Ludox TM (Dupont) was used as the source of silicon and aluminum powder (-40 mesh) was obtained from Alfa Chemicals. Precautions were taken to exclude CO_2 from the reaction mixture and all experiments were carried out in Teflon bottles.

Zeolite synthesis was done at 80-85°C in a thermostatted water bath without any stirring. Samples were removed from the same batch of the crystallizing medium at various times during the crystallization process and examined spectroscopically. The details of the spectroscopic experiments are provided later.

The Raman spectra were obtained with a Spex 1403 spectrometer controlled by a Datamate computer. Excitation of all samples was done with the 457.9 nm line of a Spectra Physics argon ion laser. The scattered light was detected with a RCA C 31034 GaAs PM tube. Slit widths were typically 6 cm^{-1}, and scan times varied between 1-3 s/cm^{-1}.

Powder X-ray diffraction patterns were obtained with a Rigaku Geigerflex D/Max 2B diffractometer with Ni filtered CuK_α source. The elemental analysis (Na, Si, Al) was done with a JEOL JXA-35 electron microprobe using zeolite A as standard (errors in Si/Al ratio are within 10% of the values reported in the paper).

Mat. Res. Soc. Symp. Proc. Vol. 111. ©1988 Materials Research Society

RESULTS AND DISCUSSION

In a typical procedure for zeolite synthesis, mixing of the aluminum and silicon reactants results in the immediate formation of an aluminosilicate precipitate, which coexists with the solution species during the entire synthesis period [6]. Most structural studies, aimed at understanding the synthesis process, are carried out on the solution and solid phases independently [3-5, 7-9]. For the solution species, there is no problem associated with such experiments. However, prior to the examination of the solid phase, it is usually washed and dried. There is some concern whether such sample pretreatment alters the structure of the solid phase. Developing probes to examine the process of zeolite synthesis insitu, is therefore, of considerable interest. We have explored the possibility of using Raman spectroscopy in this regard.

Studies on the Gel Phase - The reactant composition used in the synthesis of zeolite X was 8.6 Na_2O 0.75Al_2O_3 3SiO_2 556H_2O. At various stages of the zeolite synthesis, the solution and solid phases were separated by centrifugation and examined. The solution phase consisted primarily of monomeric and dimeric silicate anions, as evidenced by Raman bands at 450, 600, 780 and 920 cm^{-1}, throughout the crystallization process [10]. The wet solid phase, largely unperturbed, was directly examined in an anaerobic Raman spinning cell. The solid phase, prepared under these conditions, is in intimate contact with the solution species and accurately represents the gel phase present during zeolite crystallization. Figure 1 shows the evolution of the Raman spectrum of this phase during zeolite X crystallization. Initially, bands at 500, 578, 780, 860 and 1000 cm^{-1} are observed. The band at 780 cm^{-1} can be assigned to the monomeric silicate ions trapped in the gel and is observed throughout the crystallization period. No major changes occur in the first six hours of heating. The band at ~ 500 cm^{-1} begins to grow after fourteen hours of heating, and immediately thereafter, Raman bands characteristic of zeolite X crystals emerge and grow till crystallization is complete (384, 480, 515, 1000, 1090 cm^{-1}, 26 hours). It is evident from Figure 1 that the solid phase present during the initial stages of the zeolite synthesis has a well defined vibrational spectrum, and therefore, a definite structure associated with it. The determination of this structure and specific changes in it during crystal growth should provide information about zeolite formation. Unfortunately, the low S/N associated with the spectra shown in Figure 1 made it difficult to observe any Raman spectroscopic changes during the prenucleation period. The factors contributing to the low S/N are the wetness of the samples, which makes focusing very difficult and the high fluorescent background typical of these alkaline samples. Considerable improvement in S/N would be expected if the samples could be washed to remove alkali and then dried prior to the Raman studies. These experiments would also indicate if washing and drying alter the structure of the solid phase.

A series of experiments, similar to the ones described above for zeolite X synthesis were repeated, but the solid phase at each successive stage was thoroughly washed and dried at room temperature in a vacuum oven prior to the Raman spectroscopic measurements. Figure 2 shows the vibrational data as a function of crystallization time. Upon comparing with Figure 1, it is immediately obvious that the Raman spectra for the washed and unwashed samples are similar, except for the better S/N obtained from the former samples. Also, as expected, no bands due to any trapped silicate ions are observed in the washed samples. Elemental analysis was performed on the samples, and it was found that the Si/Al ratio did not alter much (within the experimental error of our measurements) during the entire crystallization period (1.2 - 1.3). The final Si/Al ratio of the zeolite was 1.2.

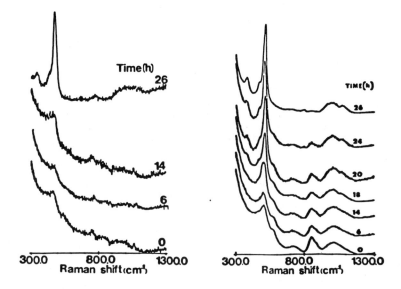

Figure 1. Raman spectra of the gel (wet solid phase) at various times during zeolite crystallization.

Figure 2. Raman spectra of the washed and dried solid phase during zeolite crystallization.

Structural Changes during Zeolite X Synthesis - Information about the structure of the solid phase present during the early stages of the zeolite synthesis can be derived from the vibrational spectrum. Raman bands at 450 (sh), 504, 575, 705, 855, 890 (sh), 1000, 1100 (sh) cm^{-1} are observed in the initial solid phase (Figure 2, OHR). Based on earlier studies of amorphous aluminosilicate solids, the following general band assignments can be made [3,4,11,12]. In the region of 300-550 cm^{-1}, the vibrational bands are assigned to motion of the oxygen atoms in the plane perpendicular to the T-O-T bond. The band between 550-600 cm^{-1} is assigned to Al-O-Si stretches arising from bonds which interconnect rings. The Al-O asymmetric stretch occurs between 700-750 cm^{-1}. The Si-O stretches due to nonbonded oxygen atoms appear between 800-900 cm^{-1}, whereas the Si-O stretches for totally polymerized segments occur between 900-1100 cm^{-1}. A notable structure marker among these bands in the prominent band between 300-550 cm^{-1}, whose frequency varies inversely with the size of the aluminosilicate rings present in the solid [13]. This correlation was developed for aluminosilicate glasses and metals, and we have adapted it to zeolites. According to this correlation, the band at 504 cm^{-1} would be indicative of four-membered aluminosilicate rings. The weak band at 450 cm^{-1} indicates the presence of higher membered and/or disordered rings. The Raman spectrum of the Ohr solid very closely resembles that of anorthite glass [14], which is thought to be made up of interconnected four-membered rings, based on X-ray radial distribution studies [15]. The first evidence of zeolite crystals in the X-ray powder diffraction patterns is observed after fourteen hours of heating, and defines the prenucleation stage of the synthesis. During this time period, the major changes in the Raman spectrum shown in Figure 2 include a decrease in intensity of the 450 cm^{-1} shoulder, narrowing of the 504 cm^{-1} band and decrease in intensity of the 855 cm^{-1} band. Based on the assignments discussed above, these spectral changes would indicate a decrease in disordered ring structures and increased polymerization of the aluminosilicate solid phase. These observations bear a striking resemblance with that reported for the zeolite A system [3]. An amorphous aluminosilicate solid consisting predominantly of four-membered aluminosilicate rings is in contact with aluminate ions in zeolite A and silicate ions in zeolite X in the initial stages of the synthesis.

Perturbations on the Zeolite X Synthesis System - In order to elaborate on the mechanistic aspects of zeolite X synthesis, we carried out a series of experiments on the aluminosilicate solid formed immediately upon mixing the aluminum and silicon reactants. This material was thoroughly washed and dried (Raman spectrum - Figure 2, 0hr). A series of cations (M=Na$^+$, K$^+$, Cs$^+$, N(CH$_3$)$_4$$^+$, N(C$_2H_5$)$_4$$^+$, N(C$_3H_7$)$_4$$^+$ and N(C$_4$H$_9$)$_4$$^+$) were exchanged into it by stirring with 1 $\underline{M}$ MCl solutions. The ion exchanged solids were then heated in the corresponding hydroxides (1.7 $\underline{M}$ MOH) and the zeolite crystals formed were analyzed by X-ray diffraction. In these sets of experiments, the starting aluminosilicate framework and the hydroxide ion concentration are identical, and the role of the cations can be directly examined. Table 1 lists the various initial zeolites that were formed. The transformation of these zeolites to subsequent forms were not investigated. All these zeolites are built up of four membered aluminosilicate rings, which, as we have noted before, is the major structural unit of the amorphous starting material. These building blocks, under the influence of the OH$^-$ ion connect to form zeolite nuclei. It is clear that the cation has a profound effect on the connectivity of the aluminosilicate framework. We propose the following model for the synthesis of these zeolites

(a) Four membered aluminosilicate rings are the simplest building block for these zeolites.

(b) These four membered rings connect to form units that are unique to a specific zeolite.

(c) The specific units for different zeolites all coexist in the gel phase during prenucleation and are readily interconvertible.

(d) The size and charge of the cation, based on steric and electrostatic factors determines which of these specific units will polymerize to form the zeolite.

A diagrammatic representation of this model is shown in Figure 3. We emphasize that this is a model and there is no spectral evidence for the intermediate building blocks shown in the figure. These structures were conceived due to their specificity for a particular zeolite and their ready interconvertibility by rearranging a few T-O-T bonds. They were also the smallest possible units that could be readily polymerized to form unique zeolite structures. Earlier work has shown that nuclei of zeolites X and P coexist during crystallization of zeolite X [16]. Kostinko [17] has shown that by controlling synthesis variables, in particular, crystallization time, zeolites A and X can be synthesized from similar compositions. The implication here is that the nuclei of these zeolites can coexist. Based on this study, the key question is why and how the cations stabilize a specific zeolite framework. The interaction of the cation with water and the aluminosilicate framework is of major importance in this regard. Many models of ion solvation have been proposed [18-22]. Based on both experimental and theoretical literature, it appears that the cations listed in Table 1 fall into three categories. For cations with high charge to radius ratios, such as Na$^+$, interaction with the aluminosilicate anion framework is favored through well ordered water molecules. As the charge to radius ratio decreases, e.g. for Cs$^+$, direct interaction with the anion is favored with disordered water molecules surrounding the cation. For NR$_4$$^+$ ions, direct interaction with the anion is still favored. However, because of the alkyl groups and the ensuing hydrophobic effect, the water molecules are well ordered around the cation. The relationship of these cation-water-anion structures to stabilization of specific zeolite nuclei is at present, unclear. The advantage of such a model lies in the suggestion of further spectroscopic experiments to study this important problem. In comparing the role of Na^+, Cs^+ and NR_4^+ in zeolite formation, a spectroscopic indicator could be the structure of the water molecule which is distinct in each of these cases. Vibrational spectroscopy of water in pure form and in the presence of ions has been extensively investigated [23-25]. The spectroscopy of water molecules trapped in the amorphous aluminosilicate gel during the zeolitization process in the presence of specific cations could provide important clues to understanding the complex mechanism of zeolite formation.

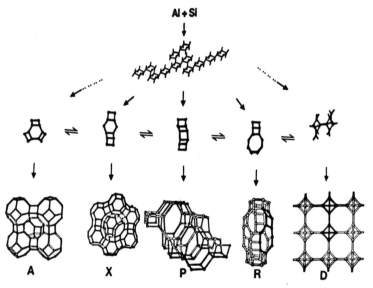

Figure 3. A model for zeolite crystallization.

Table 1
Zeolites Crystallized as a function of Cations

Cations	Zeolite
Na^+	zeolite A
K^+	zeolite R (chabazite)
Cs^+	zeolite Cs-D (edingtonite)
$N(CH_3)_4^+$	zeolite ZK-4
$N(C_2H_5)_4^+$	zeolite P (gismondine)*
$N(C_3H_7)^+$	zeolite X*
$N(C_4H_9)^+$	zeolite P

*major constituent, impurities also present

Acknowledgement - We acknowledge the support provided by the National Science Foundation (CHE 8510614).

References

1. L. B. Sand, Pure Appl. Chem., 52, 2105 (1980).
2. E. M. Flanigen, Adv. Chem. Ser., No. 121, 119 (1973).
3. P. K. Dutta, and D. C. Shieh, J. Phys. Chem., 90, 2331 (1986).
4. P. K. Dutta, D. C. Shieh, and M. J. Puri, J. Phys. Chem., 91, 2332 (1987).
5. P. K. Dutta and M. J. Puri, J. Phys. Chem., 91, 4329 (1987).
6. D. W. Breck, Zeolite Molecular Sieves, Wiley, N.Y. 1974; p. 245.
7. W. C. Beard, Adv. Chem. Ser., No. 21, 162, (1973).
8. F. Roozeboom, H. E. Robson, and S. S. Chan, Zeolites, 3, 321 (1983).
9. G. Engelhardt, B. Fahlke, M. Magi, and E. Lippmaa, Zeolites, 5, 49 (1985).
10. P. K. Dutta and D. C. Shieh, Appl. Spectrosc. 39, 343 (1985).
11. P. McMillan, Am Mineral, 69, 622 (1984).
12. S. K. Sharma, J. A. Philpotts and D. W. Mattson, J. Non Cryst. Solids, 71, 403 (1985).
13. D. W. Mattson, S. K. Sharma, and J. A. Philpotts, Am. Min., 71, 694 (1986).
14. S. K. Sharma, B. Simons, and H. S. Yoder, Jr., Am. Min., 71, 694 (1986).
15. M. Taylor, and G. E. Brown, Jr., Geochim. Cosmochim. Acta, 43, 61 (1979).
16. J. A. Kostinko, ACS Symp. Ser., 218, 3 (1983).
17. E. F. Freund, J. Cryst. Growth, 34, 11 (1976).
18. H. S. Frank, W. Y. Wen, Discussions Faraday Soc., 24, 133 (1957).
19. R. . Gurney, Ionic Processes in Solution, (Dover Publ. Co., New York, 1962).
20. P. Schuster, W. Jakubetz and W. Marius, Topics in Current Chemistry, 60, 1, (Springer-Verlag, Berlin 1975).
21. J. L. Kavanau, Water and Solute-Water Interactions, (Holden-Day, San Francisco, 1964).
22. G. Zundel, Hydration and Intermolecular Interaction, (Academic Press, New York, 1969).
23. H. G. Hertz, Angew. Chem. International Ed., 9, 124 (1970).
24. R. E. Verral, in Water - A Comprehensive Treatise, Ed. F. Franks, (Plenum Press, N.Y. 1973), p. 211.
25. T. H. Lilley, in Water - A Comprehensive Treatise, Ed. F. Franks, (Plenum Press, New York, 1973), p. 265.

THE EFFECT OF ALKALI METAL CATIONS ON
THE STRUCTURE OF DISSOLVED SILICATE OLIGOMERS

Alon V. McCormick, A. T. Bell, C. J. Radke
Center for Advanced Materials, Lawrence Berkeley Laboratory, and Department of
Chemical Engineering, University of California, Berkeley, CA 94720

ABSTRACT

By means of ^{29}Si NMR spectroscopy, it is established that the
distribution of silicate anions in alkaline silicate solutions is a moderate
function of base composition. At a fixed SiO_2 concentration and silicate
ratio, the proportion of Si present in oligomeric and cage-like structures
increases in progressing from Li to Cs hydroxide. Interactions between alkali
metal cations and silicate anions are investigated using NMR spectroscopy of
the cations; in this way the concentration of ion pairs is measured as a
function of cation size. As a result the silicate redistribution is ascribed
to cation-silicate anion pairing and to a higher selectivity for ion pairing
by large silicate anions as cation size increases.

INTRODUCTION

It is widely believed that the nucleation and growth of zeolites occur
through the reactions of dissolved silicate and aluminosilicate anions [1,2].
Supporting this view, several experimental studies have shown that the
structure of dissolved silicate species can influence the structure of solid
aluminosilicate intermediates present in the gel from which a zeolite may form
[3,4,5]. As a consequence, determination of the structure and distribution of
such anionic species has become a subject of considerable interest. Studies
utilizing ^{29}Si NMR have revealed a wide variety of silicate species in aqueous
solutions of silica at high pH [1,6-10]. Anions resembling the secondary
building units of zeolites have been observed and it has been shown that the
concentrations of these species, as well as others, are highly sensitive to
the pH of the solution [7,9]. The structures of aluminosilicates are less
well known than those of silicates but some information about the simpler
aluminosilicates has been obtained from ^{29}Si and ^{27}Al NMR [11,12].
Cation size and charge are known to influence the selectivity of zeolite
synthesis [1]. For instance, faujasite synthesis is quite specific to sodium
systems, whereas zeolite L is formed in the presence of potassium. The
effects of cation composition on the distribution of silicate species has been
examined by only a few authors. Ray and Plaisted [13], using
trimethylsilation/ chromatography, have found that the proportion of silicon
observed in oligomeric species increases as the base is changed from LiOH to
CsOH. Differences in the distribution of silicate species of Na silicate and
K silicate solutions can also be deduced from differences in the high
resolution ^{29}Si NMR spectra of K silicates [8] and Na silicates [9].
The work presented here was aimed at establishing the influence of
different alkali metal cations on the distribution of silicate anions in
aqueous solutions such as those used in zeolite synthesis. Changes in the
extent of silicate oligomerization were identified using ^{29}Si NMR
spectroscopy. The results obtained indicate that the influence of cation
composition is expressed through the formation of cation-anion pairs. Direct
evidence for the presence of such ion pairs was obtained using alkali metal
cation NMR spectroscopy.

Mat. Res. Soc. Symp. Proc. Vol. 111. ©1988 Materials Research Society

EXPERIMENTAL

Alkali metal silicate solutions were prepared from Baker analyzed SiO_2 gel, reagent grade alkali metal hydroxides, and deionized water. Stock solutions at about 3 mol% SiO_2, $SiO_2/M_2O = 3$ (M = alkali metal) were aged in polypropylene bottles for periods of several weeks, to assure that all the SiO_2 had dissolved. Samples were then formulated with extra water and base to achieve a desired composition.

NMR spectra were obtained with a Bruker AM-500 spectrometer at 99.36 MHz. About 100 70° pulses were used with relaxation delays of about 5 times the spin-lattice relaxation time, T_1. All spectra were recorded at room temperature.

RESULTS

Figures 1 and 2 show ^{29}Si NMR spectra of alkali metal silicate solutions with silicate ratios (R = $[SiO_2]/[M_2O]$) of R = 0.4 and 1.5. Each peak in these spectra represents a distinct silicon environment. In each figure, the chemical shift is referenced to that of the monomer in the Li silicate solution. The silicate solutions at R = 0.4 exhibit only three peaks, assignable to the monomer, dimer and cyclic trimer [9]. The larger cations produce higher concentrations of the cyclic trimer than do the smaller cations. The Li silicate spectrum in Fig. 1 ought not be compared directly with the other spectra, since a substantial amount of lithium inosilicate precipitated from the solution. At R = 1.5 the spectra display resonances in all of the connectivity regions except Q^4 (-34 ppm). As the cation size increases, silicon is displaced from lower connectivity states to states with Q^3_Δ and Q^3 connectivity. (Q^i indicates a silicate tetrahedron with i silicate bonds, and the subscript indicates a three-Si ring.)

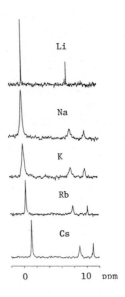

Fig. 1. ^{29}Si NMR spectra of 1.75 mol% SiO_2, R = 0.4 alkali metal silicate solutions.

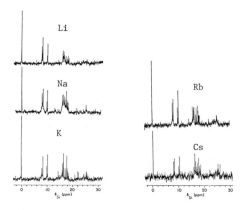

Fig. 2. ^{29}Si NMR spectra of 3.0 mol% SiO_2, R = 1.5 alkali metal silicate solutions.

The chemical shift and the linewidth of the cation NMR spectra were recorded with series of solutions with increasing cation size. Such a series of spectra is shown in Fig. 3. These spectra each display only one line due to the fast chemical exchange of the cation between the hydrated and the silicate-paired environment. These values were reduced as described in ref. 19 to reveal the trend in the concentration of paired cation with increasing cation size.

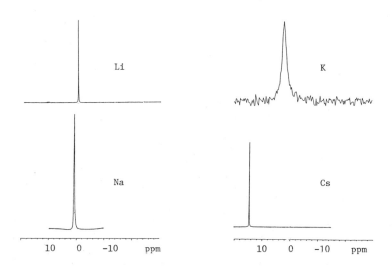

Fig. 3. Cation NMR spectra of alkali metal silicate solutions with 3 mol% SiO_2 and R = 1.5. The chemical shift reference is the fully hydrated cation in dilute chloride salt solution.

DISCUSSION

The results presented in Figs. 1 and 2 demonstrate that the distribution of silicate anions in alkaline silicate solutions is affected to a moderate degree by the choice of the alkali metal cation. The proportion of Si present in single and double rings increases progressing from Li to Cs, i.e., in the direction of increasing cation radius. The present results are consistent with those of Ray and Plaisted [13], who showed by means of trimethylsilation/chromatography that the fraction of Si present in oligomeric structures increases with increasing cation size. Also, in agreement with the trends of the present work, these authors observed that the addition of KCl to a Na silicate solution resulted in a modest increase in the degree of silicate oligomerization.

Since pH is known to influence strongly the distribution of silicate anions in alkaline solutions [7,9], one might expect the observed changes in silicate anion distribution with cation size to be a result of changes in pH. However, the pH of the solutions was not observed to increase with cation size, as might be expected from a consideration of hydroxide dissociation constants [15]. This suggests that the influence of cation composition is a consequence of direct cation-anion interactions, rather than an effect of pH.

It is well known that cation-anion pairs can form in concentrated solutions of ionic compounds [16]. The extent of ion pair formation increases with increasing ion concentration, decreasing cation size, and decreasing dielectric strength of the solvent [16]. Kinrade and Swaddle have invoked the proximity of alkali metal cations to silicates to explain anomalously short ^{29}Si spin lattice relaxation times [17].

The way in which cation-anion pairs might influence the distribution of silicate anions can be illustrated with the aid of reactions 1-3 below:

$$
\begin{array}{lll}
1. & M + M^- = D^- + H_2O \\
2. & M^- + C^+ = CM \\
3. & D^- + C^+ = CD
\end{array}
$$

Here M^- and D^- represent the anionic species $Si(OH)_3O^-$ and $(HO)_3SiOSi(OH)_2O^-$, respectively, and M and D are the corresponding neutral species $Si(OH)_4$ and $(HO)_3SiOSi(OH)_3$. C^+ is the cation and CM and CD are cation-anion pairs formed with M^- and D^-, respectively.

Reactions 2 and 3 represent the formation of cation-anion pairs. If the equilibrium constants for these reactions, K_2 and K_3, are identical, cation-anion pair formation will have no influence on the extent of dimerization at a fixed pH. On the other hand, if $K_3/K_2 > 1$, then the formation of dimers will be enhanced. The question to be posed, then, is what is the effect of cation size on K_2 and K_3, and more importantly, on K_3/K_2.

Based on Bjerrum's model of ion pairing [16], it is expected that the equilibrium constant for ion pair formation of a given anion with a cation decreases with increasing cation size. As a consequence K_2 and K_3 would both be expected to decrease in magnitude as one proceeds from Li^+ to Cs^+. Direct experimental confirmation of this trend is not available, but support for it can be inferred from the following observations. First, the precipitation of the Li inosilicate salt at solution compositions (R = 0.4) which remain stable in the presence of larger cations indicates that Li^+ cations bind to small silicate anions more strongly than do larger cations. Second, using a Born-

Haber cycle and measured heats of formation of alkali metal orthosilicates
[19], we have estimated that the heat of formation of a Li^+OSi ion pair is
25.5 kcal/mol higher than that of a K^+OSi pair in the absence of the
moderating effects of hydration. From this we would infer that K_2 for Li^+ is
larger than for K^+.

The effects of cation size on K_3/K_2 cannot be determined directly, since
the necessary thermochemical information is unavailable. Indirect evidence
for an increase in K_3/K_2 with increasing cation size can be drawn, however,
from NMR observations of the interactions between alkali metal cations and
silicate anions. These studies indicate that oligomeric anions shield the
nucleus of Cs^+ more efficiently than Na^+ and that Cs^+-silicate pairing
exhibits higher selectivity for large oligomers than does Na^+-silicate pairing
(see Fig. 4, in which large silicate anions are present at the solution
stability limit described in ref. 18). Thus we conclude that the
concentration of ion pairs increases somewhat with increasing cation size in
the presence of large silicate anions.

CONCLUSIONS

The distribution of silicate anions in alkaline silicate solutions
changes with the composition of the base. At a fixed SiO_2 concentration and
silicate ratio, the proportion of Si present in high molecular weight and
cage-like structures increases in progressing from Li to Cs hydroxide. This
trend is ascribed to cation-silicate anion pairing and to the higher
selectivity to ion pairing by large silicate anions as cation size increases.
An elementary estimate of the energetics of ion pairing and cation NMR confirm
that major differences in pairing equilibrium are to be expected as the cation
size is changed.

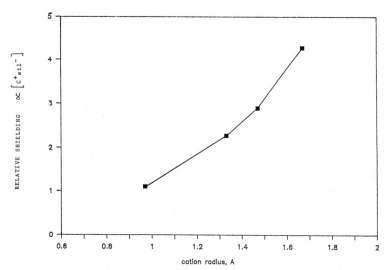

Fig. 4. Normalized shielding of the alkali metal cation due to pairing with
the silicate anion, as a function of cation size. The normalized
shielding is proportional to the concentration of cation-silicate
ion pairs [19].

REFERENCES

1. Barrer, R. M., The Hydrothermal Chemistry of Zeolites, Academic   Press, London, 1982.
2. Zhdanov, S. P., Molecular Sieves I, Adv. Chem. Ser. vol. 101, Am. Chem. Soc., 1971, p.20.
3. Derouane, E., Zeolites 2, 299 (1982).
4. Engelhardt, G., Zeolites 5, 49 (1985).
5. Wieker, W., and Fahlke, B., Zeolites (H. Drzaj, S. Hocevar, and S. Pejovnik, eds.), Elsevier, 1987, p. 161.
6. Marsmann, T., Z. Naturforschung 295, 495 (1974).
7. Engelhardt, Z., An. Allg. Chem. 418, 17 (1975).
8. Harris, R. K., and Knight, C. T. G., J. Chem. Soc, Faraday 2 79, 1525 and 1539 (1983).
9. McCormick, A. V., Bell, A. T., and Radke, C. J., Zeolites 7, 183 (1987).
10. McCormick, A. V., Bell, A. T., and Radke, C. J., to be published.
11. Dent Glasser, L. S., and Harvey, G., Proc. 6th Intl. Zeol. Conf. (A. Bisio and D. H. Olson, eds.), Butterworth, Guildford, U.K., 1984, p.925.
12. McCormick, A. V., Bell, A. T., and Radke, C. J., New Developments in Zeolite Science and Technology (Y. Murakami et al., eds.), Studies in Surf. Sci. and Cat. vol. 26, Elsevier, 1986, p.247.
13. Ray, W.H., and Plaisted, R. J., J. Am. Chem. Soc. 108, 7159 (1986).
14. Fukushima, E., and Roeder, S. B. W., Experimental Pulse NMR, Addison-Wesley, Reading, MA, 1981.
15. Gmelin Handbuch der Anorganische Chemie, Springer-Verlag, Berlin.
16. Bockris, J. O'M., and Reddy, A. K. N., Modern Electrochemistry, Plenum, New York, 1970.
17. Kinrade, S. D., and Swaddle, T. W., J. Am. Chem. Soc. 108, 7159, (1986).
18. Iler, R. K., The Chemistry of Silica, Wiley, New York, 1979.
19. McCormick, A.V., Ph.D. thesis, University of California, Berkeley, 1987.

COCATION EFFECTS IN RHODIUM-EXCHANGED Y ZEOLITES

R. K. SHOEMAKER, R. A. JOHNSON AND T. M. APPLE
University of Nebraska, Department of Chemistry, Lincoln, NE 68588-0304

ABSTRACT

Magic-angle spinning ^{13}C NMR spectra of carbon monoxide adsorbed on rhodium/Y zeolites yield information about the proportioning of CO in the various possible adsorption states; linear, bridged and dicarbonyl. The relative amounts of these adsorbed types, particularly the ratio of bridged to linear CO is influenced by the nature of the majority cations present with the rhodium. Reduced Rh-Na(+) and Rh-Li(+) zeolites form all three CO surface species, while acidic Rh zeolites, formed by the introduction of the co-cations Ca(2+) and H(+), exhibit no bridged carbonyls. The suppression of the bridged moiety results from the withdrawal of electrons from rhodium by the acid centers making the metal electron deficient (more oxidized).

Rh(I) dicarbonyl species form on all samples studied, however these species are indistinguishable from the linear monocarbonyls based solely upon the isotropic chemical shift obtained from magic-angle spinning. The number of dicarbonyl species can be quantitatively determined by the Carr-Purcell-Meiboom-Gill sequence, the powder pattern or by selective exchange experiments. At room temperature the two CO molecules in the gem-dicarbonyl appear to undergo a mutual hopping exchange. This motion is frozen out at 198K. The carbon-carbon internuclear separation in the gem-dicarbonyl is 3.3 Å.

Catalysts pre-adsorbed with ^{13}CO undergo exchange of the dicarbonyl species upon exposure at 198 K to ^{12}CO, however they also react to form $^{13}CO_2$. When exposed to CO at room temperature no CO_2 formation is detected.

INTRODUCTION

There has been great interest in the electronic nature of small metal particles in zeolites since the early work of Rabo et al..[1]   Their work showed that the turnover rate for isomerization and hydrogenolysis of neopentane over Pt/Y zeolite was greater when the cocation was acidic. Platinum in acidic zeolites behaved similarly to iridium suggesting that metals in acidic zeolites are electron deficient. Considerable discussion has arisen as to whether small particles inherently behave as if electron deficient. Gallezot et al. [2] in their study of small Pt clusters in zeolites ascribed the unusual chemistry of small particles to an electron deficiency. This "deficiency" resulted from a change in ionization potentials and not from electron transfer. A shift in the Pt L_{III} absorption edge toward higher energies has been observed when Pt is included in acidic zeolites.[3]

The interaction of CO with Group VIII metals is sensitive to the state of the metal.[4]   Analysis of solid-state ^{13}C NMR yields a wealth of information about the bonding of CO in inorganic carbonyls [5] and on supported metal catalysts [6-8]. We report ^{13}C NMR studies of CO adsorbed on Rh in Y zeolites with varying cocations. Magic-angle spinning is used to quantitatively differentiate between linear and bridged moieties. Multiple pulse experiments are used to quantify and characterize dicarbonyl species.

EXPERIMENTAL

Catalysts were prepared by ion-exchange of Na-Y zeolite obtained from Linde. Rhodium nitrate solutions were used for exchange. Each catalyst was exposed to a 0.02 M solution while stirring at room temperature. All ion exchange levels were determine by atomic emission of the supernatant solutions. Rh-Na catalysts were 4.2 wt.% rhodium. Rh-Ca catalysts were prepared by ion exchange of calcium for sodium (70% Na replacement). Rhodium was subsequently exchanged in these samples to a level of 1.8 wt.%. Rh-Li Y zeolite was prepared by lithium exchange (50% Na replacement) and rhodium exchange to 5.8 wt.%. H-Y zeolite was prepared by heating an NH_4^+ zeolite to 650 K and resulted in an exchange of 60% protons for sodium ions. Rhodium inclusion occurred to a level of 6.0 wt.%. The level of rhodium uptake appears to roughly obey an inverse dependence upon the size of the cocation.

Following exchange all catalysts were heated to 575 K for 4 hours to remove all water. Catalysts were then reduced under hydrogen at 575 K. Following reduction the catalysts were evacuated at ten degrees above the reduction temperature to a base pressure of 10-6 torr. ^{13}CO (Merck 99% enriched) was contacted with the samples on a grease-free glass vacuum system equipped with a capacitance manometer and cold-cathode gauge. Following exposure to CO the samples were evacuated at room temperature for 1 hour to remove reversibly-bound CO. Samples were then sealed under vacuum.

NMR spectra were obtained by spinning the sealed tubes in a spinner assembly described previously.[9] For all spectra shown here at least three independent samples were analyzed. NMR spectra were acquired on a spectrometer built in-house, operating at 25.16 MHz for ^{13}C. 90° pulse widths of 2.0 microseconds were used corresponding to H_1 fields of 116 G; more than a factor of ten greater than the chemical shift anisotropy. For the Carr-Purcell-Meiboom-Gill experiment composite 180° pulses were used.

RESULTS AND DISCUSSION

Representative ^{13}C NMR spectra of adsorbed CO on Rh zeolite catalysts co-exchanged with various cations are shown in Figure 1. All spectra exhibit a strong resonance (a) near 180 ppm in the deshielding direction from TMS. This peak has associated with it several spinning sidebands (s) as verified by variable spinning speed experiments. The spectra of all Na and Li Rh-Y zeolite samples studied show an additional resonance (b) at about 240 ppm relative to TMS. By comparison with work on Rh carbonyls in the solid state by Gleeson and Vaughan [5] we assign the resonance at 240 ppm to bridged CO and the resonance at 180 ppm and its associated spinning sidebands to CO bound in a terminal manner.

In the linear form the pi electrons in CO have free circulation about the C-O bond and cause a large "ring current" effect producing a shift about 80 ppm upfield of TMS when the magnetic field is along the C-O bond. When the field is perpendicular to the C-O bond the shift has a large downfield shift due to the so-called "paramagnetic" term in the chemical shift. The chemical shift anisotropy is, thus, over 400 ppm for terminally bound CO. At the magnetic field strengths used here (2.3 Tesla) this corresponds to greater than 10 kHz. Since this is greater than the spinning speeds commonly obtainable, spinning sidebands are produced for this species.

CO bonded in a bridged manner does not have a local axial symmetry and thus free electron circulation is prohibited. The chemical shift anisotropy is, therefore, greatly reduced and the isotropic value shifts to lower shielding. This yields a peak at 240 ppm, however no spinning sidebands are produced from the bridged moiety due to the smaller

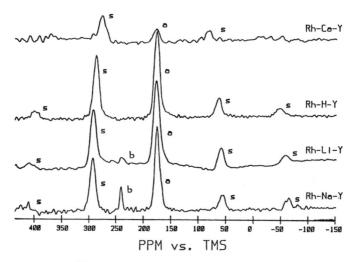

PPM vs. TMS

^{13}C NMR Spectra of CO on Rh-Y Zeolites

Figure 1

anisotropy.

The presence of a bridged peak for Rh-Li and Rh-Na Y zeolite samples and its absence in the Rh-Ca and Rh-H samples suggests that the electronic nature of the Rh in the two types of samples differs. A possible explanation for this effect is that in the acidic zeolites electron density is withdrawn from the Rh, making the Rh more oxidized. Metals in this state have stronger interactions with CO in the linear form than with CO in the bridged form due to the nature of the electron flow in the two forms.[4] Bridged CO is a better pi acid than linear CO because only electron withdrawal from the partially filled rhodium d_{xz} and d_{yz} orbitals (the z axis being the surface normal) into the vacant pi* orbitals of CO is possible. In the linear form sigma donation from the non-bonding orbital on carbon to the d_{z^2} orbital of rhodium also occurs, somewhat moderating the electron flow.

When attempts are made to analyze the spinning sideband pattern of the terminally bonded CO by the method of Herzfeld and Berger [10] a unique solution cannot be obtained. In all cases the intensity of the center band is too strong for the analysis to succeed. The peak at 180 ppm therefore appears to have more than one component. Previous work [7] on Ru in Y zeolite showed that dicarbonyls resonate near the mono-carbonyl species. We suspected that the peak at 180 ppm contained both monocarbonyl and dicarbonyl species. Duncan et al. [6] have shown that due to molecular motion the dicarbonyl species does not exhibit a large chemical shift anisotropy at room temperature. To test whether the center resonance contains a dicarbonyl species two experiments were performed. A static powder pattern was obtained both by Fourier transformation of the free-induction decay and by transformation of a free decay following a spin-echo. This powder pattern clearly showed the presence of both a broad anisotropy pattern from the linear mono-carbonyl and a partially narrowed component, presumably the dicarbonyl, both centered at 180 ppm.

A low-temperature (195K) exchange of 100% enriched CO_2 was performed.

Yates et al. [11] have shown that this procedure results in a selective exchange of ^{12}CO for dicarbonyls and leaves linear monocarbonyls on metal rafts unaffected. Following this procedure, the Rh-Y zeolite samples exhibited decreased intensity in the center band at 180 ppm relative to the associated spinning sidebands. Furthermore unique solutions to the Herzfeld-Berger procedure are obtained. The calculated values of the chemical shift tensor for the terminal Rh-monocarbonyls are presented in Table I.

Chemical Shift Components for CO on Rh-Y Zeolites

SAMPLE	δ_{iso}*	δ_{11}**	δ_{22}	δ_{33}
Rh-Li-Y				
Linear	172	370	244	-98
Bridging	239	---	---	---
Rh-H-Y				
Linear	174	356	258	-92
Bridging	---	---	---	---
Rh-Na-Y				
Linear	173	338	269	-87
Bridging	239	---	---	---
Rh-Ca-Y				
Linear	173	366	232	-79
Bridging	---	---	---	---

*All shifts in PPM relative to TMS. Positive numbers refer to decreased shielding. Error in δ_{iso} = ±2 PPM.
**Error in chemical shift anisotropy components = ±15 PPM, which represents the 95% confidence limits from statistical analysis of the Herzfeld-Berger results.

Table I

Rh-Ca and Rh-Na samples exhibit a pronounced peak at 128 ppm after ^{12}CO exchange which remains narrow even in non-spinning NMR experiments. We have previously determined this to be due to CO_2 moving about in the zeolite cage.[7] The appearance of this species in the NMR spectrum occurs even though the samples were evacuated for 1 hour after exchange of the ^{12}CO. The presence of $^{13}CO_2$ in the NMR spectrum indicates that in the Rh-Ca and Rh-Na samples the diffusion of CO_2 is too slow for all the CO_2 to escape from the zeolite cages prior to NMR analysis. Rh-H and Rh-Li samples, containing smaller cocations, show no CO_2. CO_2 presumably derives from the dissociation of the ^{12}CO and subsequent reaction of the oxygen with the preadsorbed ^{13}CO. The residual surface carbon formed is ^{12}C and invisible to NMR, while the originally adsorbed ^{13}CO is converted into CO_2. It is interesting to note that samples pre-adsorbed with ^{13}CO and left at room temperature under ^{13}CO do not produce CO_2. Perhaps the residence time of CO on one metal center is too short at room temperature for CO dissociation to occur.

The nature of the dicarbonyl species was further investigated by performing a Carr-Purcell-Meiboom-Gill (CPMG) experiment. In this experiment, spin interactions which are linear in the nuclear spin operator, such as the chemical shift and heteronuclear dipole and scalar couplings are averaged to zero. The remaining response is then dominated by homonuclear couplings, primarily dipolar. Figure 2 shows the magnetization during a CPMG experiment of a Rh-Li Y zeolite sample exposed to ^{13}CO and then outgassed. The response exhibits a pronounced oscillation

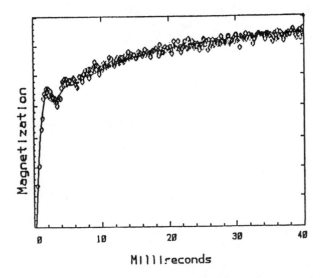

CPMG Response of Rh-Li-Y Zeolite

Figure 2

superimposed upon the normal transverse magnetization decay and has been
fit to a sum of two decaying exponentials and an oscillatory component.
This analysis indicates that 14% of the total carbon is contained in the
oscillatory component.

The "slow-beat" oscillation [8] results from the coupling of two ^{13}C
magnetic dipoles acting as a two-spin pair. In order for the oscillation
to be pronounced, the coupling between the two nearest nuclei must be much
greater than the couplings to other next nearest neighbor nuclei. This can
occur via spatial isolation of the dicarbonyl as has been suggested to
occur in supported metal systems.[4,12,13] However for the isolation to be
solely determined by spatial considerations, the separation between the
two-spin pair and neighboring spins on metal particles would have to be
greater than about 7 Å.

Another possible source of isolation of the two-spin pair is the
averaging of longer range interactions by motional processes which leave
the primary two spin interaction unchanged. Motional processes are known
to occur for the dicarbonyl species, resulting in narrowing of the
resonance line.[4] A motional narrowing of CO on Rh particles has also
been observed.[8]

When the Rh-Li-Y sample is cooled to 198K and analyzed by the CGMG
experiment at this temperature no oscillatory component remains. Warming
the sample to room temperature restores the oscillatory signal. It would
appear, therefore, that motional processes are involved in establishing the
dicarbonyl as an isolated two spin system. We suggest that the dicarbonyl
undergoes a hopping exchange among the two CO molecules and that the hop
occurs on a time scale much faster than the residence time at either site.
Motion of this type leaves the dipole-dipole coupling unchanged. However,
when combined with the motions of other next nearest CO molecules at
neighboring sites, this exchange reduces the longer-range dipolar
couplings.

Ignoring scalar coupling, the dipolar oscillation frequency is proportional to r^{-3} where r is the internuclear separation. Due to the very large chemical shift anisotropy, the ^{13}C nuclei in the gem-dicarbonyl may not behave as "like" spins with regard to their dipolar coupling. The difference in resonance frequency between the two CO molecules may range from 0 to 10 kHz at this resonance frequency. Large differences in the resonance frequency make the "spin-diffusion" term of the dipolar Hamiltonian non-secular. With this consideration, a detailed analysis to be published later yields an internuclear distance of 3.3 Å between the CO moieties of the gem-dicarbonyl. Since the species is NMR active, it is diamagnetic and must exist as $Rh(I)(CO)_2$.

REFERENCES

1.   J. A. Rabo, V. Shomaker and P. E. Pickert, Proc. 3rd Intl. Cong. Catalysis, Vol. 2 (North Holland Publishers, Amsterdam, 1965), p. 1264.

2.   P. Gallezot, J. Datka, J. Massardier, M. Primet and B. Imelik in Proc. 6th Intl. Cong. Catalysis, edited by G. C. Bond, P. B. Wells and F. C. Tompkins (The Chemical Society 2, London, 1977) p. 696.

3.   P. Gallezot, R. Weber, R. A. Dalla Betta and M. Boudart, Z. Naturforsch, 34a, 40 (1979).

4.   G. Henrici-Olive and S. Olive, Catalyzed Hydrogenation of Carbon Monoxide, (Springer-Verlag Publishers, Berlin, 1984), p. 23.

5.   J. W. Gleeson and R. W. Vaughan, J. Chem. Phys., 78, 5384 (1983).

6.   T. M. Duncan, J. T. Yates, Jr., and R. W. Vaughan, J. Chem. Phys., 73, 975 (1980).

7.   R. K. Shoemaker and T. M. Apple, J. Phys. Chem., 89, 3185 (1985).

8.   C. P. Slichter, Ann. Rev. Phys. Chem., 37, 25 (1986).

9.   R. K. Shoemaker and T. M. Apple, J. Magn. Reson., 67, 367   (1986).

10. J. Herzfeld and A. E. Berger, J. Chem. Phys., 73, 6021 (1980).

11. J. T. Yates, Jr., T. M. Duncan, S. D. Worley and R. W. Vaughan, J. Chem. Phys., 70, 1219 (1979).

12. H. F. Van't Blik, J. B. A. D. Van Zon, T. Huizinga, J. C. Vis, D. Konigsberger and R. Prins, J. Chem. Phys., 87, 2264 (1983).

13. J. L. Robbins, J. Phys. Chem., 90, 3381 (1986).

THE CHEMISTRY OF DETRITAL ALUMINUM IN ZEOLITES
STUDIED BY MULTINUCLEAR NMR SPECTROSCOPY AND OTHER TECHNIQUES

RICHARD H. JARMAN*
Exxon Research and Engineering Co., Rte. 22 East, Annandale, NJ 08801

ABSTRACT

High resolution ^{27}Al NMR spectroscopy has been used to study the dealumination of the A-type zeolite ZK4. The aluminum removed from the lattice during calcination undergoes reversible changes in coordination when treated in acidic and basic media. ^{29}Si NMR spectroscopy reveals that the changes are confined to the detrital aluminum species and do not affect the zeolite framework. Thermogravimetric analysis data combined with compositional data available from ^{29}Si NMR spectra have been used to probe the state of charge of the detrital aluminum species and the nature of defects present in the lattice.

INTRODUCTION

The gross features of the thermal decomposition of zeolites containing ammonium ions in the presence of water vapour have been well established by the use of many experimental techniques including, in particular, high-resolution NMR spectroscopy[1-9]. These techniques permit the chemistry of the framework and non-framework species to be followed virtually independently. At elevated temperatures, some fraction of the tetrahedal framework aluminum atoms that are formally associated with protons, usually, but not necessarily, provided by ammonium ions, are removed from the zeolite framework to form detrital (or extra-lattice) aluminum species. In this way, the silicon-aluminum ratio of the zeolite framework increases. The concomitant increase in the thermal stability of the lattice and modification of the acidic properties confer upon the dealuminated zeolite its utility in catalytic processes. The dealumination procedure appears to be strongly enhanced, if not totally dependent upon, the presence of water vapour. This can be explained by the water molecule engaging in nucleophilic attack on the aluminum atom in the Al-OH-Si group. Note that the dealumination process induced by calcination of a protonated zeolite differs from that induced by treatment of the zeolite with $SiCl_4$ vapour[10]. In the latter case, the driving force for the reaction is provided by the high Al-Cl bond strength and hence the presence of framework protons is no longer a prerequisite.

The general features of the dealumination process are well established. The resulting framework composition depends upon the calcination temperature, the degree of ion-exchange and, to a lesser extent, the water vapour pressure. However, the chemistry of the detrital aluminum species is not well understood, although this aspect has been the subject of study using IR spectroscopy[11]. In the ^{27}Al NMR experiment, the appearance of a resonance with a chemical shift in the range characteristic of octahedral Al generally signals the presence of detrital Al[1,4-6]. However, recent evidence suggests that the detrital Al can, under some circumstances, adopt tetrahedral coordination[12-14]. During progressive dealumination of faujasite, the octahedral peak in the ^{27}Al spectrum does not increase in intensity as the framework Si/Al ratio increases, suggesting that some part of the detrital Al adopts a non-octahedal coordination[15].

* Present address: Amoco Corporation, P.O. Box 400, Naperville, IL 60566

The detrital aluminum species may be present either as neutral or charged entities;in the latter case they may act as non-framework cations in neutralizing the framework charge due to the remaining framework aluminum.The chemistry of the detrital aluminum may have considerable influence on the catalytic properties of the zeolite in so far as, for example, the zeolite acidity may be controlled or modified by its presence.

In this paper we hope to provide further insight into the chemistry of detrital aluminum by the combined use of thermogravimetric analysis and high-resolution ^{29}Si and ^{27}Al NMR spectroscopy. The A-type zeolite ZK-4 was chosen for this study. The structure of the material is well characterized and, like faujasite, the framework contains only one crystallographically distinct framework site. Therefore, interpretation of the ^{29}Si NMR spectra is not complicated by the presence of overlapping peaks due to contributions from different framework sites. Furthermore,unlike faujasite, the sodium ion population can be completely exchanged by ammonium ions with ease, thus avoiding any complications due to residual sodium ions. In addition, as will be shown later, the detrital aluminum is entirely trapped within the zeolite cages and hence the material can be subjected to successive treatments without altering the overall composition. The enhancement of the activity towards cracking of hydrocarbons after calcination of zeolite ZK-4 at elevated temperature as re-ported in the patent literature[16] had suggested that dealumination occurs during the calcination of ammonium exchanged materials. High-resolution ^{29}Si NMR spectroscopy has confirmed that this is the case[17]. This is not surprising in view of the common structural features of ZK-4 and faujasite and that the process of dealumination appears to be quite general to all zeolite structures.

EXPERIMENTAL

Samples of zeolite ZK-4 in the composition range 2.5< Si/Al <2.9 were prepared using an adaptation of the method of Kerr[18] as described in detail in ref 19. Chemical analysis was performed using inductively coupled plasma emission spectroscopy. Before performing dealumination experiments, ammonium-exchanged forms were prepared by stirring 2 g of zeolite with 20 g of NH_4Cl in 100 cm^3 of distilled water made up to pH 9 with aqueous ammonia solution. One exchange was sufficient to remove greater than 95% of the sodium ion population (the residual sodium was less than .2% by weight). Complete exchange was effected by two treatments. The as-prepared zeolites always contain a considerable concentration of tetramethylammonium ion which is trapped in the alpha and beta cages[20]. These ions are removed by calcination above about 350°C. Ammonium-exchanged samples were also prepared from starting materials from which the tetramethylammonium ions had been removed by calcination at 500°C. This procedure itself causes a small degree of dealumination.

Ammonium-exchanged materials were calcined in air for one hour at a temperature between 520°C and 550°C. Sample geometry was kept constant for all the experiments, the sample depth being about 3 mm. Although the calcinations were performed without additional water vapour, that generated by evolution of the zeolitic water content is generally considered sufficient to perform "self-steaming" of the zeolite. After calcination, the materials were characterized by powder X-ray diffraction, TGA and NMR spectroscopy. Some of the dealuminated samples were subjected to further treatments with ammonium ion-exchange solutions and calcinations in air. In some cases, the ammonium chloride solution was not made basic but used in acidic form at pH 6.

^{29}Si and ^{27}Al NMR spectra were obtained using a JEOL FX200 spectrometer with magic-angle sample spinning,operating at a field strength of 47 kG (200 MHz proton). The corresponding ^{29}Si and ^{27}Al frequencies were 39.5 MHz and

52.0 MHz respectively. Powder X-ray diffraction patterns were obtained on a Siemens D500 automated diffractometer. Refined lattice parameters were obtained from the data using a least-squares iterative procedure. Elemental analysis was performed with a Jarrel-Ash Atomcomp III inductively-coupled plasma emission spectrometer. Thermogravimetric analysis was performed in air using a Dupont 1090 Thermal Analyzer. A heating rate of 10°C per minute up to 1000°C was used.

RESULTS AND DISCUSSION

The ^{29}Si NMR spectra show that the framework Si/Al ratio increases upon calcination of the ammonium form of ZK-4, figure 1. The relative intensities of the peaks due to Si coordinated by fewer Al atoms increase in the spectra of the dealuminated samples. The chemical shifts of the peaks remain almost constant however. The values of the framework Si/Al ratio were calculated from the relative peak intensities using the expression,

$$Si/Al = 4\Sigma I_n/\Sigma nI_n \qquad (1)$$

where I_n is the intensity of the Si(nAl) peak. Some uncertainty attaches to the values because of the poor resolution of the peaks due to n=3 and n=2. In the spectra of the calcined materials, the resolution of these peaks improves after the material is treated with ammonium chloride solution as shown by figures 1c and 1d.

Figure 1   Figure 2

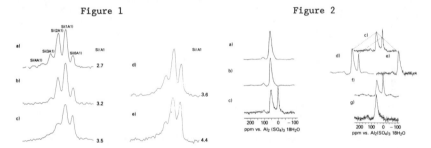

Figure 1:   a as-prepared zeolite; b, calcined, ammonium-exchanged; c, b after calcination; d, c after ammonium-exchange; e, four cycles of calcination and ammonium exchange.

Figure 2:   a, as-prepared zeolite; b, ammonium-exchanged; c, b after calcination; d, c after acidic ammonium exchange; e, c after basic ammonium exchange; f, d or e after calcination; g, basic ammonium exchange after four cycles of calcination and ammonium exchange.

The Si/Al ratio of the material prior to calcination was about 2.7. After one calcination the ratio was between 3.4 and 3.6. Calcination of the as-prepared material at 500°C also causes some dealumination, figure 1b. After four calcinations the ratio increased to between 4.0 and 4.4. More framework aluminum is removed during the first step than during subsequent steps under the same conditions.

Powder X-ray diffraction data reveal that changes in the unit-cell dimension occur during the dealumination procedure. The refined lattice parameter of the as-prepared material was 12.079(1) A Ammonium exchange causes a contraction of .026 A Calcination of the as-prepared zeolite followed by ammonium ion exchange causes a larger contraction to 12.0Q7(1) A The values of a_o for the dealuminated materials are 11.972(1) A. The shrinking of the unit-cell during dealumination is consistent with the increase in the framework Si/Al ratio,since the length of the Si-O bond is less than that of the Al-O bond.

The relative intensities of the diffraction peaks change during the course of the various treatments. The strongest peak in the diffraction pattern of the as-prepared zeolite is hkl = 311. After calcination (and removal of the tetramethylammonium ions) the strongest peak becomes hkl = 111. The change in intensities of the reflections after this treatment is due most probably to removal of the the tetramethylammonium ions. There are only very small differences between the diffraction patterns of the freshly calcined materials and ammonium exchanged materials. The integrated intensity of the 111 reflection does not diminish significantly with successive cycles of calcination and ammonium exchange. This suggests that the integrity of the lattice remains intact throughout the course of several treatments.

Representative ^{27}Al NMR spectra of the dealuminated samples are shown in figure 2. Ammonium ion-exchange has no effect on the spectrum of the parent, figures 2a and 2b. After calcination, a peak at about 2 ppm vs $Al_2(SO4)_3$.18H$_2$O appears in the spectrum in addition to the broader peak at about 54 ppm, as shown in figure 2c. By analogy with previous work, they are assigned to aluminum in octahedral and tetrahedral coordination to oxygen respectively.

The nature of subsequent treatment has a significant effect on the appearance of the ^{27}Al NMR spectra. After treatment with acidic ammonium chloride, the "octahedral" resonance diminishes in intensity relative to the tetrahedral peak, figure 2d; after treatment with basic ammonium chloride, the "octahedral" peak vanishes, figure 2d. These changes are completely reversible, since a spectrum very similar to that of the first calcined material is observed after calcination of either of these materials, figure 2f. Treatment of this material with basic ammonium chloride again causes loss of the octahedral resonance, figure 2g. All of the spectra shown were obtained using the same rf power and pulse width. The effect of changing these parameters on the relative intensities of the "octahedral" and "tetrahedral" resonances was not examined in this work. In none of the cases are the relative intensities of the two resonances consistent with the values predicted from the ^{29}Si NMR spectra.

Chemical analysis reveals that the overall Si/Al ratio of the materials remains constant during the cycles of calcination and ion-exchange. Therefore, none of the aluminum removed from the lattice during the dealumination is removed from the zeolite pores during subsequent chemical treatment. The residual sodium content is in all cases extremely low.

The observed ratios of Al_{oct}/Al_{tet} are taken from the integrated intensities of the ^{27}Al NMR spectra. The calculated values are obtained from the relation,

$$Al_{oct}/Al_{tet} = (1-R_1/R_2) / R_1/R_2 \qquad (2)$$

where R_1 and R_2 are the Si/Al ratios of the parent and calcined zeolites respectively, as shown in Table I. In the case of the freshly calcined materials, the intensities of the octahedral resonance are greater than that expected from the change in framework composition. The origin of this discrepancy lies in the fact that ^{27}Al is a quadrupolar nucleus with spin 5/2 which is therefore subject to interactions with the electric field gradient at the nucleus.

Table I

Sample	R_2	Al_{OCT}/Al_{TET}(OBS)	Al_{OCT}/Al_{TET}(CALC)	NH_3/Al
a	3.2	0	0.19	0.68-0.80
b	3.6	0	0.33	0.50-0.54
c	3.7	0.28	0.37	0.22-0.38
d	3.5	0.49	0.30	
e	3.4	0.72	0.26	
f	4.4	0.69	0.63	
g	4.0	0	9.48	0.18-0.29

$$R_1 = 2.7$$

Samples: a) calcined plus basic ammonium exchange, b) calcined plus basic ammonium exchange, c) calcined twice plus acidic ammonium exchange, d) calcined once, e) calcined once, f) calcined four times, g) calcined four times plus basic ammonium exchange.

Since the absolute intensities of the spectra were not carefully monitored, it is not possible to determine from the present spectra whether the aluminum lost from the octahedral resonance after treatment with ammonium ion solutions has been broadened beyond detection or has adopted a tetrahedral coordination, the chemical shift of which coincides with the framework aluminum. Clearly, the ^{29}Si NMR spectra show that the aluminum does not reenter the framework and the elemental analysis shows that it does not leave the sample.

It is tempting to propose that there exists in the solid an equilibrium similar to that in aqueous solution,

$$Al(H_2O)_6^{3+} + 4\ OH^- = Al(OH)_4^- + 6\ H_2O \quad (3)$$

If so, when treated in a basic medium, the detrital aluminum adopts a tetrahedral environment. Then it would be predicted from the above equation that the octahedral species were cationic and the tetrahedral species were anionic. To some degree, support for this hypothesis can be obtained from analysis of the TGA data.

The ammonium contents derived from the TGA are shown in Table I as ratios of NH_3/Al. The ratios for samples treated in base are higher than for those treated in acidic ammonium chloride. Therefore, the chemical composition of the zeolite as well as the structure of the detrital aluminum is affected by the pH of the medium. In both cases, the ammonium contents are insufficient to counterbalance the framework negative charge, which implies that the detrital aluminum may well carry a net positive charge, but one that is higher for materials treated in acid media.

It is also possible that the ammonium exchange under acid conditions does not effectively replace all the framework hydroxyls with ammonium ions. However, it appears that, in the presence of ammonia, the equilibrium,

$$NH_4^+(cage) = NH_3\ (cage) + H^+(lattice) \quad (4)$$

lies strongly to the left[21], implying that the presence of ammonia alone is sufficient to convert the framework hydroxyls into ammonium ions.

124

REFERENCES:

1. J. Klinowski, J. M. Thomas, C. A. Fyfe and G. C. Gobbi, Nature, $\underline{296}$, 553 (1982).
2. G. Englehardt, U. Lohse, A. Samoson, M. Magi, M. Tarmak and E. Lippma, Zeolites, $\underline{2}$, 59 (1982)
3. I. E. Maxwell, W. A. van Erp, G. R. Hays, T. Couperus, R. Huis and A. D. H. Clague, J. Chem. Soc., Chem Commun., 523 (1982).
4. C. A. Fyfe, G. C. Gobbi, J. S. Hartman, J. Klinowski and J. M. Thomas, J. Phys. Chem., $\underline{86}$, 1247 (1982).
5. J. Klinowski, J. M. Thomas, C. A. Fyfe, G. C. Gobbi and J. S. Hartman, Inorg. Chem., $\underline{22}$, 63 (1983).
6. D. Muller, W. Gessner, H. J. Behrens and G. Scholer, Chem. Phys. Lett., $\underline{79}$, 59 (1981).
7. D. Freude, T. Frohlich, H. Pfeifer and G. Scheler, Zeolites, $\underline{3}$, 171 (1983).
8. G. Engelhardt, U. Lohse, V. Patzelova, M. Magi and E. Lippmaa, Zeolites, $\underline{3}$, 233 (1983).
9. G. Engelhardt, U. Lohse, V. Patzelova, M. Magi and E. Lippmaa, Zeolites, $\underline{3}$, 239 (1983).
10. See for example H. K. Beyer and I. Belenykaja in Catalysis by Zeolites, edited by B. Imelik, Elsevier Sci. Pub. Co. (1980).
11. R. D. Shannon, K. H. Gardner, R. H. Staley, G. Bergeret, P. Gallezot and A. Auroux, J. Phys. Chem., $\underline{89}$, 4778 (1985).
12. D. R. Corbin, R. D. Farlee and G. D. Stucky, Inorg. Chem., $\underline{23}$, 2920 (1984).
13. D. Freude, J. Haase, H. Pfeifer, D. Prager and G. Scheler, Chem. Phys. Lett., $\underline{114}$, 143 (1985).
14. A. Samoson, E. Lippmaa, G. Englehardt, V. Lohse and H. G. Jerschkewitz, Chem. Phys. Lett., $\underline{134}$, 589 (1987).
15. G. J. Ray, B. L. Meyers and C. L. Marshall, Zeolites, $\underline{7}$, 307 (1987).
16. R. L. Wadlinger, E. J. Rosinski and C. J. Plant, U.S. patent, 3,375,205 (1968).
17. C. A. Fyfe, G. J. Kennedy, G. T. Kokotailo and C. T. DeSchutter, J. Chem. Soc., Chem. Commun., 1093 (1984).
18. G. T. Kerr, Inorg. Chem., $\underline{5}$, 1537 (1966); G. T. Kerr, Canadian Patent, 817,322 (1969).
19. R. H. Jarman, M. T. Melchior and D. E. W. Vaughan, A.C.S. Symposium Series 218, edited by G. D. Stucky and F. G. Dwyer, 267 (1983).
20. R. H. Jarman and M. T. Melchior, J. Chem. Soc., Chem. Commun., 414 (1984).
21. J. B. Uytterhoeven, L. G. Christener and W. K. Hall, J. Phys. Chem., $\underline{69}$, 2117 (1965).

IMPROVEMENT AND USE OF EMPIRICAL ESTIMATES OF ^{29}Si CHEMICAL SHIFTS IN ZEOLITES

J. M. NEWSAM, M. T. MELCHIOR and R. A. BEYERLEIN*
Exxon Research and Engineering Company, Route 22 East, Annandale, NJ 08801
*Present Address Amoco Oil Company, Amoco Research Center, PO Box 400, Naperville, IL 60566

ABSTRACT

^{29}Si chemical shifts, δ, are for aluminosilicate zeolites expressed as δ_{Si-4Al} (which can be estimated from the T-O-T angles observed crystallographically) plus terms involving b_1 and b_2, the quantitative influences of first and second neighbor aluminum atoms. The former, b_1, is assumed to be the sum of a fixed electrostatic component, α, and a framework-specific angular term, β ($0 < \beta < 0.9$ ppm), that reflects the generally smaller Si-O-Al angles compared to Si-O-Si. The compositional dependence of δ is approximated by an equal influence, b_2, per added second neighbor aluminum atom, the concentration of which is assumed to vary linearly with composition. Values of δ_{Si-nAl} ; n=0-3 for representative L (LTL-framework) and omega (MAZ) zeolites calculated on this basis agree well with those observed. Applying this description of the compositional dependence to as-synthesized and dealuminated CSZ-1 zeolites enables the measurement of framework aluminum contents in this four T-site case.

INTRODUCTION

Extraction of the invaluable information on aluminum distributions (and hence framework compositions) that is contained in the ^{29}Si nmr spectra of zeolites requires that the measured intensity be correctly apportioned to the various (Si-nAl; n = 0-4)$_i$; i = 1-m components (where m is the number of crystallographically distinct silicon sites). That is, the chemical shifts of these various components (and, where significant overlap occurs, the peak shapes and widths) must be known with reasonable accuracy. This has, in general, not been the case and quantitative analyses of the ^{29}Si nmr spectra of aluminosilicate zeolites were possible only for relatively simple systems with one (m = 1) or, at most, two inequivalent silicon sites. Considerable progress has been made, however, in assessing quantitatively the influence of first and second neighbor aluminum atoms [1], and the dependence of the ^{29}Si chemical shifts on the local framework geometry [2-7]. We describe here how these developments can be exploited, firstly, in determining framework compositions for CSZ-1 zeolites (a four T-site case) and, secondly, in calculating ^{29}Si chemical shifts for representative L (LTL-framework) and omega (MAZ) zeolites.

METHOD

In sodium aluminosilicates with the faujasite (FAU) framework, the average effect of adding a single aluminum atom to the first neighbor shell of a ^{29}Si nucleus (converting it, for example, from Si-0Al to Si-1Al) is to change its chemical shift, δ, by 6.17 ppm [1]. At typical compositions the observed separation between Si-0Al and Si-1Al peaks is, however, considerably smaller as a result of the compositional dependence of δ. The chemical shifts, $δ_{Si-nAl}$; n=0-4, vary approximately linearly with framework aluminum content [1]. This dependence can be reproduced by assuming that a change in chemical shift of 0.70 ppm accompanies the replacement of a second neighbor silicon atom by aluminum, and that the average concentration of second neighbor aluminum atoms varies linearly with composition. This latter assumption is known to be only approximate, because a high degree of local order persists over a broad composition range in FAU-materials [8]. Further, the assumption of a constant effect per second neighbor aluminum atom, irrespective of its distance or mode of bond connectivity to the central silicon atom is simplistic. More refined models involve allowing for known local Si - Al ordering effects [8], or for a 'direct' contribution for each aluminum atom that scales as some function of the Si - Al distance [9]. Nevertheless, given estimated experimental errors on reported ^{29}Si chemical shifts of, typically, some ± 0.1 - 0.2 ppm, the simple assumptions outlined above are sufficient to reproduce the observed compositional dependence of the reported ^{29}Si chemical shifts in the FAU and LTA frameworks [1].

This description of the compositional dependence of the ^{29}Si chemical shifts can be exploited directly as a means of determining the framework compositions of as-prepared and dealuminated CSZ-1 zeolites. These materials are closely related structurally to zeolite Y (FAU), but differ in having rhombohedral rather than cubic symmetry [10]. As a result of this reduction in symmetry, the single degenerate T-site (T = Tetrahedral species, Si or Al etc.) observed in zeolite Y is replaced by four inequivalent T-sites in CSZ-1 materials. The ^{29}Si nmr spectra of highly dealuminated CSZ-1 zeolites thus comprise four peaks of approximately equal intensities [11], contrasting with the single peak observed for similarly dealuminated zeolite Y. CSZ-1 zeolites with lower framework Si:Al ratios give heavily overlapped, unresolvable ^{29}Si nmr spectra. In order to extract framework aluminum contents from such spectra the compositional dependences for $δ_{Si-nAl}$ deduced for the FAU-framework (using exactly the same form and parameters), were combined with estimated peak intensities, I_n; n=0-4, taken from FAU-materials, and a chemical shift calibration point provided by a non-dealuminated CSZ-1 zeolite of known framework composition. A plot of ^{29}Si spectrum centroid position (in ppm) against framework composition was thus deduced.

An observed centroid for a given material is readily measured from the ^{29}Si nmr spectrum, as it is effectively equivalent to the half-area point. Framework aluminum contents for materials in this four T-site case were thus derived from the measured centroid positions by reference to the calculated centroid curve. The results for both as-prepared and dealuminated materials agree well with chemical analysis values (by inductively coupled plasma emission spectroscopy (ICPES)) when detrital aluminum is absent. For representative samples in which significant detrital aluminum is retained, total aluminum contents obtained by ICPES agree reasonably with the sum of the framework aluminum content determined from the ^{29}Si nmr spectrum centroids, plus the non-framework aluminum content estimated by ^{27}Al nmr [11].

In principle, this same analysis can be applied to zeolites other than LTA and FAU-framework materials (and systems that are related structurally). Unfortunately, however, other zeolite structures (excepting sodalites (SOD-framework)) cannot, as yet, be synthesized over a sufficiently wide range of Si:Al ratios as required for similarly detailed analyses of the compositional dependence of the ^{29}Si chemical shifts. We must therefore consider the transferability of the method and parameters deduced for the LTA and FAU frameworks.

Crystallographic data for a wide range of silicates and aluminosilicates and theoretical results demonstrate that the expectation Si-O-Al angle ($\omega_E(Al) = 137°$) is slightly smaller than the expectation Si-O-Si angle ($\omega_E(Si) = 144°$) [12]. In sodium aluminosilicates with the FAU-framework, the observed variation of the mean T-O-T angles with composition indicates a mean change ($\omega(Si) - \omega(Al)$) of about 7° [13], consistent with this difference between the expectation values. Given the dependence of the ^{29}Si chemical shifts on the T-O-T angles [2-7], it is therefore reasonable to suggest that the influence of a first neighbor aluminum atom on the chemical shift of a central ^{29}Si nucleus can be described as the sum of two distinct (but, perhaps, experimentally inseparable) components, namely a direct, electrostatic deshielding effect, and a geometrical component that reflects this small change in coordination geometry. The change in ^{29}Si nmr chemical shift (between, say, Si-0Al and Si-1Al) corresponding to the expected reduction in angle is calculated as some +0.89 ppm (the difference between the values of $\bar{\rho}$ calculated for T-O-T angles of 144° and 144-(7/4)° [7]; $\bar{\rho}$ is defined below). In the FAU-framework, the total change in chemical shift per added first neighbor shell aluminum of $b_1 = 6.17$ ppm is, therefore, apportioned $\alpha = 5.28$ ppm to the electrostatic component, and $\beta = 0.89$ ppm to the geometrical term.The electrostatic component, α, is, to first order, expected to be independent of the framework topology. The geometric component, β, on the other hand, might be expected to vary somewhat, depending on the framework flexibility. That is, the full expected reduction in the Si-O-T angle on placing an aluminum atom in the first neighbor shell may not be realized in frameworks of intrinsically lower flexibility. Thus we expect, generally, $0 < \beta < 0.9$ ppm, and we can write the calculated chemical shift at a framework Si:Al ratio of R, $\delta_n(R)$ as

$$\delta_n(R) = \delta_4 - (4-n)(\alpha + \beta) + A_n(R) \qquad (1)$$

where $\qquad A_n(R) = M(n) \times b_2 \times (2 / (R + 1)) \qquad (2)$

where $M(n)$ is the maximum number of second neighbor sites that can, subject to Loewenstein's rule [14], be occupied by aluminum for a given number of first neighbor aluminum atoms, n. The values $M(n)$ depend on the framework topology. The reference point (δ_4) is the chemical shift for the Si-4Al component (which is effectively independent of framework composition). The quantitative influence of a single aluminum atom in a second neighbor site, b_2, was determined to be 0.70 ppm in FAU and LTA-framework materials [1].

RESULTS AND DISCUSSION

Recent structural work on a dehydrated potassium zeolite L [15] provides mean T-O-T angles of T4(Si1)-O-T = 142.0(3)° and T6(Si2)-O-T = 142.9(2)°. Comparison with earlier powder X-ray diffraction results, with data for analogous gallosilicates [16], and with data for other zeolite L materials [17] suggests that these angles vary little on hydration/dehydration or with framework (and certain nonframework) cation substitution. The similarities between the local geometries of the two silicon sites are such that the ^{29}Si nmr spectrum can be well-fitted when treated as a single site case [15,18]. From the individual T-O-T angles, values for $\overline{\rho}$,

$$\overline{\rho} = 1/4 \, \Sigma_i \, (\cos \theta_i / \cos \theta_i - 1) \; ; \; i = 1 - 4 \qquad (3)$$

of 0.439(Si1) and 0.442(Si2) are calculated for the two sites. Values for the corresponding chemical shifts, δ_4, of -86.66 ppm and -87.20 ppm (with respect to tetramethylsilane (TMS)) respectively are then taken from a recent plot of δ_4 vs. $\overline{\rho}$ [7]. Using the mean of these two values in equation (1) (with $\alpha = 5.28$, $b_2 = 0.70$ ppm as above) and initially assuming $\beta = 0$, the calculated chemical shifts for a material with Si:Al = 2.89 differ from the observed values by -2.8 (n = 0), -1.8 (n = 1), -1.2 (n = 2) and -1.0 ppm (n = 3) (Table I). Based on the above argument these differences should each correspond to (n-4) x β. The consistency is good and these data provide a mean value of $\beta = 0.67$ ppm. The fact that β is less than 0.89 ppm is in line with the available structural data for LTL-framework zeolites [15-17] which indicate less variability in the framework geometry than is observed in the

FAU-framework [13]. The chemical shifts calculated using this value of β are compared with those observed in Table I.

It should be stressed that this analysis of data for only a single composition cannot yield unique values of b_2 and β. The factor $M(n)/(4-n)$, obtained on combining equations (1) and (2), is nearly independent of n. The quantity $\Delta_n = \{\delta_n(R) - \delta_4\}/(4-n)$ is therefore nearly constant and the magnitudes of β and b_2 are highly correlated. Using $\delta_4 = -86.9$ ppm, as above, calculated values for Δ_n are 5.1, 5.1, 5.1 and 5.4 ppm for n = 0, 1, 2 and 3 respectively (equations (1) and (2)). Excluding the small n=3 peak for which the chemical shift is least precise, and averaging over the remaining data yields $\Delta_n = (\alpha + \beta) - (1.16 \pm 0.06)b_2$. Thus for $\alpha = 5.28$ ppm and $0 < \beta < 0.9$, as discussed above, the second neighbor contribution is $0.15 < b_2 < 0.91$ ppm. Any value for b_2 within this range, coupled with the appropriate value of β, will provide predicted chemical shifts essentially identical to those listed in Table I. To separately determine appropriate values of b_2 and β requires data for other compositions.

The assignment of the ^{29}Si nmr spectra of synthetic MAZ-framework zeolites (exemplified by zeolites omega and ZSM-4) has been discussed on several occasions [19-22]. Similar to zeolite L, the MAZ-framework contains two crystallographically distinct T-sites (labelled T4 and T6). Although in zeolite L the Si-nAl components from the two sites are nearly coincident, in zeolite omega the two sets of components are separated by a considerable margin, some 8ppm. Typical spectra for materials with Si:Al = 3.1 - 4.2. contain 5 or 6 maxima and have given rise to two alternative assignments in which the Si-2Al peak for the T6 site is taken to coincide, respectively with the Si-1Al peak [19,20], or the Si-0Al peak for T4 [21]. Unfortunately, no structural data for synthetic aluminosilicate MAZ-framework materials are yet available. Single crystal X-ray structure analyses of both dehydrated [23] and hydrated [24] natural mineral samples have been reported, but those results are unlikely to model the synthetic materials accurately (see, e.g., [7]).

For data at only a single composition, the respective separations between the Si-0Al and Si-1Al peaks provide initial estimates of β_{T4} and β_{T6} (assuming, say, $b_2 = 0.7$ ppm). For the spectrum of a MAZ-framework aluminosilicate with Si:Al = 3.1 reported earlier [20], using these initial estimates in equations (1) and (2) enabled the assignment of the six pronounced maxima in the full spectrum. Optimized values for β_{T6} and β_{T4} were then derived from the six observed peak positions by least-squares, keeping b_2 fixed at 0.7 ppm (because of the high degree of correlation mentioned above). The agreement between observed peak positions and those calculated using the derived values was reasonable (although poorer than that for the zeolite L spectrum discussed above), and the same constants also yielded reasonable estimates for the peak positions observed in the spectrum of a MAZ-framework material with a somewhat lower aluminum content, Si:Al = 4.12.

Limiting ^{29}Si chemical shifts of -106.0 ppm (T4) and -114.4 ppm (T6) in a highly dealuminated omega zeolite have, however, also been reported [25], extending significantly the data range and, at least approximately, enabling the resolution of the first and second neighbor influences. Although the exact composition of this material was not known, it is assumed that these chemical shifts apply to a pure SiO_2 composition. Assigning the six maxima in the resolution enhanced spectrum of the material with Si:Al = 4.12 as above, least-squares optimized values of $\beta_{T4} = 0.32$, $\beta_{T6} = 0.87$ and $b_2 = 0.40$ were then obtained. The chemical shifts computed using these values are compared with those observed in Table I and Figure 1. The agreement is excellent. Chemical shifts were also calculated for a framework composition Si:Al = 3.1 using exactly the same parameters, but changing only the framework composition in the computation of A(n) (Equation (2) above). The agreement with the observed peak positions is again good (Table I).

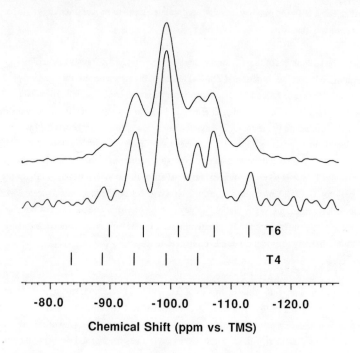

Chemical Shift (ppm vs. TMS)

Figure 1. ^{29}Si nmr spectrum of a synthetic sample of aluminosilicate MAZ (Si:Al = 4.1) measured with magic-angle sample spinning on a JEOL FX200B spectrometer operating at 4.7T (upper). The lower portion is a resolution enhancement of the spectrum achieved by applying a negative exponential to the free induction decay signal before Fourier transformation. Empirical estimates of the peak positions of each of the Si-nAl components for both the T4 site (lower) and T6 site (upper) are marked by the vertical bars (see Table I).The chemical shift scale is drawn referenced to TMS.

TABLE I

Observed and Calculated ^{29}Si Chemical Shifts (given as ppm versus TMS)

LTL

		n = 4	n = 3	n = 2	n = 1	n = 0
	obs[a]	--	-92.3	-97.1	-102.2	-107.4
	calc[b]	-86.9	-92.0	-97.2	-102.4	-107.3

MAZ

Si:Al = 4.1

		n = 4	n = 3	n = 2	n = 1	n = 0
T6	obs[c]	--	--	--	-107.2	-112.9
	calc[d]	-89.8	-95.5	-101.3	-107.2	-112.9
T4	obs[c]	--	-88.7	-94.0	-99.3	-104.4
	calc[e]	-83.6	-88.7	-94.0	-99.3	-104.5

Si:Al = 3.1

		n = 4	n = 3	n = 2	n = 1	n = 0
T6	obs[f]	--	--	--	-106.9	-112.4
	calc[g]	-89.8	-95.4	-101.1	-106.9	-112.5
T4	obs[f]	--	-88.6	-93.7	-98.8	-103.5
	calc[h]	-83.6	-88.6	-93.8	-99.0	-104.1

[a] Spectrum accumulated using 20% DSS as an internal standard. Chemical shifts derived by full-matrix least-squares profile analysis assuming gaussian peak shapes [18].

[b] Calculated according to equation (1) using δ_{Si-4Al} = -86.93, α = 5.28, b_2 = 0.70, β = 0.67 ppm with mean maximum numbers of aluminum second neighbors, M(n), of 0 (n = 4), 2.5(3), 4.5(2), 6.5(1) and 9.5(0).

[c] Peak positions taken from the spectrum illustrated in Figure 1 after resolution enhancement to slightly less extent than illustrated in the figure. The peak positions agree reasonably with those reported earlier for spectra obtained from commercial Linde omega zeolite of similar framework composition [20,21].

[d] Calculated according to equation (1) using δ_{Si-0Al} = -114.40 (at Si:Al = ∞), α = 5.28, b_2 = 0.395, β = 0.87 ppm with M(n) = 0(n = 4), 3(3), 5(2), 7(1), 10(0).

[e] As (d), but δ_{Si-0Al} = -106.00 (at Si:Al = ∞), β = 0.32 ppm.

[f] Original spectrum for this material was reported in [20]. Listed peak positions were measured from the original spectrum after resolution-enhancement to the same degree as for spectrum of the Si:Al = 4.1 material.

[g] Calculated according to equation (1) using identical parameters to (d), but using the different framework composition.

[h] Calculated according to equation (1) using identical parameters to (e), but using the different framework composition.

The optimized values for the constants β_{T4} and b_2 differ from similar results for the FAU and LTA-frameworks. The reduced first-neighbor influence for T4 suggests, in the context of the present methodology, less variability in the Si_{T4}-O-T angles with composition than is observed in the FAU-framework [13]. An optimized value for b_2 which differs from that initially derived for the FAU and LTA-frameworks is not unexpected. The value of $b_2 =$ 0.7 ppm for the FAU and LTA frameworks was, as above, based on the (simplest) approximations both of a linear change in the concentration of second neighbor aluminum atoms with framework composition, and of an equal influence for each such aluminum in a second neighbor site. The individual contributions of aluminum atoms beyond the first neighbor shell are, however, expected to depend on their disposition with respect to the central ^{29}Si nucleus (which is controlled by the framework topology). Local aluminum ordering effects which, say, selectively populate the nearer second neighbor sites will give rise to non-linear changes in the ^{29}Si chemical shifts with composition and limit the usefulness of the assumption of linear behavior. Aluminum ordering effects are compounded in the cases of the LTL and MAZ-frameworks by the unequal partitionings of aluminum between the two inequivalent T-sites in the structures [15,21,26]. Such ordering effects may, indeed, be manifested directly in the ^{29}Si nmr spectra of the MAZ-framework materials in the suggestion of structure in the Si-0Al (T6) peak (such as apparent in the resolution enhanced spectrum shown in Figure 1).

The present analysis (and another recent report [22]) indicates that the Si-2Al peak for T6 coincides with neither the Si-1Al peak [19,20], nor the Si-0Al peak [21] for T4. The resolution enhancement of the spectrum of the Si:Al = 4.1 material (Figure 1) reveals that the position of the Si-2Al component for T6 is indeed close to the computed position. Apart from chance matching of other factors, there is, of course, no reason why a peak for a silicon nucleus on one crystallographic site should coincide with any peak from another silicon nucleus at a crystallographically distinct site. The earlier simplified assignments were necessitated by the inability to calculate peak positions, and by the occurrence of less resolvable features in the observed spectrum than there are known contributing components. The accuracy to which the observed peak positions are reproduced by the present calculation is good, and is significantly better than that described earlier based on a composition-independent model for the ^{29}Si chemical shifts [22]. The limitations of such a model are well illustrated by the significant changes in peak positions that accompany a change in framework Si:Al ratio from 3.1 to 4.1 to ∞ (Table I). A more extended discussion of the ^{29}Si nmr spectra of MAZ-framework materials will be given elsewhere.

The present empirical description of the factors influencing ^{29}Si chemical shifts thus reproduces quite well observed peak positions in these LTL- and MAZ-framework materials. As discussed above, the use of an angular component, β, in forming the first neighbor aluminum influence appears justified, and gives generally consistent results. To further corroborate this description, however, it will be necessary to examine in a more quantitative fashion the relationship between empirically derived values of β and the relevant structural parameters. In particular, it will be interesting to determine whether the implied difference between the variabilities of the geometries of the T4 and T6 sites in the MAZ-framework is supported by structural data.

The precision of the ^{29}Si chemical shifts computed on the basis of the present empirical correlations is probably not yet sufficient to enable accurate peak intensities for complex spectra in the general case to be extracted by full spectrum profile analysis [18] assuming *exactly* the calculated peak positions. However, by constraining peak positions to values close to the computed values in the curve-fitting process, the combination of empirical estimates of peak positions with a flexible full profile analysis procedure may become a powerful general means of determining aluminum distributions in the more complex zeolites.

CONCLUSION

The quantitative description of the compositional dependence of the ^{29}Si chemical shifts derived for FAU and LTA framework materials has been applied successfully to determining framework compositions in CSZ-1 zeolites, a four T-site case. Combining this compositional dependence with estimates of geometrical contributions provides a general framework for estimating ^{29}Si chemical shifts empirically. The agreement between observed and calculated chemical shifts for representative zeolite L and omega materials is good, and similar methods can therefore presumably be applied to other zeolites.

Acknowledgements

We thank R. H. Jarman for providing the sample of aluminosilicate zeolite omega used in producing the resolution enhanced spectrum.

References

1. J. M. Newsam, J. Phys. Chem. 89, 2002-2005 (1985).
2. J. V. Smith and C. S. Blackwell, Nature 303, 223-225 (1983).
3. J. M. Thomas, J. Klinowski, S. Ramdas, B. K. Hunter and D. T. B. Tennakoon, Chem. Phys. Lett., 102, 158-162 (1983).
4. S. Ramdas and J. Klinowski, Nature, 308 (1984) 521-523.
5. G. Engelhardt and R. Radeglia, Chem. Phys. Lett., 108, 271-274 (1984).

6. R. Radeglia and G. Engelhardt, Chem. Phys. Lett., 114, 28-30 (1985).

7. J. M. Newsam, J. Phys. Chem. 91, 1259-1262 (1987).

8. M. T. Melchior in Intrazeolite Chemistry, edited by G. D. Stucky and F. G. Dwyer (ACS Symp. Ser. 218, American Chemical Society Washington DC 1983) pp. 243-265.

9. L. P. Aldridge, R. H. Meinhold and D. M. Bibby poster presented at 7th. Int. Zeolite Conf., Tokyo Japan, August 1986 and private communication 1987.

10. M. M. J. Treacy, J. M. Newsam, R. A. Beyerlein, M. E. Leonowicz and D. E. W. Vaughan, J. Chem. Soc. Chem. Comm., 1986, 1211-1214.

11. R. A. Beyerlein, J. M. Newsam, M. T. Melchior and H. Malone, J. Phys. Chem., accepted for publication (1987).

12. see, e.g., K. L. Geisinger, G. V. Gibbs and A. Navrotsky, Phys. Chem. Minerals, 11, 266-283 (1985).

13. J. M. Newsam, A. J. Jacobson and D. E. W. Vaughan, J. Phys. Chem. 90, 6858-6864 (1986).

14. W. Loewenstein, Amer. Mineral., 39, 92-96 (1954).

15. J. M. Newsam, J. Chem. Soc. Chem. Comm., 1987, 123-124.

16. J. M. Newsam, Mater. Res. Bull., 21, 661-672 (1986).

17. J. M. Newsam, presented at ACS National Meeting, New Orleans LA, September 1987 and in preparation 1987.

18. J. M. Newsam, M. T. Melchior and H. Malone, in "Proc. Symp. New Developments in Heterogeneous Oxide Catalysts, Including New Microporous Materials" Eds. K. R. Poeppelmeier and S. T. Wilson, Solid State Ionics in press (1987)

19. C. A. Fyfe, G. C. Gobbi, G. J. Kennedy, C. T. DeSchutter, W. J. Murphy, R. S. Ozubko and D. A. Slack, Chem. Lett., 1984, 163-166.

20. R. H. Jarman, A. J. Jacobson and M. T. Melchior, J. Phys. Chem., 88 5748-5752 (1984).

21. C. A. Fyfe, G. C. Gobbi, G. J. Kennedy, J. D. Graham, R. S. Ozubko, W. J. Murphy, A. Bothner-by, J. Dadok and A. S. Chesnick, Zeolites, 5, 179-183 (1985).

22. J. Klinowski and M. W. Anderson, J. Chem. Soc. Faraday Trans. I, 82, 569-584 (1986).

23. R. Rinaldi, J. J. Pluth and J. V. Smith, Acta Cryst., B31, 1603-1608 (1975).

24. E. Galli, Cryst. Struct. Comm., 3, 334-339 (1974).

25. J. M. Thomas, J. Klinowski, S. Ramdas, B. K. Hunter and D. T. B. Tennakoon, Chem. Lett. 1983, 1547-1550.

26. J. Klinowski, M. W. Anderson and J. M. Thomas, J. Chem. Soc. Chem. Comm. 1983, 525-526.

HIGH RESOLUTION ELECTRON MICROSCOPIC STUDY OF THE GROWTH OF DENSER PHASES ON ZEOLITE Y AND ZEOLITE L.

H.W. Zandbergen,
National Center for Electron Microscopy, Lawrence Berkeley Laboratory, University of California, Berkeley, CA 94720, USA, and Gorlaeus Laboratories, State University Leiden, P.O. Box 9502 Leiden, The Netherlands.

Abstract

High resolution electron microscopy (HREM) is a powerful tool in materials research. This is especially true in the study of zeolites because most of the questions concerning the properties of zeolites require a detailed characterization of the local structure [1]. The potential of HREM is illustrated with studies on the growth of denser phases on Zeolite Y and Zeolite L.

Introduction

With the development of a new generation of commercial high resolution electron microscopes, a point resolution below 0.2 nm can be achieved, which is close to the technical and theoretical limits of this technique.

Therefore, it has become possible to record electron micrographs that contain structural detail of atomic scale, which thusfar, could only be attained in a statistical form by X-ray diffraction methods. Especially in materials science, the importance of high resolution electron micrographs for structure determination is growing tremendously.

Zeolites are particularly suitable for high-resolution electron microscopical investigation since their pore structure often allows an unambiguous interpretation of the images, especially when taken with the electron beam parallel to one of the pore channel directions. This will be explained below.

Zeolites are very important in a number of industrial processes. The key to their importance is their pore system which allows only molecules up to a specific size to enter the structure.

Further insight in the processes involved in the synthesis of zeolites could open the way to new and tailor-made zeolites and better performing zeolites. In this respect electron microscopy can be of great importance, because it can give very useful information in a number of key areas concerning zeolite synthesis:

Nucleation;

Defect characterization;

Characterization of growth mechanisms of denser phases;

Structure determination;

Identification of small metal particles inside the zeolite matrix.

As an example the growth of denser phases will be discussed in this paper.

Experimental

High resolution electron microscopy was carried out with a JEOL JEM200CX electron microscope, operating at 200kV, and equipped with a 10° double tilt goniometer. Specimens for electron microscopy were prepared by suspending unpowdered material in ethanol; a few droplets of the suspension were put on copper grids with holey carbon support films.

Zeolites are often very susceptible to electron beam damage. Unless special precautions are taken it is impossible to select, tilt into the orientation of interest and take photographs. Ion exchange of the large cations by UO_2^{2+} or Li^+ was found to give a significant increase in stability (10 times or more); also replacement of the bulk of Al ions by Si ions can result in decreased beam damage [1,2]. This last method did not always result in much greater stability, whereas the ion exchange did. Stabilisation methods have the disadvantage that structural changes can be introduced. Consequently, the non-stabilized material has to be investigated too.

A very important piece of equipment that allows a reduction in radiation damage is an image intensifier. Using this extension to the electron microscope it is possible to work with an electron beam approximately 10 times less intense. Since the radiation damage was found to be exponentially related to the electron beam intensity, it is evident that the use of an image intensifier allows a much longer investigation of a given crystal. Drift is one of the most serious problems in the recording of high resolution images. This problem can be overcome by taking micrographs which are strongly underexposed, or by using the images from the image intensifier and correct for the drift. The first technique needs additional image processing and is unsuitable for non-periodic images such as grain boundaries and defects.

Image processing, which allows one to extract more information from the electron micrographs, is becoming a very important technique in high resolution electron microscopy. The images are sampled into a network of points; in each point the contrast is measured and digitized. Then the images are fed into a computer and processed using image processing algorithms.

Applications of image processing are manyfold but with respect to the study of zeolites, the applications are mainly concerned with image enhancement and filtering.

If the periodic high resolution image of a zeolite is obscured by non-periodic noise, such as radiation damage, which often occurs during examination in the electron microscope, the Fourier transform of the image, i.e. the diffraction pattern, consists of sharp, periodic Bragg spots, corresponding to the periodic image and diffuse contrasts from the noise.

By selecting only the sharp, periodic reflections and reverse Fourier transforming, a filtered image is obtained in which the non-periodic noise has disappeared. Another way of improving the signal/noise ratio for periodic objects is the averaging method in which the images of different unit cells are superimposed. In this way the signal/noise ratio increases with the square root of the number of superimposed images.

Denser phases

A very serious problem in the synthesis of zeolites is the transformation of the object zeolite to denser phases with much smaller or no pores. In numerous synthesis mixtures these denser phases already start to be formed before all the reaction mixture is transformed into the zeolite of interest. The effect of a denser phase on the sorption properties of the zeolite strongly depends on the zeolite itself and on the way the denser phase is formed. In cases where the denser phase grows on a specific crystallographic surface of the zeolite, the effect of small amounts of denser phase can be disastrous, especially for one-dimensional pore systems.

HREM can give important information on the formation of denser phases in several ways:

• The growth of denser phases on specific crystallographic surfaces of the zeolite of interest and their mutual orientation can be studied.

• By combining HREM with elemental analysis (EDX), the chemical composition around the interface of the zeolite and the denser phase can be determined.

• The effect of additives on the growth rate of denser phases on specific crystallographic surfaces can be established.

• It can be investigated how crystalline material of the zeolite of interest is consumed for the formation of denser phases.

Two examples of the presence of a denser phase will be given. In the first place, the growth of Zeolite P on Zeolite Y; the latter being a zeolite with a cubic unit cell and a three-dimensional pore system. Secondly, the growth of the zeolite Erionite on Zeolite L. Both these zeolites have a hexagonal structure with a one-dimensional pore system, but the diameter of the pores is 0.42 and 0.70 nm respectively.

Zeolite P on Zeolite Y

Zeolite Y is widely used as a cracking catalyst [3]. The main efforts in the research on this zeolite are the synthesis of better performing Zeolite Y by increasing the Si/Al ratio or by introduction of small cations, such as Zn, Co or Fe, thus possibly creating catalytically active sites for specific reactions. Also post-synthesis processing are studied: for instance, reactions with steam and SiF_4 to increase the Si/Al ratio [3], ion exchanges with rare earth ions to increase the thermostability [3], and impregnation with metal complexes to obtain small metal particles inside the zeolite lattice [2,4].

In general, it is difficult to obtain a 100% yield of pure Zeolite Y from the reaction mixture. Often impurities of Zeolite P and W are frequently observed. When the proportion of Zeolite P remains below about 3%, its existence has no significant effect on the performance of the product. However, as soon as Zeolite P starts to be formed, its growth can be very rapid. At that moment it is better to stop the synthesis, even when the reaction mixture still contains considerable amounts of starting material.

Several reaction products containing Zeolite Y with a few percent of Zeolite P were investigated. In a number of these samples, the growth of Zeolite P was found to start at one of the corners of the octahedrally shaped crystals of

Zeolite Y, as can be seen in figure 1b.

With HREM on pure Zeolite Y crystals, it is observed that these corners often contain material which does not show a pattern of Zeolite Y, as can be seen in figure 1. Because no regular pattern is obtained, the material at the corner is thought to be amorphous.

Such growth phenomena, starting at the corner of the Zeolite Y octahedra, are not observed in all samples. This indicates that the growth of Zeolite P is more complex and that growth, starting on a corner, is only one of the possible mechanisms.

Erionite on Zeolite L

Another example of intergrowth of a denser phase is Erionite on Zeolite L. Zeolite L has a one-dimensional pore system and blocking of the pores will already occur if the two opposite planes of entrance ([001] planes in the case of Zeolite L) contain a very thin layer of contamination.

In figure 2, an example is given of Erionite grown on the [001] surface of Zeolite L. Growth of Erionite was only observed on the basal planes of the Zeolite L crystals. This indicates that all Erionite crystals will block the one-dimensional pores on one side, which influences the behaviour of this Zeolite L batch.

The amount of Erionite contamination in Zeolite L , as determined by X-ray powder diffraction, was only a few percent but the performance of this Zeolite L was significantly inferior than that of a pure Zeolite L. However, it cannot be concluded that the poor performance arises only from the presence of the contamination. Differences in composition and shape of the zeolite crystals can have similar effects.

An important observation in this respect is the presence of differently oriented Zeolite L on the exterior of the Zeolite L crystal. In figure 2 a lattice spacing of 1.60 nm corresponding to the d_{100} distance is observed in the more central part of the crystal, whereas on the exterior a spacing of 0.41 nm corresponding with d_{220} is present. This type of intergrowth is also reported by Terasaki et al [5]. Although this phenomenon is also observed for crystals from batches of pure Zeolite L, in this specimen it was found to occur much more frequently. Since such a defect will also lead to a blocking of the pores, the poor performance of this zeolite will also partly be caused by these defects. It is not known whether the intergrowth of [110] oriented Zeolite L on [100] oriented Zeolite L has a promoting effect on the growth of Erionite.

An interesting feature in figure 2 is the extension of Erionite into the crystal of Zeolite L. The basal planes of almost all Zeolite L crystals, and certainly of the crystals in this batch, are either flat or stepped increasing in height towards the center of the basal plane. Because a hole in the center is very unlikely, it is almost certain that the Zeolite L crystal is partly consumed by the Erionite crystal. During the growth of the Erionite crystal, the interface between Erionite and Zeolite L will move into of the Zeolite L crystal. Another explanation is the continuation of the growth of Zeolite L next to the growth of Erionite, but this is less plausible because, at this stage of the synthesis, the growth rate of Zeolite L is far smaller than that of Erionite.

Determination of the Denser Phase

The easiest way to indentify a contaminant or a specific zeolite is to take a series of diffraction patterns of one crystal. Since the electron beam intensity required to take diffraction patterns is much less than that for HREM imaging, this can be carried out without strong radiation damage of the zeolite. The most straightforward techniques would be conventional electron diffraction in an orientation which shows reflections of higher order Laue zones. In this case the zero order Laue zone reflections provide information on the cell dimensions in two dimensions, whereas the positions of the circles of higher order Laue zones gives information on the continuation of the reciprocal lattice in the direction perpendicular to the zero order Laue zone. Unfortunately higher order Laue zone reflections are often not present, due to even minor irradiation damage. For this reason one has to take a number of diffraction patterns from the same crystal and use the different zero order Laue zones. Knowing their mutual orientation a three dimensional diffraction pattern can be constructed. The easiest way is to rotate the crystal about a direction, parallel to a short reciprocal lattice vector and showing to be a mirror plane or showing to have systematically absent reflections. Doing this the chance of rotating about one of the unit cell axes or a very specific direction (e.g. {111} for a cubic system) is very high, which simplifies the determination of the unit cell and its symmetry. From the different diffraction patterns, the plane perpendicular to the rotation axis can be constructed. Then any plane in reciprocal space can be constructed. From this the smallest reciprocal unit cell, describing all the reciprocal lattice points can be constructed. Sometimes this unit cell has to be transformed to a larger one, because of the symmetry elements present (e.g. monoclinic to C-centered orthorhombic). Often about 4 differently oriented diffraction patterns are sufficient to determine the three dimensional unit cell provided the crystal is not multiply faulted.

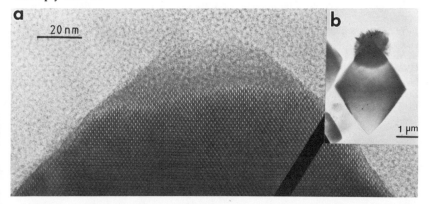

Figure 1. High resolution image (a) of the corner of an octahedrally shaped crystal of Zeolite Y. The top part of the crystal near the corner does not show the image of the zeolite lattice and is believed to be amorphous. (b) shows the initial growth of Zeolite P on the corner of an octahedron of Zeolite Y.

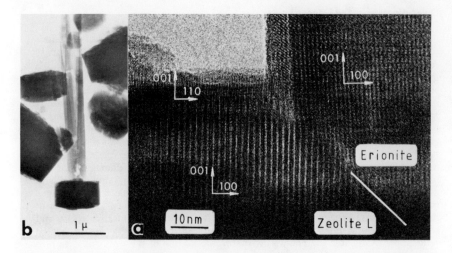

Figure 2 (a) High resolution image showing the growth of Erionite on the basal plane of Zeolite L. It can also be seen that near the surface (110) oriented Zeolite L is grown on top of (100) oriented Zeolite L. Both intergrowths will deteriorate the performance of the zeolite. (b) shows a low resolution image of the needle shaped crystal of Erionite on top the Zeolite L crystal.

References

1. J.M. Thomas, Proc. 8th Intern. Cong. on Catalysis Berlin (1984) Verlag Chemie, pag 1.
2. H.W. Zandbergen, C.W.R. Engelen and J.H.C. van Hooff, Appl. Catal. 25, 231, (1986)
3. R.M. Barrer, Hydrothermal chemistry of zeolites, Academic Press, London (1982)
4. N.I. Jaeger, P. Ryder and G. Schulz-Ekloff, Structure and Reactivity of Modified Zeolites, ed P.A. Jacobs, Elsevier, Amsterdam (1984) pag. 294.
5. O. Terasaki, J.M. Thomas and S. Ramdas, J. Chem. Soc. Chem. Commun. 1984, 216.

STRUCTURE-SENSITIVE VIBRATIONS IN ZEOLITES
AS STUDIED BY RAMAN SCATTERING

R. G. Buckley*, H. W. Deckman, J. M. Newsam, J. A. McHenry, P. D.
Persans** and H. Witzke
Exxon Research and Engineering Company, Clinton Township, Route 22
East, Annandale, NJ   08801
*Physics and Engineering Laboratory, DSIR, Private Bag, Lower Hutt,
New Zealand
**Rennsalaer Polytechnic Institute, Physics Department, Troy, NY
12180-3590

ABSTRACT

In this paper we demonstrate the usefulness of a nearest neighbor
force model for studying the structural dependence of the lattice
dynamics of zeolites. Interpreting Raman scattering in the context of
this model confirms earlier empirical vibrational assignments in
zeolites and underlines the importance of short range forces in
determining the frequencies of certain zeolite vibrational modes. The
model describes fully the low frequency bond bending mode of A_1
symmetry, although long range forces are shown to be required to model
complete zeolite vibrational spectra.

INTRODUCTION

Zeolites are typically aluminosilicates with complex framework
structures that are constructed from corner-shared SiO_4 and AlO_4
tetrahedra. They are of widespread technological importance in
catalysis, sorption and ion-exchange applications. These properties
reflect particular zeolite structural characteristics and techniques,
such as Raman scattering, that probe the lattice vibrations are,
therefore, important both as characterization tools, and as potential
means of developing an understanding of the dynamics of zeolite pore
structures under process conditions.
   The frequencies of modes in the infrared vibrational spectra of
zeolites have for many years been used to provide indications as to the
nature of certain local framework features [1,2]. Such empirical
correlations have been of value because the complexity of zeolite
structures has prevented detailed group-factorial analyses such as are
required for complete spectral assignment and interpretation. Progress
is being made towards improving and extending mode assignment methods
[3-11], although the full quantitative description of zeolite
vibrational spectra remains, perhaps, a still distant goal. Raman
spectra of zeolites were first reported in 1972 [12-14], but
experimental problems, largely related to sample fluorescence [6], have
until recently [15,16] hampered a more widespread exploitation of this
versatile technique. In common with the infrared studies, correlations
between Raman mode frequencies and zeolite structural features have
been essentially empirical.
   In this communication we outline the manner in which a simple
lattice dynamic model developed to describe lattice vibrations in
glasses [17-19] can also be used to assist interpretation of the Raman
spectra of zeolites and to confirm the empirical mode assignments of
Flanigen et al [1,2]. The use of the model is demonstrated by
reference to features observed in the Raman spectra of lithium, sodium
and potassium chloride sodalites, and of cesium zeolite ZK-5.

EXPERIMENTAL

Raman spectra were measured with a Spex 1877B triple spectrograph-monochromator and a Princeton Applied Research 1420 multichannel detector. Data were collected over the frequency range 200 cm^{-1} to 1200 cm^{-1} at a resolution of 5 cm^{-1}. Spectra were recorded in a backscattering configuration with the laser power at the sample usually below 100 mW over approximately 10^{-3} cm^2.

DYNAMIC MODEL AND RESULTS

Displayed in Figure 1a is the Raman spectrum of dehydrated Cs-ZK5. Applying the empirical scheme of Flanigen [1,2], the following 'internal tetrahedra' mode assignments can be made: The peaks at 1145 cm^{-1}, 784 cm^{-1} and around 460 cm^{-1} are assigned to asymmetric stretch, symmetric stretch and bond bending modes respectively. Flanigen [1,2] has also made 'external linkage' assignments such that the 600 cm^{-1} and 345 cm^{-1} modes would be assigned to a double ring structure and pore opening respectively. The complexity of zeolite structures is a considerable deterrent to a full calculation of the vibrational density of states. Our approach has, therefore, been to employ a simple model for the dynamics, following earlier work on AX$_2$ glasses [17]. For

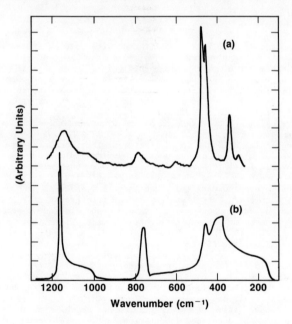

Fig. 1a   The Raman spectrum of dehydrated cesium zeolite ZK-5. The peaks at 1145 cm^{-1}, 784 cm^{-1} and around 460 cm^{-1} are assigned to asymmetric stretch, symmetric stretch and bond bending respectively.

1b   The vibrational density of states calculated using the nearest-neighbor model [17] with force constants $\alpha = 476$ Nm^{-1} and $\beta = 65$ Nm^{-1}.

these systems the density of states was modelled by a calculation based on a Bethe lattice with nearest-neighbor forces only, and a potential energy given by

$$V = \frac{1}{2} (\alpha-\beta) \sum_{<ij>} \left| (\bar{u}_i - \bar{u}_j) \cdot \bar{r}_{ij} \right|^2 + \frac{1}{2} \beta \sum_{<ij>} (\bar{u}_i - \bar{u}_j)^2 \qquad (1)$$

where α and β are the central and non-central force constants respectively, $\bar{u}_i$ the displacement, $\bar{r}_{ij}$ is a unit vector and $<ij>$ implies a sum over nearest-neighbor pairs. The four-fold Bethe lattice employed here incorporates neither long range order nor any closed rings. This model has the advantages that an exact solution can be found and that the local zeolite environment is precisely described. As is shown below, the model indicates the importance of short range forces. The solution involves a tetrahedral atom (assumed to be Si) at a site of tetrahedral symmetry with a single inter-tetrahedral angle about the bridging oxygen, in an isotropic network with only nearest-neighbor forces.

In the limit of a zero non-central force constant, $\beta = 0$, the calculation reduces to the central force model of Sen and Thorpe [18], which predicts modes corresponding to asymmetric stretch, symmetric stretch and bond bending. Galeener [19] has shown that these modes are observed in AX_2 glasses although the central force model was unable to predict the mode frequencies accurately. Barrio, Galeener and Martinez [17] demonstrated that this situation can be improved by including non-central forces.

In the present work, we have found that for (crystalline) zeolites the inclusion of non-central forces (i.e. $\beta \neq 0$) in a similar nearest-neighbor model enables the prediction of the prominent Raman modes. Shown in Figure 1 is the correspondence between the calculated vibrational density of states and the measured Raman spectra of Cs-ZK5. The values of the force constants α and β used in this calculation were 476 Nm^{-1} and 65 Nm^{-1} respectively, which are within the range of values reported in the literature [3,4]. The ability of this simple model to predict the prominent vibrational modes of zeolite ZK-5 in this region indicates the dominance of local structure and nearest-neighbor forces in dictating the lattice dynamics. It confirms the empirical assignments of Flanigen [1,2]. Further, it provides some physical insight as to the nature of the lattice vibrations as it details their sensitivity to central and non-central forces, and to the inter-tetrahedral angle. For example, the bond bending mode seen in Figure 1a shows some structure. Cesium zeolite ZK-5 has four crystallographically inequivalent framework oxygen atoms giving four unique inter-tetrahedral angles [20]. These angles are grouped (143°, 145°, 146°, 150° [20]) in such a way as to account for the observed peak structure.

A more rigorous test of the model is provided by the chloride sodalites, where changing the non-framework counter cations in the sequence Li^+ - Na^+ - K^+ results in an increase in the (single) inter-tetrahedral angle from 126° to 138° to 155° respectively [21]. The Raman spectra for this series (Figure 2) clearly display a strong dependence of mode frequencies on the inter-tetrahedral angle, θ. With increasing frequency the modes can be identified as bond bending and symmetric stretch. However, the two modes in the asymmetric stretch region constitute a TO-LO splitting [22]. The symmetric stretch is only weakly split and is thus not resolvable; the bond bending mode is of A_1 symmetry and hence single [22]. The nearest-neighbor force model, of course, does not predict these splittings because the Coulomb forces responsible have not been included.

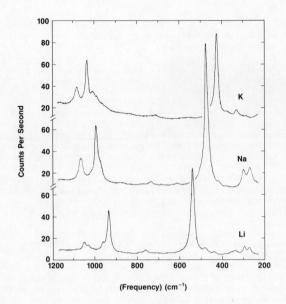

Fig. 2   Raman spectra for the chloride sodalite series; Li, Na and
         K. Note that the spectra are on a displaced vertical scale.

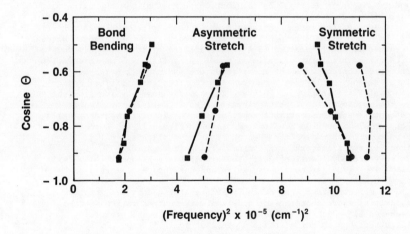

Fig. 3   The cosine of the inter-tetrahedral angle plotted against
         the square of the mode frequency for Li, Na and K chloride
         sodalite. The circles are experimental points. The boxes
         show the results of the nearest-neighbor model with force
         constants $\alpha = 360$ Nm^{-1} and $\beta = 70$ Nm^{-1}. The lines are to
         guide the eye.

In the limit $\beta = 0$, a linear relationship is expected between cosine (θ) and the square of the mode frequency [18,19]. The data in Figure 3 for the chloride sodalite series remarkably deviate only slightly from linear behavior (although a different force constant would be extracted for each mode). Also shown in Figure 3 are the results of optimizing α and β so as to best reproduce the observed shift in the low frequency bond bending mode. An excellent fit is obtained. The shift in the bond bending mode is chosen for the fit as this mode is not influenced by Coulomb forces and is hence fully within the scope of the present model. This Raman active mode has A_1 symmetry [22]. We suggest that the nearest-neighbor model will predict fully the behavior of this mode in both the present and other zeolite systems. The values of α and β obtained from the fit are 360 Nm^{-1} and 70 Nm^{-1} respectively which fall within the range of literature values [3,4]. The estimated value of α for sodalite is less than the estimated value for ZK-5 as expected [1,2], reflecting, at least partly, the lower aluminum content of cesium ZK-5 compared to the chloride sodalites.

The shifts of the stretching modes are not well reproduced by the present model, although modes are predicted to fall in the correct spectral regions. For instance, as shown in Figure 3 the model predicts the symmetric stretch to fall within 10% of the observed mode frequency and the asymmetric stretch to lie between the TO and LO modes (as expected for a model that does not include Coulomb forces). This represents the limit of the nearest-neighbor force model which is insufficient to describe fully the Raman spectra of zeolites, in particular the behavior of modes of symmetry other than A_1.

SUMMARY

The application of a nearest-neighbor force model has been shown to predict the position and the inter-tetrahedral angle dependence of the Raman active bond bending mode of A_1 symmetry. The frequency of this mode in other zeolite systems can probably be interpreted to yield local structural information. The success of this model in describing the dynamics of zeolites allows for considerable physical insight into the nature of the lattice vibrations. However, the inability of the model to predict the high frequency stretching modes with reasonable accuracy indicates that a more detailed model in which longer range forces can be incorporated is required for a complete description of the spectra. When care is taken to minimize fluorescence arising from structural defects, impurities or carbonaceous material it proves relatively straightforward to obtain good Raman spectra from zeolitic materials. Compared to other structure-sensitive techniques, such as X-ray diffraction, nmr or neutron scattering, Raman scattering has proved to be experimentally convenient and offers the possibility of obtaining structure-sensitive information within a short time scale.

REFERENCES

1. E. M. Flanigen, H. Khatami and H. A. Szymanski in Molecular Sieve Zeolites 1, Eds. E. M. Flanigen and L. B. Sand (Adv. Chem. Ser. 101, Amer. Chem. Soc., Washington, DC) 1971, pp. 201-229.
2. E. M. Flanigen in Zeolite Chemistry and Catalysis, Ed. J. A. Rabo, (ACS Monograph No. 171, Amer. Chem. Soc., Washington, DC) 1976, pp. 80-118.
3. C. S. Blackwell, J. Phys. Chem., 1979, 83, 3251-3257.
4. C. S. Blackwell, J. Phys. Chem., 1979, 83, 3257-3261.
5. A. Miecznikowski and J. Hanuza, Zeolites, 1985, 5, 188-193.

6. Y. S. Kong, K. T. No and M. S. Jhon, Bull, Korean Chem. Soc., 1985, 6, 57-60.

7. S. V. Gorainov and A. P. Shebanin, Tr. Inst. Geol. Geofiz., Akad. Nauk SSSR, Sib. Otd., 1985, 610, 101-110.

8. K. T. No, D. H. Bae and M. S. Jhon, J. Phys. Chem., 1986, 90, 1772-1780.

9. P. Walther, Zeit. Chem., 1986, 26, 189-190.

10. P. Walther, Zeit. Chem., 1986, 26, 222-223.

11. A. P. Maroni, 1986, private communication.

12. J. J. P. M. de Kantner, I. E. Maxwell and P. J. Trotter, J. Chem. Soc., Chem. Commun., 1972, 733-734.

13. B. D. McNicol, G. T. Pott and K. R. Loos, J. Phys. Chem., 1972, 76, 3388-3390.

14. C. L. Angell, J. Phys. Chem., 1973, 77, 222-227.

15. P. K. Dutta and B. Del Barco, J. Chem. Soc., Chem. Commun., 1985, 1297-1299.

16. P. K. Dutta and B. Del Barco, J. Phys. Chem., 1985, 89, 1861-1865.

17. R. A. Barrio, F. L. Galeener and E. Martinez, Phys. Rev. B, 1985, 31, 7779-7787.

18. P. N. Sen and M. F. Thorpe, Phys. Rev. B, 1977, 15, 4030-4038.

19. F. L. Galeener, Phys. Rev. B, 1979, 19, 4292-4297.

20. J. B. Parise, R. D. Shannon, E. Prince and D. E. Cox, Z. Kristallogr., 1983, 165, 175-190.

21. B. Beagley, C. M. B. Henderson and D. Taylor, Mineral. Mag., 46, 459-464 (1982).

22. J. Ariai and S. R. P. Smith, J. Phys. C: Solid State Phys., 1981, 14, 1193-1202.

SYNTHESIS AND STRUCTURAL CHARACTERIZATION OF A HYDRATED
GALLOSILICATE ZEOLITE WITH THE NAT-FRAMEWORK

D. XIE*, J. M. NEWSAM**, J. YANG*† AND W. B. YELON*
*University of Missouri Research Reactor, Columbia, MO 65211, USA
**Exxon Research and Engineering Company, Route 22 East, Annandale, NJ 08801,
USA
†Permanent address: Institute of Atomic Energy, Beijing, People's Republic
of China

ABSTRACT

Gallosilicate NAT-framework zeolites have been synthesized from
$Na_2O:TMA_2O:Ga_2O_3:SiO_2:H_2O$ gels and characterized by chemical analysis and
powder X-ray diffraction. The structure of a hydrated material (Orthorhombic,
Fdd2, a = 18.423(4) Å, b = 18.826(3) Å, c = 6.652(1) Å) was determined using
powder neutron diffraction. The framework Si:T ratio, Si-Ga segregation and
non-framework cation and water configurations are similar to those in the
analogous aluminosilicates. The geometrical effects of framework gallium
substitution are consistent with those observed for other zeolite framework
topologies.

INTRODUCTION

Although the structural characteristics of aluminosilicate zeolites have
been studied extensively [1,2], it is only relatively recently that the structural consequences of framework T-atom hetero-substitutions (T = Tetrahedral
species, Si, Al or Ga etc.) have begun to be explored. The effect on unit cell
constants of, for example, Ga [3,4], Fe [5] and B [6] substitution, and complete
structural details for ABW [7], FAU [8], LTL [9,10], MAZ [11] and SOD-framework
[12,13] gallosilicates have been reported. Our detailed understanding of
gallium substituent effects such as might enable the prediction of gallosilicate
zeolite properties remains, however, somewhat sketchy [4]. In order to extend
our understanding of the structural chemistries of gallosilicate zeolites, we
have examined the syntheses and structures of a series of gallosilicates
adopting a range of framework topologies. We present here results for the NAT-
framework. Precise structural details for the isotopological natural mineral
aluminosilicates natrolite [14-19], scolecite [20,21], mesolite [22], and
gonnardite [23] and for the related, non-zeolitic phase $Rb_2Ga_2Ge_3O_{10}$ [24] are
already available [1]. The synthesis of gallosilicate NAT-framework materials
has previously been described by Ponomareva et al. [25].

SYNTHESIS

Gallosilicate NAT-framework zeolites were encountered in crystallizations
from sodium-tetramethylammonium (TMA) aluminosilicate gels (Table I). Gallia
(Ga_2O_3; Ingal) was added to an NaOH solution, heated for 30 min., the required
amount of 25 wt% TMAOH solution (RCA Corp.) added and the mixture (which
remained cloudy), heated for another 30 min. The mixture was allowed to cool
and the silica component (Ludox, HS-40) added while stirring. The mixture was
allowed to cold age at 28°C for 1 d, heated without stirring at 100°C for 1 d,
and then at 95°C with shaking for 24 d (the mixture was sampled every 3-5 days
throughout). The product was identified as a NAT-framework gallosilicate based
on the similarity of the powder X-ray diffraction (PXD) profile to that of an
aluminosilicate NAT-framework standard [1]. The synthesis was subsequently
repeated with similar results (Table I). The products were characterized by
PXD, inductively-coupled plasma emission spectroscopy (ICPES), thermogravimetric
analysis and ^{29}Si nmr (results from the latter two techniques are not included

Table I

Gallosilicate NAT-framework Zeolites: Syntheses and Products

Sample Number	137D	140
Syntheses Compositions		
TMA_2O	2.049	2.049
Na_2O	1.476	1.476
Ga_2O_3	1.000	1.000
SiO_2	6.053	6.053
H_2O	310	270
Conditions		
	28°C 1d/100°C 1d/95°C shaker, 24d	95°C shaker, 85d
Products		
Si:Ga	1.661	1.572
Na:Ga	0.95	0.96
a_o Å[a]	18.42(4)	18.41(5)
b_o Å[a]	18.66(2)	18.70(4)
c_o Å[a]	6.632(6)	6.68(2)

[a]Least-squares optimized lattice constants based on, typically, 22
reflection positions to 70° in 2θ (Phillips analog diffractometer;
Cu Kα, λ = 1.5418 Å).

here). The least squares optimized lattice constants derived from the PXD
(Table I) and powder neutron diffraction data (Table II) are in reasonable
agreement with the values a = 18.40(1) Å, b = 18.80(1) Å, c = 6.66(1) Å
reported by Ponomareva at al. [25].

POWDER NEUTRON DIFFRACTION

Powder neutron diffraction (PND) data were collected from a fully hydrated
sample on the powder diffractometer at the Missouri University Research Reactor
Facility [26]. Approximately 2g of sample was contained in a 6.35 mm outside
diameter, 0.002" walled vanadium can. A wavelength of 1.2891 was selected from
the (220) planes of a Cu monochromator at a take-off-angle of 60.6°. Data from
four 25° spans of the linear position sensitive detector were each accumulated
over some 6 hrs at 298(5)K and combined to yield the diffraction profile
5 < 2θ < 105°, rebinned in 0.1° steps. As a result of calibration drifts, the
data measured by the extreme end portions of the detector were less reliable
than those measured closer to the detector center. The 1° end sections of data
for each setting were therefore excluded, leading to 2° gaps in the measured
profile at 30°, 55°, 80° and 104°.

The PND data were analyzed by full matrix least squares Rietveld refinement
[27] using a locally modified version of Rietveld's original code. The background was treated by linear interpolation between a set of estimated points
that were constrained to lie on a smooth curve. The background estimate was
updated periodically during the refinement process. The observed background
is high, reflecting the large incoherent scattering from the protons of the
sorbed water. The peak shapes were described by the Voigt function [28,29].

Previous results for aluminosilicate natrolite [16] provided the starting point for structure refinements. Space group Fdd2 (No. 43 [30]) was assumed and confirmed by the subsequent analyses (see below). The final overall and atomic parameters are listed in Tables II and III. Selected separations and angles are given in Table IV. The final observed and calculated diffraction profiles are shown in Figure 1. For the final results presented in the tables, the number of framework gallium atoms was constrained to satisfy the known chemical composition and the water site was fixed at full occupancy. Allowing the T2 and T3 site occupancies to vary without constraint yielded values within, at most, two esd's of the known composition. The results indicate effectively complete gallium segregation onto the T3 site. There was no evidence for gallium substitution at the T1 site and, as indicated in Table III, the refined gallium population at the T2 site is within two esd's of zero. The occupancy of the water site, when allowed to vary, converged to 16.1(8). Representations of the structure of this gallosilicate NAT-framework zeolite are shown in Figures 2 and 3.

DISCUSSION

Although the starting gel compositions from which gallosilicate NAT-framework zeolites were crystallized had Si:Ga = 3.0 (Table I), the crystallized zeolites have Si:Ga ratios close to 1.5. Although we have not studied in detail the boundaries to the gel composition field from which gallosilicate NAT materials crystallize, the earlier work of Ponomareva at al. [25] also indicates that products have Si:Ga ~ 1.5, irrespective of the starting composition. The natural mineral aluminosilicates natrolite, scolecite and

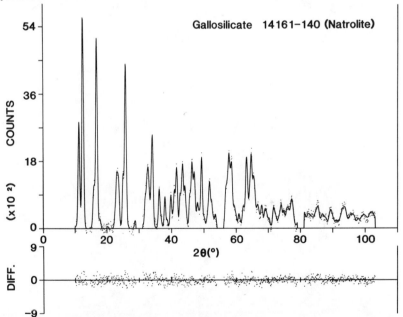

Figure 1.   Powder neutron diffraction (PND) profiles for gallosilicate NAT-
framework zeolite: final observed (points), calculated (continuous
line) and difference (lower) profiles after background subtraction
and structure optimization.

mesolite also have Si:Al = 1.5, implying that this is a preferred composition
for the NAT topology. Although partial aluminum disorder has been reported
in one natrolite sample [17], this composition generally corresponds to full
Al(Ga)-occupancy of the T3 site and full Si occupancy of sites T1 and T2. The
apparent stability of this composition may therefore reflect the character of
the summed electrostatic (and short-range) interactions for the framework constituents. The framework composition also dictates, for charge balance, the
required number of non-framework cations. Although the natural occurrence of
both sodium (natrolite) and calcium exchanged materials (mesolite and scolecite) with this topology might be considered as evidence against a major
influence for the non-framework cations in stabilizing the Si:Al(Ga) = 1.5
composition, the possibility of geochemical synthesis in the sodium form,
followed (for mesolite and scolecite) by subsequent calcium ion exchange can
probably not be excluded. Comparison of the present synthesis conditions with
the results of Ponomareva et al. [25] implies that the presence of tetramethylammonium cations in the gel is unnecessary for the crystallization of
gallosilicate NAT species.

Structural data for a synthetic NAT-framework zeolite have not been
reported previously. However, the present results can be compared in detail
with those for the aluminosilicate NAT-materials natrolite [14-19], scolecite
[20,21], mesolite [22], gonnardite [23] and the non-zeolitic phase $Rb_2Ga_2Ge_3O_{10}$
[24]. In the series natrolite-mesolite-scolecite, complete replacement (in,
at least, a figurative sense) of Na_6 in natrolite by $(Ca + H_2O)_3$ yields scolecite. Partial replacement to a composition $Na_2(Ca + H_2O)_2$ yields mesolite,
which corresponds structurally to regularly alternating layers of scolecite
and natrolite. From a structural standpoint we can therefore argue that a
material with composition $Na_4(Ca + H_2O)_1$ would also have a tripled unit cell,
$b' = 3b_{natrolite}$, and that other discrete compositions, with more extended
supercells, should also exist. Single crystal X-ray diffraction of a gonnardite sample [23], shows net Na-Ca disorder, although distinct natrolite and

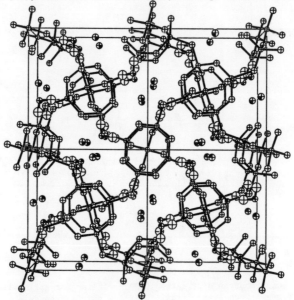

Figure 2.   ORTEP Representation of the structure of gallosilicate
NAT-framework zeolite.

Figure 3.   Stereoview of the structure of gallosilicate NAT-framework zeolite in polyhedral representation [32], viewed approximately along [001].

scolecite sheets may be present in a non-regularly repeating fashion. Gonnardite (and tetranatrolite [31]), however, have significantly higher aluminum contents and Si:Al disorder.

For the gallosilicate NAT materials studied here no evidence for deviation from the a = 18.4 Å, b = 18.8 Å, c = 6.6 Å cell with Fdd2 symmetry is seen in the PXD or PND patterns. The gallosilicate has a slightly larger cell than aluminosilicate natrolite (V_{Ga} = 2307 Å^3, V_{Al} = 2256 Å^3 [16], $\Delta V/V$ ~ 2.3%), with more pronounced expansions in the b and c directions. The accuracy of the refined lattice constants is, of course, limited by the accuracy to which the incident neutron wavelength, λ, is defined. The MURR powder diffractometer [26] is operated effectively in single wavelength mode, and the value λ = 1.2891(1) Å has been derived from cross calibrations based on a large number of data sets from a wide range of materials.

Compared to aluminosilicate natrolite [14-19], atomic coordinate shifts in the gallosilicate are small. The observed bond lengths and angles (Table IV) are also similar in the two structures. The mean Si-O bond lengths of 1.616(17) Å [T1], 1.641(18) Å [T2] compare with values of 1.621(4) Å [T1] and 1.625(5) Å [T2] in aluminosilicate natrolite [16]. The mean Ga(T3)-O distance, 1.802(17) Å, is close to the value of 1.797(2) observed in $Rb_2Ga_2Ge_3O_{10}$ [24]. The mean T-O-T angles for the three T-sites are 140.3(9)° [T1], 133.6(7)° [T2] and 133.0(7)° [T3] (Table IV). The corresponding values reported for aluminosilicate natrolite [16] are 143.0(3)° [T1], 137.4(3)° [T2] and 136.4(3)° [T3]. Similar decreases in mean T-O-T angles are observed following gallium incorporation into other zeolite frameworks [7-13], an effect which can be rationalized in terms of the non-bonded radii [4]. Although only one of the three T-sites in the present gallosilicate NAT material has significant gallium occupancy, all of the bond angles about framework oxygen atoms are more acute than in the aluminosilicate. Thus the Si-O5-Si angle of 141(2)° in the gallosilicate compares with a value of 144.8(6)° in the aluminosilicate [16]. The consistent aluminum and gallium preference for the T3 site contrasts with the behavior observed in gallosilicate MAZ [11] and LTL [10] framework zeolites, where random gallium partitioning between the two T-sites is observed.

The configuration of the sorbed water molecules is closely similar to that determined in the single crystal neutron diffraction studies of aluminosilicate NAT materials [16,19,21]. The proton positions are far better determined than in the most careful X-ray study, despite the use of powder rather than single crystal data. As above, the water site is observed to be fully populated.

Table II

Final Overall Parameters
with estimated standard deviations in parentheses

Temperature	296(5)K
Sample Weight, g	2.00
Data acquisition time	24 h
Data range: 2θ (°); d-spacing (Å)	5.0-104.0; 14.8-0.82
Number of data points	918
Number of contributing reflections	594
Number of atomic variables	39
Total number of variables	48
a Å	18.423(4)
b Å	18.826(3)
c Å	6.652(1)
Halfwidth Parameters (°^{2})	
U	1.04(13)
V	-0.78(10)
W	0.37(2)
Gamma	0.02(2)
zeropoint °	-0.118(5)
R_I	5.42
R_P	9.67
R_{WP}	9.56
R_E	8.35
Residual $(R_{WP}/R_E)^2$	1.31

Table III

Final Atomic Parameters, with estimated standard deviations in parentheses

Atom	Site	x	y	z	B ($Å^2$)	Occn.
T1(Si)	8a	0.0	0.0	0.0	1.2(1)	8.0 [0.0]
T2(Si)	16b	0.1543(9)	0.2112(10)	0.610(5)	–	14.5(6) [1.5(6)]
T3(Ga)	16b	0.0380(7)	0.0933(7)	0.609(5)	–	2.0(6) [14.0(6)]
O1	16b	0.0183(8)	0.0699(7)	0.867(6)	1.8(1	16.0
O2	16b	0.0693(6)	0.1850(7)	0.604(6)	–	16.0
O3	16b	0.1015(8)	0.0352(8)	0.497(6)	–	16.0
O4	16b	0.2064(8)	0.1510(8)	0.708(5)	–	16.0
O5	16b	0.1792(7)	0.2301(7)	0.381(6)	–	16.0
Na1	16b	0.2185(14)	0.0271(11)	0.603(8)	2.5(4)	15.5
OW1	16b	0.0558(7)	0.1905(7)	0.110(6)	1.5(2)	16.0
H1	16b	0.0506(14)	0.1468(13)	0.029(6)	4.0(3)	–
H2	16b	0.1023(16)	0.1942(13)	0.160(6)	–	–

Occupancy numbers expressed as population per unit cell. For the T-sites, the numbers in brackets are the refined gallium populations.

CONCLUSION

Powder neutron diffraction analyses of a hydrated synthetic gallosilicate zeolite with the NAT-framework reveal structural characteristics quite similar to those of analogous natural mineral NAT-framework aluminosilicates. The manner in which the framework adjusts to accommodate gallium incorporation is consistent with that observed previously for other framework topologies.

Table IV

Selected Separations (Å) and Angles (°) with
Estimated Standard Deviations in Parentheses

[TO4] Tetrahedra					
T-O Distances		**O-O Distances**		**O-T-O Angles**	
T1-O1	1.62(2)(2x)	O1-O1	2.72(3)	114(2)	
T1-O5	1.61(2)(2x)	O1-O5	2.58(4)(2x)	106(1)	
		O1-O5	2.62(4)(2x)	108(1)	
Mean T1-O	1.616(17)	O5-O5	2.71(3)	115(2)	
		Mean	2.638(14)	Mean 109.5(7)	
T2-O2	1.64(2)	O2-O3	2.57(3)	102(2)	
T2-O3	1.66(4)	O2-O4	2.70(2)	111(2)	
T2-O4	1.62(3)	O2-O5	2.65(3)	108(2)	
T2-O5	1.63(5)	O3-O4	2.75(2)	114(2)	
		O3-O5	2.71(5)	110(2)	
Mean T2-O	1.641(18)	O4-O5	2.69(4)	111(2)	
		Mean	2.677(14)	Mean 109.4(8)	
T3-O1	1.81(5)	O1-O2	2.94(4)	108(2)	
T3-O2	1.82(2)	O1-O3	2.98(5)	113(2)	
T3-O3	1.77(3)	O1-O4	3.00(5)	112(1)	
T3-O4	1.81(3)	O2-O3	2.97(2)	112(1)	
		O2-O4	2.81(3)	101(1)	
Mean T3-O	1.802(17)	O3-O4	2.94(2)	111(2)	
		Mean	2.939(14)	Mean 109.4(7)	
T-O-T Angles		**H-O Distances**		**H-OW-H Angles**	
T1-O5-T2	141(2)	H1-OW	0.99(4)	111(3)	
T1-O1-T3	139(2)	H2-OW	0.92(4)		
T2-O2-T3	126(1)	H1-O1	1.90(4)		
T2-O3-T3	135(1)	H2-O5	2.15(4)		
T2-O4-T3	132(1)				
Mean T1-O-T	140.3(9)				
T2-O-T	133.6(7)				
T3-O-T	133.0(7)				

[NaO6] Polyhedra			
Na-O Distances		**OW-H-O Distances**	
Na-O2	2.60(5)	OW-O1	2.87(3)
Na-O2	2.50(5)	OW-O5	2.99(4)
Na-O3	2.27(3)		
Na-O4	2.45(3)	**Further OW-O Distances**	
Na-OW	2.36(5)		
Na-OW	2.40(5)	OW-O2	3.30(6)
		OW-O4	3.42(4)
Mean Na-O	2.451(2)	OW-O3	3.46(4)
Mean Na-OW	2.381(3)		
Mean Na-O	2.431(2)		

ACKNOWLEDGMENTS

We would like to thank J. Dunn for help with the gallosilicate zeolite
syntheses.

REFERENCES

[1]W. M. Meier, D. H. Olson, "Atlas of Zeolite Structure Types" (Structure
Commission of the International Zeolite Association, 1978; available from
Polycrystal Book Service, Pittsburgh, PA; second edition in preparation
(1987).
[2]W. J. Mortier, "Compilation of Extra-framework Cation Sites in Zeolites"
(Butterworths, Surrey, 1982).
[3]G. Kuhl, J. Inorg. Nucl. Chem. 33, 3261-3268 (1971).
[4]J. M. Newsam and D. E. W. Vaughan, in: New Developments in Zeolite Science
Technology, Eds. Y.Murakami, A.Iijima and J.W.Ward (Kodansha, Tokyo, 1986)
pp. 457-464.
[5]R. Szostak, T. L. Thomas, J. Chem. Soc. Chem. Commun. 113-114 (1986).
[6]B. L. Meyers, S. R. Ely, N. A. Kutz, J. A. Kaduk, E. van den Bossche, J.
Catal. 91, 352-355 (1985).
[7]J. M. Newsam, J. Chem. Soc. Chem. Comm. 1295-1296 (1986).
[8]J. M. Newsam, A. J. Jacobson, D. E. W. Vaughan, J. Phys. Chem. 90, 6858-6864
(1986).
[9]P. A. Wright, J. M. Thomas, A. K. Cheetham, A. K. Nowak, Nature 318, 611-614
1985.
[10]J. M. Newsam, Mater. Res. Bull. 21, 661-672 1986.
[11]J. M. Newsam, R. H. Jarman, A. J. Jacobson, Mater. Res. Bull. 20, 125-136
(1985).
[12]L. B. McCusker, W. M. Meier, K. Suzuki, S. Shin, Zeolites 6, 388-391 (1986).
[13]J. M. Newsam, J. D. Jorgensen, Zeolites, in press (1987).
[14]W. H. Taylor, C. A. Meek, W. W. Jackson, Zeit. Kristallogr. 84, 373-398
(1933).
[15]W. M. Meier, Zeit. Kristallogr. 113, 430-444 (1960).
[16]F. Pechar, W. Schafer, G. Will, Zeit. Kristallogr. 164, 19-24 (1983).
[17]K. F. Hesse, Zeit. Kristallogr. 163, 69-74 (1983).
[18]A. Kirfel, M. Orthen, G. Will, Zeolites 4, 140-146 (1984).
[19]G. Artioli, J. V. Smith, A. Kvick, Acta Cryst. C40, 1658-1662 (1984).
[20]L. Falth, S. Hansen, Acta Cryst. JB35, 1877-1880 (1979).
[21]A. Kvick, K. Stahl J. V. Smith, Zeit. Kristallogr. 171, 141-154 (1985).
[22]G. Artioli, J. V. Smith, J. J. Pluth, Acta Cryst. C42, 937-942 (1986).
[23]F. Mazzi, A. O. Larsen, G. Gottardi, E. Galli, N. Jb. Miner. Mh. 5, 219-228
(1986).
[24]K. H. Klaska, O. Jarchow, Zeit. Kristallogr. 172, 167-174 (1985).
[25]T. M. Ponomareva, N. P. Tomilov, A. S. Berger, Geokhimiya 6, 925-931 (1974).
[26]C. W. Tompson, D. F. R. Mildner, M. Mehregany, J. Sudol, R. Berliner, W. B.
Yelon, J. Appl. Cryst. 17, 385-394 (1984).
[27]H. M. Rietveld, J. Appl. Cryst. 2, 65-71 (1969).
[28]M. Ahtee, L. Unonius, M. Nurmela, P. J. Suorrti, Appl. Cryst. 17, 352-357
(1984).
[29]W. I. F. David, J. C. Mathewman, J. Appl. Cryst. 18, 461-466 (1985).
[30]International Tables for Crystallography Volume A, (D. Riedel, Dordrecht,
Holland, 1983).
[31]E. Krogh Andersen, M. Dano, O. V. Petersen, Medd. Gronland 181, 1-19 (1969).
[32]R. Fischer, J. Appl. Cryst. 18, 258 (1985).

ELECTRON MICROSCOPY STUDY OF MORPHOLOGY AND COMPOSITION OF A FERRISILICATE CATALYST

Roseann Csencsits*, Ronald Gronsky*, Vinayan Nair** and Rosemarie Szostak***
*Center for Advanced Materials, Materials and Chemical Sciences Division, Lawrence Berkeley Laboratory, and Materials Science and Mineral Engineering Department, University of California, Berkeley, CA 94720
**Union Carbide Company, Tarrytown, NY
***Zeolite Research Program, Georgia Tech Research Institute, Georgia Institute of Technology, Atlanta, GA 30332

ABSTRACT

The effects of various synthesis conditions on the structure and composition of ferrisilicate analogs of zeolite ZSM-5 were considered. Scanning electron microscopy (SEM) was used to determine the particles size distributions and morphologies. Particle sizes vary from tenths of a micron to several microns, depending on degree of agitation during crystal growth, while morphology is additionally dependent on the concentration of iron in the gel during crystallization.

X-ray emissive spectroscopy (XES) performed in the transmission electron microscope (TEM) was used to determine their composition variation. The distribution of iron amongst the crystals is more homogeneous if the gel is stirred and it does not depend on particle size.

INTRODUCTION

Iron supported on silica or alumina is an active, flexible catalyst for the conversion of synthesis gas (CO and H_2) to hydrocarbons and water (Fischer-Tropsch synthesis) [1]. In the late 1970's, researchers at Mobil synthesized a shape selective zeolite, ZSM-5, that was excellent for converting methanol to "gasoline range" hydrocarbons with high yield [2,3]. Mobil researchers also tested the ability of the ZSM-5 to control the products of the Fischer-Tropsch reaction by reacting CO and H_2 over the physical mixture of catalytically active iron and ZSM-5 powder [4,5]. The results were very promising. Throughout the eighties, researchers worldwide have been trying to optimize the distribution of Fe *inside* the ZSM-5 channels to produce a better Fischer-Tropsch catalyst [6-9].

Ferrisilicate analogs of the ZSM-5 zeolite may be directly synthesized from ferrisilicate gels in a manner which differs slightly from the alumino-silicate ZSM-5 [9]. The resultant white, crystalline ferrisilicate is referred to as FeZSM-5 in the as-synthesized form. Thermal treatment removes the organic crystal-directing agent and moves some of the framework iron into non-framework sites producing the calcined form of the molecular sieve FeZSM-5 [10-12]. Subsequent hydrothermal treatment moves more of the iron out of the molecular sieve framework [10-12]. This second iron oxide phase can then be on the surface or inside the pores of the ZSM-5 structure.

To optimize the Fischer-Tropsch catalyst, homogeneity in the particle size distribution and in the distribution of catalytic iron throughout the particles is desired. Electron microscopy, with its high spatial resolution, plays an important role in the physical characterization of these catalysts. Scanning electron microscopy (SEM) is used to characterize the molecular sieve particle sizes and morphologies as a function of preparation conditions. X-ray emissive spectroscopy (XES) performed in the

Figure 1 - SEM image of ferrisilicate molecular sieve in the as-synthesized form, grown from a stirred gel with Si/Fe-ratio $\simeq$ 25. Bar = 1.5 μm.

transmission electron microscope (TEM) is used to determine the inter-particle composition variation (i.e., Si/Fe) of the ferrisilicates.

SCANNING ELECTRON MICROSCOPY

The particle sizes and morphologies as a function of various preparatory conditions, thermal and hydrothermal treatments were studied using SEM. Samples of FeZSM-5 in the as-synthesized form showed marked differences in particle size as well as morphology depending on whether the gel was stirred or not stirred during crystallization. When the gel was stirred the particles were generally less than 1μm diameter and appeared to be spherical and complex-shaped aggregates of smaller crystallites, see figure 1. Particles grown from unstirred gels varied in size as well as morphology, see figure 2. Most particles appeared to be 2-5μm diameter aggregates of smaller elementary crystallites; however, some single, twinned and inter-grown crystals were observed. Decreasing the iron concentration in the gel from Si/Fe $\simeq$ 25 to Si/Fe $\simeq$ 100 resulted in more spherical particle aggregates in the stirred samples and larger particle aggregates in the unstirred samples; in the unstirred batch some of the spherical particles were greater than 5 μm in diameter. Images of thermally and hydrothermally treated forms of the molecular sieves were identical in size and morphology to the as-synthesized forms.

Figure 2 - SEM image of ferrisilicate molecular sieve in the as-synthesized form, grown from an unstirred gel with Si/Fe-ratio $\simeq$ 25. Bar = 4.5 μm.

X-RAY EMISSIVE SPECTROSCOPY

The inter-particle iron distribution in the ferrisilicate molecular sieves has been studied using X-ray emissive spectroscopy in the transmission electron microscope. For the purposes of this discussion, the word "particle" is used to describe the particle aggregates shown in figure 3, not to describe the individual crystallites making up the agglomerate. The effects of gel iron concentration, stirring during crystallization, particle size, as well as thermal treatment were considered. The stirred samples were in the as-synthesized and calcined forms and had Si/Fe-ratios $\simeq$ 25, 50 and 100; the unstirred samples were in the as-synthesized and calcined forms and had Si/Fe-ratio $\simeq$ 25. The Si/Fe-ratios were determined by atomic absorption spectroscopy and closely matched the Si/Fe-ratio of the gel. TEM specimens consisted of uniformly thin (90-100nm) sections of the ferrisilicate particles embedded in an acrylic resin, (figure 3); they are prepared by microtomy, described in detail elsewhere [13]. Experiments were carried out in a Phillips 400T TEM/STEM operated at 100kV accelerating voltage. X-rays were detected with a Kevex, beryllium-window, X-ray detector; spectra were collected for 300 seconds, (livetime) to minimize statistical counting uncertainty. Probes were formed, as best possible, to equal the size of the particle probed without illuminating adjacent particles. This choice of probe size could, at times, have resulted in under-sampling of the particle edges. This would result in an incorrect assessment of the inter-particle iron distribution only if the iron concentration at the edges of a particle differed greatly from its internal concentration. Work is in progress to address this question; preliminary results indicate that it is not a significant effect.

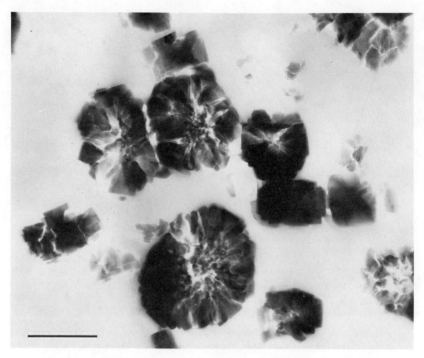

Figure 3 - TEM image of microtomed thin section of as-synthesized FeZSM-5, embedded in an acrylic resin. The ferrisilicate was grown from a stirred gel with Si/Fe-ratio ≃ 100. Bar = 500 nm.

Energy dispersive X-ray spectra (EDS) of all FeZSM-5 samples showed silicon and iron peaks corresponding to the ferrisilicate molecular sieve particles, and copper peaks resulting from the copper support grids. In some spectra, a small chromium peak was observed; this was probably due to a small amount of chromium contamination from the stainless steel crystallization vessel.

On average, 20 spectra from various particles were collected for each sample; this was to insure that the spectra were representative of the bulk samples. For each spectrum, a "hand-fit" background was subtracted and the total number of X-ray counts in 200 eV wide windows for silicon and iron K lines were recorded. The number of the counts in the silicon window divided by the number of counts in the iron window was taken as the Si/Fe-ratio for this study. Absorption and fluorescence corrections were not necessary since the specimens met the thin-film conditions (<150nm) [14]. In the near future, this relative measure of Si/Fe-ratio will be converted to actual Si/Fe-ratios by calibration of the X-ray detector for oxides containing silicon and iron. Currently, the numerical values of Si/Fe-ratios, determined by XES, are artificial in that they do not correspond to those determined by the atomic absorption, however their relative values are important and are used in the present study.

A summary of the relative Si/Fe-ratios and their percent standard deviations is shown in table I. In this case, the standard deviation is used as a measure of the spread in the data points for the given sample.

Table I - Si/Fe-ratios for various FeZSM-5 samples.			
Form	Si/Fe-ratio by AA	Relative Si/Fe-ratio by XES	percent std. dev.
stirred, as-synthesized	27	16.6	4.2
calcined		16.2	7.4
stirred, as-synthesized	46	27.2	4.4
calcined		26.0	10.
stirred, as-synthesized	88	50.2	16.
calcined		51.2	48.
unstirred, as-synthesized	25†	16.8	20.
calcined		12.0	18.
† Si/Fe-ratio of the gel, not measured by AA.			

The Si/Fe-ratio of FeZSM-5 increases as the concentration of iron in the crystallizing gel decreases; this is in agreement with the atomic absorption measurements. The effect of particle size on the distribution of iron among the FeZSM-5 particles can be seen best by comparing the as-synthesized, stirred and unstirred samples. As stated above, the particle sizes of the stirred samples are 1 μm or less, while in the unstirred samples, particle diameters are greater than 1 μm. The Si/Fe-ratios measured by XES for both the stirred and unstirred samples of FeZSM-5 are the same, thus it can be concluded that the distribution of iron is independent of particle size. Although the average Si/Fe-ratios are equal, the spread of the data is larger in the unstirred sample, indicating a more homogeneous distribution of iron with stirring during crystallization.

For all stirred samples, thermal treatment does not change the average Si/Fe-ratio, within a standard deviation. The particle-to-particle variations are larger as indicated by the increase in percent standard deviation; this suggests that there may be iron migration occurring during thermal treatment. In the case of the unstirred sample, there appears to be an increase in iron concentration with heat treatment; work is in progress to determine if this is generally true or an anomaly.

In the sample with the very high Si/Fe-ratio, stirring does not insure homogeneity of the inter-particle iron concentration. For the very low iron ferrisilicate, the spread in the data is three and one half times that of the samples containing moderate and high iron levels, indicating less homogeneity of iron distribution. Thermal treatment of the high Si/Fe-ratio sample results in greater inhomogeneity of the iron distribution amongst the particles, as compared to the other samples. This enhanced inhomogeneity could possibly be due to easier migration of some of the iron to the particle surfaces during heat treatment, for samples with very low iron concentrations. In samples with higher iron concentrations, a second iron oxide phase may form inside the ZSM-5 particles and reduce the migration of the iron to the outside of the particle.

CONCLUSIONS

SEM has shown that stirring the gel during crystal growth results in small (.5μm - 1μm) molecular sieve particle aggregates. Crystal growth without agitation produces some single crystals in addition to the micron or larger sized particle aggregates. In the stirred batches, the particle morphology changes from regular spheres to irregular spheres and cubes as the iron content increases; this is analogous to the effect of increasing the aluminum content in ZSM-5 zeolites. XES has shown that the Si/Fe-ratio of the as-synthesized FeZSM-5 particle aggregates is independent of particle size and stirring, although the homogeneity of the inter-particle iron distribution does improve slightly with stirring. As expected, thermal treatments do not significantly change the average Si/Fe-ratios of the FeZSM-5; however, the particle to particle variations do increase, particularly in very low iron samples.

ACKNOWLEDGMENTS

The FeZSM-5 samples were prepared at Georgia Institute of Technology in the Zeolite Research Program; electron microscopy was performed at the Electron Microscopy Laboratories at the College of Letters and Sciences, University of California at Berkeley and at the Materials and Chemical Sciences Division at Lawrence Berkeley Laboratory. This work has been supported by the Director, Office of Energy Research, Office of Basic Energy Sciences, Materials Sciences Division, U.S. Department of Energy, under Contract DE-AC03-76SF00098 and through the Pittsburgh Energy and Technology Center, under Contract DE-AC22-86TC90007.

REFERENCES

[1] Roger A. Sheldon, *Chemicals From Synthesis Gas*, (D. Reidel Publishing Company, Dordrecht, Holland, 1983), pp. 64-75.

[2] C.D. Chang and A.J. Silvestri, J. Catal. **47**, 249 (1977).

[3] G.T. Kokotailo, S.L. Lawton, D.H. Olson and W.M. Meier, Nature **272**, 437 (1978).

[4] C.D. Chang, W.H. Lang and A.J. Silvestri, J. Catal. **56**, 268 (1979).

[5] P.D. Caesar, J.A. Brennan, W.E. Garwood and J. Ciric, J. Catal. **56**, 274 (1979).

[6] L.E. Iton, R.B. Beal and D.T. Hodul, J. Molec. Catal. **21**, 151 (1983).

[7] A.N. Kotasthane, V.P. Shiralkar, S.G. Hegde and S.B. Kulkarni, Zeolites **6**, 396 (1986).

[8] A. Shamsi, V.U.S. Rao, R.J. Gormley, R.T. Obermyer, R.R. Schehl and J.M. Stencel, Appl. Catal. **27**, 55 (1986).

[9] R. Szostak and T.L. Thomas, J. Catal. **100**, 555 (1986).

[10] R. Szostak, V. Nair and T.L. Thomas, J. Chem. Soc. Faraday **83**, 487 (1987).

[11] V. Nair, PhD Thesis, School of Chemical Engineering, Georgia Institute of Technology, Atlanta, GA, USA.

[12] A. Meagher, V. Nair and R. Szostak, Zeolites (in press).

[13] R. Csencsits, C. Schooley and R. Gronsky, J. Electr. Microsc. Tech. **2**, 643 (1985).

[14] C.E. Lyman, P.W. Betteridge and E.F. Moran, *Intrazeolite Chemistry* , (American Chemical Society, Washington, D.C., 1983), pp.199-215.

STRUCTURAL AND HIGH RESOLUTION STUDIES OF B-ZEOLITE CATALYSTS USING TEM AND FOURIER TRANSFORM PROCESSING TECHNIQUES

MIGUEL JÓSE-YACAMÁN, J.G. PÉREZ-RAMÍREZ AND D.R. ACOSTA.
IFUNAM. Apdo Postal 20-364 01000 Mexico, D.F.
J.P. GILSON AND G.C. EDWARDS. W.R. Grace & Co Davison Chemical Division. Washington Research Center 7379 Route 32 Columbia, Maryland 21044, U.S.A.

Introduction

Zeolite Beta, synthesized in 1967 (1), is the first example of a large pore, high silica zeolite ($SiO_2O_3 > 20$) prepared in the presence of an organic template (Tetraethylammonium Hydroxide). Zeolite NU-2, whose synthesis has been reported in 1982 (2) seems to be related to the zeolite Beta. Although Zeolite Beta has been claimed as useful catalyst in hydrocracking (3), cracking (4), dewaxing (5), hydrotreating (6), shale oil (7) and Fisher-Tropsh products (8) conversions and as a selective adsorbent in the separation of alkylaromatics (9); this zeolite has not been extensively characterized in the open literature.

The original patent (1) describes zeolite Beta as having cubic unit cell of $\alpha_o = 1.204$ nm. This result is however disputed by Breck (10). More recently, Martens et. al. (11) have shown, by a careful analysis of the product distribution on n-decane hydroconversion, that the void structure of zeolite Beta is similar to that of zeolite L and consists of pores with 12 membered rings and lobes. Martens et al. (12) also concluded that Pt/H-Beta behaves like an ideal bi-functional catalyst in the isomerization and hydrocracking of $nC_{10} - nC_{17}$ alkanes.

In order to shed some light on the structure of this interesting zeolite, lattice imaging using Fourier Transform Electron Microscopy (FTEM) has been undertaken.

Experimental

The zeolite Beta has been synthesized using a modification of the procedure outlined in reference (1). The X-ray powder pattern of the calcined material (two hours at 550°C), Figure 1, agrees very well with the published data (1). However a full determination of the structure based on X-ray data is not possible. We applied a number of additional techniques in order to determine the detail structure of this material. A JEOL 100-CX fitted with an STEM attachment and a JEOL 100-C fitted with an ultra high resolution pole piece were used to obtain selected area, micro-diffraction (from areas 25 nm in size) and high resolution images. Occasionally a 1200 CX high resolution microscope was used. Images were digitalized using a scanning microdensitometer or a TV-camera an then processed in an Innovion system coupled to an IBM-AT computer. Fourier transforms were obtained and displayed on the monitor. Images were reconstructed using different kinds of filters. Great care was taken to avoid extra periodicities on the final image that can be introduced during the processing. For instance all the "double" periods introduced by the FT operation were eliminated from the final image. Other techniques such as differential thermal analysis were used.

It was noted that the sample was decomposed during the irradiation on the TEM. In several high resolution pictures amorphous zones were observed at different stages of the degradation process. This phenomena is caused principally by radiolytic process.

Results

DTA:

Figure 2 shows the result of the DTA analysis of the β-Zeolite.
As can be observed two peaks at 333°C and 429°C are present.
The corresponding energies of the peaks are 30.21 cal/gr and
34.43 cal/gr respectively. These data suggests that two phases
are produced on the β Zeolite. A preliminary study using
differential scanning calorimetry confirms the results of DTA.

Fourier Transform

It has been shown by Tomita et. al. (13) that in the case
of a not very thin sample and with dynamic scattering taking
place, the Fourier Transform (FT) of a high resolution image is
directly related to the diffraction pattern of the sample. In
fact the distance of the spots appearing on the Fourier Transform is equal to g (the reciprocal lattice vector). Therefore
the distance between spots on the FT is equal to the distance
between spots in a diffraction pattern. Figure 3 shows an example of FT of a β-Zeolite. A rectangular two-fold pattern can be
observed. It is particullarly conspicuous on the pattern "extra
diffuse" periodicities in one direction. Carefull examination
of the spot positions reveals that sometimes there are some
differences between the peak in orthogonal positions. Additionally the angle between spots in the FT is not 90° but 86° or
95°. The main periodicities observed on the FT are 1.204 nm,
1.127 nm, and 1.154 nm.

Reconstructed and High Resolution Images

Images can be recovered by transform using filters to select frequencies. Figure 4 shows an image formed using all
frequencies observed in Fig.3. On the other hand Fig. 5b) shows
the result of reconstructing the original image (Fig. 5a) but
now using only the central spot and the "diffuse spots" arrowed
in Fig. 3. This type of technique was applied in a somewhat
different way by Thomas and coworkers (14-15) to study Zeolite
L, and has proved its usefullness to obtain structural information from Zeolitic materials.

The image in Fig.4 clearly shows that there are linear defects
which defines micro-domains of slightly distorted square ar-
rays. The micro-domains are rotated -10°. This is also appreciated in the high resolution pictures shown in figures 6 and 7.
In such pictures it is clear that the material is conformed by
platelets which are displayed. The figure 7 shows the two domains of 1.156 nm periodicities forming angles of about 80°.
Note that in the image, some amorphous regions are observed,
which are the result of the electron irradiation. The longer

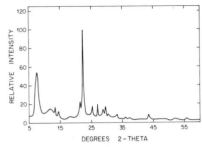

Fig. 1 X-ray diffraction pattern

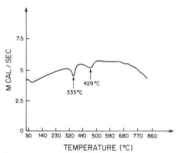

Fig. 2. Differential thermal analysis of β-zeolite.

Fig. 3. Fourier transform
of a high resolution image
of a β-zeolite.

Fig. 4. Reconstructed
image including all the
frecuencies observed in
Figure 3

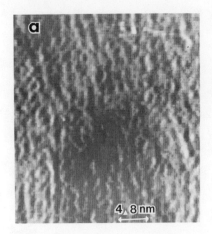 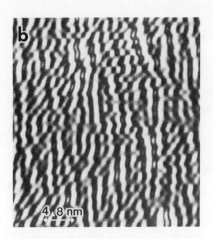

Fig. 5 Reconstructed images of β-Zeolite
a) original b) image using the central spot and the diffuse
portion of the FT.

 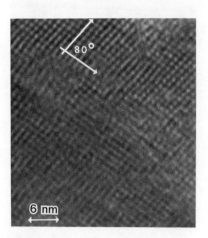

Fig. 6 High resolution images
(non reconstructed) of a β-Zeolite
showing the layer structure of the
material.

Fig. 7 High resolution
image of β-Zeolite show
ing two periodicities
making an angle of 80°.
Note some amorphous re-
gions.

the observation time, the larger the regions. The contrast observed in figure 5b) is basically diffraction contrast, the
image strongly resemble that of planar faults which are fully
documented on the literature (16). This picture strongly supports the idea that β-Zeolite is a heavy faulted material.

Diffraction Results

Individual but irregular shaped crystals were localized to
obtain selected area diffraction patterns from different
regions of the samples at several tilting angles. An example of
the tilting sequence is shown in Fig. 8. The tilting axes was
closed to a cubic [(100)] axis. It was possible in many cases,
to index the pattern as a simple cubic with a spacing $a_0 = 1.204$
nm as originally reported for this zeolite (1). However many
extra spots corresponded to a super lattice with $a' = 2 a_0$ and
$a' = 3a_0$ Many complicated patterns formed from several crystals
were avoided. The spots showed also splitting in other sets of
patterns; the value of 1.252 nm was obtained from this patterns. It should be noted that many of the distances observed
for the cubic parameter are related by a factor Cos 10.5 which
corresponds to twice the average angle observed between different periodicities on the image.

A very interesting example can be observed in the figure 9
in which two SADP of the same area are shown. Fig. 9a) can be
fully indexed as a cubic structure with $\alpha_0 = 1.204$ nm and extra
spots at positions 1/3, and 2/3 suggesting twins. After tilting by 14°, Fig 9b) shows now a rectangular structure with a ≠
c forming angles of 84° and the relation c = α Cos (10.5°).
Therefore it appears that the β-Zeolite has at least in two dimensions a cubic twinned structure. The high concentration of
planar defects creates a micro-domain structure, in which domains are deviated from orthogonallity by −10°. A superstruc-
ture with na_0 period is also observed.

High resolution electron micrographs strongly suggest a
channel structure for beta zeolite, but a full similarity with
other zeolites like that claimed by Martens et al. (12) between
beta and L zeolites has not been found.

Conclusions

From High Resolution and processed images it can be appreciated
the presence of linear defects and micro-domains in the beta
structure.
Beta zeolite is something less sensitive than other zeolites to
electron irradiation in the microscope and a full study of the
degradation process is in progress.
Electron diffraction studies reveals a cubic structure but additionally, extra spots reveals a superlattice structure with a
parameter 2 or 3 times the cell unit parameter.

Acknowledgements

The authors are indebted to Mr. Francisco Ruiz-Medina and to
Mr. Roberto Hernandez for technical help with the TEM. To Mr.
Samuel Tehuacanero for the image processing and to Mr. Oel
Guzman from I.M.P. for DTA determination.

166

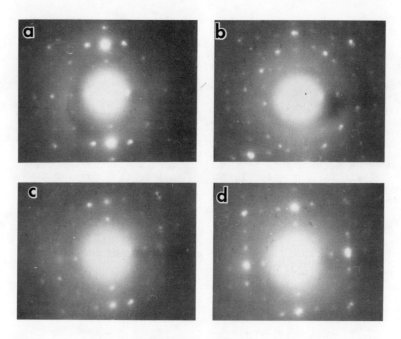

Fig. 8 Selected area diffraction patterns in different tilting
angles with respect to a [100] cubic axis
a) 0°, b) 4°, c) 15°,d) 20°. The patterns show 1/3
and 2/3 extra spots.

Fig. 9. Two SADP of a ß-Zeolite in a) 0° showing 1/3 and 2/3
extra spots and b) after 14° showing asymmetric distances.

REFERENCES

(1) Wadlinger, R.L., Kerr, G.T., Rosinski, E.J., US Pat. 3,308,069 (1967), assigned to Mobil Oil Corp.
(2) Whittman, T.V., Eur. Pat. Appl. 55,046 (1982),assigned to ICI PLC.
(3) Angevine, P.J., Mitchell, K.M., Oleck, S.M. Shih, S.S. US Pat. 4,612,108 (1986), assigned to Mobil Oil Corp.
(4) Kennedy, C.R., Ware, R.A., Eur. Pat. Appl. 186,447 (1986), assigned to Mobil Oil Corp.
(5) Chu, P., Yen, J.H., US Pat. 4,601,993 (1986), assigned to Mobil Oil Corp.
(6) Angevine, P.J., O:leck, S.M., Mitchell, K.M., Shih, S.S., Eur. Pat. Appl.180 354 (1986), assigned to Mobil Oil Corp.
(7) Angenine, P.J., Kuehl, G.H.m Mizraki, S., US Pat. 4,5,21,297 (1985), assigned to Mobil Oil Corp.
(8) Chen, N.Y., Haag, W.O., US Pat. 4,584,424(1986), assigned to Mobil Oil Corp.
(9) a)   Barthomeuf, D.M., US Pat. 4,584,424(1986) assigned to Exxon Research & Engineering Co.
    b)   Barthomeuf, D.M., Rosenfeld, D.D., US Pat. 4,554,398 (1985), assigned to Exxon Research & Engineering Co.
(10) Breck, D.W., "Zeolite Molecular Sieves: Structure Chemistry & Use, John Wiley & Sons, New York, 1974, p. 309.
(11) Martens, J.A., Tielen, M., Jacobs, P.A.. Weitkamp, J., Zeolites 4, 98, 1984.
(12) Martens, J.A., Perez-Pariente, J., Jacobs, P.A., Proc. Int. Symp. Zeolite Catalysis, Siofok, Hungary, 1985, pag, 487.
(13) M. Tomita, H. Hashimoto, T. Ikuta, K. Endoh and Y. Yokota. Ultrascopy 16, 9 (1985).
(14) O. Terasaki, J. M. Thomas and G.R. Millward. Proc. of Roy. Soc. of Lond. A 395 153 (1984).
(15) R. Millward, J. Thomas and R. Glaeser. J. Chem. Soc. Comm. 520, 962 (1985)
(16) See for instance; Modern Diffraction and imagining Techniques in Materials Science by S. Amelinckx, R. Gevers, G. Renault and J. Van Landuit. North Holland Publish Co. 1970.

CHARACTERIZATION OF ERIONITES WITH A SUPPORTED
METALLIC PHASE BY TEM,IR,AEM AND AUGER METHODS

D.R. ACOSTA,J.SALMONES,J.NAVARRETE AND G. JUAREZ
Instituto Mexicano del Petróleo,IB.BP.
Eje Lázaro Cárdenas 152;A.P. 14-805,México D.F., MEXICO

INTRODUCTION

Catalytic isomerization of paraffines of low molecular
weight is a very important process in petrochemical refination
and zeolites like mordenites and erionites are used frequently
as supports, due to   acid properties, neccesary   for a dual
action in hydroisomerization processes(1-3). In the present
work, we analyze the behavior of erionites interchanged with Pt
at low weight concentrations, using test reaction for n-pentane
hydrodesintegration. At several stages of the process, small
quantities of the samples were studied using TEM,AEM, IR, ESCA
and Auger methods.

EXPERIMENTAL

Commercial erionites were prepared for preactivation in a
vacuum line during 16 hours. Interchange was realized using
chloroplatinic acid solution with deionized water.the platinum
dispersion was obtained with athermal "in situ" treatment in or
der to avoid variations of pressure and temperature. Catalytic
activity was tested in an integral fixed bed reactor. In the
reactor, erionite samples, interchanged and not interchanged,we-
re observed in a kinetic regime. Temperatures range   from
200 °c to 500° C at atmospheric pressure.

RESULTS

Figure 1 shows the results of IR measurements, which were
realized varying temperature and vacuum conditions.It can be se
en, the disappearance of a band at 1399 cm^{-1} after temperature
increased. This is connected with missing   template agent(4)
and in our case, this   suggests   that   nitrate ions are missing. In the same figure,the band at 1635 cm^{-1},
related to the vibration of the group OH, has disappeared.this
suggests   that total extinction of zeolitic water has
ocurred.
Figure 2 shows the results obtained in a N_2 atmosphere and
besides similar results of figure 1, it can be appreciated the
appearance of a band at 2360 cm^{-1} which is associated to the
presence of CO_2 as a consequence of the oxidation reaction of
the organic radical of templant agent.Auroux et al.(5) from
studies realized in ZSM-5, proposed the de composition of template agent due to releasing of protons which cause an increase
of acidity in the zeolite. In our case the presence of CO_2 suggests a total oxidation and the presence of nitrogen oxides in
the lattice.

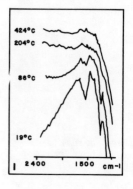

Figures 1 and 2.
Infra Red spectra
at different tem-
perature conditions.
See details in the
text.

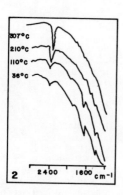

Quantitative analysis of the elemental composition at the surfaces in the samples with different thermal treatments was done using Auger spectroscopy. The results are shown in table 1.It can be observed that the ratio Si/Al is smaller in thermal treated samples compared to the non treated.Evidence not presen ted here, reveals an enrichment of Al concentration at surface during the activation. This has not been explained but it seems that migration from the bulk has ocurred.

Analytical and conventional electron microscopy was perfor med in a 100 KV side entry electron microscope,equipped with an EDS system.Results of the elemental analysis are shown in table II for the two samples studied. Si/Al ratio variations from these measurements were observed and related with the acidity of the erionites and from here with the catalytic activity. Pt dis persed in the erionites was not detected with this technique and its presence on the surfaces of erionites was detected using ESCA and Auger techniques.All selected area diffraction patterns were consistant with erionites but local variations in the c/a ratio were observed.This suggests local variations of compositi on in the crystallites, but its effect has not been correlated to catalytic properties. Bright and dark field imaging was us- ed to identify the small metallic particles.The size distribu-- tion and the dispersion determined from these images are shown in figure 4.

Table I. Auger Analysis

Sample	Si/Al	SiO$_2$/Al$_2$O$_3$
EMT/A	1.3628	0.7714
EM/A	1.7481	1.0099
EMT/B	0.9912	0.5660
EM/B	1.6262	0.9265
EMT/C	1.3447	0.7612
EM/C	1.5487	0.8766

EMT = Erionite/Metal/Treated
EM = Erionite/Metal

Table II. EDS Analysis

Sample 1 EL-Line	KF	AT%	El W%	PK
Si-K	1.00	77.47	76.31	11342
Al-K	1.25	24.36	23.27	2284
Fe-K	1.30	1.80	3.57	501
K -K	0.95	0.00	0.00	0
Sample 2				
Si-K	1.00	70.04	65.07	92
Al-K	1.25	20.52	18.38	20
Fe-K	1.30	7.68	14.27	15
K -K	0.95	1.76	2.28	3

A part of thecatalytic activity of the samples studied is shown in table III.It can be clearly observed the variations

of activity with metal concentrations and temperature treatments.
Samples were also studied after each catalytic test in order to
determine possible variations of structural characteristics.

Table III. Metal effect on erionite selectivity

Temperature	Cat.	Metal	%Weight	Select. I-C5	Select. C4	Select. C3
240 °C	ELZ1	Pt	0.28	4.05	30.20	48.92
	ELZ3	Pt	0.095	1.08	10.40	15.40
	ELZ5	Pt	0.25	3.78	25.70	30.72
	ELZ	--	-----	0.07	2.34	3.49
360 °C	ELZ1	Pt	0.28	28.37	24.78	25.93
	ELZ3	Pt	0.095	0.92	0.89	11.31
	ELZ5	Pt	0.25	22.40	20.70	21.88
	ELZ	--	-----	0.07	2.45	3.69
400°C	ELZ1	Pt	0.28	16.76	17.89	10.76
	ELZ3	Pt	0.095	5.41	6.78	3.79
	ELZ5	Pt	0.25	15.40	16.71	9.88
	ELZ	Pt	-----	0.07	2.45	3.69

Work conditions: Vapor Pressure = 174 mm Hg
Weight = 0.2 gr.
LHSV = 1.049 gC5/seg-g.Cat.
H_2/HC = 2.354

Figures 3a,3b and 3c are electron micrographs that show
the erionite crystallites with supported metallic particles.In
3a and 3b we have a sequence of two pictures, where an interesting fact is observed:the particle arrowed has behind it a trace
similar to a channel and this makes think that the particle was
moving during TEM observations.In figure 3b we have the same zo
ne of 3a but in another focus condition(less than 9 nm) and the
particle and the trace are not so visible as in figure 3a.This
phenomena was observed in several crystallites and resembles
the case of metallic particles forming channels on graphite.Pic-
ture 3c shows clearly the particle arrowed and its trace.More
detailed observations are being realized in order to correlate
thermal treatment, electron irradiation and particle motion
in or on the crystallites.Figures 4a to4d show the particle
size distribution in four of the samples studied.

CONCLUSIONS

This work shows that erionites with a metallic phase are
selective to the production of linear hydrocarbons,C_3 and C_4
for instance.The samples showed a strong dependence on metallic
concentration with respect to selectivity and activity.The behavior of particles with traces, seems to be unusual and
there is not a full explanation for it.

ACKNOWLEDGMENT

Technical assistance of G. Arciniega is here recognized.

Figures 3a and 3b.It can be observed a particle with a trace behind it, similar to a channel.

Figure 3c. Several small particles with clear traces are here visible .

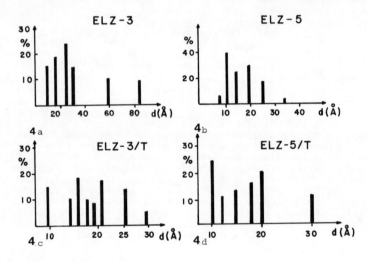

Figures 4a to 4d.- The size distribution and dispersion of metallic phase observed in bright field conditions is shown

REFERENCES

1.- Benesi,H.A.;U.S.Patent 3,190,939 (1965)
2.- Benesi,H.A.;U.S.Patent 3,527,835,(1960)
3.- Adams,C.E. et al.;U.S.Patent 299,153,(1967)
4.- Auroux,A. et al.;J.Chem.Soc.Far.Trans. 75(11)2544,1979
5.- Lyman,Ch.;"Intrazeolite Chemistry"ACS Series, 1983

COMPARATIVE STUDIES OF THE DEGRADATION OF
SEVERAL ZEOLITES UNDER ELECTRON IRRADIATION

D.R. ACOSTA,O.GUZMAN,P.DEL ANGEL AND J.DOMINGUEZ
Instituto Mexicano del Petróleo,I.B.P.,A.P. 14-805,MEXICO D.F.
M.J.YACAMAN: IFUNAM, A.P.20-364,01000 MEXICO D.F.,MEXICO.

INTRODUCTION

High resolution electron microscopy has proven to be a powerful technique to determine structural characteristics of zeolites (1-2),symmetry variations and identification of several
kind of defects.together with ideal projected potential images,
the microscopist usually finds in electron micrographs the influence of electro-optical parameters and alterations of the
crystallinity of the material under electron irradiation. One
of the purposes of this work is to contribute to the understanding
of the degradation process of zeolites under electron irradia--
tion in the electron microscope and in this way,discriminate
when it is possible,what is reliable information recorded in
the images obtained in high resolution conditions.

Due to radiolytic,knock-on and thermal processes, zeolites
under electron irradiation, change from crystalline to amorphous state following several steps. The simplest way to study the
se transformations is to obtain selected area diffraction pat--
terns and record them as a function of the time and from this,
detect blurring and loss of diffraction spots. With high resolu
tion electron microscopy we can use real space images to detect
and follow the transformation process.

EXPERIMENTAL

We observed ZSM-5,Calcium A,Y Faujasite and Beta zeolites
in a 100KV top entry electron microscope equipped with a pole
piece of C_s = 1.77 mm and tungsten filament and a 100 KV electron microscope with STEM equipment,pole piece of C_s =2.3 mm
and LaB_6 filament for beta zeolite observations.
In all the cases the samples were under thermal treatment in a
vacuum line during several hours before microscope observation
and the samples were mounted in a nitrogen atmosphere in order
to improve the stability of the zeolites in the microscope. In
some cases focal series and time series in high resolution conditions were obtained in order to follow the degradation.

RESULTS

ZSM-5 Zeolite

Figure 1 shows a part of a ZSM-5 crystallite where a high
degree of crystallinity can be observed together with some zones where loss of crystallinity is in the initial steps of the
process like that arrowed. This tells us that in this zeolite
crystalline and amorphous states coexist and that there is not
an abrupt change from one state to the another

Figure 1b shows part of another ZSM-5 crystallite after 25 seconds in high resolution conditions.Besides the ordered entrance of tunnels, it can be observed that several of these tunnels present missing parts,clearly visible . The point to consider here is that the missing parts are in most of the cases, following particular directions, which can be indicative of the process: This begins at specific places of the rings that conform to the tunnels and some loss of mass is ocurring; after this, mesopores of about 1.3 nm to 1.7 nm are created and a chain process of this type conduce to a generalized amorphous state.

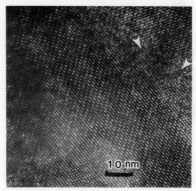 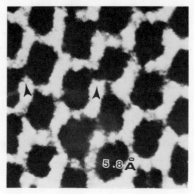

Figure 1a. A part of a ZSM-5 crystallite showing a high degree of crystallinity and some amorphous zones.

Figure 1b. A projection of tunnels are showed in this ZSM-5 crystallite sight showing incompleted and connected rings.

Calcium A Zeolite

This cubic zeolite, a= 1.227 nm, under electron irradiation,showed several phenomena before total degradation; first, an apparent size decrease was observed in single crystallites. This was interpreted as a consequence of bending processes. In another series, it was observed that the degraded zone, was moving from the border to the centre of the crystallites (3). Figure 2a shows zones where rows of tunnels, the white circles, are interrupted or where some tunnels have disappeared.Plane deviations and dislocations are common at the first stages of the degradation process of this zeolite.In some cases, supercages are observed, similar to those reported by Thomas et al.(4). In some crystallites the generation of circular zones was observed as a function of irradiation time.Figures 2b and 2c are two of a series of five pictures, where the generation of the circular can be appreciated.This can be connected with the presence of surface barriers and mass losses as has been suggested by Karger et al.(5). In our case the samples did not receive any hydrothermal treatment before microscope observations.

 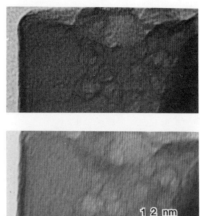

Figure 2a. Calcium A zeolite showing broken planes and dissapeared tunnels.

Figures 2b and 2c. Part of a sequence,where circular zones are generated and the border arrowed,propagates to the centre

Y Faujasite

This zeolite also undergoes notable changes before the final amorphous state.It could be observed in focal and time series how the progressive disappearance of tunnels cause broken and interrupted rows of planes. Figures 3a and 3b show the disappearance of planes and the generation of a circular zone, still in the case that the sample did not receive any hydrothermal treatment.Figure 3c shows a zone where there is an amorphous island and interrupted planes are clearly visible.

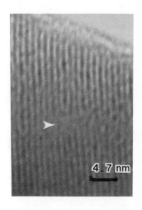  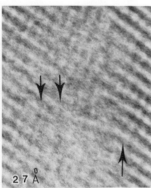

Fig.3a Some broken planes are clearly visible

Fig.3b A circular zone is observed.

Fig.3c An amorphous zone and interrupted planes are showed in this sight.

Beta zeolite

This zeolite with apparent cubic structure, a= 1.204 nm, pre sents several interesting aspects in the degradation process. Figure 4a shows in a very clear way the crystallinity of this material together with some distortions and alterations of the initial order of this material.figure 4b shows another stage of the process where the rows of planes are interrupted in certainlocal groups.this zeolite presents a longer life under elec tron irradiation with respect to those previously mentioned in this work.

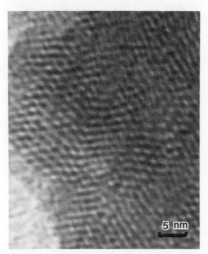 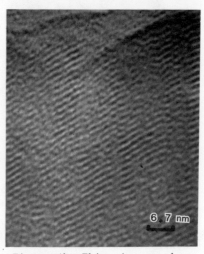

Figure 4a. High resolution sight of beta zeolite showing some degree of disorder.

Figure 4b. This picture shows local groups of ordered planes with non crystallinity between each pair of groups.

CONCLUSIONS

Zeolites ZSM-5,Calcium A,Y Faujasite and beta, are all sensitive to electron irradiation in the microscope.Except the beta zeolite,the degradation process is similar for the zeolites mentioned in this work. Quantification of the phenomenon is the next step and it will be reported in a forthcoming paper.

REFERENCES

1.- Thomas,J. et al.:"Intrazeolite Chemistry"ACS Series,1983
2.- Domínguez,J.M. et al.:J.of Cat. 83,450,1983
3.- Acosta,D.R.: EMSA Procc. 1986,page 852
4.- Bursill,L.-Thomas,J.:J.Phys.Chem. 85,3007,191
`5.- Karger,J. et al.:Zeolites 6,146,1986

ACKNOWLEDGEMENT

We thank the technical assistance of Gilberto Arciniega

ON THE PROPAGATION OF TWIN–FAULT–INDUCED STRESS IN PLATELET FAU–FRAMEWORK ZEOLITES

M. M. J. TREACY*, J. M. NEWSAM*, D. E. W. VAUGHAN*, R. A. BEYERLEIN*#,
S. B. RICE* and C. B. DE GRUYTER^fl
*Exxon Research and Engineering Company, Clinton Township, Annandale, NJ 08801
#Present Address, Amoco Research Center, PO Box 400, Naperville, IL 60566
^flExxon Chemical Holland, Botlekweg 121, PO Box 7225, 3000 HE, Rotterdam, The Netherlands

ABSTRACT

A model describing the propagation of a rhombohedral distortion in platelet CSZ–1 zeolites is presented. It is proposed that internal stress gradients grown into CSZ–1 platelets at synthesis are responsible for this distortion of the cubic FAU framework, where the spacings of 111 planes parallel to the platelet surfaces are elongated relative to the $\{11\bar{1}\}$ planes. The presence of inhomogeneities is suggested by the presence of thin bands of twin faults which are invariably observed near the central layers of each platelet. Elastic modelling confirms that the effects of any stress associated with such twin faults will be most pronounced in the thinnest platelets, where the effects of elastic relaxation are minimal, and where the width of the fault zone relative to the platelet thickness is maximal. Platelet CSZ–3 and Y-type zeolites, which are considerably thicker, are therefore not expected to show significant rhombohedral distortion despite the presence of similar twin fault bands.

INTRODUCTION

Extended defects, predominantly of the crystallographic twinning type, are common in zeolite structures. The presence of such defects is unsurprising in view both of the close similarities, on a local scale, between different but related zeolite framework topologies [1,2], and of the kinetically-controlled conditions under which zeolites are generally crystallized [3,4]. Pore structures, and hence sorption characteristics, are generally altered by the presence of such twin planes, and the topological effects of twinning can provide information about related known, or new, zeolite framework structures. As a result, the gross effects of such twinning operations have been discussed quite frequently [5]. The more subtle effects of such extended defects have, however, been little explored, even though such effects can apparently produce noticeable changes in the macroscopic properties [6]. We discuss here the manner in which stress, that might be associated with twin planes, is propagated in platelet crystallites of the faujasite type [1]. A preliminary communication introducing some of the ideas discussed here has already appeared [7].

A wide range of materials adopting the FAU-framework is known [1,2,8,9]. Zeolites CSZ-1 and CSZ-3 are cesium-containing high silica polymorphs that crystallize as platelets from $Cs_2O.Na_2O.SiO_2.Al_2O_3.H_2O$ gels [10,11]. CSZ-3 is observed at relatively low levels of cesium [10]; CSZ-1 at higher Cs concentrations [11]. Analysis of these materials by various techniques such as powder X-ray diffraction, ^{29}Si magic angle spinning (m.a.s.) n.m.r. spectroscopy [12], selected area electron diffraction and structure imaging shows that CSZ-3 possesses the cubic framework of faujasite [10], whereas CSZ-1, although topologically similar, is distorted and has rhombohedral rather than cubic symmetry [6]. The precise origin of this rhombohedral distortion is still uncertain. In the present paper we demonstrate how internal stress gradients (such as might be associated with growth twins), are propagated throughout platelet crystallites of the CSZ-1 and CSZ-3 type, leading potentially to net crystallographic distortions of magnitudes similar to those that are actually observed in these materials.

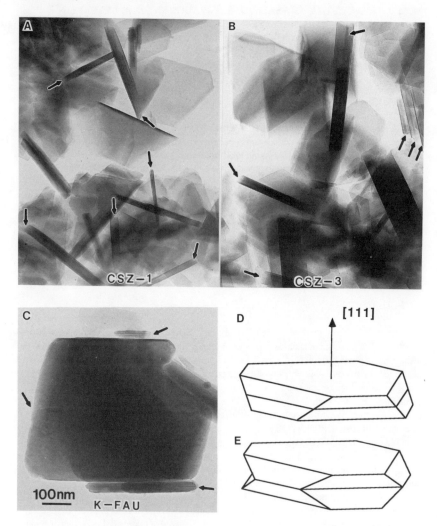

Figure 1. Bright field images of a) zeolite CSZ-1; b) zeolite CSZ-3; c) potassium zeolite Y (labelled K–FAU). CSZ-1 crystallites measure typically 300 nm wide by 40 nm thick. CSZ-3 crystallites are typically 300 nm by 80 nm. Both CSZ–1 and CSZ–3 have a tabular, or platelet, morphology, as shown in d). Frequently, as a result of twinning, crystallites exhibit the twinned tabular morphology shown in e). Arrows indicate planar faults near the crystal center layers.

Bright-field images obtained from representative samples of zeolites CSZ-1, CSZ-3 and of a potassium zeolite Y (labelled subsequently K–FAU), supported on amorphous carbon films, are shown in figure 1. CSZ-1 and CSZ-3 invariably grow with a tabular, or platelet, morphology (figure 1(d)) with a cubic [111] axis (or rhombohedral [001] axis) normal to the platelet. The average platelet diameters are similar in the two materials, about 300 nm. However, CSZ-1 is thinner; the ratio of platelet thickness to platelet diameter is typically about 1:8 in CSZ-1 and 1:4 in CSZ-3. In K–FAU, ratios range between 1:4 and about 1:1.

All three platelet FAU-framework materials contain faults on $(111)_{cubic}$ (or $(001)_{rhomb.}$ in the case of CSZ–1) planes near the platelet centers (see arrows in figure 1). The crystal habit thus frequently adopts a twinned tabular morphology as shown in figure 1(e). The width of the fault band, c, is roughly the same in all materials, and thus the ratio of the width of the fault band to the platelet thickness, c/b, is roughly doubled in CSZ-1 relative to CSZ-3.

Figure 2(a), is a $[1\overline{1}0]$ structure image of an uncharacteristically squat CSZ–1 crystallite (the larger diameter platelets are unstable when lying edge–on and rapidly twist and buckle under the influence of the electron beam), showing that the defect layer contains twins on the cubic (111) planes. Twinning occurs commonly in faujasite [13-15] and faujasite related materials [16].

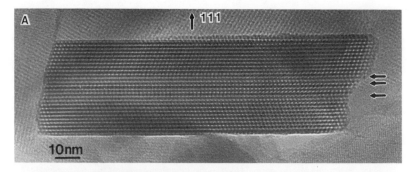

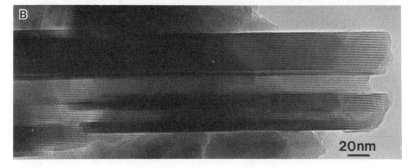

Figure 2. a) $[1\overline{1}0]$ structure image of a squat CSZ-1 crystallite, revealing twinning on (111) planes (cubic indexing, (001) rhombohedral) near platelet centers. b) Edge view of an atypical CSZ–1 crystallite showing a marked distinction between the central and outer layers. The 111 spacings of the central layer are ~5% expanded relative to the outer layers.

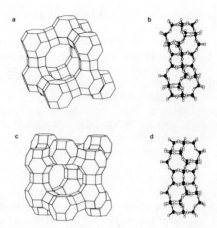

Figure 3. (a) Unfaulted faujasite framework in polyhedral representation. T–sites (T ≡ tetrahedral species, Si or Al) correspond to polyhedron vertices. (b) Connectivity of sodalite cages across double 6–ring prism in faujasite. Filled circles represent T-sites, and open circles represent oxygen atoms. (c) Framework in vicinity of twin plane passing through double 6–ring prisms. (d) Connectivity of sodalite cages across twin plane bisecting double 6–ring prisms.

The structure at the twin interface is illustrated in figure 3. The faujasite structure is based on that of diamond, space group Fd$\overline{3}$m, with a cuboctahedron, or sodalite cage centered on each lattice point (Figures 3a and 3b). The rhombohedral distortion that occurs in CSZ-1 is such that the cubic (111) plane spacings (stacking along the platelet normal) are elongated relative to the {111} plane spacings inclined to the platelet normal [6]. The resulting rhombohedral angle, θ, which corresponds in the undistorted cubic case to the angle of 60° between [110] and [101] (and, equivalently, [110] and [011], and [011] and [101]), is found to be typically in the range 59.5° – 60°. When a twin plane, that bisects the hexagonal prisms, is present, the sodalite cages become related not by inversion centers, but rather by mirror planes (Figures 3c and 3d). Relative to unfaulted faujasite, the sodalite cages are thus rotated 180° before connection. Regular, recurrent twins of this type would generate a new, hexagonal structure [17] labelled "structure 6" by Breck [18] with space group P6$_3$/mmc, and $a \approx$ 17.4 Å, $c \approx$ 28.3 Å. This had initially been proposed as the structure adopted by zeolite CSZ-1 [11], a suggestion supported by interpretations of earlier electron diffraction and lattice image data [19]. However, as indicated in figure 2(a), the density of faults in the ABC... type cubic stacking in CSZ-1 crystallites is always low, and no incidence of recurrence has been observed in CSZ-1 samples [6].

Both the cubic and rhombohedrally distorted FAU-framework structures share a three-fold axis along the cubic [111] (or rhombohedral [001]) directions. The hexagonal variant, "structure 6" [17,18] ($\overline{3}$), would have 6-fold symmetry along this same direction. Electron diffraction, to first order, is sensitive only to the projected structure, and the symmetries of both the cubic FAU framework and hexagonal Breck's 'structure 6' appear to be 6–fold in projection. In less beam–sensitive materials, convergent beam electron diffraction could be used to establish the true character of the symmetry about the platelet normal direction. However, convergent beam diffraction involves exposing the samples to intense focused irradiation, which destroys zeolite structures rapidly, and prevents the ready application of such techniques in the present case.

Views down the [111] axis invariably show "bend contours" which reveal strikingly the three–fold symmetry characteristic of the cubic (ABC...) type of stacking in the FAU-framework. Figure 4 shows a bright-field image of a CSZ-1 platelet with its cubic [111] axis aligned closely parallel to the beam direction. The positions of the three <776> poles (which are inclined at about 4° to the [111] pole) imply a radius of curvature of the diffraction planes of about 2.5 μm. The corresponding radius of curvature for CSZ-3 crystals is estimated to be about 10 μm to 20 μm. On tilting the crystallites relative to the electron beam, the [111] pole shifts position in such a way that the platelet always appears to be curved toward the beam direction, irrespective, presumably, of whether the platelet is lying on the electron beam entrance or exit side of the carbon film (see figures 5(a) and 5(b)).

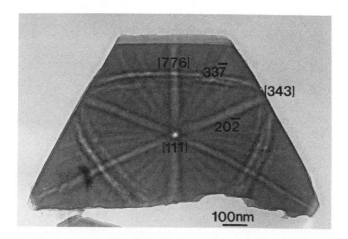

Figure 4. Bright-field image of CSZ-1 platelet aligned with [111] axis (cubic indexing, (001) rhombohedral) close to beam direction. Crystallite is strained as revealed by the bend contours, The angle between the [111] and [776] poles is ~ 4°, implying an effective radius of curvature of 2.5 μm.

Edge views of the platelets such as those in figure 2, do not reveal any detectable net curvature and the "bend contours" observed in figure 4 are apparently not associated directly with platelet curvature. However, careful measurement of the lattice spacings and relative interplanar angles reveal that they are generally not constant across the crystallites. For a rhombohedral distortion that was uniform throughout the crystallite, no shear strain gradients would be expected. Since CSZ-3 and K–FAU, to some extent, exhibit such bend contours (but with much reduced effective radii of curvature) it is plausible that the rhombohedral distortion, along with the anomalous "bend contours", derive from the presence of shear strain gradients within the platelets, possibly associated with the twin planes. Strain is common in other platelet systems, particularly clay minerals where it is thought to be primarily caused by lattice substitution or layer disorder(see ref [20] for a review).

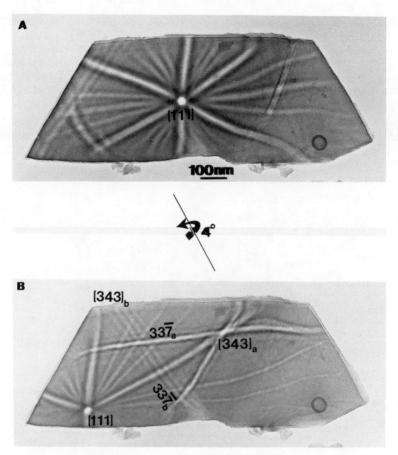

Figure 5. Bright-field images of CSZ-1 platelet aligned with [111] axis close to beam direction.Tilting the specimen and noting the direction of movement of the [111] pole reveals that the curvature in the diffracting planes is such that the surface facing the beam is always contracted. This is irrespective of whether or not the platelet is on the electron beam entrance or exit surface of the support film, implying that the central layers are expanded relative to the upper and lower layers. Note the faint set of $33\overline{7}_b$ contours positioned at 60° to the strong $33\overline{7}_a$ contours. "a" contours are from the upper platelet layers and the weaker "b" contours from the lower layers.

The possible presence of stress at the twin boundaries, or, expressed in another way, at the interface between the hexagonal and cubic structures, is unsurprising (see, e.g. [21,22]). There are several mechanisms by which this stress might arise. Compositional changes are known to give rise to differing lattice constants in faujasite-type materials [1,8,9]. Although X-ray microanalyses indicate that the Si/Al ratio is uniform throughout CSZ-1 crystallites (equal to $2.6 \pm 2\%$ for the samples studied here), the variabilities in the Cs^+ cation concentrations across the crystallites are difficult to measure, since these cations are highly mobile under electron irradiation. Even in the absence of cesium compositional gradients close to the twinned region, however, it is quite likely that some stress is associated with the twinning operation in the framework itself (the local environments of the atoms in the immediate vicinity of the twin differ from those in the bulk – figure 3). Were there not a significant activation barrier associated with the twinning operation, a more widespread occurrence of twin planes might be expected.

An effective model of the rhombohedral distortion in CSZ–1, based on internal stress gradients, must be able to explain why CSZ–3 and FAU, with similar types of faults, do not exhibit any measurable rhombohedral distortion. To do this convincingly we need to model the elastic behavior of thin platelets in the presence of internal stresses.

ELASTIC MODEL OF THE RHOMBOHEDRAL DISTORTION

A qualitative description of distortions in a platelet arising from internal stresses can be obtained by examining the behavior of a two dimensional plate whose dimensions are equivalent to a rectangular cross section of the three dimensional platelet. We take the x direction to be along the platelet diameter and the y direction to be through the platelet thickness. Stresses in the z direction are ignored, simplifying the problem to that of plane stress.

The origins of the internal stresses in the plate are taken to be due to gradients in unit cell parameter throughout the thickness of the platelet (although the general validity of the model is not limited to this assumption). For simplicity, we consider a gradient which is symmetric about the central layer. Figure 6 shows the situation for the particular case when the platelet is constructed as a simple sandwich with a central layer of material having different unit cell constant to that of the outer slabs. The interface in unit cell constant between the outer and inner layers in this case is abrupt. If all layers have identical numbers of cubic unit cells along x, then the layers, if cut free from the plate, will have different lengths; a_1 for the inner layer, of width c; and a_2 for the outer layers which are of width $(b-c)/2$. This gives a relative strain, ε_0, of

$$\varepsilon_0 = \frac{a_1 - a_2}{a_2} \qquad (1)$$

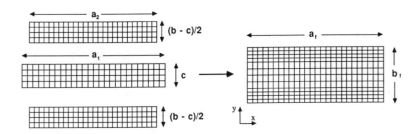

Figure 6. Simple "Sandwich" model for rhombohedral distortion. The central layer has different unit cell constant relative to the outer layers. Distortions in the majority layers will dominate crystallite properties.

No Elastic Relaxation

If relaxation of internal stresses at the plate edges is ignored, the internal stress distribution is given simply by that stress $S_{xx}(y)$ which needs to be applied to the ends along the x direction so as to suppress this relative strain, as shown in figure 6(b). When this external stress is removed, but now with the layers connected, the plate will attempt to return to its previous configuration, with accompanying relaxations at the surfaces. The driving force for this response is $-S_{xx}(y)$. Ignoring these relaxations, the internal stress distribution is given by

$$\sigma_{xx} = S_{xx}(y) = -\frac{(1 - (c/b))/(1 - \mu)}{\left(1 + (1 - c/b)\varepsilon_0\right)} E\,\varepsilon_0 \tag{2 a}$$

in the central layer ($|y| < c$), and by

$$\sigma_{xx} = S_{xx}(y) = \frac{(c/b)/(1 - \mu)}{\left(1 + (1 - c/b)\varepsilon_0\right)} E\,\varepsilon_0 \tag{2 b}$$

in the outer layers ($c < |y| < b$). Here E is Young's modulus for the material of the plate.

A net rhombohedral distortion is seen to be a natural consequence of these stresses, even when viewed in the context of this simple model. In the example of figure 6, the central layer is compressed along the x direction, and the two outer layers are expanded along the x direction so that they have the same length, a_f. At equilibrium, the total force integrated along the ends is zero. Compressing the central layer also expands it along y by an amount related to Poisson's ratio μ, and, conversely, stretching the outer layers shrinks them along y.

In an isotropic face–centered cubic material, with the [111] direction parallel to y, such compressions and expansions give rise to a rhombohedral distortion. The respective layer dimensions become

$$a_f = \frac{a_1}{\left(1 + (1 - c/b)\varepsilon_0\right)} = \frac{(1 + \varepsilon_0)\,a_2}{\left(1 + (1 - c/b)\varepsilon_0\right)} \tag{3 a}$$

$$c' = c\left\{1 + \frac{(1 + \mu)}{(1 - \mu)}\frac{(1 - c/b)\varepsilon_0}{\left(1 + (1 - c/b)\varepsilon_0\right)}\right\} \tag{3 b}$$

$$(b - c)' = (b - c)\left\{1 - \frac{(1 + \mu)}{(1 - \mu)}\frac{(1 - c/b)\varepsilon_0}{\left(1 + (1 - c/b)\varepsilon_0\right)}\right\} \tag{3 c}$$

The distortion in the central layer has a rhombohedral angle distorted below 60°, and in the outer layers the rhombohedral angle is distorted above 60°. The rhombohedral angles (in degrees) can be written as,

$$\theta_1 = \frac{360}{\pi}\,tan^{-1}\left\{\left\{\frac{8}{3}\left(\frac{c'/c}{a_f/a_1}\right)^2 + \frac{1}{3}\right\}^{-1/2}\right\} \tag{4 a}$$

$$\theta_2 = \frac{360}{\pi} \tan^{-1} \left\{ \left\{ \frac{8}{3} \left(\frac{(b-c)'/(b-c)}{a_f/a_2} \right)^2 + \frac{1}{3} \right\}^{-1/2} \right\} \tag{4 b}$$

in the inner and outer layers respectively.

Elastic Relaxation

In practice, elastic relaxation near the surfaces of the platelet will modify significantly the above results. Relaxation may be modelled using the principle of least work, following the approach of Timoshenko and Goodier [23] .

To calculate the relaxation we must consider the full response of the plate when the suppressing stress $S_{xx}(y)$ is removed from the ends. The response stress term is $-S_{xx}(y)$. The boundary conditions for the elastic response of the plate are therefore;

For $x = \pm a$

$$\sigma_{xy} = 0 \qquad \sigma_{xx} = -S_{xx}(y)$$

For $y = \pm b$

$$\sigma_{xy} = 0 \qquad \sigma_{yy} = 0 \tag{5}$$

Since we have a simply connected boundary, the stress distribution is independent of the elastic constants of the plate, and the strain energy can be written in terms of the stress function ϕ

$$U = \frac{1}{2E} \iint \left\{ \left(\frac{\partial^2 \phi}{\partial y^2} \right)^2 + \left(\frac{\partial^2 \phi}{\partial x^2} \right)^2 + 2 \left(\frac{\partial^2 \phi}{\partial x \partial y} \right)^2 \right\} \, dx \, dy \tag{6}$$

where

$$\sigma_{yy} = \frac{\partial^2 \phi}{\partial x^2} \qquad \sigma_{xx} = \frac{\partial^2 \phi}{\partial y^2} \qquad \sigma_{xy} = -\frac{\partial^2 \phi}{\partial x \partial y} \tag{7}$$

The correct value for the stress function ϕ is that which minimizes the elastic energy. Following the approach of Timoshenko and Goodier [23], the stress function can be written as a series of the type

$$\phi = \phi_0 + (x^2 - a^2)^2 (y^2 - b^2)^2 \sum_{n=0}^{N} \sum_{m=0}^{M} \alpha_{nm} x^{2n} y^{2m} \tag{8}$$

where α_{nm} are constants. The factor $(x^2 - a^2)^2 (y^2 - b^2)^2$ ensures that the boundary conditions will always be satisfied provided

$$S_{xx}(y) = -\frac{\partial^2 \phi_0}{\partial y^2} \tag{9}$$

Only even powers in y are considered in equation (8) since we assume a symmetrical distribution of layers. Odd terms must be included for asymmetrical layer distributions where overall platelet curvature is present, as in the case of a bimetallic strip. The polynomial form of the stress function greatly reduces computational effort since the differentiations and integrations in equation (6) can be carried out by hand. The coefficients α_{nm} can be calculated from the conditions for minimum energy, namely

$$\frac{\partial U}{\partial \alpha_{nm}} = 0 \tag{10}$$

These give a set of simultaneous equations from which the coefficients α_{nm} can be calculated, namely

$$\sum_{n=0}^{N}\sum_{m=0}^{M}\alpha_{nm}a^{(2n+2m+6)}\left[A_{nmpq}\left(\frac{b}{a}\right)^{(2m+2)} + A_{mnqp}\left(\frac{b}{a}\right)^{(2m+6)} + 2B_{np}B_{mq}\left(\frac{b}{a}\right)^{(2m+4)}\right]$$

$$= -\frac{4}{b}\int_{-b}^{b} S(y) \sum_{Q=0}^{2}\frac{(2q-2Q+4)(2q-2Q+3)(-1)^{Q}}{(p+5)(p+3)(p+1)(2-Q)!\,Q!}\left(\frac{y}{b}\right)^{(2q-2Q+2)} dy \tag{11}$$

The constants are given by

$$A_{nmpq} = \sum_{K=0}^{2}\sum_{L=0}^{2}\frac{(2m-2K+4)(2m-2K+3)(2q-2L+4)(2q-2L+3)}{(2m+2q-2K-2L+5)}$$

$$\times \frac{4(-1)^{(K+L)}}{K!\,(2-K)!\,L!\,(2-L)!}\sum_{M=0}^{4}\frac{4!\,(-1)^{M}}{(2n+2p-2M+9)\,M!\,(4-M)!} \tag{12}$$

and

$$B_{mq} = \sum_{K=0}^{2}\sum_{L=0}^{2}\frac{4(2m-2K+4)(2q-2L+4)(-1)^{(K+L)}}{(2m+2q-2K-2L+7)\,K!\,(2-K)!\,L!\,(2-L)!} \tag{13}$$

From the coefficients α_{nm} the response stress field can be calculated using equations (7) and (8). The strain suppression term $S_{xx}(y)$ must then be added to the solution for σ_{xx}, since in the fully relaxed state there is zero stress on the outer surfaces. This is equivalent to ignoring ϕ_0 in the expression for the stress function, equation (8). The strain field in the plate can then be calculated by the use of Hooke's law. As shown by Timoshenko and Goodier [23], for a simple parabolic stress distribution, the strain field can be accurately calculated using very few terms, upper limits of $N = 1$ and $M = 1$ being sufficient (giving up to four coefficients a_{00}, a_{10}, a_{01} and a_{11}). This is a natural consequence of the polynomial expansion of the stress function (equation 8). For the sandwich case of figure 6, however, more terms $(N, M > 10)$ are needed to provide an adequate match at the abrupt layer interface, giving more than 100 coefficients α_{nm}.

RESULTS AND DISCUSSION

Figures 7(a) and 7(b) show graphically the calculated effects of relaxation on a platelet of aspect ratio $b/a =$ 0.25, with a (hypothetical) parabolic distribution of unit cell dimensions of the form $a(y) = a_0 + \varepsilon_0 E(1 - (y/b)^2)$. The strain amplitudes (somewhat exaggerated for graphic purposes) are $\varepsilon_0 = + 0.1$ in figure 7(a) and $\varepsilon_0 = - 0.1$ in figure 7(b). Note how the plane bending varies across the plate, being most pronounced towards the ends. It is the shear strain gradients across the platelets which give rise to the bend contours so strikingly visible in figure 4. The bend contours provide a measure of the relaxation within the platelet. They are not necessarily a measure of (dishlike) platelet curvature, as is usually assumed when such contours are observed.

This model also explains why the upper and lower layers of the platelet, when they are in a twin orientation relative to each other (as in figure 4), do not give rise to two mutually inverted sets of three–fold symmetric contours, resulting in a six–fold star. The upper and lower layers are bent in opposite senses. The crystallographic inversion that reflects the mirror symmetry of the twinning is thus counteracted by an inversion in the direction of the bend contours. Careful inspection of the three sets of $33\overline{7}$ contours in figure 4 reveal that they are doubled, with contributions from the upper and lower layers almost, but not quite, superimposing. In instances where the upper and lower layers are not in a twin relation, such as is probably the case in figure 5, a six–fold star, rather than a triangle, would be expected. However, when such stars are observed, the two sets of bend contours are not of equal intensity. The reason for this is not clear, although it may be simply due to the upper and lower layers being of unequal thickness. It is conceivable that dynamical

a

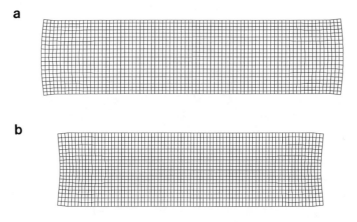

b

Figure 7. Calculated distortions in a plate of aspect ratio 0.25, due to a (hypothetical) parabolic internal strain gradient arising from a gradation in unit cell dimension of the form $a(y) = a_0 + \varepsilon_0(1 - (y/b)^2)$, with a) $\varepsilon_0 = + 0.1$, and b) $\varepsilon_0 = - 0.1$. Note the curvature of the planes, particularly near the plate edges, and that the sense of curvature is of opposite sign in the upper and lower halves. The rhombohedral angle is less than $60°$ near the platelet upper and lower surfaces in a), and near the central layers in b).

electron diffraction effects may be playing a role where, in the presence of anomalous absorption, the upper and lower crystal layers do not necessarily play an equal role in generating the contrast (much in the same way that contrast from inclined stacking faults is generated [24]). The sense of curvature of the diffraction planes in the platelet, revealed by the tilting experiment shown in figure 5, implies that the central layers of the platelets are expanded relative to the outer layers. This is in agreement with the observations of figure 2(b) where 111 lattice measurements indicated an expansion in the central layers.

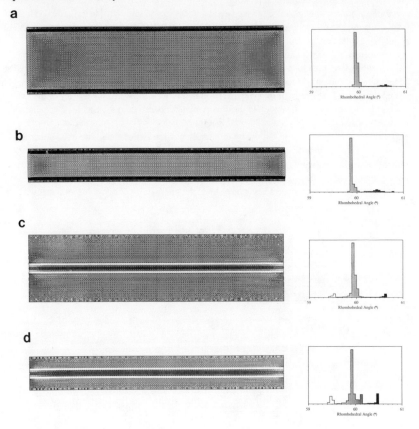

Figure 8. Plots of the distortion for the "sandwich" case, for crystals with same length a, but different thicknesses b. Case 1; Central layer is expanded by 1% ($\varepsilon_0 = 0.01$). (a) $b/a = 0.25$, and (b) $b/a = 0.125$. Outer layers are the same width, $(b - c)/2$, in each case, thus $c/b = 0.9$ and 0.8 respectively. Case 2; Central layer is contracted by 1% ($\varepsilon_0 = -0.01$). (c) $b/a = 0.25$, and (d) $b/a = 0.125$. Inner layers are the same width in each case, thus $c/b = 0.1$ and 0.2 respectively. Histograms of resulting rhombohedral distortions are plotted for each case, indicating the bias in all cases plotted toward rhombohedral angles less than 60°. In all plots the darkest areas have the highest rhombohedral angles, and the lightest areas have the lowest.

Figure 8 illustrates the propagation of the calculated rhombohedral distortion for sandwich–type plates, of aspect ratios $b/a = 0.125$ and 0.25. The plate diameter a is constant, but the plate thickness b is doubled in the latter case. The absolute widths of the minority sandwich layers, are held the same in each case, (ie. if $c \gg (b - c)$ then $c/b = 0.8$ and 0.9 respectively; or, if $c \ll b$ then $c/b = 0.2$ and 0.1 respectively). All α_{nm} up to $(N + M) = 13$ are included, giving 105 coefficients. Histograms of area fraction vs rhombohedral distortion are shown next to each figure.

Provided c is significantly greater than $(b - c)$, then the central portion of the crystal, which has a rhombohedral angle $\theta < 60°$, also forms the bulk of the crystal. However, if one layer type overwhelmingly dominates, then the distortions induced by the minority layer are too feeble to have any measurable impact on the bulk structural properties. Furthermore, when $b/a \geq 1$, relaxation further reduces the impact of the minority layer. This is expected to be the case for the K–FAU sample shown in figure 1(c), where the effect of the twinned region at the center would be negligible. As the platelet becomes thinner (ie b/a decreases), elastic relaxation diminishes and the volume fraction of the minority component (whose thickness is held constant) is increased. This can be seen by comparing figures 8(a) $(b/a = 0.125, c/b = 0.8)$ and figure 8(b) $(b/a = 0.25, c/b = 0.9)$. The rhombohedral distortion in the central region dominates in each case, the rhombohedral angle being less then $60°$. The effect is larger in the thinner crystallite (figure 8(a)).

The results are similar if the inner layer is the minority component. In this case the properties of the outer layers will dominate (figures 8(c) and 8(d). However, the shear strains (and hence the effective radii of curvature of the bend contours) will be opposite to those determined from figure 5. If the volume fractions of the inner and outer layers are comparable, then the crystal properties would be essentially bimodal. Splitting of powder X–ray diffraction peaks would, for example, be expected. Peaks may be split into more than two components due to the effects of elastic relaxation. Multiple peaks are seen in the rhombohedral angle histograms in figure 8.

In both variations of the 'sandwich' model, whether the minority contracting layer is the outer, or the inner layer, a relative strain of about 4% $(\varepsilon_0 = -0.04)$ when $b/a = 0.125$ and $(b - c)/b = 0.9$ (or $c/b = 0.1$ for the thinner inner layer), will produce a rhombohedral angle of $59.5°$ in the majority layer, and shear strain gradients of magnitude comparable to those implicated in the observed bend contours. Of course, all the parameters in the model interact; smaller strains being needed if the minority layer is thicker. However, it should be noted that even if Si/Al variations across the platelet had not been negligible, and indeed had ranged from 1.0 to ∞, they still would not have been enough to produce strains of the magnitude observed here [1,8,9].

It is important to note, however, that although the twin layers are the obvious suspects as sources of internal stress, the success of the model presented here is not contingent on any particular source for the internal stress. The twinning may be symptomatic of a more subtle stress–inducing variation within the platelet, such as cation concentration or location within the framework. Another possibility is the effect of framework termination at the surfaces. It is conceivable that hydroxyl terminations, or T–O–T cross linkages, put zeolite surfaces under a tensile stress, or "surface tension". This tension might be different on the 111, 110 and 100 surfaces. Such anisotropies in surface tension could also induce distortions, again, the strongest effects being observed in the thinnest platelets.

CONCLUSION

The possible presence of internal stress gradients explains why CSZ–1 platelets show the rhombohedral distortion, whereas related, similarly twinned materials such as CSZ–3 do not. CSZ–1 platelets are thinner than CSZ–3, whereas the width of the fault bands is the same in each case. There is less elastic relaxation of internal stresses in thinner platelets, and the distortions induced by the fault layers can be proportionately larger. Any distortion present in CSZ–3 is thus expected to be small, and probably within the limits of experimental error.

ACKNOWLEDGMENT

The authors would like to thank Dr. A. J. Jacobson of Exxon for his support and help during the course of this work.

REFERENCES

1. W. M. Meier and D. H. Olson, Atlas of Zeolite Structure Types, (Structure Commission of the International Zeolite Association; available from Polycrystal Book Service, Pittsburgh PA, 1978. Second edition in preparation, 1987)
2. J. V. Smith, Chem. Rev. in press, 1988.
3. D. W. Breck, Zeolite Molecular Sieves: Structure, Chemistry and Use (Wiley and Sons, London) 1974 (reprinted R. E. Krieger, Malabar FL, 1984).
4. R. M. Barrer, Hydrothermal Chemistry of Zeolites (Academic Press, London) 1982.
5. J. M. Thomas, Proc. Eighth. Int. Congr. Catalysis Vol. I (Verlag-Chemie) 31, 1984.
6. M. M. J. Treacy, J. M. Newsam, R. A. Beyerlein, M. E. Leonowicz and D. E. W. Vaughan, J. Chem. Soc., Chem. Commun., 1986, 1211.
7. M. M. J. Treacy, J. M. Newsam, D. E. W. Vaughan, R. A. Beyerlein, and S. B. Rice, in Proc. 1987 Analytical Electron Microscopy Workshop, San Fransisco CA (San Francisco Press, CA) in press, 1987.
8. J. V. Smith, Molecular Sieve Zeolites - I, edited by R. F. Gould (American Chemical Society; Washington, DC) Adv. Chem. Ser. No. 101, 1976, 109.
9. W. J. Mortier, Compilation of Extra-framework Sites in Zeolites, (Structure Commission of the International Zeolite Association; Butterworth, Guildford UK), 1982.
10. D.E.W. Vaughan and M. G. Barrett, U.S. Pat. 4,333,859, 1982.
11. M. G. Barrett and D. E. W. Vaughan, U.S. Pat., 4,309,313, 1982.
12. R. A. Beyerlein, J. M. Newsam, M. T. Melchior and H. Malone, J. Phys. Chem., accepted for publication, 1987.
13. J. M. Thomas, M. Audier and J. Klinowski, J. Chem. Soc., Chem. Commun., 1981,1221.
14. M. Audier, J. M. Thomas, J. Klinowski, D. A. Jefferson and L. A. Bursill, J. Phys. Chem., 86,1982, 581.
15. J. M. Thomas, S. Ramdas, G. R. Millward, J. Klinowski, M. Audier, J. Gonzalez-Calbet and C. A. Fyfe, J. Solid State Chem., 45, 1982, 368.
16 G. T. Kokotailo and J. Ciric, Molecular Sieve Zeolites - I, edited by R. F. Gould (American Chemical Society; Washington, DC) Adv. Chem. Ser. No. 101, 1976, 109.
17. P. B. Moore and J. V. Smith, Mineralog. Mag., 35, 1008, 1964.
18. D. W. Breck, reference 3, pp 56 - 58, 1974.
19. G. R. Millward, J. M. Thomas, S. Ramdas and M. T. Barlow, Proceedings of 6th International Zeolite Conference, edited by D. Olson and A. Bisio (Butterworths, Surrey) 1984, 793.
20. G. W. Brindley, in X–Ray Structures of Clay Minerals and their X–Ray Identification, Edited by G. W. Brindley and G. Brown, Min. Soc. (London), ch 2, 1980.
21 D. J. H. Cockayne, M. L. Jenkins and I. L. F. Ray, Philos. Mag., 24, 1971, 1383.
22. M. M. J. Treacy, J.M. Gibson and A. Howie, Phil. Mag. A, 51, 1985, 389.
23. S. P. Timoshenko and J. N. Goodier, Theory of Elasticity , (New York: McGraw–Hill), pp 58 – 262, 1970.
24. P. B. Hirsch, A. Howie, R. B. Nicholson, D. W. Pashley and M. J. Whelan, Electron Microscopy of Thin Crystals, (Krieger, New York) (1977), see chapters 8, 9 and 12.

COMPOSITION VARIATION BETWEEN Cu-Ni CRYSTALLITES USING TEM AND EDS

Pankaj K. Sinha and Timothy S. Cale
Department of Chemical, Bio and Materials Engineering
Arizona State University, Tempe, Arizona 85287

ABSTRACT

The distribution of nickel and copper between crystallites supported on high surface area silica (Cabosil HS5) has been studied using a Philips 400 series transmission electron microscope equipped with super twin lenses, a field emission gun and EDAX energy dispersive X-ray spectroscope. The same Ni-Cu catalysts have previously been characterized and used in cyclopropane hydrogenolysis kinetics. The good intercrystallite homogeneity found in a well reduced sample having 31% nominal copper validates a critical assumption made in these previous studies. Limited data on unreduced and partially reduced samples also highlight the need to carefully control the degree of reduction in kinetic and characterization studies.

INTRODUCTION

Multimetallic catalysts have received considerable attention recently [1-3]. Surface-bulk heterogeneity, a surface composition different from the nominal, has been predicted for many such systems [4,5], and is an important factor in the performance of a catalyst. A cherry-like model has been discussed by Sachtler and van Santen [1] to help visualize the phenomenon. According to this model, a highly enriched surface can form at low copper content in samples having large crystallites; however, a higher concentration is required for highly dispersed samples. Indeed, in a magnetic, adsorptive, and magneto-adsorptive study of the same Ni-Cu catalyst samples used in the present work, Cale and Richardson [6] concluded that mass balance limited surface segregation of Cu occurs. The assumption that the overall composition of each crystallite was the same formed the basis for estimating the surface composition of this set of samples. These surface composition estimates were used to interpret cyclopropane hydrogenolysis kinetics over these samples [7]. The areal rates of ring opening and fragmentation, as well as the selectivity to fragmentation, decrease with increasing surface fraction of copper. The kinetic results are consistent with an ensemble size of two for ring opening and three for fragmentation, if the distribution of metals on the surface is assumed random. The present work was undertaken to check the validity of assuming intercrystallite composition homogeneity, as well as to gain insight into the assumption of surface randomness.

Mat. Res. Soc. Symp. Proc. Vol. 111. ©1988 Materials Research Society

EXPERIMENTAL

Two samples having 12% (JA) and 31% (JC) copper concentrations were taken as representative of the series of catalysts having nominal Cu concentrations ranging from 0 to 40% relative to total metal [6]. All of the catalysts have about 25% metal and were prepared by homogeneous precipitation-deposition using urea, a technique known to yield catalysts having a fairly uniform crystallite size [8].

Sample JC-PR (intermediate reduction) was reduced at 400 °C for 4 hours in hydrogen. A 400 °C for 20 hours, 600 °C for 4 hours and 800 °C for 1 hour reduction was used for JC-R (thoroughly reduced). Each catalyst sample was dispersed on a titanium grid for transmission electron microscopy (TEM). A Philips 400 electron microscope equipped with super twin lenses, a field emission gun and an EDAX energy dispersive spectroscope (EDS) was used for elemental analyses. Samples were mounted on a liquid nitrogen cooled specimen holder and analyzed using a 10 nm probe size. All spectra were obtained by focussing at the edge of a particle, in order to avoid absorption effects. The copper fractions, with respect to total metal, were computed from the areas under the peaks corresponding to nickel and copper K-alpha lines in the EDS spectra. The high resolution imaging was performed on a JEOL-200CX.

RESULTS

The compositions determined by EDS for the samples match those reported by Cale and Richardson [6]. The reproducibility of composition between runs at the same point (data points 1-3 and 4-6 for JC-R in Table III) was within $\pm$ 3% (abs.) of the average concentrations at those points. The remaining data were obtained by probing different points of the respective samples.

Analysis of the JC-R sample supports the assumption of intercrystallite composition homogeneity used by Cale and Richardson [6,7]. The sample displayed the presence of individual crystallites which were distinct and stable under a focussed electron beam. It is estimated that one or two crystallites lie within each probed volume. As indicated in Table I, the variation in composition obtained by probing crystallites in different regions of the catalyst is about the same as the reproducibility of the measurements. Another feature of interest is the presence of Moire fringes. Horizontal lattice fringes in a relatively large crystallite are shown in Fig. 1. Moire fringes on a part of the crystallite may indicate an overlapped crystalline phase, probably a copper rich phase; however, these may also occur due to a faulted crystal.

Electron microscopy of JC-PR (see Fig. 2) did not reveal individual crystallites. A few of the darker spots at the edge of the support, marked P in the figure, were probed. Upon focussing the beam at one of these darker spots, it dispersed and subsequently was not visible. This instability has also been reported for triosmium clusters [9]. A second scan at the same location yielded a few percent lower copper concentration than the first scan. In addition, the copper concentration in this sample varied significantly over a set of measurements at several of the dark spots. Note the wide spread in copper concentration data for the JC-PR presented in Table II. Focussing the beam on a large black region, marked B, of the sample yields an average concentration of 27% (Sl. No. 4, JC-PR), close to the concentration predicted by other techniques.

The results of EDS analysis of the unreduced 12% copper sample (JA) indicate good spatial compositional uniformity. This is expected, considering the catalyst preparation method [8].

TABLE I				TABLE II				TABLE III			
JA				JC-PR				JC-R			
Sl No	Cu % metal	Sl No	Cu % metal	Sl No	Cu % metal	Sl No	Cu % metal	Sl No	Cu % metal	Sl No	Cu % metal
1	8.8	7	10.4	1	16.5	1	27.8	7	24.0	13	24.0
2	9.0	8	9.8	2	46.3	2	28.6	8	21.4	14	28.0
3	9.7	9	9.2	3	37.1	3	24.6	9	21.9	15	20.0
4	12.4	10	10.6	4	27.2	4	22.9	10	24.4	16	26.5
5	10.4	11	11.9	$\overline{x} = 31.8$		5	20.8	11	21.8	17	25.8
6	12.7	12	11.6	$\sigma_{n-1} = 12.8$		6	20.6	12	24.9	18	26.3

$$\overline{x} = 10.5$$
$$\sigma_{n-1} = 1.4$$

$$\overline{x} = 24.1$$
$$\sigma_{n-1} = 2.5$$

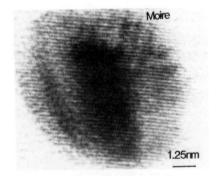
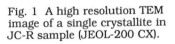

Fig. 1 A high resolution TEM image of a single crystallite in JC-R sample (JEOL-200 CX).

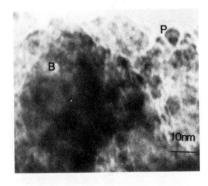

Fig. 2 A TEM image of JC-PR sample (Philips 400 ST/FEG).

CONCLUSIONS

EDS results on a representative Ni-Cu catalyst sample (JC-R) support the assumption of intercrystallite composition homogeneity which was used in the interpretation of cyclopropane hydrogenolysis activity/selectivity patterns [7]. In addition to more EDS experiments, HREM studies are planned in order to explain the Moire fringes and perhaps to test the surface randomness assumption.

Data on unreduced and partially reduced samples, though limited, suggest that the distribution of the two metals goes through a maximum in non-uniformity upon reduction when compared to the JC-R data. This was also suggested by Datye and Schwank [10] for the Ru-Au system. This suggests that compositional and kinetic data should be compared between samples having the same degree of reduction, as is reasonable. Note that JC-R had about the same degree of reduction as in the earlier studies [6,7]. The data also suggests that an intermediate reduction may not form stable crystallites. Instead, the two metals may cluster separately. This suggests that activity/selectivity studies of catalyst systems should be done on samples with the same degree of reduction. A systematic investigation of the distribution of a metal as a function of reduction may answer some important questions regarding the mechanisms of multimetallic crystallite formation. This would help explain catalyst performance trends.

ACKNOWLEDGEMENT

The authors acknowledge the NSF/ASU HREM facility at Arizona State University, where the TEM work was performed.

REFERENCES

1. W. M. H. Sachtler, and R. A. van Santen, Advances in Catalysis and Related Subjects, 26, (Academic Press, New York, 1977), p. 69.
2. G. Leclercq, Appl. Catal., 27(2), 299, (1986).
3. R. L. Moss, and R. Whalley, Advances in Catalysis and Related Subjects, 22, (Academic Press, New York, 1975), p. 115.
4. T. Hashizume, T. Sakurai, H. W. Pickerin, J. Physique, 47, 381, (1986).
5. Y. Takasu and H. Shimizu, J. catal., 29, 479 (1973).
6. T. S. Cale and J. T. Richardson, J. Catal., 79, 378 (1983).
7. T. S. Cale and J. T. Richardson, J. Catal., 94,289 (1985).
8. J. A. van Dillen, J. W. Geus, L. A. M. Hermans, and J. van der Meijden, Proc. Sixth International Congress on Catalysts, 677 (1976).
9. J. Schwank, L. F. Allard, M. Deeba and B. C. Gates, J. Catal., 84, 27 (1983).
10. A. K. Datye and J. Schwank, International Catalysis Congress in Berlin, 587 (1984).

THE EFFECT OF SUPPORT COMPOSITION ON PLATINUM CRYSTALLITE AGGLOMERATION IN OXYGEN

MICHAEL F. MORRISSEY, SHOKO KITAZUMI, AND ROBERT F. HICKS
University of California, Los Angeles, Dept. of Chemical
Engineering, 5531 Boelter Hall, Los Angeles, CA   90024

ABSTRACT

The effects of temperature of oxygen exposure on metal dispersion were determined for platinum deposited on Al_2O_3, ZrO_2, 12 wt% Y_2O_3-ZrO_2, and MgO. Platinum on zirconia and on magnesia showed higher thermal stability than platinum on alumina.

INTRODUCTION

Precious metals deposited on monolithic ceramic supports are used to catalytically oxidize hydrocarbons and CO (1-3). The most notable application is automotive exhaust emission control. These catalysts must operate in oxidizing or reducing atmospheres at temperatures between 500 and 1000 ^{O}C, where catalyst deactivation due to metal crystallite agglomeration is a problem. Many studies have been conducted on platinum crystallite agglomeration (5-13). However, most of this work has focussed on the agglomeration of Pt on alumina supports (4,6-11). Relatively few studies have examined the effect of different supports (5) or of different promoters (12,13).

In this report, we discuss the progress we have made in studying the effects on support composition on Pt crystallite agglomeration. Our main interest has been the interaction of Pt with zirconia and yttria-stabilized zirconia. The reason for investigating Y_2O_3-ZrO_2 is to determine whether the oxygen vacancies present in the oxide interact with the Pt and stabilize higher metal dispersions (14).

EXPERIMENT

The catalyst supports used in this study were: Degussa gamma-Al_2O_3 (100 m^2/g); Morton Thiokol alpha-Al_2O_3 (160 m^2/g); Morton Thiokol alpha-Al_2O_3 calcined at 1050°C for 48 hours (60 m^2/g); Degussa experimental ZrO_2 (42 m^2/g); Zircar ZrO_2 calcined at 1050°C for 48 hrs (4 m^2/g); Zircar 12 wt% Y_2O_3/ZrO_2 calcined at 1050°C for 48 hrs (4 m$_2$/g); and Mallincrodt light MgO calcined at 1050°C for 48 hrs (3 m^2/g). The platinum was deposited on these supports by ion exchange and by incipient wetness impregnation of either H_2PtCl_6 or $Pt(NH_3)_4Cl_2$. After the deposition step, the samples were dried at 105°C overnight, pelletized to 32-60 mesh size, and stored in a desiccator. The platinum weight loadings were determined by inductively coupled plasma analysis. The method of preparation of each catalyst sample is summarized in Table 1.

Mat. Res. Soc. Symp. Proc. Vol. 111. ©1988 Materials Research Society

Table 1.

Sample Designation and Method of Preparation

Sample	Support	Salt Solution	Method	Platinum Loading (wt%)
$Pt/Al_2O_3(A)$	Degussa Al_2O_3	$Pt(NH_3)_4Cl_2$	Incip. Wetness	0.5
$Pt/Al_2O_3(B)$	M.T. Al_2O_3	H_2PtCl_6	Ion Exchange	0.8
$Pt/Al_2O_3(C)$	calcined Al_2O_3	H_2PtCl_6	Incip. Wetness	0.4
$Pt/ZrO_2(A)$	Degussa ZrO_2	$Pt(NH_3)_4Cl_2$ NH_4OH	Ion Exchange	0.3
$Pt/ZrO_2(B)$	Degussa ZrO_2	$Pt(NH_3)_4Cl_2$	Incip. Wetness	0.5
$Pt/ZrO_2(C)$	calcined Zircar ZrO_2	H_2PtCl_6	Incip. Wetness	0.5
$Pt/YZrO_2$	calcined Zircar 12wt% Y_2O_3/ZrO_2	H_2PtCl_6	Incip. Wetness	0.5
Pt/MgO	calcined MgO	H_2PtCl_6	Incip. Wetness	0.5

The thermal stabilities of the catalysts were investigated by measuring the Pt dispersion after heating in oxygen. The catalysts were heated in pure O_2 for 2 hours to the desired temperature, oxidized for 2 hours at this temperature, reduced in pure H_2 for one hour at 300°C, and then cooled in 2×10^{-5} torr vacuum for six hours to room temperature. Once at room temperature, the platinum dispersion, D_{Pt} = exposed metal/total metal, was measured by H_2 titration of preadsorbed O_2, according to the method of Boudart (15):

$$Pt-O_s + 3/2 \; H_2 \quad = \quad Pt-H_s + H_2O \qquad (1)$$

A volumetric chemisorption apparatus was used for this purpose. The oxidation, reduction, and dispersion measurement sequence was repeated at successively higher oxidation temperatures until D_{Pt} decreased to a low value.

RESULTS AND DISCUSSION

Shown in Figure 1 is the dependence of platinum dispersion on oxidation temperature for 0.8 wt% $Pt/Al_2O_3(B)$ and 0.3 wt% $Pt/ZrO_2(A)$. Both of these catalysts were prepared by ion exchange, so that the initial dispersion of Pt is 100%.

For $Pt/Al_2O_3(B)$, the dispersion begins decreasing at 600°C, and falls to 7.5% after an 800°C oxidation. A TEM micrograph of $Pt/Al_2O_3(B)$ after 800°C oxidation reveals metal crystallites of approximately 100 angstroms in diameter. These results for $Pt/Al_2O_3(B)$ are in agreement with previous work, which showed that Pt crystallites rapidly agglomerate to large particles when exposed to O_2 above 500°C (9-11,13).

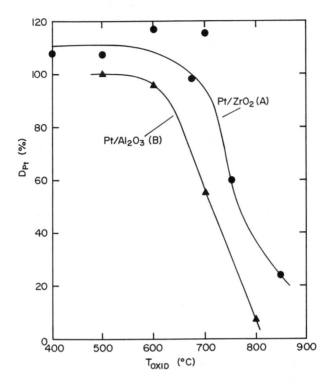

Figure 1. The dependence of the platinum dispersion on the oxidation
temperature for 0.8wt% Pt/Al_2O_3 (B) and 0.3wt% Pt/ZrO_2 (A).

The thermal stability of Pt/ZrO_2(A) differs markedly
from that of Pt/Al_2O_3(B). The initial dispersion of
Pt/ZrO_2(A) shown in Figure 1 is anomalously high, most likely
because of H_2 adsorption by the support. Nevertheless, it
is not until 700°C that the percentage of metal exposed on
Pt/ZrO_2(A) starts to decrease. This is roughly 100°C higher
than the corresponding decrease observed for Pt/Al_2O_3(B).
The decline in the dispersion of Pt on Pt/ZrO_2(A) at
temperatures above 700°C may or may not be due to platinum
crystallite agglomeration. Simultaneous with the decline in
D_{Pt} is a loss of ZrO_2 surface area due to sintering. The
dependence of ZrO_2 surface area on temperature is shown in
Figure 2. During sintering, the metal crystallites may
become buried in the oxide, and thereby lower the value of
D_{Pt}. TEM micrographs of the Pt/ZrO_2(A) samples are being
prepared to determine whether it is Pt particle growth and/or
ZrO_2 sintering which causes the decline in metal surface
area.

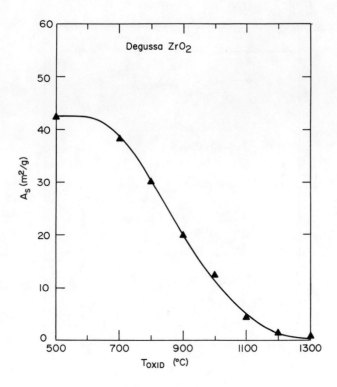

Figure 2. The dependence of the tap density and the BET surface area on the oxidation temperature for the Degussa zirconia support.

    The effects of oxidation temperature on catalysts with lower metal dispersions are summarized in Table 2.   The supports used for these catalysts were calcined at 1050°C for 48 hours and have low surface areas as described above. Perusal of Table 2 reveals no differences in the thermal stability of platinum deposited on calcined Al_2O_3, ZrO_2, or Y_2O_3-stabilized ZrO_2. The Pt dispersions of these catalysts decline as the temperature is raised above 500°C, falling to a minimum value of about 15%.   Platinum deposited on MgO shows improved low temperature stability, exhibiting a constant dispersion of 30% up to 700°C. This finding is in agreement with the earlier study of Wanke (5).   However, above 700°C the dispersion of Pt on Pt/MgO decreases to 7%. Thus, there appears to be little effect of support composition on the thermal stability of Pt at high temperatures on supports which have been previously calcined at 1050°C.

Table 2.

Effect of Catalyst Support and Oxidation Temperature on Platinum Dispersion

Catalyst	D_{Pt} (%) after O_2 Exposure at:					
	500°C	600°C	700°C	800°C	900°C	1000°C
$Pt/Al_2O_3(C)$	26	--	21	--	12	17
$Pt/ZrO_2(C)$	55	--	25	--	18	16
$Pt/YZrO_2$	32	19	12	12	15	15
Pt/MgO	29	34	30	22	9	7

CONCLUSIONS

These experiments show that the stability of Pt crystallites in O_2 is influenced by the composition of the support. However, the pretreatment history of the support also has a large effect which obscures the dependence on support composition. Further experiments are underway to better define the role of support composition and structure on the rate of platinum crystallite agglomeration at high temperatures.

REFERENCES

1. J.P. Kesselring, W.V. Krill, H.L. Atkins, R.M. Kendall, R.M. Kelley, and J.T. Kelley, Design Criteria for Stationary Source Catalytic Combustion Systems (EPA-600/7-79-181, 1979).

2. J.H. Conklin, D.M. Sowards, and J.H. Kroehling, in: Kirk Othmer Encyclopedia of Chemical Technology, **Vol. 9**, 511 (Wiley, New York, 1980).

3. R. Prasad, L.A. Kennedy, and E. Ruckenstein, Catal. Rev. -Sci. Eng. **26**, 1 (1984).

4. S.E. Wanke, and P.C. Flynn, Catal. Rev. -Sci. Eng. **12**, 93 (1974).

5. S.E. Wanke, in: Sintering and Heterogeneous Catalysis, Eds., G.C. Kuczynski, A.E. Miller, and G.A. Sargent, 223, (Plenum, New York, 1984).

6. P. Wynblatt, Acta Metall. **24**, 1175 (1976).

7. E. Ruckenstein and D.B. Dadyburjor, J. Catalysis **48**, 73 (1977).

8. Y.F. Chu and E. Ruckenstein, J. Catalysis **55**, 281 (1987).

9. F.M. Dautzenberg and H.B.M. Wolters, J. Catalysis **51**, 26 (1978).

10. H.C. Yao, P. Wynblatt, M. Sieg, and H.K. Plummer, in: Sintering Processes, Ed., G.C. Kuczynski, 561, (Plenum, New York, 1980).

11. W.G. Rothschild, H.C. Yao, and H.K. Plummer, Langmuir **2**, 588 (1986).

12. H.C. Yao, Appl. Surface Sci. **19**, 398 (1984).

13. Y.F.Y. Yao, J. Catalysis **87**, 152 (1984).

14. M.G. Sanchez and J.L. Gazquez, J. Catalysis **104**, 120 (1987).

15. J.E. Benson and M. Boudart, J. Catalysis **4**, 704 (1965).

STRUCTURAL TRANSFORMATIONS IN PLATINUM SUPPORTED ON
γ-ALUMINA STUDIED BY X-RAY ABSORPTION SPECTROSCOPY

H.J. ROBOTA, M.J. COHN, A.Z. RINGWELSKI, AND R.A. EADES
Allied-Signal Engineered Materials Research Center, Des Plaines, IL,
60017-5016

ABSTRACT

Pt particles in a highly dispersed Pt/γ-Al$_2$O$_3$ catalyst have been found to undergo ordering following the removal of chemisorbed oxygen with hydrogen at room temperature. The very small Pt particles are found to be disordered following an initial reduction at 500°C. Treating the ordered phase of Pt at 500°C in hydrogen is found to restore the disordered phase.

INTRODUCTION

The structure of very small metal particles present in supported catalysts is believed to play an important role in determining catalyst activity and selectivity. In particular, Pt supported on either alumina or silica has been extensively studied by a multitude of catalytic and physical methods. Previous x-ray absorption studies have been conducted from a variety of perspectives. General studies have compared particle properties for Pt supported on alumina with Pt supported on silica [1]. Other studies have been concerned with the structure of supported Pt catalysts for specific applications, such as detoxification of automotive exhaust [2]. More recently, the influence of adsorbed molecules [3] and even the intimate contact between the Pt particle and its support [4] have begun to be addressed.

In our recent studies of room temperature reactions of gases adsorbed on small Pt particles, we have observed dramatic changes in the structure of the particles by x-ray absorption spectroscopy (XAS). Most models of such very small particles assume either a perfectly crystalline or pseudo crystalline morphology. If disorder is present, it is assumed to be in the form of slight relaxation of surface atoms from the normal interatomic distances. We believe the changes we observe are caused by ordering of initially severely disordered Pt particles. Ordering of the small particles appears to occur when chemisorbed oxygen is reacted off by hydrogen at room temperature. These ordered particles can be returned to the disordered state by heating in hydrogen to 500°C.

EXPERIMENTAL

The catalyst contained 0.38% Pt supported on γ-Al$_2$O$_3$ prepared by aqueous impregnation of a H$_2$PtCl$_6$ precursor. The impregnated Al$_2$O$_3$ was dried in flowing air, oxidized at 525°C and reduced at 565°C. Following this initial reduction, the catalyst was passivated and stored under N$_2$. Three separate methods were used to determine catalyst dispersion: pulse titration with CO; desorption of strongly held hydrogen; and titration of chemisorbed oxygen. Each method yields results consistent with a catalyst of unit dispersion.

XAS measurements were made using fluorescence detection on beam line X-18B at the National Synchrotron Light Source located at Brookhaven National Laboratory. Spectra were collected at -185°C following a series of gas treatments. Initially, the catalyst was reduced in situ in flowing H$_2$ at 500°C, purged with He and cooled to -185°C for measurement. Next, the catalyst was warmed to 28°C and exposed to about 250 cm^3 of flowing dry

air in a one-minute period and again purged with He and cooled to -185°C.
Chemisorbed oxygen was removed from the Pt surface by warming to 28°C in He
followed by a 250cc dose of flowing H_2 for about one minute followed by
another He purge and measurement. A final measurement was made after again
heating the catalyst to 500°C in H_2. The final measurement was made at
-185°C in flowing H_2.

RESULTS AND DISCUSSION

The normalized extended fine structure, designated as χ, was isolated
using a procedure similar to that described by Cook and Sayers [5]. Figure
1 shows the χs observed at three stages in the sequence of treatments with
the corresponding k^3 weighted Fourier transforms shown in Figure 2. (k
is the wave vector magnitude of the outgoing photoelectron created in the
core absorption process.) Following the initial reduction and He purge,
weak EXAFS characteristic of Pt-Pt scattering pairs is detected. The
signal amplitude merges with the background at about $k=10A^{-1}$. The corresponding Fourier transform (FT) reveals a single feature consistent with
the Pt-Pt pairs expected for the average Pt coordination environment in a
small, fully reduced catalyst particle. Following removal of chemisorbed
oxygen by a short exposure to hydrogen at 28°C, a dramatic qualitative
difference is immediately detected in χ. Clear EXAFS oscillations are
now evident over a much wider energy range. Also, oscillations at lower
k values have slightly greater amplitude. The Fourier transform is also
substantially different than observed following the initial reduction.
The Pt-Pt nearest neighbor peak has grown considerably in amplitude. This
is to be expected since the oscillatory signal now extends to much higher

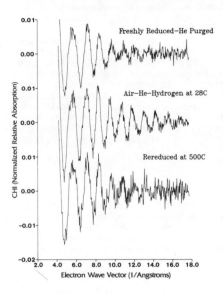

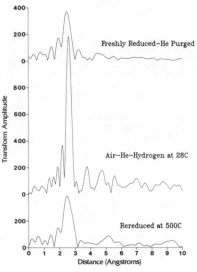

Figure 1. The normalized EXAFS
obtained following the indicated
gas treatments.

Figure 2. The radial structure
functions obtained by performing k^3
Fourier transforms of the EXAFS.

k values and the k^3 weighting amplifies this difference. Also, additional new coordination features are now visible at distances in accord with the structure of metallic Pt. Finally, following another treatment in H_2 at 500°C, a signal and FT very similar to the initial measurement are observed. Again, the χ amplitude is no longer visible beyond about $k=10A^{-1}$ and the FT exhibits only a single feature related to nearest neighbor Pt-Pt pairs.

Before considering quantitative evaluation of these data, some qualitative discussion is warranted. The χ observed following initial reduction in H_2 at 500°C is in good agreement with previous _in situ_ studies of highly dispersed, supported Pt [1,3,4]. These earlier studies used catalysts with Pt loadings near 1% and measured dispersions near unity. The χ amplitude observed for the simply reduced catalyst is somewhat smaller than previous measurements with complete damping of the signal at a lower value of k, consistent with either Pt particles containing fewer atoms or particles of a given size which are more disordered. Following removal of chemisorbed oxygen by H_2 at 28°C, χ is very similar to previous measurements. Coordination shells as distant as the fourth nearest neighbors become clearly visible following this simple gas treatment. If this signal were observed in an isolated experiment, the immediate qualitative conclusion would be that Pt sintering has occurred, producing particles of an intermediate size. However, the conditions under which the transformation has taken place are extremely mild and the catalyst is known to exhibit unit dispersion following high temperature hydrogen reduction. This leads us to suggest that the catalyst initially contains Pt particles which are extremely disordered. Removal of the chemisorbed oxygen at 28°C by H_2 results in the formation of Pt particles with a more crystalline, ordered morphology. This suggestion is supported by the results of the final high temperature treatment in H_2. Assume for purposes of discussion that sintering was the cause of the observed differences in particle structure. It follows that either further sintering or a more complete ordering of these large particles should result from high temperature treatment. Actually quite the opposite is observed. Rather than retaining the crystalline structure induced at 28°C, the particles revert to a morphology consistent with that observed following the initial reduction. Redispersion of Pt particles by treatment in hydrogen has never been observed, so it is unlikely that a change in particle size is responsible for the changes in the observed structure.

An attractive explanation for such a morphological change involves interaction of the reduced Pt atoms in the very small particles with the support. If Pt atoms spread over the support surface following high temperature treatment to produce raft-like structures, it is conceivable that this interaction could be broken by chemisorption of oxygen. Then, when the adsorbed oxygen is removed at 28°C by H_2 treatment, insufficient thermal energy is available to overcome a barrier to forming the raft-like structure. As a result, the Pt atoms might assume a more crystalline structure. Finally, by heating these small crystalline particles to 500°C, the raft-like morphology could be recovered.

Unfortunately, this scenario is not supported by experimental observations. In such a raft-like morphology, the average Pt atom would be forced to interact with several oxygen ions in its nearest neighbor environment. This additional coordination should be observable in the experimental χ. Recently, exactly this type of interaction has been claimed for Pt and Rh supported on a variety of substrates [4,6]. The method used to reveal the weaker Pt-O interaction in the presence of the dominant Pt-Pt coordination relies on phase and amplitude correction of χ for the Pt-Pt pair prior to Fourier transformation. This method has been applied in the study of these catalysts. No evidence for coordination of Pt by any species other than Pt is found in the fully reduced catalyst. Following exposure of the catalyst to air, the chemisorbed oxygen is clearly visible in addition to the Pt-Pt coordination.

It is generally assumed that EXAFS can be analyzed within the assumption of small disorder. As long as the disorder is small, its effect can be included as an exponential damping factor similar to the Debye-Waller term found in x-ray scattering. However, for cases with greater disorder, the simple damping description is no longer valid [7,8]. Rather, an explicit analysis shows that a disorder related phase shift arises in addition to amplitude reduction greater than predicted by the exponential term. It has been shown that this disorder induced phase shift can be observed by superimposing the amplitude and imaginary components of a phase and amplitude corrected Fourier transform. In the absence of residual phase terms, the imaginary component will peak positively and symmetrically in coincidence with the amplitude maximum. An asymmetric imaginary component is found when the disorder exceeds the small disorder limit [9]. Exactly this behavior is observed when analyzing the data for this catalyst. Following either the initial or final high temperature H_2 treatment, the imaginary component of the FT is sharply shifted in maximum from the amplitude maximum with severe asymmetry in the shape. Following the room temperature removal of chemisorbed oxygen, where the more crystalline Pt FT is observed, the imaginary component is nearly perfectly in coincidence with the amplitude maximum and is quite symmetric about the peak. These observations are qualitatively in accord with a transformation from a disordered phase to a crystalline ordered phase and back again.

Quantitative analysis of EXAFS data when a material is suspected of being severely disordered is problematic. We have chosen to use analysis procedures which apply for the small disorder limit since extensive modeling of the disorder has not been attempted and the data do not lend themselves to alternative possible analyses. Table I gives the analyzed bond lengths, coordination numbers, and simple relative disorder parameter obtained by curve fitting the isolated nearest neighbor Pt-Pt signal by Fourier filtering. Even under ideal conditions, these parameters are not strictly linearly independent [10]. When severe disorder is present, coupling of parameters is even more severe. Normally, one expects some coupling of the bond length with E_0 and coupling of the amplitude with the disorder term.

These couplings become even more severe when additional phase shifts and amplitude reductions are present [8]. Consider first the apparent Pt-Pt bond lengths. Following high temperature H_2 treatment, the bond distances are considerably shorter than the 2.775Å distance in Pt metal. Such contracted distances have been previously observed and attributed to a real interatomic distance. In light of the severe disorder present in the particles under these conditions, it appears far more likely that a disorder phase shift induces an artifactual bond contraction. This may even still be responsible for the short distance found in the more ordered crystalline Pt. The highly disordered state of the particles is clearly evident even in the relative disorder term. The values observed for this catalyst are substantially larger than values near 0.003 found in other studies. When disorder is sufficiently large to produce additional phase terms, damping of the scattering amplitude beyond that predicted by the Debye-Waller term will also be present. This will result in coordination numbers which are artificially large. A coordination number of near 9 is completely inconsistent with particles exhibiting unit dispersion as measured for this catalyst, even when the conventional error estimate of 20% is included. In order to account for the added amplitude damping, a much larger Debye-Waller factor must be used. In order to compensate for this excess damping at low k values, the coordination numbers must be increased. Thus, using coordination numbers to estimate particle sizes can be of questionable validity unless it can be guaranteed that a small disorder approximation can be used.

CONCLUSIONS

Extremely small Pt particles supported on γ-Al$_2$O$_3$ undergo reversible disorder-order transformations in response to gas treatments and temperature. Since high temperature treatment in H$_2$ produces disordered particles, it appears that the disordered form is thermodynamically preferred. No evidence for enhanced interaction with the substrate to produce raft-like structures in the disordered form is observed.

Table I

Catalyst Treatment	Coordination Number	Bond Length (Angstroms)	Relative Disorder
Freshly reduced-He Purged	9.73	2.692	0.00667
Air-He-H$_2$ at 28°C	9.27	2.742	0.00311
Rereduced at 500°C	9.83	2.718	0.00486

REFERENCES

1. G.H. Via, J.H. Sinfelt, and F.W. Lytle, J. Chem. Phys. 71, 690 (1979)

2. R.W. Broach, J.P. Hickey, and G.R. Lester, presented in the Symposium on Advanced Catalyst Characterization at the Annual Meeting of the Materials Research Society, November 1982; H.J. Robota, R.W. Broach, J.W.A. Sachtler, and S.A. Bradley, Ultramicroscopy, 22, 149 (1987)

3. F.W. Lytle, R.B. Greegor, E.C. Marques, D.R. Sandstrom, G.H. Via, and J.H. Sinfelt, J. Catal. 95, 546 (1985)

4. D.C. Koningsberger and D.E. Sayers, Solid State Ionics 16, 23 (1985)

5. J. W. Cook and D.E. Sayers, J. Appl. Phys. 52, 5024 (1981)

6. J.B.A.D. van Zon, D.C. Koningsberger, H.F.J. van't Blik, and D.E. Sayers, J. Chem. Phys. 82, 5742 (1985); D.C. Koningsberger, J.H.A. Martens, R. Prins, D.R. Short and D.E. Sayers, J. Phys. Chem. 90, 3047 (1986)

7. P. Eisenberger and G.S. Brown, Sol. Stat. Comm. 29, 481 (1979)

8. G. Bunker, Nucl. Instrum. Methods 207, 437 (1983)

9. E.D. Crozier and A.J. Seary, Can. J. Phys. 59, 876 (1981)

10. P.A. Lee, P.H. Citrin, P. Eisenberger, and B.M. Kincaid, Rev. Mod. Phys. 53, 769 (1981)

THE REACTIONS OF C_2H_2 AND CH_3C_2H ON Ag POWDER

PAUL B. DORAIN* AND JOSEPH E. BOGGIO**
*Amherst College, Amherst, MA 01002
**Fairfield University, Fairfield, CT 06430

ABSTRACT

Ag powder is activated by pulsing it with NO_2 gas which forms fresh Ag microclusters. This powder is then exposed to subsequent pulses of C_2H_2 or CH_3C_2H (3.7% in N_2). The surface enhanced Raman scattering (SERS) spectra show dramatic intensity variations due to rapid changes in adatom concentration. Normalization of these time-dependent SERS spectra to the background scattering intensity, which is proportional to the adatom concentration, provides spectra which represent adsorbate coverage if major surface reconstruction does not occur. The temporal development of the SERS spectra of C_2H_2 shows rapid degradation with no evidence for adsorbed species. In contrast, propyne reacts more slowly, as evidenced by the behavior of the intensity at $1980 cm^{-1}$ due to adsorbed -C_2-. The reactions observed are in accord with the models developed by Barteau and Madix[1] and Vohs, Carney, and Barteau[2]. Exposure to both alkynes results in the appearance of SERS active NO, an adsorbant not previously observed at room temperature. Ellipsometric measurements are consistent with the presence of a carbon overlayer, which may stabilize the NO and render the system inert to further chemical reaction.

INTRODUCTION

The reactions of acetylene on oxygen-covered Ag metal have been elucidated in the last few years using ultra high vacuum (UHV) spectroscopic techniques. A review of Ag surface chemistry has been published recently[3]. At low temperatures, the surface oxygen abstracts a H from C_2H_2 to form adsorbed HC_2 and OH. As the temperature is increased, the adsorbates recombine to form C_2H_2, which is desorbed, and C_2 species, which form surface C matrices upon further heating.

Previous studies of the SERS spectra of acetylene adsorbed on Ag are limited to pure Ag surfaces in UHV systems. The spectra of C_2H_2 on cold evaporated Ag films at 120K are characterized by peaks at $171(\nu_{Ag-C_2H_2})$, $635(\nu_4, \Pi_g)$, $756(\nu_5, \Pi_u)$, $1934(\nu_2, \Sigma_g^+)$, $3245(\nu_3, \Sigma_u^+)$, and $3317(\nu_1, \Sigma_g^+) cm^{-1}$[4]. The strongest characteristic, the peak at $\sim 1934 cm^{-1}$, is assigned to the $C{\equiv}C$ stretching mode. In gaseous C_2H_2, this mode occurs at $1974 cm^{-1}$.

Recently, the reaction mechanism of NO_2/N_2O_4 gas pulses on oxygen-covered Ag metal powder has been reported[5]. The oxygen-covered surface reacts with NO_2 to form $AgNO_3$, and the Ag metal forms $AgNO_2$. As time evolves, the $AgNO_2$ undergoes an autocatalytic reaction to form Ag, $AgNO_3$, and NO gas. The Ag imbedded in the $AgNO_3$ migrates to sites where Ag_n microclusters are formed. These microstructures are a necessary part of the SERS enhancement mechanism. They are also highly reactive towards additional pulses of gases, even though they are imbedded in a $AgNO_3/AgNO_2$ matrix. This paper reports the results of the reactions of C_2H_2 and CH_3C_2H gas pulses with Ag_n microstructures prepared by exposing a Ag surface to a pulse of NO_2/N_2O_4 gas at room temperature. It is shown that: 1) the reactions occur rapidly and are in accord with the results obtained under UHV conditions, 2) the alkynes form additional Ag microstructures which augment the intensity of the SERS signal. 3) the intensity of the SERS background increases several fold, indicative of a corresponding increase in the surface concentration of adatoms. The implication is that the background signal may be used to monitor the contribution of the charge transfer mechanisms to the total SERS intensity.

EXPERIMENTAL

The SERS experimental apparatus, described previously[5], permits the measurement of Raman scattering spectra over a range of 100-2600cm^{-1} in 1s intervals with a S/N of $\sim$ 100. Gas pulses are typically 5s wide and the volume is adjusted to correspond to an exposure of 5L for the Ag powder used (0.6m^2/gr). No attempt was made to pre-clean the Ag powder surface, which is exposed to ambient conditions and presumed to be covered with oxygen and carbon compounds.

The ellipsometric measurements are also described earlier[5]. Evaporated Ag mirrors, $\sim$ 1000A thick, were exposed to NO_2/N_2O_4 prepared from Cu and concentrated HNO_3. The gases were dried by passage over $CaSO_4$ and carried into the ellipsometric cell by N_2 gas. A $AgNO_3/Ag$ matrix $\sim$ 80A thick was thereby generated. Propynyl lithium was reacted with water and the resulting propyne gas was dried and carried into the cell by N_2 gas.

RESULTS

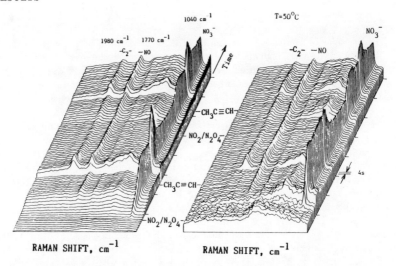

RAMAN SHIFT, cm^{-1} RAMAN SHIFT, cm^{-1}

Figure 1: a) SERS spectrum using a sequence of NO_2/N_2O_4 followed by $CH_3C\equiv CH$. λ=514nm, P=10mw, and Δt=4s. No signal messaging. b) Spectrum after normalization to the background intensity at the highest frequency shift shown for each frame.

The time-dependent SERS spectra of propyne on Ag powder are shown in Figures 1a and 1b. Figure 1a is the spectrum as recorded with an optical multichannel detector. Figure 1b shows the same data, but each frame is normalized to the average photon count at $\sim 2100\text{cm}^{-1}$, just beyond the portion of the spectrum arising from the $C\equiv C$ vibrational mode at 1900cm^{-1}.

Consider the intensity of the peak at 1040cm^{-1} due to the symmetric stretch of NO_3^- adsorbed on Ag[5]. In the first step, NO_2/N_2O_4 gas forms $AgNO_2$ which covers the Ag surface. The $AgNO_2$ undergoes autocatalytic decomposition into $AgNO_3$, Ag, and NO. As this process occurs, Ag microclusters are formed, causing an increase in surface "roughness" and a corresponding increase in the the SERS signal. There are two primary mechanisms that control the SERS enhancement. One is the electromagnetic enhancement, which relies upon the presence of Ag particles about 20nm in radius[6]. The other mechanism is due to a charge transfer process which depends upon the formation of Ag adatoms, or clusters of a few Ag atoms, on the surface. The SERS enhancement factor is $\sim 10^4$ for the electromagnetic mechanism and ~ 100 for the charge transfer mechanism[7].

Figure 1a shows that the background signal increases after the addition of NO_2. It is generally accepted that this scattering is due to electron-hole recombination processes, similar to those of the charge transfer enhancement mechanism, that occur when a metal is irradiated with laser light[8]. The intensity of the background may serve as a measure of the contribution of the charge transfer mechanism to the overall SERS enhancement. Thus, if the overall gain is

$$G(t) = G_{EM}(\omega_L,\omega)G_{CT}(\omega_L,\omega)G_{other}(\omega_L,\omega)\Theta(t),$$

then

$$G(t)/G_{CT}(\omega_L,\omega) = G_{EM}(\omega_L,\omega)G_{other}(\omega_L,\omega)\Theta(t)$$

where $G_{EM}(\omega_L,\omega)$ is the electromagnetic enhancement factor, $G_{CT}(\omega_L,\omega)$ is the adatom enhancement factor, $G_{other}(\omega_L,\omega)$ is any other enhancement mechanism, and $\Theta(t)$ is the coverage[9]. If the background for a given spectral frame is measured at a wavelength away from any vibrational peaks, and if this background is used to normalize the rest of that spectral frame, the result is a SERS spectrum arising from electromagnetic enhancement alone. Figure 1b is an example of such a spectrum.

This analysis is true only if the heat of reaction does not reconstruct the Ag surface, destroying the 20nm scale "roughness". Figures 2a and 2b illustrate the effect of surface reconstruction on the SERS spectra with a pulse sequence of N_2O/N_2O_4 followed by C_2H_2. Although the overall spectrum (Figure 2a) shows dramatic changes in the intensties of both the background and the peaks, the electromagnetic spectrum (Figure 2b) is nearly featureless. Thus, it appears that the intensity fluctuations are largely reflections of the transient concentration of adatoms.

The reactions of $CH_3C\equiv CH$ with Ag have been studied extensively[2]. An oxygen-covered Ag metal surface is a strong base that not only abstracts the acidic H to form $CH_3C\equiv C(a)$, but also forms the propargyl radical, $CH_2C\equiv CH(a)$, and OH(a). At 300K, these molecules dissociate on the surface to form adsorbed C and H. The H(a) reacts with OH(a) to form H_2O.

Figure 1b shows SERS spectra obtained as these reactions occur on Ag microclusters. After the initial pulse of NO_2/N_2O_4, the rapid growth of the NO_3^- peak at 1040cm^{-1} occurs, as well as a slow growth of a peak at 1771cm^{-1}. Upon exposure of this surface to a pulse of propyne, a strong peak appears at 1979cm^{-1}. This peak is assigned to the stretching mode of C≡C[4]. A strong increase in the nitrate and carbon signals from ∼ 1300cm⁻¹–1500cm⁻¹ also occurs, indicating that some of the $CH_3C\equiv C$ degrades. Finally, the peak at 1771cm⁻¹ has grown strong and remains with a nearly constant intensity while the intensities of the C≡C and NO_3^- peaks decrease.

The following sequence of reactions is in agreement with those observed under UHV conditions and can provide an explanation of the spectra shown in Figure 1. These reactions occur on the same Ag microclusters responsible for the SERS enhancement. Upon exposure to CH_3CCH, two reactions occur immediately:

$$CH_3CCH(g) + O(a) \longrightarrow CH_3CCAg(a) + OH(a)$$

and

$$CH_3CCH(g) + AgNO_3 \longrightarrow CH_3CCAg(a) + H_2O + NO_2.$$

The second reaction explains the decrease in the intensity of the SERS signal due to NO_3^- upon initial addition of CH_3CCH. The growth of the CH_3CCAg peak at 1979cm⁻¹ is rapid, but because the CH_3CCAg slowly decomposes to form adsorbed C and H, this C≡C bond signal nearly disappears before the addition of a second pulse of NO_2/N_2O_4 causes a regeneration of the surface.

The peak at 1771cm⁻¹ is assigned to a nitrosyl group by comparison of known frequency shifts of nitrosyl-metal compounds. It is apparently stabilized by the presence of the acetylide decomposition products. The peak does not appear unless acetylide is present. Though NO is produced during the autocatalytic decomposition of $AgNO_2$–the reaction that initially forms the Ag microclusters–no NO(a) was ever observed[5]. With each pulse of CH_3CCH, this peak shifts by ∼ +5cm⁻¹ then returns to 1771cm⁻¹.

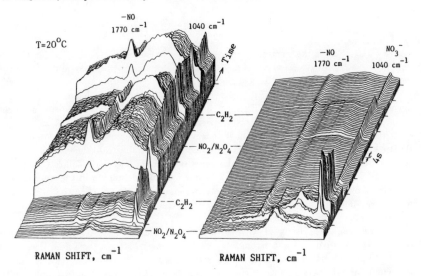

Figure 2: a) SERS spectrum obtained during an experiment identical to that of Figure 1 except that C_2H_2 was used in place of $CH_3C\equiv CH$. b) The same spectrum after normalization.

Figure 2 shows spectra obtained for the reaction of acetylene with Ag microclusters produced, again, by a pulse of NO_2 gas on Ag powder. The reactions are noteably different from those of propyne. No peak arising from the C≡C moeity is observed. Copious adatoms are produced photochemically as $AgNO_3$ reacts with both acidic hydrogens on acetylene. Also, acetylene reacts with the oxide-covered Ag forming adsorbed OH and carbon. The only peak in Figure 2 resulting from the acetylene reaction is the NO peak at $1771cm^{-1}$. As before, this peak shifts in frequency when acetylene is added but, in this case, the shift is $+25cm^{-1}$. It appears that the addition of acetylene causes a positive polarization of adsorbed NO. The reason for the change in surface charge is not clear, but perhaps depends upon the presence of an adsorbed C overlayer, rapidly produced in the chemical reaction.

Spectra of reactions involving deuterated acetylene and propyne showed no isotopic shifts except for a CH peak at $\sim 3000cm^{-1}$, which is commonly observed.

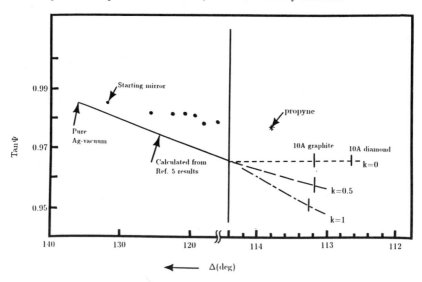

Figure 3: The experimental ellipsometric results for the growth of a film on a Ag mirror after treatment with NO_2 gas followed by CH_3CCH compared with three different models of the surface.

Figure 3 shows the computed ellipsometric angles, Ψ and Δ, for a diamond layer, a transparent graphite layer, and absorbing graphite layers, all formed on a Ag substrate with an 80A $AgNO_3/Ag$ matrix overlayer. The following optical constants were used to perform the ellipsometric calculations: $N_{Ag} = 0.082 - 3.61i[10]$, $N_{AgNO_3/Ag} = 3.311 - 0.1068i[5]$, $N_{DIAMOND} = 2.4235[11]$, $N_{GRAPHITE} = 1.918$, $1.918 - 0.5i$, and $1.918 - i$, and $N_{AMBIENT(AIR)} = 1$. The experimental points to the left of the solid vertical line correspond to exposing the Ag mirror to NO_2/N_2O_4. The first point represents the Ag mirror with its intrinsic film of oxygen and carbon compounds. The point to the right of the vertical line shows the result of exposure to propyne. Once the carbonaceous overlayer formed, it did not seem capable of further growth with continued exposure to propyne. The propyne exposure was ~ 70 (490mL at STP) times as great as the NO_2/N_2O_4 exposure (7mL at STP). Nevertheless, the carbon layer formed by degradation of the propyne was only on the order of 5A. These observations are in accord with the SERS results reported here and with the reaction scheme that has been proposed by previous investigators.

CONCLUSION

The results presented in this paper indicate that normalization of SERS spectra may be a useful procedure for the study of surface chemical reactions. Additional results from a variety of surface chemical measurements are necessary to support this conclusion. It was also shown that the reactions of alkynes on Ag powder at room temperature are rapid and passivate the surface to further chemical reaction due to formation of a carbon overlayer. Finally, as previously suggested[5], it appears that the addition of NO_2/N_2O_4 forms catalytically active Ag_n microstructures which are useful in the study of Ag surface chemical reactions.

REFERENCES

1. M.A. Barteau and R.J. Madix, Surf. Sci. 115, 355 (1982).

2. J.M. Vohs, B.A. Carney, and M.A. Barteau, J. Am. Chem. Soc. 107, 7841 (1985).

3. R.J. Madix, Science 233, 1159 (1986).

4. K. Manzel, W. Schultze, and M. Moskovits, Chem. Phys. Lett. 85, 183 (1982).

5. P.B. Dorain and J.E. Boggio, J. Chem. Phys. 84 (1), 135 (1986).

6. M. Moskovits, Rev. Mod. Phys. 57, 783 (1985).

7. X. Jiang and A. Campion, Chem. Phys. Lett. 140, 95 (1987).

8. E. Burstein, Y.J. Chen, C.Y. Chen, S. Lundquist, and E. Tosatti, S. S. Comm. 29, 567 (1979).

9. A. Otto, in Light Scattering in Solids, Vol. IV, edited by M. Cardona and G. Guentheradt (Springer-Verlag, NY, 1984) p.

10. R.H. Muller, R.F. Steiger, G.A. Somerjai, and J.M. Morabito, Surf. Sci. 16, 234 (1969).

11. A.I.P. Handbook, 3rd ed., (McGraw-Hill, NY, 1972), Section 6, page 27.

12. Lange's Handbook of Chemistry, 11th ed., edited by J.A. Dean, (McGraw-Hill, NY, 1973), Section 10, page 258.

Research supported by ONR.

STRUCTURE AND EPITAXY OF SILVER/GOLD MICROCLUSTERS ON MGO

L. D. Marks, Department of Materials Science, Northwestern University, Evanston, IL 60208, M.C. Hong, H. Zhang and B. K. Teo, University of Illinois at Chicago, Department of Chemistry, Chicago, IL 60680.

ABSTRACT

We report observations by high resolution electron microscopy of model catalysts produced by impregnating MgO smoke particles with inorganic clusters. With the use of very low beam currents the substrates are of sufficiently low noise that we have been able to image cleanly both single atoms and very small clusters of size 1-2nm in sufficient detail to determine their atomic structure. Two types of metal structures are observed: single atoms decorating atomic steps and in a few cases two-dimensional surface rafts, and a population of mainly single crystal with a few multiply twinned particles. The single crystal particles are pseudomorphically epitaxed on the MgO substrates.

INTRODUCTION

There has for a long time been extensive studies of supported metal catalysts by electron microscopical techniques with a view towards detailing the atomic structure of the metal and measuring the particle size distribution in order to explain the catalytical activity. Almost without exception the major emphasis of the work has concentrated on larger particles, generally larger than 2nm in diameter which has been in many respects a practical limit on the resolution. Whilst larger particles can be imaged without any problems (e.g. [1]), noise due to amorphous contamination layers or amorphous substrates has made smaller sizes almost undetectable [2] except for the STEM Z contrast technique which can image single atoms in favourable cases [3]. In addition, a number of recent papers [4-6] have reported structural fluctuations in small clusters in the microscope which implied that the particles would not remain stationary for long enough to be imaged, a process which we believe to be due to energy transferred by inelastic scattering of the incident electron beam [7,8].

We have recently been attempting to push the limits to which meaningful images of small particles can be obtained by carefully choosing our experiments. To attain this the experiment must be chosen so that:
a) Noise from the substrate must be exceedingly small so that it does not interfere with the images of the small particles.
b) The specimens must be clear of amorphous layers such as carbonaceous contaminats.
c) The metal particles must be stable on the substrate.

In this paper we report results on MgO smoke particles which satisfy criteria a) above, taken under conditions free from contamination (b) and at very low beam currents (c). We have been successful in cleanly imaging single atoms and very small clusters in sufficient detail to be able to identify their atomic structure.

Mat. Res. Soc. Symp. Proc. Vol. 111. ©1988 Materials Research Society

EXPERIMENTAL METHOD

Clusters of $Au_{13}Ag_{12}$ and $Au_{18}Ag_{20}$ prepared as described elsewhere [9-10] stabilised by triphenyl phosphide ligands were dissolved in methyl chloride and a drop of the resultant solution was allowed to dry on MgO smoke particles produced by collecting the smoke from burning Mg directly onto pyrolised copper grids. These specimens were then examined in a Hitachi H-9000 under minimum exposure conditions with 8 to 12 second exposures. The point is to minimise electron beam changes to the particle structures, and experimentally it was noted that higher beam fluxes lead to shape changes and even net particle motion along the surface as reported by other authors [4-6]. Due to the clean turbomolecular pumping system of the instrument no contamination was observed over extended periods of time even without a liquid nitrogen cold finger.

RESULTS

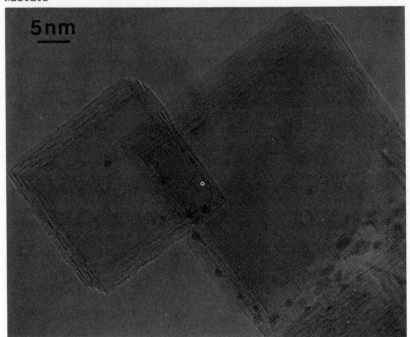

Figure 1. Typical field of view showing particles and MgO.

A typical field of view is shown in Figure 1, cubic MgO particles with steps, a few particles and some very faint 'black dots'. Image simulations indicated that a single gold atom with no substrate would give approximately a 20% contrast signal under the conditions that we have used provided that there is no strong diffraction from the substrate which would completely obscure the single atoms. This will be an overestimate since the image will be degraded by inelastic scattering from the MgO substrate, and a value of 10% would seem more appropriate. This implies that we can identify the faint dots as single atoms, and Figure 2 shows a larger magnification image of surface steps and single atoms. Experimentally, it is apparent that the atoms are decorating the surface steps of the MgO cubes, which are of course preferential adsorption sites.

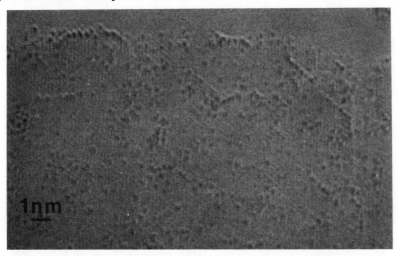

Figure 2. Micrograph showing single black dots which are identified as single atoms, decorating surface steps. At the top the stronger contrast is due to two dimensional surface rafts.

In addition to single atoms and rafts, somewhat larger clusters ranging in size from 1 to 3nm were observed, for instance the single crystals in 3a and b and the multiply twinned particles shown in Figures 3c and d. The dominant particles were the single crystal, pseudomorphically epitaxed on the substrate, as would be expected considering that the lattice mismatch is only ~4%. The particles were also in many cases preferentially adhering to either surface steps or the bottom of larger surface steps, as shown in Figure 4.

One additional observation that is important is that in regions where there was a higher concentration of particles there was also in many cases an amorphous looking carbonaceous layer as shown in Figure 5. The significance of this we will return to in the discussion.

DISCUSSION

The main conclusions of our work are threefold, one technical concerning the limits of electron microscopy of small particles, the second catalytic and the last with respect to the observed small particle structures. Firstly it should be clear that with the appropriate choice of conditions there really is no major problem in imaging down to the atomic level. Indeed, in our experience single atom imaging is somewhat

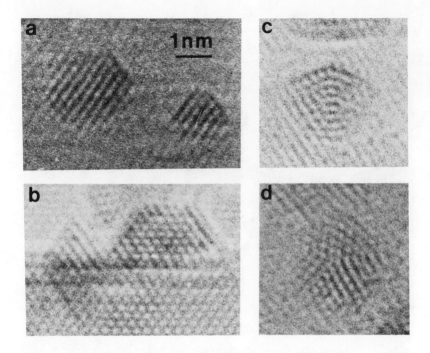

Figure 3. Images of a) and b) single crystal particles, with the epitaxy relative to the MgO shown more clearly in b) where the MgO [110] zone axis structure can be seen; c) Icosahedral multiply twinned particle and d) decahedral multiply twinned particle.

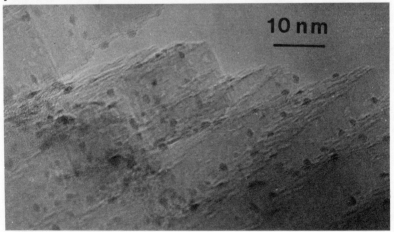

Figure 4. Image showing particles residing at the bottom of larger steps

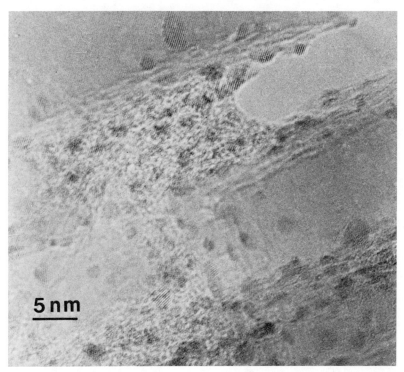

Figure 5. Image showing carbonaceous residue in regions of higher particle density.

easier than imaging the small 1 nm clusters since the clusters must be correctly oriented with respect to the beam in order to image along the atomic columns. It should be repeated that the appropriate substrates and the use of very low beam currents are very important.

The second point concerns the activity of catalyst systems such as these. It is hard to be quantitative, but in many cases we would suspect that as much as 20% of the exposed atoms were in the form of single atoms, and this could well be of major catalytic significance. However, it should at the same time be acknowledged that whereas single atoms are consistent with Ostwald ripening processes, we do not completely understand why they are arising. The small particles show quite strong epitaxy with (100) faces parallel to the (100) faces of the magnesium oxide, with large boundary regions indicating that the gold/magnesium oxide interface energy is quite large and this may assist the initial clusters seperation from the triphenyl phosphide ligands. The observation of carbonaceous regions associated with higher particle density regions does indicate that the phenyl groups at least have not vanished. Where exactly in this process the single atoms come in is currently being explored further.

The final concerns the observation that the major population was single crystals, not multiply twinned particles which free from a substrate would have a lower energy. The strong epitaxy explains why there is a low fraction of multiply twinned particles, since these do not have substantial (100) facets and hence are energetically disfavoured relative to the single crystals.

ACKNOWLEDGEMENTS

One of us (LDM) would like to acknowledge support from the Petroleum Research Foundation on grant number 17899-G5.

REFERENCES

1. L. D. Marks, Ultramicroscopy, _18_, 445 (1985).
2. K. Heinemann and F. Soria, Ultramicroscopy, _20_, 1 (1986).
3. M. J. Treacy, private communication
4. S. Iijima and T. Ichihashi, Japanese J. Appl. Phys., _24_, L125 (1985); Phys. Rev. Letts, _56_, 616 (1986).
5. S. Iijima, J. Electron Microscopy, _34_, 249 (1985).
6. J. O. Bovin, R. Wallenberg and D. J. Smith, Nature, _317_, 47 (1985).
7. J. Dundurs, P. M. Ajayan and L. D. Marks, Phil Mag in press
8. P. M. Ajayan and L. D. Marks, these proceedings
9. B. K. Teo and K. Keating, J. Am. Chem. Soc., _106_, 2224 (1984).
10.B. K. Teo, M. C. Hong, H. Ahang and D. B. Huang, Angew. Chem. Int. Ed. Engl., _26_, 897 (1987); J. C. S. Chem. Comm., submitted.

MÖSSBAUER ANALYSIS OF IRIDIUM IN Pt-Ir/SiO2 CATALYSTS

AECL-9678

J.A. Sawicki*, F.E. Wagner** and J.H. Rolston*

*Chalk River Nuclear Laboratories, Chalk River, Ontario KOJ 1JO, Canada
**Department of Physics, Technical University of Munich, D-8046 Garching, Federal Republic of Germany

ABSTRACT

The state of iridium in highly dispersed bimetallic Pt-Ir catalysts supported on amorphous silica was studied by ^{193}Ir Mössbauer spectroscopy, after deposition via ion exchange, as well as after calcination, reduction in hydrogen, and on exposure to carbon monoxide. Characteristic patterns and relative contributions of adsorbed $[Ir(NH_3)_5Cl]^{2+}$ complexes, $IrCl_3$, IrO_2 and Pt-Ir alloy particles have been identified. The composition of bimetallic particles indicates a strong tendency for segregation of Pt and Ir.

INTRODUCTION

Bimetallic catalysts consisting of Pt-Ir clusters are applied in petrochemical processing as hydrocarbon conversion catalysts [1-4]. In the past, microstructural characteristics of Pt-Ir clusters have been studied by X-ray diffraction [2], scanning electron microscopy [3], extended X-ray absorption fine structure (EXAFS) [4], and also, indirectly, through Mössbauer spectroscopy on ^{57}Fe-decorated catalyst specimens [5]. Recently, we have demonstrated the usefulness of Mössbauer spectroscopy of ^{193}Ir in the characterization of the local structure and chemical forms of iridium in such catalysts [6]. Specimens of silica-supported Pt-Ir catalysts containing ~5-10 wt% Ir have been successfully studied. In order to better assess the merits of the method, in the present paper we have extended the sensitivity of the measurements to catalysts with much smaller contents of iridium (~1 wt%).

EXPERIMENTAL TECHNIQUES

Specimens

A substrate of amorphous silica granules was prepared from a uniform slurry of Cab-O-Sil (EH-5) powder in double distilled water (~8 mL of water per gram of Cab-O-Sil). The material was suction-filtered, oven-dried at 150°C, calcined at 400°C for 4 h and then lightly crushed and screened to 5-10 mesh. Platinum and iridium were adsorbed as complex ions via ionic exchange from aqueous solutions of chloropentammine iridium (III) chloride $[Ir(NH_3)_5Cl]Cl_2$ and chlorotetrammine platinum (II) chloride $Pt(NH_3)_4Cl_2 \cdot H_2O$. During exchange, the pH of the solution was adjusted to 11 with NH_4OH. After deposition, lasting from 10 to 20 h, the material was washed with distilled water until free of Cl and NH_4OH and then dried in air at temperatures between 40°C and 150°C for 2 to 24 h. Calcinations were carried out in air and reductions were performed in flowing hydrogen.

Metal areas (dispersion) of the catalysts were determined by hydrogen chemisorption at 25°C and 200°C using a static volumetric apparatus. Prior to the measurements each specimen had been outgassed and reduced at 150°C.

The average particle sizes, as determined by hydrogen chemisorption and small angle X-ray diffraction, were in the range of 3 to 5 nm. Chemisorption of carbon monoxide was examined by the FTIR (Fourier Transform Infrared) technique, using a Bomem spectrophotometer. The kinetics of CO oxidation were measured as a function of the total gas flow at a pressure of 110 kPa at 32°C and for relative humidities of 30 and 100%, with a single-pass flow reactor and a precision infrared CO/CO_2 dual gas analyzer.

Mössbauer measurements

The Mössbauer spectra of 73.0 keV transition in ^{193}Ir were measured at 4.2 K in transmission geometry, using a Ge detector and a ^{193}Os metal source ($T_{\frac{1}{2}}$=30 h) irradiated in a thermal neutron flux of 2×10^{13} n/cm^2·s. The quadrupole splitting in the source was assumed to be 0.48 mm/s [7]. The absorbers contained ~1 g/cm^2 of the catalyst. The spectra were fitted with sets of Lorentzian lines with common widths. Spectra for 10 wt% Ir and 9 wt% Pt-1 wt% Ir specimens, together with the fitted curves, are shown in Figs. 1 and 2. The Mössbauer parameters obtained by least squares fits of the spectra are given in Table 1.

RESULTS AND DISCUSSION

The Mössbauer spectra of as-adsorbed specimens (Figs. 1 and 2) exhibit a symmetric quadrupole doublet with the isomer shift very close to that found for $[Ir(NH_3)_5Cl]Cl_2$ (IS=-1.63 mm/s) [8]. However, the quadrupole splitting measured for the adsorbed complexes (QS=1.54-1.66 mm/s) is considerably smaller than for the crystalline compound (QS=2.06 mm/s) [8]. Iridium complexes appeared to be stable upon thermal treatment, for instance at 150°C in air, despite the already black coloration of the specimen.

Calcinations at 350-500°C resulted in formation of IrO_2 clusters, represented in the Mössbauer spectra by a quadrupole doublet with the parameters rather close to the values found for bulk IrO_2 (IS=-0.93 mm/s; QS=2.73 mm/s) [7,8].

The Mössbauer spectra of reduced specimens exhibit single broad lines characteristic of iridium metal (IS~ 0 mm/s) or Ir rich Pt-Ir alloy (IS=-0.15 mm/s).

A small single-line component with an isomer shift IS=-1.94 mm/s, observed in the specimens reduced at relatively low temperatures, can be tentatively ascribed to $IrCl_3$, in accordance with the literature data [9]. $IrCl_3$ may be formed as an intermediate thermal decomposition product of the initially deposited $[Ir(NH_3)_5Cl]^{2+}$ ions.

The composition of Pt-Ir alloys can be determined from the Mössbauer spectra, since the isomer shift in binary $Pt_{1-x}Ir_x$ alloys varies linearly with x from -0.63 mm/s at x=0 to zero at x=1 [10]. On the basis of the determined compositions, it can be concluded that iridium in bimetallic clusters occurs in regions where platinum is depleted, which points to a strong tendency for segregation of both metals during heat treatments or during deposition.

In our most recent experiment, the degree of segregation vs. nominal metal composition has been only slightly lessened for the catalysts prepared by sequential deposition of metals, as can be seen by comparing data B5 and B7 in Table 1. In addition, the Mössbauer spectra of the

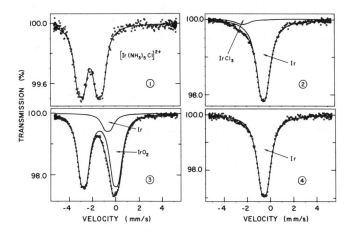

Figure 1. Mössbauer spectra of ^{193}Ir in 10 wt% Ir/Cab-O-Sil specimens: (1) as-adsorbed and dried at 50°C; (2) calcined at 500°C; (3) reduced at 200°C; (4) calcined at 500°C and reduced at 300°C.

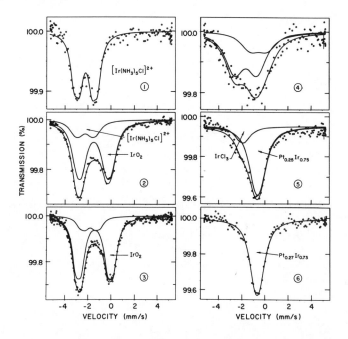

Figure 2. Mössbauer spectra of ^{193}Ir in 9 wt% Pt-1 wt% Ir/Cab-O-Sil specimens: (1) as-adsorbed and dried at 50°C; (2) calcined at 350°C; (3) calcined at 450°C; (4) reduced at 200°C; (5) calcined at 350°C and reduced at 200°C; (6) calcined at 450°C and reduced at 300°C.

Table 1. ^{193}Ir Mössbauer results for bimetallic 9 wt% Pt-1 wt% Ir/Cab-O-Sil catalysts. The first column gives information on the time and temperature of calcination and reduction. The next three columns represent, respectively, the isomer shift with respect to metallic iridium (IS), quadrupole splitting (QS) and the FWHM linewidth (W). The last column summarizes the conclusions about the form of iridium as drawn from the Mössbauer data.

Specimen	IS mm/s	QS mm/s	W mm/s	Form of iridium
A. 10 wt% Ir				
1. as-adsorbed	-1.61(1)	1.66(1)	0.84(3)	Ir complex
2. calc. 500°C/2h	-0.83(1)	2.78(1)	0.87(1)	IrO_2
	-0.14(1)	-		11% Ir metal*
2. red. 300°C/2h	+0.00(1)	-	1.11(2)	Ir metal
	-1.81(7)	-		6% $IrCl_3$*
4. calc. 500°C/2h; red. 300°C/2h	-0.01(1)	-	1.07(2)	Ir metal
B. 9 wt% Pt-1 wt% Ir				
1. as-adsorbed,				
dried at 50°C	-1.64(2)	1.54(3)	0.80(5)	Ir complex
dried at 150°C	-1.59(3)	1.43(5)	0.78(8)	Ir complex
2. calc. 350°C/5h	-0.92(2)	2.56(3)	1.10(3)	IrO_2
	-1.67	-		22% Ir complex*
3. calc. 450°C/2h	-0.88(1)	2.63(3)	1.07(4)	IrO_2
	-2.20	-		11% Ir complex*
4. red. 200°C/12h	-1.12(7)	1.99(11)	1.77(20)	Ir complex
	+0.01(19)	1.50		30% Ir metal*
5. calc. 350°C/5h; red. 200°C/12h	-0.12(4)	-	1.30(13)	$Pt_{2.25}Ir_{0.75}$
	-1.48(14)	-		22% $IrCl_3$*
6. calc. 450°C/2h; red. 300°C/3h	-0.13(2)	-	1.20(5)	$Pt_{0.27}Ir_{0.73}$
7. calc. 350°C/5h; red. 200°C/5h; kept in air**	-0.18(1)	-	1.16(10)	$Pt_{0.32}Ir_{0.68}$
8. calc. 350°C/5h; red. 200°C/5h; kept in CO**	-0.17(1)	-	1.16(10)	$Pt_{0.32}Ir_{0.68}$

* Area in Mossbauer pattern; true content can be stated only when f-factors are known.
** Catalysts prepared by sequential deposition of Pt and Ir.

specimens kept in air (B7) and in CO (B8) did not indicate any significant
influence of the surroundig gas on the state of iridium.

CONCLUSIONS

Mössbauer spectroscopy of ^{193}Ir emerges as a new tool for the
determination of the chemical and metallurgical structure of iridium in
catalysts. As has been shown, one can determine the valence state of
iridium, its chemical form and the composition of alloy clusters at iridium
contents as small as ~1 wt%. In future work, even lower Ir contents
should be accessible. It is to be noticed that the extraction of similar
information from the EXAFS data [4,11], would be much more difficult, due
to a weak signal from iridium and close proximity of the Pt and Ir_{III}
edges.

Acknowledgements

The authors would like to thank Mrs. L.M. Paterson and Mr. D.E. Clegg
for the preparation of the catalysts, as well as Mr. K.D. McCrimmon and Mr.
J.E. Winegar for supplementary analyses of the specimens. They also thank
Dr. E.A. Symons and Dr. K. Marcinkowska for helpful comments.

References

1. J.H. Sinfelt, Rev. Mod. Phys. 51, 569 (1979).

2. J.H. Sinfelt and G.H. Via, J. Catal. 56, 1 (1979).

3. K. Fager, D. Hay and H. Jaeger, J. Catal. 96, 154 (1981); ibid 96, 170
   (1981).

4. J.H. Sinfelt, G.H. Via and F.W. Lytle, J. Chem. Phys. 76, 2779 (1982);
   Cat. Rev. Sci. Eng. 26, 1 (1984).

5. R.L. Garten and J.H. Sinfelt, J. Catal. 62, 127 (1980).

6. F.E. Wagner, J.A. Sawicki and J.H. Rolston, Hyperfine Interactions, in
   press.

7. F.E. Wagner, Hyperfine Interactions 13, 149 (1983).

8. W. Potzel, U. Wagner and H.H. Schmidtke, Chem. Phys. 4, 294 (1974).

9. F.E. Wagner, W. Potzel, U. Wagner, H. Schaller, and P. Kienle, Phys.
   Letters 25B, 253 (1971).

10. F.Y. Fradin, B.D. Dunlap, G. Shenoy and C.W. Kimball, Intern. J.
    Magnetism 2, 415 (1971).

11. F.W. Lytle, J. Non. Cryst. Solids 61-62, 1359 (1984).

RAMAN SPECTRA OF ION INTERCALATED VERMICULITES

H.X. JIANG,[*] S.A. SOLIN,[*] H. KIM[**] and T.J. PINNAVAIA[**]
*Department of Physics and Astronomy
**Department of Chemistry
Center for Fundamental Materials Research
Michigan State University, East Lansing, MI 48824-1116

ABSTRACT

The Raman spectra of ion intercalated Vermiculites have been studied. The torsional mode frequencies exhibit a shift to higher energy with increasing basal spacing d. Using a van der Waals type force model, and the nearest neighbor approximation, we have successfully derived the formula for the torsional mode frequency as functions of basal spacing d and other physical parameters. The calculation results are quite consistent with the experimental observations.

INTRODUCTION

Clay Intercalation Compounds (CIC's) have recently drawn attention in solid state physics because of their interesting physical properties. [1-3] One of the most important features of CIC's is that they can be used as an ideal quasi-two-dimensional system with which to investigate 2-D physical and chemical processes. One of the common CIC's is vermiculite which is classified as a 2:1 layered silicate due to its structure. [4, 5] In the interlayer galleries of vermiculite as well as other clays, chemical reactions can be selective, specific, and quite distinct compared with the corresponding reaction in free space. [6]

It is always important to understand the physical properties of CIC's for different intercalated ions. The most important properties for CIC's are layer rigidity, the structure (such as basal spacing), and the distributions of intercalated ions in the galleries all of which are heavily influenced by the guest - guest and guest - host interactions. Extensive work on graphite intercalation compounds [7, 8] and other layered solids has shown that Raman spectroscopy is an effective tool with which to study these interactions.

In this paper, we present the results of the torsional Raman mode of ion intercalated vermiculites. The detailed results and composition dependence of the Raman frequency of this mode of $[(CH_3)_4N^+]_{1-x}$ $[(CH_3)_3NH^+]_x$-Vermiculites will be published elsewhere. [9] The distinct behaviors of these systems suggests the importance of the ion size. The different force constants between guest ions and the basal oxygens also plays an important role. Using a van der Waals type force model and the nearest neighbor approximation, we have successfully derived a formula for the Raman frequency of the torsional mode as a function of basal spacing d, force constants and other physical parameters.

EXPERIMENTAL

For this study, M^+-V(ermiculites) samples were made from Mg-Vermiculite using an ion exchange method [10] with $M^+=Rb^+$, Cs^+, $(CH_3)_3NH^+$ and $(CH_3)_4N^+$, respectively. The samples were made of powder in which each crystallite was a few microns in size. The samples are well ordered and water free. Highly oriented films were prepared from powdered natural vermiculite. The water was removed by heating the samples in an oven for 12 hours at a temperature of 100 ^{o}C. Film samples formed on glass slides exhibited mosaic spreads of $\Omega\sim5^o$ [2] indicating an oriented morphology with the silicate layers parallel to the slide surface.

Raman spectra were measured using the 5145 Å line of an argon laser, a Spex 1402 double spectrometer and a photon counting detection system. All measurements were made with the scattered light collected at 90° to the direction of propagation of the laser light which was incident at an angle of about 45° to the sample plane and polarized in the plane of incidence.

RESULTS AND DISCUSSIONS

The Raman frequencies of the torsional mode of M^+-V [$M^+=Rb^+$, Cs^+, $(CH_3)_3NH^+$ and $(CH_3)_4N^+$)] are at about 106 to 118 cm^{-1}. The torsional mode is an interlayer mode of the C_s symmetry group and is the vibration of oxygens in the basal plane. [1] A few other high frequency modes have also been observed which correspond to intralayer modes of the host layer. These high frequency modes will not be discussed here. In Fig. 1, we plot the Raman shift as a function of basal spacing d for M^+-V with M^+ = Rb^+, Cs^+, $(CH_3)_3NH^+$, and $(CH_3)_4N^+$, respectively. The basal spacing d was measured from x-ray experiments. [1, 2] In the figure, the Raman frequencies are normalized to the frequency of Rb-V, which is $106 cm^{-1}$, $\nu_{norm} = \nu/\nu_{Rb}$ and d_{norm} is the normalized basal spacing distance. $d_{norm} = (d_{obs} - d_{min})/(d_{max}$

$- d_{min}$) where d_{obs} is the observed basal spacing, and d_{min} and d_{max} are the basal spacing of Rb-V and $(CH_3)_4N^+$-V, which are 10.23Å and 13.3Å, respectively.

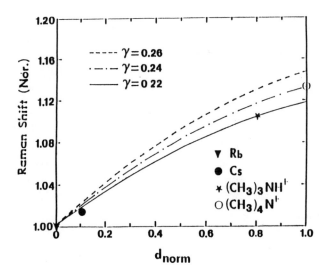

Fig. 1. Calculation results of Raman shift of torsional mode (normalized) as a function of basal spacing (normalized) of M^+-V with three different values of Y, the ratios of force constants between oxygen-guest ion to oxygen-oxygen. The experimental results are shown in the figure with M^+ being Rb^+, Cs^+, $(CH_3)_3NH^+$ and $(CH_3)_4N^+$, respectively. The basal spacings d were measured from x-ray experiments.[1],[2]

From Fig. 1, we can see that the Raman frequency of the torsional mode increases with the basal spacing d, which increases with the size of the guest ions. The connection between the Raman frequency and the basal spacing d is very clear. This implies that the basal spacing d is one of the most important parameters for the torsional mode. As we will see, the force constants between oxygen-oxygen and guest ions-oxygen are also important for the torsional mode.

The Raman frequency of the torsional mode as a function of the basal spacing and other physical parameters can be calculated under the following assumptions. First, we assume that the interaction between the oxygen-oxygen and oxygen-guest ion are   of the van der Waals type.
[11] Second, the guest ions are located in the centers of the hexagonal

pockets of the Kagome lattice of the bounding oxygen planes. This should be true if we consider the minimum energy of the system or the equilibrium position of the guest ions. One should notice that when the system vibrates in the torsional mode, the position of each oxygen in the basal plane is determined uniquely, since we know the eigenvector of the mode (guest ions are motionless in the torsional mode).

Under the nearest neighbor approximation, the relation between the Raman frequency of the torsional mode, basal spacing and other physical parameters can be derived as

$$\tilde{v}^2 = (-2b^{n-1}/m)(\alpha/4\pi^2 c^2)\{3 + Y(1 + h^2/b^2)^{(n-3)/2}(n + h^2/b^2)\}, \quad (1)$$

where $h=(d - 6.42/2$ (Å) is the distance between the center of the guest ion and the basal plane and 6.42Å is the distance between the top and bottom layers of the basal planes in each period. In Eq.(1), b is the lattice parameter, $b = 2.67$Å, m is the mass of the oxygen atom, c is the speed of light, and $n = -7$ represents the van der Waals type force. The parameters α and $\beta < 0$ correspond to the attractive force constants between oxygen-oxygen and oxygen-guest ion while $Y = \beta/\alpha$ is the ratio of these force constants.

Equation (1) connects the Raman shift of the torsional mode with the physical parameters of M^+-V. From the parameters b, h, and m, we can calculate the Raman shift of torsional mode as a function of the basal spacing. In Fig.1, we have plotted the calculated results of Eq.(1), Raman shift of the torsional mode (normalized), $\tilde{v}_{norm}$, as a function of basal spacing d (normalized), d_{norm}, for three different values of Y.

The calculated results are quite consistent with experimental observations. For a fixed value of Y, the Raman shift of the torsional mode increases with basal spacing d. The force constant for the oxygen-guest ion interaction is in the order of about 0.2 to 0.3 of the oxygen-oxygen interacton. The same method can also be used for other vibrational modes of CIC's. By comparing calculation and experimental results for other modes, we can derive the force constants between oxygen-oxygen and oxygen-guest ion in CIC's.

ACKNOWLEDGMENTS

We gratefully acknowledge helpful discussions with Dr. S. Mahanti, S. Lee, W. Jin and Y. B. Fan. This work was supported by NSF-MRG grant DMR 85-141545 and is part by the MSU Center for Fundamental Materials Research.

REFERENCES

1. B.R. York, S.A. Solin, N. Wada, R. Raythatha, I.D. Johnson, and T.J. Pinnavaia, Solid State Comm., 54, 475 (1985).

2. S. Lee, H. Kim, S.A. solin, and T.J. Pinnavaia, Chemical Physics of Intercalation, edited by A. P. Legrand, (Nato ASI, 1987), (to be published).

3. N. Wada, R. Raythatha, and S. Minomura, Solid State Comm. 63, 783 (1987).

4. R.E. Grim, Clay Mineralogy, (Mcgraw-Hill, 1968)

5. G.W. Brindly and G. Brown (eds.), Crystal Sturctures of Clay Minerals and Their x-ray Identification, (Minerological Society, London, (1980).

6. T.J. Pinnavaia, Science 220, 365 (1983).

7. S.A. Solin, P. Chow, and H. Zabel, Phys. Rev. Lett. 53, 1927 (1984).

8. S.A. Solin, Adv. Chem. Phys. 49, 455 (1982).

9. S.A. Solin, H.X. Jiang, H. Kim and T.J. Pinnavaia, to be published.

10. H. Kim and T. J. Pinnavaia, unpublished.

11. C. Kittel, Introduction to Solid State Physics, 5th edition (John Willey and Sons, Inc.), p.79.

STRUCTURAL AND DYNAMICAL PROPERTIES OF INTERCALATED LAYERED SILICATES

W. Jin, S. D. Mahanti, S. A. Solin and H. C. Gupta[*]
Department of Physics and Astronomy, Center for Fundamental Materials
Research, Michigan State University, East Lansing, MI. 48824.

INTRODUCTION

Structural and dynamical properties of intercalated solids in general [1], and layered silicates of the type A_xB_{1-x}-Vermiculite in particular, are of both fundamental and practical interest. In these systems, two types of ions A and B with different ionic radii occupy the space between two silicate layers. On the fundamental side, one is interested in studying the average interlayer spacing as a function of (1) the concentration x of the large ion, (2) sizes and compressibilities of the intercalated ions and (3) the transverse rigidity of the silicate layers. In addition, one is interested in the dynamic properties of these solids. On the practical side, when the size difference between the two intercalants is large, one obtains pillared clays which are characterized by widely spaced silicate layers that are propped apart by sparsely distributed larger interlayer cations (sometimes referred to as pillars) [2]. The enormous free volume of accessible interior space that is derived from such an open structure has significant practical application in the field of catalysis and sieving.

In the first part of this paper we discuss our results of a systematic analysis of the average normalized interlayer spacing $d_n(x)$ in a single layer of a model ternary of the type $A_xB_{1-x}V$ obtained using both simulation techniques and elastic spring models. The latter takes into account the effect of different compressibilities of the intercalants and transverse rigidity of the silicate layers on the average interlayer spacing. In the second part, we discuss the dynamics of these systems, in particular the zone center vibrational modes.

STRUCTURAL STUDIES

A model monolayer system with finite layer rigidity has been studied by computer simulation [3]. For simplicity we assume that the intercalant ions are hard-spheres. Starting from a 2-dimensional triangular lattice of lattice constant a_0 representing a single gallery between two silicate layers and with each lattice site occupied by a B ion of height d_B, we randomly replace the B ions with A ions with height $d_A > d_B$. The height of a cell within a healing length λ of the A ion is also increased to d_A. A second ion in this region does not affect already expanded cells but expands unexpanded cells within λ of its location. Random replacement of the B ions continues to saturation. If we define $f(x)$ as the fraction of cells with height d_A, then the normalized interlayer spacing $d_n(x) = f(x)$; where $d_n(x) = (d(x)-d(0))/(d(1)-d(0))$. The simulation results for $d_n(x)$ are shown in Fig. 1 for several different healing lengths λ. Note that the nonlinearity in $d_n(x)$ is a collective effect associated with long range interaction between the large ions through the distortion fields. Using effective medium type arguments [4], we can describe the normalized spacing $d_n(x)$ for a monolayer by the equation

$$d_n(x) = [1 - (1 - x)^p],$$ (1)

where p is a layer rigidity parameter given by p = Z+1 for our "lattice-gas" simulation model; Z is the number of neighboring sites that are puckered by the insertion of an isolated A ion as shown in Fig. 1.

The interlayer spacing of silicates with compressible intercalants will be different from those for the hard-sphere case. To study the effect of ionic compressibility on $d_n(x)$, we consider a "spring model" [5]. For the moment we assume that the layer is infinitely rigid against transverse distortion, i.e. a flat layer without gallery height fluctuations from site to site. This is a generalization of a model used by Dahn et al. [6] for binary intercalation compounds. The elastic energy of a monolayer intercalated with A and B ions can be written as

$$E = \frac{1}{2} k_o(d - d_o)^2 + \frac{1}{2} k_A(d - d_A)^2 x + \frac{1}{2} k_B(d - d_B)^2(1 - x) , \quad (2)$$

where d_o is the pure host gallery height and k_o is the effective spring constant of the pure host; k_A and k_B are effective spring constants of the local galleries containing intercalants A or B. These spring constants are directly related to the compressibitili of A and B ions in the gallery which in turn depend upon guest-host interaction in the solid. In general, k_o is quite small compared to k_A and k_B, thus d_o does not play an important rôle in the x-dependence of $d_n(x)$. The average interlayer spacing is obtained by minimizing the energy in Eqn. (2). One then has

$$d(x) = \frac{k_o d_o + k_A d_A x + k_B d_B(1 - x)}{k_o + k_A x + k_B(1 - x)} . \quad (3)$$

Fig. 2 shows the normalized spacing $d_n(x)$ for different ratios of k_B/k_A and a fixed $k_o/k_A(=0.01)$. We see that when $k_A \gg k_B$ the result is similar to the hard-sphere simulation results, as we expect.

A simple generalization of the monolayer "spring model" that can take into account the finite layer rigidity is given by

$$E = \frac{1}{2} k_o \sum_{i=1}^{N} (d_i - d_o)^2 + \frac{1}{2} \sum_{i=1}^{N} k_i(d_i - d_{io})^2 + \frac{1}{2} \sum_{\langle ij \rangle} k_H (d_i - d_j)^2 , \quad (4)$$

where k_i is k_A or k_B, d_{io} is d_A or d_B according to the type of intercalants at site i, and k_H is related to the transverse rigidity of the silicate layers. In contrast to the model discussed in the last paragraph, the present model allows for fluctuations of the local interlayer spacing from site to site. Note that in the "floppy" limit, i.e. when $k_H = 0$, we get a linear behaviour in $d_n(x)$ (which is called the Vegard's law). In the limit $k_H \to \infty$, this model reduces to the previously discussed rigid-layer model. On the other hand, it is easy to show that if $k_A = k_B$, $d_n(x)$ also shows a Vegard's law behaviour independent of the transverse rigidity of the layer.

For a given x and a particular distribution of A and B ions inside the gallery, the minimum energy configuration can be obtained by a conjugate gradient method [7]. For simplicity we have looked at a single chain and the normalized d-spacing for this system with different values of k_H are given in Fig. 3. As expected, the results for large k_H are similar to those for large healing lengths.

DYNAMICAL PROPERTIES

The transverse rigidity of the silicate layer which determines the healing length λ discussed earlier can be probed by studying the vibrational properties of these layered silicates. Recently, York et al. [8] have carried out Raman measurements in $Cs_{1-x}Rb_xV$. In particular they have measured the x-dependence of the torsional mode frquency ω_+. This mode consists of a collective rotational motion of SiO_4 tetrahedra about vertical Si-O bonds. We have used a simple radial and angular force constant model and a virtual crystal approximation to study the x dependence of ω_t. In addition, the same force constants have been used to calculate the elastic anisotropy in these layer silicates.

For the purpose of understanding the vibrational spectra, the single silicate sheet can be considered as a combination of two tetrahedral units of SiO_4 such that the oxygen atoms at the basal plane are shared to give Si_2O_5. The structure of anhydrous vermiculite has an octrahedral brucite-like sheet between two Si_2O_5 (actually there is a finite ammount of Al - Si substitution) sheets and an M^+ ion linking two different tetrahedral sheets. There are 22 atoms per unit cell. For lattice dynamics studies we associate a central (α) and an angular (α') force constant pair with each bond. The angular forces are represented by the method proposed by de Launey [9]. The dominant force constants used in our calculation are α_{Si-O}, α'_{Si-O}, α_{O-O}, α'_{O-O}, α_{Mg-O}, α'_{Mg-O}, α_{M^+-O}, α'_{M^+-O}. The values of the force constant that fit the experimental data rather well are (in units of 10^4 dyn/cm) 35.50, 10.95, 1.50, -2.70, 3.2, 0.4, 0.42, 0.268 (M =Rb) and 0.33, 0.54 (M =Cs) respectively. We have found that q_1= 0 vibrational frequencies of the Si_2O_5 layer which are above 500cm^{-1} do not change in the presence of the octahedral sheet and the M^+ ion. On the other hand, the mode frequencies less than 300cm^{-1} change by a factor of nearly 2. The torsional mode frequency agrees closely to the experimental value of York et al. [8] for CsV and RbV.

York et al. [8] find a large deviation from linear behavior for the torsional mode frequency ω_t vs. x in $Cs_xRb_{1-x}V$. Since the gallery cation M^+ is at rest in this particular mode for the pure system and nearly so for the ternary, their nonlinear x-dependence must be attributed to the force constants. We find that to fit the experimental data, the central force constant α_{M^+-O} decreases whereas the angular force constant α'_{M^+-O} increases with increasing Cs substitution for Rb. This observation is important because the increasecd role of the angular force constant in CsV can be quatlitatively understood if it is noted that the force between M^+ ion and the basal oxygen plane can have significant three-body component whose strength increases with an increase in the polarizability of the cation.

To see whether the force constant model used to explain the q = 0 phonon frequencies can account for the elastic anisotropy, the elastic constants C_{11} and C_{33} for RbV and CsV have been calculated. For simplicity $C_{11} = C_{22}$ is assumed. For the actual system with monoclinic symmetry, there is however a small difference between C_{11} and C_{22}. The elastic anisotropy factor $r = C_{33}/C_{11}$ has been calculated from the dynamical matrix in the leading order approximation. Using lattice constant a = 5.34 $\overset{o}{A}$ for both RbV and CsV and d = 10.23 $\overset{o}{A}$ for RbV and 10.54 $\overset{o}{A}$ for CsV, we obtain r = 0.2 for RbV and 0.4 for CsV. Although experimental measurments of r are not available in these systems, they are consistent with the Brilloion measurementsin muscovite by Vaughan and Guggenheim[10]. These authors find r = 0.324.

This work was supported by the National Science Fundation Material Research Group grant DMR 85-14154 and in part by the Center for Fundamental Materials Research at Michigan State University.

*Permanent Address: Physics Department, Indian Institute of Technology, New Delhi, India.

234

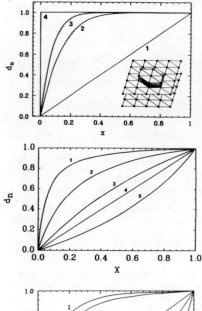

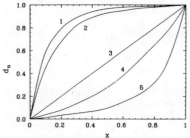

Fig. 1. Monolayer Triangular Lattice Computer simulations of $d_n(x)$ for several values of the healing length , and rigidity parameter p. The solid lines are from Eq. (1) of the text with (1) p = 1, λ= 0; (2) p = 7, $\lambda = a_0$; (3) p = 13, $\lambda = \sqrt{3}\, a_0$; and (4) p = ∞, $\lambda = \infty$. Insert - The puckered region of a triangular lattice with $\lambda = a_0$. In this case the number of expanded sites is p = Z + 1 = 7 whwre Z is the number of nearest neighbors.

Fig. 2. $d_n(x)$ in the monolayer model for different values of k_B/k_A : (1) 0.05, (2) 0.2, (3) 0.6, (4) 1.0 and (5) 2.0

Fig. 3. Inter-layer spacing $d_n(x)$ in the monolayer model for $k_A = 1.0$ and several values of k_B and k_H : (1) $k_B = 0.05$, $k_H = 10$, (2) $k_B = 0.05$, $k_H = 1$, (3) $k_B = k_A$ and any k_H, (4) $k_B = 10$, $k_H = 10$, (5) $k_B = 10$, $k_H = 50$.

REFERENCE

[1] S. A. Solin and H. Zabel, Adv. Phys, in press.
[2] T. J. Pinnavaia, Science, 220, 365 (1983)
[3] H. Kim, W. Jin, S. Lee, T. J. Pinnavaia, S. D. Mahanti, and S. A. Solin, Phys. Rev. Lett, (submitted).
[4] W. Xia and M. F. Thorpe, Phys. Rev. A, (submitted).
[5] W. Jin and S. D. Mahanti, (submitted).
[6] J. D. Dahn, D. C. Dahn and R. R. Haering, Solid State Comm. 42, 179 (1982).
[7] R. Fletcher, Practical Methods of Optimization, Vol. 2, (1981).
[8] B. R. York, S. A. Solin, N. Wada, R. H. Rayathatha, I. D. Jhonson, and T. Pinnavaia, Solid State Comm. 54, 475 (1985).
[9] J de Launey, The Theory of Specific Heats and Lattice Vibrations, Solid State Physics, Vol. 2, ed. F. Seitz and D. Turnbull, (Acad. Press, N.Y.), 219-303 (1956).
[10] M. T. Vaughan and S. Guggenheim, Jour. of Geophysical Research 91, 4657 (1986).

CHARACTERIZATION OF ZEOLITES BY GAS ADSORPTION AT LOW PRESSURES

AGUSTIN F. VENERO* AND J.N. CHIOU*
*Omicron Technology Corp., 160 Sherman Avenue, Berkeley Heights, N.J. 07922

ABSTRACT

The nitrogen and argon adsorption isotherms of three different zeolites, calcium A, ZSM-5, and sodium Y were obtained at 77.3 and 87.5 K. Using a continuous volumetric technique high resolution data was obtained in the Henry's Law region of the isotherm at relative pressures below 10^{-3} P/Po. The data for the two adsorbates indicates that nitrogen, due to its quadrupole moment, shows a high degree of localized adsorption, which makes it difficult to differentiate among zeolites of different pore sizes. Argon, on the other hand, can be used to identify zeolites of different pore sizes. The argon data was used in conjunction with the Horvath-Kawazoe model to obtain a pore size distribution of the zeolite.

Introduction

Nitrogen and argon were selected as adsorbates for two reasons. First, both gases have been used extensively [1] in the characterization of porous solids and therefore there is a large body of literature dealing with their properties as pertaining to physical adsorption. Secondly, the presence of a strong quadrupole moment in nitrogen and its absence in argon makes these gases suitable for the study of the effects of the quadrupole moment on the properties of the adsorption isotherms and on the characterization of zeolites by gas adsorption. Properties of these gases relevant to adsorption are given in Table I.

TABLE I
PROPERTIES OF NITROGEN AND ARGON

	Nitrogen	Argon
Boiling Point, K	77.3	87.4
Molecular diameter, A	3.0	2.89
Polarizability, cm^3	1.74×10^{-24}	1.6×10^{-24}
Quadrupole moment, cm^3	0.31×10^{-24}	-----

Experimental

The zeolites selected for this study were calcium A, ZSM-5, and sodium Y. Samples of calcium A, and sodium Y were provided by H. Hillery of Union Carbide. The ZSM-5 sample was supplied by E. Bowes of Mobil Corporation. Prior to obtaining the adsorption isotherms, the samples were outgassed at 350°C to a vacuum of 10^{-4} torr. In the case of ZSM-5 the sample was received in the ammonium form and it was calcined in air following the instructions of E. Bowes [2] of Mobil Corporation. The adsorption isotherms were determined using the Omnisorp 360 from Omicron Technology Corporation. This gas adsorption instrument uses a continous volumetric technique that allows data acquisition starting at relative pressures around 10^{-6} P/Po. A schematic of this instrument is given in Figure 1.

236

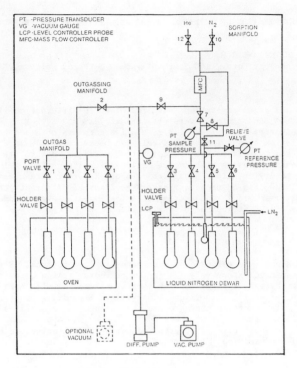

Figure 1
SCHEMATIC OF OMNISORP 360

The nitrogen adsorption isotherms are shown in Figure 2 in terms of the
normalized volume adsorbed, w/wo (volume adsorbed divided by volume adsorbed
at P/Po=0.5) versus the log of the relative pressure. The curves show the
characteristic "S" shaped curves of microporous materials predicted by the
vacancy solution model of Dubinin [3], and in the case of ZSM-5 a second step
around 0.1P/Po can be observed, as described by Unger and Muller [4].
Although differences in the isotherms can be observed, specially in the first
transition region (lower part of the S curve), one should expect larger
differences in view of the large range in pore diameters (from 0.5nm in
calcium A to around 0.75nm in sodium Y). If the shape of the isotherm were
to be primarily a function of the pore diameter, then one should expect
the first transition region of the isotherm to occur at pressure values
proportional to the size of the pore, that is, the smaller the pore, the
lower the pressure at which the transition occurs. Obviously, this is not
the case with the nitrogen isotherms: calcium A with the smallest pore
diameter has the highest transition pressure (around 3×10^{-5} P/Po) while ZSM-
5 with about the same pore diameter shows the lowest (around 4×10^{-6}).

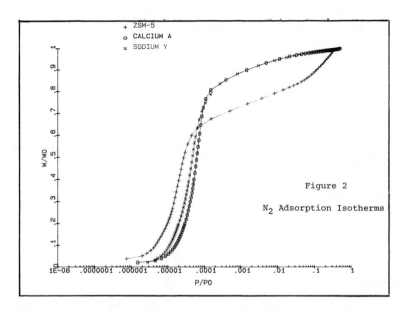

Figure 2

N_2 Adsorption Isotherms

In the case of the argon isotherms, shown on figure 3, a different picture emerges. The first transition region correlates with the pore diameter of the zeolite. In Table 2 the pressure at which the first transition region occurs for nitrogen and argon is given together with the publised values of the pore diameters for the different zeolites. The transition pressures were estimated by extrapolation of the linear regions below and above the transition and taking the point of convergence of these two lines as the transition pressure.

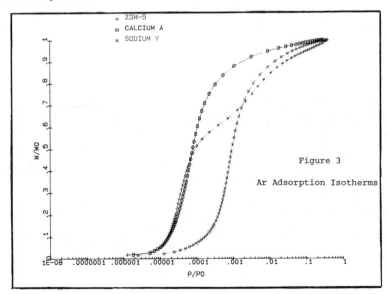

Figure 3

Ar Adsorption Isotherms

TABLE II
Transition pressures for nitrogen and argon isotherms.

Sample	Nitrogen	Argon	Pore Diameter (nm)
Ca A	3.2×10^{-5}	1.5×10^{-5}	0.5
ZSM-5	1.0×10^{-5}	1.0×10^{-5}	0.56
NaY	2.0×10^{-5}	3.6×10^{-4}	0.74

The data for argon correlates well with the accepted effective pore diameters. In the case of ZSM-5 and calcium A, the resolution is poor because the small differences in pore diameters (0.06nm) is probably within the experimental uncertainties.

Another observation that can be made from the argon isotherm data is that differences among the different zeolites disappear as the pressure increases, to the point that at relative pressures above 10^{-2} P/Po it is very difficult to distinguish among the different samples.

Pore size distribution

The argon data was used in conjuction with the Horvath-Kawazoe model [5] to obtain an effective pore size for the different zeolites.

The H-K model was developed primarily for molecular sieve carbons but has been used for zeolites by Seifert and Emig [6,7].

The model correlates the pressure-volume data of the isotherm to an effective pore diameter using the Lennard-Jones potential functions. The correlating expression is given by:

$$RT\ln(P/Po) = K \frac{N_1 A_1 + N_2 A_2}{r^{-4}(1-d)} \times \left[\frac{r^{-4}}{3(1-d/2)^3} - \frac{r^{-10}}{9(1-d/2)^9} - \frac{r^{-4}}{3(d/2)^3} + \frac{r^{-10}}{9(d/2)^9} \right]$$

where K is Avogadro's number, N is the number of molecules per unit area, A is the Lennard-Lones constants obtained by the Kirwood-Muller expressions, l is the sum of the diameters of adsorbent and adsorbate and r is the distance between a gas atom and the surface at zero interaction energy. The subscripts 1 and 2 refer to the adsorbent and adsorbate, respectively.

The procedure consists in choosing suitable values of the effective pore diameter $(1-d_1)$ between 0.35 and 1.4 nm and the values of l substituted in equation 1 and the corresponding P/Po value obtained. Then using the isotherm data, the pore size distribution is obtained by plotting the pore size distribution function (the derivative of W/Wo with respect to the pore radius) versus the pore radius.

The data for the different samples is presented in figure 4.

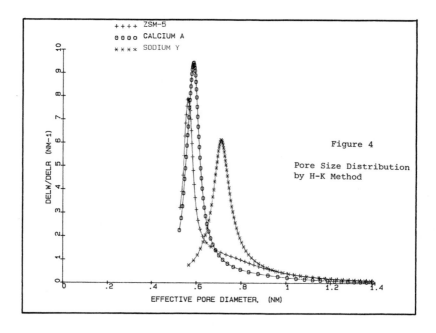

Figure 4

Pore Size Distribution
by H-K Method

TABLE III

Comparison of pore diameters obtained from H-K model and accepted values.

| Zeolite | Pore diameter (nm) | |
	H-K Model	Accepted values
Ca A	.59	.50
ZSM-5	.57	.56
Na Y	.71	.74

Conclusions

From the results of the adsorption experiments using argon and nitrogen, we can arrive at the following conclusions:

1. Nitrogen is unsuitable for the pore size characterization of zeolites. We attribute this shortcoming to the strong interaction of nitrogen with zeolites due to nitrogen's large quadrupole moment.

2. Argon interacts less strongly with zeolites, and therefore, can be used successfully in ranking zeolites on the basis of pore sizes.

3. The Horvath-Kawazoe model in conjunction with the argon adsorption data predicts well the effective pore sizes of the zeolites studied.

References

1. S.J. Gregg and K.S.W. Sing, Adsorption, Surface Area and Porosity, 2nd edition, Academic Press, New York, 1982
2. E. Bowes, Private Communication
3. M.M. Dubinin, Molecular Sieves - II, ACS Symposium Series 40, J.R. Katzer,Editor, 1977.
4. K.K. Unger and U. Muller, IUPAC-Symposium on Characterization of Porous Solids, Reprints, Bad Soden, W. Germany, 1987
5. G. Horvath and K. Kawazoe, J. Chem. Eng. Japan, 16(6), 470-475 (1983)
6. J. Seifert and G. Emig, Chem. Ing. Tech. 59(6), 475-485 (1987)
7. J. Seifert, Diploma Thesis, University Erlangen-Nuremberg, 1985

SMALL-ANGLE X-RAY SCATTERING OF CARBON-SUPPORTED METAL CATALYSTS

J. GOODISMAN*, M.G. PHILLIPS** AND H. BRUMBERGER*
*Syracuse University, Dept. of Chemistry, Syracuse, N.Y. 13244-1200
**SUNY College at Cortland, Dept. of Chemistry, Cortland, N.Y. 13045

ABSTRACT

The slit-smeared intensity of X-radiation scattered from carbon-supported metal catalysts behaves as $c_1 h^{-1} + c_2 h^{-3}$ for higher h-values in the small-angle region ($h=4\pi\lambda^{-1}\sin\theta$, λ=X-ray wavelength, θ=half of scattering angle, c_1, c_2 are constants). For three-phase systems with sharp phase boundaries, uniform electron-densities in the metal and void phases, and lamellar electron-density fluctuations in the carbon phase, the small-angle X-ray scattering (SAXS) can be described in terms of inter- and intraphase electron-density correlation functions, and can be shown to yield the observed behavior. Further assumptions are required to separate these contributions and calculate interfacial specific surfaces. SAXS measurements on Pt, Ru, Pd, and Rh/C are presented.

INTRODUCTION

The small-angle X-ray scattering of carbon-supported metal catalysts in the region of larger h-values differs from that predicted by the Porod law for systems of internally homogeneous phases with sharp boundaries [1-3]. One-dimensionally ordered regions displaying lamellar electron-density fluctuations in the carbon phase, averaged over spatial orientations, have been shown to account for the experimentally observed intensity behavior [4,5]. We present a theoretical analysis of the scattering from such systems which separates the inter- and intraphase fluctuations, and permits the calculation of all surface areas with the help of appropriate structural models of the catalyst. The theory is applied to scattering measurements of 5 w/o Pt, Ru, Pd, and Rh/C. This approach does not require the reduction of the system to one of two phases by filling the pores with a liquid of equal electron density to that of support [6].

THEORY

The scattered intensity, $I(h)$, for a spatially isotropic sample is, in terms of the correlation function $\gamma(r)$, given by

$$I(h) = I_e \langle \eta^2 \rangle V \int_0^\infty 4\pi r^2 \gamma(r) \frac{\sin hr}{hr} \, dr \qquad (1)$$

where $\langle \eta^2 \rangle$ is the mean-square electron density fluctuation, V the irradiated volume, and I_e, the scattering by a single electron, is essentially constant for constant incident intensity. If $P_{ij}(r)$ is the probability that a line segment r has one end in phase i and the other in phase j, and if n_i is the electron density of the ith phase (ϕ_i being its volume fraction), then $\gamma(r)$ is expressed as follows [7]:

$$\langle\eta^2\rangle\gamma(r) = \sum_{i,j} P_{ij}(r)\langle[n_i(u) - \bar{n}][n_j(u+r)-\bar{n}]\rangle \qquad (2)$$

with

$$\bar{n} = \sum_i \phi_i n_i \;;\; \langle\eta^2\rangle = \overline{n^2} - \bar{n}^2 = \frac{1}{2}\sum_{i,j}(n_i - n_j)^2\phi_i\phi_j \qquad (3)$$

The triangular brackets indicate an average over all points **u** such that **u** is in phase i and **u+r** in phase j; the average is required because of possible intraphase density fluctuations. For a phase i without internal density fluctuations, n_i is a constant. The internal density fluctuations occur in phase 1 and are not correlated with the other phases. Then

$$\langle\eta^2\rangle\gamma(r)=P_{11}(r)\{\langle[n_1(u)-n_1][n_1(u+r)-n_1]\rangle + \langle(n_1- \bar{n})^2\rangle\}$$

$$(4)$$

$$\begin{array}{ll} (j\neq1) & (j\neq1) & (i\neq1)\ (j\neq1) \\ + \sum_j P_{1j}(r)(n_1-\bar{n})(n_j-\bar{n}) + \sum_j P_{j1}(r)(n_j-\bar{n})(n_1-\bar{n}) + \sum_i \sum_j P_{ij}(r)(n_i-\bar{n})(n_j-\bar{n}) \end{array}$$

with the average being performed over points **u** such that **u** and **u+r** are both in phase 1. n_1 is now the average electron density in phase 1. The correlation function can then be written as a sum of intraphase and interphase terms:

$$\gamma(r) = \frac{P_{11}(r)\langle[n_1(u)-n_1][n_1(u+r)-n_1]\rangle + \sum_{i,j} P_{ij}(r)(n_i-\bar{n})(n_j-\bar{n})}{\langle\eta^2\rangle} \qquad (5)$$

Note that the P_{ij} depend only on the magnitude of **r** in the spatially isotropic system.

From (3) and conditions which the P_{ij} obey, such as

$$\sum_j P_{ij}(r) = \phi_i \text{ and } P_{ij}(0) = \delta_{ij}\,\phi_i \qquad (6)$$

one can write [7]

$$\sum_{i,j} P_{ij}(r)(n_i-\bar{n})(n_j-\bar{n}) = \frac{1}{2}\sum_{i,j}(n_i-n_j)^2[\phi_i\phi_j-P_{ij}(r)] \qquad (7)$$

The interphase term, γ_2, in γ is thus

$$\gamma_2 = \frac{\sum_{i,j}(n_i-n_j)^2(\phi_i\phi_j-P_{ij})}{\sum_{i,j}(n_i-n_j)^2 \phi_i\phi_j} \qquad (8)$$

which is identical to that for a system of internally homogeneous phases. The interphase surface area S_{ij} obeys [7]

$$S_{ij}/4V = [dP_{ij}/dr]_{r=o} \qquad (9)$$

We turn to the intraphase term. The average of $[n_1(u)-n_1][n_1(u+r)-n_1]$ approaches zero for large r and becomes $n_1^2 - n_1^2$ for $r = 0$; n_1^2 is the average squared electron density in phase 1. At any point u, density fluctuations will be correlated only in one direction, so that

$$P_{11}(r)\langle[n_1(u)-n_1][n_1(u+r)-n_1]\rangle = (\overline{n_1^2}-n_1^2)\ P_1(r)\gamma_1(z) \qquad (10)$$

where z is the component of r in the stacking direction of the lamellae, and $\gamma_1(0) = 1$. $P_1(r)$ is the probability that the points u and $u+r$ lie in the same lamellar stack; $P_1(r)$ should resemble $P_{11}(r)$ for small r, but fall below it for large r, since P_{11} includes the probability that u and $u+r$ are in different (uncorrelated) regions in the same phase. In averaging u over the sample, we will then also have to average over the orientations of the z-axis relative to the scattering vector.

The scattered intensity contributions of the inter- and intraphase portions of $\gamma(r)$ are, respectively,

$$I_2(h) = I_e\int dr\ e^{ih\cdot r}\ \gamma_2(r) = I_e\int_0^\infty 4\pi r^2\ \frac{\sum_{i>j}(n_i-n_j)^2(\phi_i\phi_j-P_{ij})}{\langle\eta\rangle^2}\ \frac{\sin hr}{hr}\ dr \qquad (11)$$

and

$$I_1(h) = I_e\ \frac{\overline{n_1^2}-n_1^2}{\langle\eta^2\rangle}\ \left\langle\int dr\ e^{ih\cdot r}\ P_1(r)\gamma_1(z)\ \right\rangle \qquad (12)$$

After averaging over all orientations of z, one obtains for the total scattered intensity

$$I = I_1+I_2 = \alpha\sum_{i>j}\int_0^\infty rh^{-1}\ \frac{(\phi_i\phi_j-P_{ij})(n_i-n_j)^2}{\langle\eta^2\rangle}\ \sin hr\ dr$$

$$+ \beta\int_0^\infty h^{-1}P_1(r)\ \sin hr\ dr \qquad (13)$$

Here, $\alpha = 4\pi I_e$ and $\beta = 2\pi I_e\ (\overline{n_1^2} - n_1^2)\int\gamma_1(z)dz/\langle\eta^2\rangle$. For "infinite-slit" conditions [8], the slit-smeared intensity $\tilde{I}(h)$ is given by

$$\tilde{I}(h) = \int_0^\infty J_0(hr)\left[\ \alpha r\frac{\sum_{i>j}(n_i-n_j)^2(\phi_i\phi_j-P_{ij})}{\sum_{i>j}(n_i-n_j)^2\phi_i\phi_j} + \beta P_1(r)\ \right]dr \qquad (14)$$

Boundary conditions on P_1 and the P_{ij} are $P_1(0) = \phi_1$, $P_1(\infty) = 0$, $P_{ij}(0)=\phi_i\delta_{ij}$ and $P_{ij}(\infty) = \phi_i\phi_j$.

Upon successive integrations by parts of eq.(14), being careful to remove singular terms in the integrand and using the appropriate boundary conditions, one arrives at the required intensity relationship:

$$\tilde{I}(h) = \beta\phi_1 h^{-1}+ \left\{\ \sum_{i>j}\alpha_{ij}P_{ij}'(0) - \tfrac{1}{2}\beta P_1''(0)\ \right\}h^{-3} + O(h^{-5}), \qquad (15)$$

where

$$\alpha_{ij} = \alpha \; (n_i - n_j)^2 / \sum_{i>j} (n_i - n_j)^2 \phi_i \phi_j. \qquad (16)$$

Thus, if $\tilde{I}(h) = c_1 h^{-1} + c_2 h^{-3}$ for large h (within the small-angle region), c_1 can be identified with $\beta\phi_1$, i.e. it arises only from the intraphase electron density fluctuations, whereas c_2 involves contributions from interphase and intraphase terms.

EXPERIMENTAL DETAILS AND DATA ANALYSIS

SAXS measurements on four commercial catalyst samples (5 w/o Pt, Ru, Pd, and Rh on activated carbon; Spex Industries) were made with CuK_α radiation, using a Kratky collimation system in the "infinite-slit" geometry [9], and a one-dimensional position-sensitive detector with multichannel analyzer. Skeletal and bulk densities (d_s and d_B respectively) were obtained by He and Hg displacement. Catalyst properties are summarized in Table I. A typical plot of $\tilde{I}h^3$ vs. h^2 is shown for Pd/C in Fig. 1.

Because of the appreciable scatter in the high-angle data, where intensities can be quite low, we work with integrals of the intensity over h. This procedure tends to smooth out statistical errors in individual intensity points. It was convenient to use the quantities

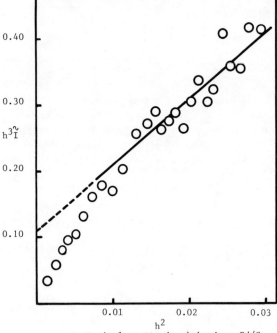

Fig. 1 Typical scattering behavior; Pd/C

Table I

CATALYST CHARACTERIZATION

Catalyst Sample	d_s	$d_B{}^b$	$d_B{}^a$	$\emptyset_2$	$\emptyset_3$	n_3	$\langle n^2 \rangle$
5% Pt/C	2.57	0.935	0.408	0.636	0.00219	8.544	0.434
5% Ru/C	2.34	0.992	0.415	0.576	0.00413	5.310	0.399
5% Pd/C	2.21	0.900	0.437	0.593	0.00371	5.242	0.386
5% Rh/C	2.38	0.929	0.388	0.610	0.00372	5.466	0.387

d_s is skeletal density (gcm^{-3}). $d_B{}^b$ is bulk density with pores of diameter $> 1\mu$ filled. $d_B{}^a$ is bulk density corresponding to displacement liquid at atmospheric pressure, with pores of diameter $> 15\mu$ filled. Densities were measured by Porous Materials, Inc., Ithaca, NY. n_3 = electron density of metal phase in moles of electrons per cm^3.

$$M(H) = \int_0^H \tilde{h}I(h)dh = \int_0^\infty drr\left[\sum_{i>j}\alpha_{ij}(\emptyset_i\emptyset_j - P_{ij}) + \beta P_1/r\right] \int_0^H dhhJ_0(hr)$$

$$= H \int_0^\infty dr\left[\sum_{i>j}\alpha_{ij}(\emptyset_i\emptyset_j - P_{ij}) + \beta P_1(r)^{-1}\right]J_1(Hr) \tag{17}$$

The analysis of the SAXS data for each of the four metal/C samples was carried out using the limiting slope and intercept from plots of $h^3\tilde{I}$ vs. h^2, and the integrals of eq.(17) from $h = 0$ to H-values corresponding to 30, 60, 90, 120, 150, and 180 channels ($h \approx 7.1 \times 10^{-4}$ N;N = no. of channels). In order to analyze the data and extract surface areas from them, a number of assumptions are required. From the scattering curves, only a single function of distance can be derived by inverse Fourier transformation. Although one may write this function as a sum of two terms, i.e. as

$$\sum_{i>j}\alpha_{ij}(\emptyset_i\emptyset_j - P_{ij}) + \beta P_1 r^{-1} = \alpha P_2 + \beta P_1 r^{-1} \tag{18}$$

(which also defines the quantity P_2 used later) there is no experimental way of separating them. One requires additional information, even though the value of β can be determined from the large-h behavior of $\tilde{I}$. It is certainly incorrect to subtract the asymptotic term, $\beta\emptyset_1 h^{-1}$, from $\tilde{I}(h)$ and expect the remainder to represent interphase fluctuations only [4]. The stack height distribution (which determines P_1) in lamellar, heat-treated carbon black follows an exponential law [10]; it is therefore appropriate to assume, here, that

$$P_1 = \emptyset_1 e^{-Br} \tag{19}$$

We postulate that

$$P_2 = e^{-Br}(1 + Cr + Ar^2), \tag{20}$$

since one has the conditions $P_1(0) = \emptyset_1$, $P_2(0) = 1$; $P_1 \to 0$ and $P_2 \to 0$ as $r \to \infty$.

Because P_1 should resemble P_2 for small r but fall below it for larger r, C and A are expected to be positive, but not so large as to make $P_2'(0)$ positive. The total intensity, in terms of P_1 and P_2, is then

$$\tilde{I}(h) = \int_0^\infty dr J_0(hr) \, [\beta \, P_1(r) + \alpha r P_2(r)] \tag{21}$$

and the integrals from h=0 to h=H are

$$M(H) = H \int_0^\infty dr J_1(Hr) \, [\beta P_1 r^{-1} + \alpha P_2] \tag{22}$$

With the forms for P_1 and P_2 in eqs.(19) and (20),

$$M(H) = \beta\phi_1 [(H^2 + B^2)^{\frac{1}{2}} - B] +$$
$$\alpha \, [1 - B \, (H^2 + B^2)^{-\frac{1}{2}} + CH^2(H^2 + B^2)^{-3/2} + 3ABH^2(H^2 + B^2)^{-5/2}] \tag{23}$$

M(0)=0, while, for large H,

$$M(H) \to \beta\phi_1(H - B) + \alpha \, (1 - BH^{-1} + CH^{-1} + \dots) \tag{24}$$

The coefficient of H in eq.(24), $\beta P_1(0) = \beta\phi_1$, is the coefficient of h^{-1} in the large-h form for $\tilde{I}(h)$, and is obtained as the slope of a plot of $h^3\tilde{I}$ vs. h^2.

The intercept on this plot is the coefficient of h^{-3} in the asymptotic form of $\tilde{I}(h)$, which, as shown above, is equal to $-\frac{1}{2}P''_1(0) - P'_2(0)$:

$$\text{Intercept} = -\tfrac{1}{2}\beta\phi_1 B^2 - \alpha \, (-B + C). \tag{25}$$

Therefore, starting from the slope and intercept, we determine α:

$$\alpha = (\text{intercept} + \tfrac{1}{2} \, \beta\phi_1 B^2)(B - C)^{-1} \tag{26}$$

The remaining parameters, A, B, and C, are determined by minimizing the deviation between calculated and experimental moments,

$$\sum_{j=1}^{6} \, [M_{calc.}(30jh_0) - M_{exp}(30jh_0)]^2 \tag{27}$$

where h_0 is the range of h corresponding to one channel.

SURFACE AREAS

For determination of surface areas, the quantity of interest is $P'_2(0)$, or $-(B-C)$. According to eqs.(9), (14), and (21),

$$-4P'_2(0) = \sum_{i>j} \, (n_i - n_j)^2 \, (S_{ij}/V) \, / \, \langle n^2 \rangle \tag{28}$$

Similar values (~0.15-0.16) are found for three of the catalysts, but for Rh/C the value is only 0.0679.

It is clear that to extract surface-to-volume ratios for the three individual surfaces S_{12}, S_{13}, and S_{23}, additional assumptions (based on a model of the system) are required. These would constitute relations between the three surface areas or P_{ij}'s. Even if measurements for the metal-free support were available [11] one would have to ensure that the support morphology in the catalysts is unchanged by the preparation procedure, and also make an assumption for the ratio S_{23}/S_{13}.

As shown by equation (28), the SAXS and density measurements can determine the sum

$$(n_1-n_2)^2 (S_{12}/V) + (n_1-n_3)^2(S_{13}/V) + (n_2-n_3)^2 (S_{23}/V) = -4P_2'(0) \langle n^2 \rangle \quad (29)$$

Since all surfaces are positive, this gives an upper bound on the individual S_{ij}/V, although, because of the small size of S_{13} and S_{23} relative to S_{12}, only the bound on S_{12}/V is useful. The upper bounds on S_{12}/V for the four samples, Pt, Ru, Pd and Rh/C, obtained by putting $S_{13} = S_{23} = 0$ in equation (29), are 0.0576, 0.0470, 0.0466, and 0.02095 A^{-1}. Since S_{ij}^*, the specific surface, is $(S_{ij}/V) 10^4(d_B^a)^{-1}$ (d_B^a is the bulk density obtained by atmospheric pressure mercury pycnometry), the upper bounds on the specific surfaces are calculated to be 1411, 1132, 1060, and 540 m^2g^{-1}, respectively.

In order to calculate all surfaces S_{ij}^*, a specific model of the catalyst morphology must be invoked. In various previous applications, we have found cell models very useful and capable of giving good agreement between SAXS and BET surface areas [12-14]. Uncorrelated cell models with a single cell size appear appropriate for these systems, which have extremely high surface areas and a high degree of dispersity for all phases. For such models [14], all the P_{ij} are products of $\phi_i\phi_j$ with a single function of r, and from eq.9, it follows that

$$S_{ij} = \phi_i\phi_j S, \quad (30)$$

S being a constant. Equations (29) and (30) permit determination of S:

$$-4P_2'(0) = S/V, \quad (31)$$

and the S_{ij} are found from (30); specific surfaces are reported in Table II, and are in good agreement with BET areas except for Rh/C, whose SAXS area is substantially lower than that determined by gas adsorption. Because of space limitations, we only quote the results for this type of model, though others can be used. It is in any case clear that some additional assumptions, concerning P_1 and P_2 and the separation of P_2 into the P_{ij}, are always required to arrive at all of the specific surface areas from SAXS data. We have shown that usable S_{ij}^* can be calculated from experimentally reasonable assumptions concerning P_1 and P_2, in conjunction with an appropriate morphological model of the catalyst, and a single SAX scattering curve for each sample.

Table II
SPECIFIC SURFACES

Sample	$B(A^{-1})$	$C(A^{-1})$	$A(A^{-2})$	$-4P_2'(0)(A^{-1})$	S_{12}^*	S_{13}^*	S_{23}^*
					$(m2\ g^{-1})$		
5% Pt/C	0.0417	0.001166	0.000284	0.1621	914	3.1	5.5
5% Ru/C	0.0375	0.000610	0.000123	0.1477	861	6.2	8.4
5% Pd/C	0.0379	0.000016	0.000362	0.1516	830	5.2	7.6
5% Rh/C	0.0191	0.002120	0.000048	0.0679	412	2.5	3.9

1 support   2 voids   3 metal

ACKNOWLEDGEMENT

The support of the National Science Foundation, under Grant No. CPE-793779, is gratefully acknowledged.

REFERENCES

1. G. Porod, Kolloid-Z. 124, 83 (1951).
2. P. Debye, H.R. Anderson, Jr. and H. Brumberger, J. Appl. Phys. 28, 679 (1957).
3. S. Ciccariello, J. Goodisman and H. Brumberger, J. Appl. Cryst., in press.
4. R. Perrett and W. Ruland, J. Appl. Cryst. 1, 308 (1968).
5. W. Ruland, J. Appl. Cryst. 4, 70 (1971).
6. R. G. Jenkins, P.L. Walker, Jr., A. Linares-Solano, F. Rodriguez-Reinoso and C. Salinas-Martinez de Leca, Carbon 20, 185 (1982).
7. J. Goodisman and H. Brumberger, J. Appl. Cryst. 4, 347 (1971).
8. O. Kratky, G. Porod and Z. Skala, Acta Phys. Austriaca 13, 76 (1960).
9. O. Kratky, in Small-Angle X-Ray Scattering, edited by H. Brumberger (Gordon and Breach, Science Publishers, New York, 1967).
10. S. Ergun and T.J. Gifford, J. Appl. Cryst. 1, 313 (1968).
11. J. Goodisman, H. Brumberger and R. Cupelo, J. Appl. Cryst. 14, 305 (1981)
12. H. Brumberger, F. Delaglio, J. Goodisman and M. Whitfield, J. Appl. Cryst. 19, 287 (1986).
13. H. Brumberger and J. Goodisman, J. Appl. Cryst. 16, 83 (1983).
14. J. Goodisman and N. Coppa, Acta Cryst. A37, 170 (1981).

THREE-STATE LATTICE-GAS MODEL OF H-S ON Pt(111)

PER ARNE RIKVOLD*, JOSEPH B. COLLINS**, G. D. HANSEN***,
J. D. GUNTON** and E. T. GAWLINSKI**
*Department of Physics, Florida State University, Tallahassee, FL 32306-3016
**Department of Physics and Center for Advanced Computational Science,
Temple University, Philadelphia, PA 19122
***ChemLink Industrial/Petroleum Chemicals Division, Malvern, PA 19355

ABSTRACT

We consider a three-state lattice-gas with nearest-neighbor interactions on a triangular lattice as a model for multicomponent chemi- and physisorption. By varying the lateral interaction constants between the adsorbate particles, this model can be made to exhibit either enhanced adsorption or poisoning (inhibited adsorption). We discuss here the conditions on the interaction constants that lead to poisoning. We present the results of a ground-state calculation and detailed numerical study of the phase diagram for a set of interactions that exhibits poisoning. We calculate the phase diagrams and adsorption isotherms by the finite-size scaling transfer-matrix method. We consider the result as a simple model for the coadsorption of Sulphur and Hydrogen on a Platinum (111) surface, with interaction constants estimated from experimental data. The resulting adsorption isotherms are in good agreement with experimental results.

INTRODUCTION

Much attention has recently been given to the effects of lateral interactions on multicomponent adsorption at interfaces [1]. Such interactions may give rise to numerous ordered adsorbate phases separated by either continuous or discontinuous phase transitions. The resulting adsorption isotherms differ dramatically from the multicomponent Langmuir form

$$\theta_i = \exp(\beta\mu_i)/(1 + \sum_j \exp(\beta\mu_j)) \tag{1}$$

where θ_i and μ_i are, respectively, the coverage and chemical potential for the i-th adsorbate species. The inverse temperature is $\beta = 1/(k_B T)$. Among the scientifically and technologically important effects that may be induced by lateral adsorbate-adsorbate interactions are enhanced and inhibited adsorption. In enhanced adsorption, the adsorption of one species is facilitated by the presence in small amounts of a coadsorbate. This effect is an important aspect of corrosion and dispersion control technologies [2]. It is also one aspect of promotion in heterogeneous catalysis [3], [4],[5]. In inhibited adsorption small amounts of a coadsorbate make the surface less attractive to the adsorbate, promoting desorption. Inhibited adsorption is closely related to the poisoning of solid catalysts [6],[7],[8].

We have studied the effects that varying the lateral interactions has on the surface phase diagrams and adsorption isotherms of a simple, well defined, statistical mechanical model. This model system, which exhibits many of the interesting phenomena associated with multicomponent, monolayer adsorption including enhanced adsorption and poisoning, is a three-state lattice-gas model with nearest-neighbor lateral interactions. This simplified model of multicomponent adsorption yields phase diagrams and highly nonlinear adsorption isotherms in good qualitative agreement with experiments [3]-[8]. The theoretical methods we use to study the model, finite-size scaling and numerical transfer matrix calculations, have in the past proved to be efficient tools in the theoretical study of surface phase transitions [9]-[12]. Readers interested in details about the transfer-matrix calculation, and the model's phase diagrams and relationships to other statistical-mechanical models, are referred to the paper by Collins, et al.[13].

THREE-STATE LATTICE-GAS MODEL

In this model, a three-state, nearest-neighbor lattice gas, adsorbate particles of two species, A and B, may occupy the sites of a two-dimensional lattice representing the adsorbent surface. The adsorption state of the i-th lattice site is given by the local concentration variables c_i^A and c_i^B. The local concentration of A, c_i^A, equals unity if the site is occupied by an A particle and vanishes otherwise. The local concentration of B, c_i^B, behaves analagously. Any site can be occupied by at most one particle, so that $c_i^A c_i^B = 0$. If the model is interpreted to represent adsorption from a binary gas, a site that is occupied by neither A nor B is considered vacant. If it is alternatively interpreted to represent adsorption from a solution of A and B in a solvent, then the third state represents a site occupied by a solvent particle. The latter physical situation is considerably more complicated than the former and, although encouraging experimental results have recently been reported [7],[14], we do not expect it to be more than qualitatively described by this simplest three-state model. For simplicity we restrict ourselves to the case of only nearest-neighbor interactions on the surface. This simplification in addition to truncating the two-body interactions, excludes any effects of unequal particle sizes. The lattice-gas Hamiltonian for this model (in the grand-canonical ensemble) is

$$H_{LG} - \mu_A \theta_A N - \mu_B \theta_B N = -\phi_{AA} \sum_{<i,j>} c_i^A c_j^A - \phi_{AB} \sum_{<i,j>} [c_i^A c_j^B + c_i^B c_j^A] - \phi_{BB} \sum_{<i,j>} c_i^B c_j^B$$
$$- \mu_A \sum_i c_i^A - \mu_B \sum_i c_i^B , \qquad (2)$$

where $\sum_{<i,j>}$ is the sum over all nearest-neighbor bonds, and $\sum_i$ is the sum over all lattice sites. The change in chemical potential when one X particle is removed from the bulk phase and adsorbed on the surface is $-\mu_X$ (defined to include the binding energy relative to the bulk phase). The total number of lattice sites is N, and $\theta_X = N^{-1} \sum_i c_i^X$ is the surface coverage by species X. The interaction energies ϕ_{AA}, ϕ_{BB}, and ϕ_{AB} describe effective interactions between particles adsorbed on the surface. In general they depend on the substrate, and their determination from first principles would demand quantum-mechanical calculations beyond the scope of the present work [15]. These interaction energies, in general, may bear little or no relation to the interaction energies between the same particles in the bulk phase. The sign convention is such that $\phi_{XY} > 0$ denotes an effective attraction and $\mu_X > 0$ denotes a tendency for adsorption in the absence of adsorbate-adsorbate interactions. In terms of a liquid solution interpretation, all the interaction strengths and chemical potentials are considered relative to the state in which both the adsorbent surface and the adsorbate particles are completely solvated. We have determined the model interaction constants from experimental chemical data in such a way as to produce phase diagrams and adsorption isotherms in qualitative agreement with experiments. A discussion of our assignment of interaction constants is given later.

We have chosen for our model the triangular lattice, corresponding to the various sets of adsorption sites on the {111} plane of a three-dimensional face-centered cubic lattice, e.g., platinum.

ORDERED STATES AND GROUND-STATE CALCULATION

Those ordered states on a triangular lattice that can be reached by second-order phase transitions are determined by Landau-Lifshitz group-theoretical arguments [16]. We only sketch the results here, and readers interested in further detail are referred to Ref. [13]. The ordered ground states are denoted $(\sqrt{3} \times \sqrt{3})$ and correspond to a separation of the lattice into three equivalent sublattices. The primitive unit cell contains three sites, one from each sublattice. Disordered states are denoted (1x1). The boundaries between the regions in the five-dimensional parameter space that correspond to particular ground states are obtained by pairwise equating the ground-state energies. The complete set of ground states together with their corresponding values of Q, P, θ_A, θ_B, and energy per lattice site are listed in Ref. [13]. We choose the notation for identifying ground states as $(1 \times 1)_P^Q$ for disordered phases and $(\sqrt{3} \times \sqrt{3})_P^Q$ for the ordered phases.

POISONING EFFECTS

Among the many interesting and technologically important phenomena that can occur in systems with multicomponent adsorption are enhancement and inhibition of the adsorption of a species, A, through its interactions with a co-adsorbate, B. An understanding of these effects in solutions is important in designing efficient programs for, e.g., corrosion control, and control of dispersion and sedimentation of solid suspensions through controlled adsorption of appropriate substances [2]. In heterogeneous catalysis, both at solid-gas and solid-liquid interfaces, these same effects are of great importance for the action of reaction modifiers, such as enhancers and poisons [6]. The interaction constants in the three-state lattice-gas model studied here can be chosen so as to produce either phenomenon. Although the model may be too simple to be directly applicable to the above-mentioned technological problems, it is of interest to determine the sets of interaction constants for which these effects are displayed. We introduce some general considerations for the case of poisoning.

With the phrase "inhibition (poisoning) of the adsorption of A by B" we understand the existence of a direct transition from a phase in which $\theta_A \approx 1$, $\theta_B \approx 0$ to one in which $\theta_A \approx 0$, $\theta_B \ll 1$, with the additional requirement that the transition can be crossed by a small increase in μ_B at constant μ_A. We do not claim this definition to be unique, but we believe that it contains the essential features of the phenomenon. In the lattice-gas model these conditions are satisfied by the first-order phase transition between $(1\times1)_1^1$, $(\theta_A = 1, \theta_B = 0)$, and $(\sqrt{3}\times\sqrt{3})_{-1/3}^{1/3}$, $(\theta_A = 0, \theta_B = 1/3)$. Our examination of the full set of ground-state diagrams reveals that the range of parameters in the antiferromagnetic case is restricted by the intervention of the $(\sqrt{3}\times\sqrt{3})_{\pm1/3}^1$ phases. The regions in which poisoning may occur are

$$\phi_{AB} > 0 \ \underline{and} \ \tilde{\phi}_{BB} < 0 \ \underline{and} \ \tilde{\phi}_{AA} > 2 \tag{3a}$$

or

$$\phi_{AB} < 0 \ \underline{and} \ \tilde{\phi}_{BB} < 0 \ \underline{and} \ \tilde{\phi}_{AA} > Max[-\frac{1}{2}, 1/3\tilde{\phi}_{BB}] , \tag{3b}$$

where $\tilde{\phi}_{XX} = \phi_{XX}/\phi_{AB}$. In no case is mutual poisoning allowed. These inequalities, (3a) and (3b), correspond, respectively, to the diagonally hatched regions in Figs. 1a and 1b.

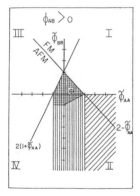
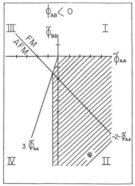

Figure 1.
The plane of reduced coupling constants $\tilde{\phi}_{AA} = \phi_{AA}/|\phi_{AB}|$ and $\tilde{\phi}_{BB} = \phi_{BB}/|\phi_{AB}|$.
a. $\phi_{AB} > 0$. The diagonally hatched region corresponds to inhibited adsorption (poisoning) of A by B. The vertically hatched region corresponds to enhanced adsorption of A by B, and the cross-hatched region corresponds to mutual enhancement.
b. $\phi_{AB} < 0$. The diagonally hatched region corresponds to inhibited adsorption (poisoning) of A by B.

It is worth noting that the regions in the $\tilde{\phi}_{AA}$, $\tilde{\phi}_{BB}$ plane corresponding to enhancement and poisoning do not overlap.

The main physical point to note is that the BB interactions must be repulsive, $(\phi_{BB} < 0)$, (as has also been noted by Reynolds, et al., for a lattice-gas model with AB nearest-neighbor exclusion [6]), in order to ensure a low-coverage phase of B. Furthermore, the AA interactions must be sufficiently non-repulsive to allow the formation of the $\theta_A = 1$ phase at sufficiently low μ_A for the transition into $(\sqrt{3}\times\sqrt{3})_{-1/3}^{1/3}$ to occur, $(\phi_{AB} \leq 0, \tilde{\phi}_{AA} > Max[-\frac{1}{2}, 1/3\tilde{\phi}_{BB}])$, or to avoid the transition being destroyed by the intervention of the $(\sqrt{3}\times\sqrt{3})_{1/3}^1$ phase, $(\phi_{AB} > 0, \tilde{\phi}_{AA} > 2)$.

DISCUSSION OF INTERACTION STRENGTHS

For detailed numerical study at finite temperatures we have chosen a set of adsorbate-adsorbate interaction parameters illustrative of poisoning. We have chosen values that are of the correct order of magnitude to represent the interactions between nearest-neighbor, chemisorbed (local binding energy > 60 kJ/mole) adparticles[1], and in particular we have considered one of the lattices of three-fold degenerate adsorption sites on a close-packed substrate, such as the (111) surface of a face-centerd cubic transition metal. Possible adsorbate particles include single donor atoms (such as H and the alkali metals, which often function as catalytic promoters), single acceptor particles (such as S or O, which often function as catalytic poisons), and small molecules (such as CO or HCl). The literature in the field of heterogeneous catalysis is vast. We have based our estimates of effective interaction constants on experimental and theoretical results from some of these sources, in particular Ref. [6] for S-CO, and Refs. [4] and [5] for K-CO. We also have referred to theoretical calculations by Dreyssee, et al.[19], for Re/W(110) (not a triangular lattice), and by Muscat[15] for H on various transition metal (111) surfaces. Results that are particularly encouraging for the application of simple lattice-gas models to interfaces between solids and aqueous electrolytes have recently been obtained by Schardt, et al.[14], for Pb+HCl+H_2O, and by Protopopoff and Marcus[7] for H+S+H_2O, both on Pt(111). However, we must emphasize that the parameters chosen are not intended to describe particular systems in complete detail, but only qualitatively. In return we expect that the qualitative aspects of our results, especially the adsorption isotherms, should apply to a far wider class of systems.

The system H+S/Pt(111) in aqueous solution may provide a simple physical realization of two-component adsorption with poisoning on a triangular lattice[7],[8], although there is some disagreement as to whether H and S adsorb on the same or on different lattices of three-fold hollow sites [20]-[22]. In an attempt to provide a simple model of this system, we have estimated effective nearest-neighbor lattice-gas interactions consistent with experimental results. The H-H and H-S interactions we have estimated from thermodynamic data in Ref. [7] as ϕ_{AA}=+2 kJ/mol, and ϕ_{AB}=−3 kJ/mol, respectively. The attractive nature of the effective H-H interaction is confirmed by the absence of ordered phases of H/Pt(111) [20],[21]. The S-S interaction we have estimated as the strongly repulsive ϕ_{BB}=−16 kJ/mol from the maximum critical temperature, 653K, of the ($\sqrt{3}$x$\sqrt{3}$) S phase reported in Ref. [22]. We present the results of our model for this system below.

TRANSFER MATRIX RESULTS AT FINITE TEMPERATURES - PHASE DIAGRAMS AND ADSORPTION ISOTHERMS

We have obtained thermodynamic data for the model at finite temperatures and finite chemical potentials using the transfer matrix method [23] and finite-size scaling [24]. We provide our numerical data for phase-transition surfaces in graphical form with accompanying descriptions for our sample system whose parameters allow for inhibited adsorption.

Figure 2 shows the ground-state diagram for H-S on Pt(111). Figure 3 shows a three dimensional view of the shape of the field-temperature phase diagram. The h and d coordinates are those shown in Fig. 2. Here there are two sets of double-humped surfaces; one extending from large ($\mu_A + \mu_B$), and the other extending from large $−\mu_A$. Their cross sections quickly approach their asymptotic values. Extending from the large $−\mu_B$ region there is a first-order "wall" dividing the $(1x1)_0^0$ phase from the $(1x1)_1^1$ phase.

Sample isotherms for our model of H+S/Pt(111) at T=1.22, corresponding to room temperature, are shown in Fig. 4. The temperature is above the asymptotic critical temperature for the $(1x1)_1^1$-to-$(1x1)_0^0$ transition, but below the asymptotic maximum critical temperatures for the ($\sqrt{3}$x$\sqrt{3}$) phases. The steep slopes exhibited by the isotherms around μ_B =+0.8 do not correspond to a phase transition. Although there are considerable finite-size effects in this region, finite-size extrapolation by the methods discussed in Ref. [13] does not predict finite coverage discontinuities, confirming the absence of a first-order transition. The θ_B isotherm shows a series of four second-order transitions at approximately μ_B = 4, 20, 28, and 44, as expected from the ground-state diagram and asymptotic critical surfaces. The initial slope of the coadsorption curve is ≈ -7 which corresponds to the hard hexagon model [6],[18]. This leads us to interpret the steep slopes exhibited by the isotherms around μ_B =+0.8 as representing local fluctuations into a hard-hexagon configuration.

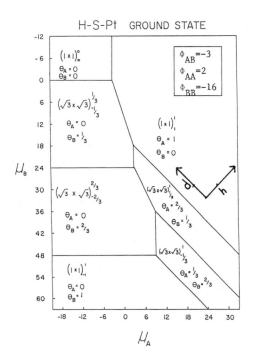

H-S-Pt GROUND STATE

$\Phi_{AB} = -3$

$\Phi_{AA} = 2$

$\Phi_{BB} = -16$

Figure 2.

Ground-state diagram for H-S on Pt(111) which exhibits inhibited adsorption (poisoning) of A by B.

Chemical potentials and temperature are given in units of the quantity

$$|J| = |\Phi_{AA} - 2\Phi_{AB} + \Phi_{BB}|/4$$

$$= 2.0 \text{ KJ/mole}.$$

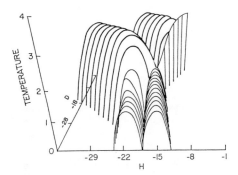

Figure 3.

A three-dimensional view of the the shape of the field-temperature phase diagram for a model similar to ours for H-S on Pt(111) where the axes are the same as in a. There are two sets of double-humped surfaces; one extending from large $(\mu_A + \mu_B)$, and the other extending from large $-\mu_A$. Extending from the large $-\mu_B$ region there is a first-order "wall" dividing the $(1 \times 1)_0^0$ phase from the $(1 \times 1)_1^1$ phase.

254

Figure 4.

Isotherms for our model of H+S/Pt(111) ($\tilde{\phi}_{AA} = 2/3$, $\tilde{\phi}_{BB} = -16/3$) at T=1.22, corresponding to room temperature. This temperature is above the asymptotic critical temperature for the $(1x1)_1^1$-to-$(1x1)_0^0$ transition, but below the asymptotic maximum critical temperatures for the ($\sqrt{3}x\sqrt{3}$) phases.

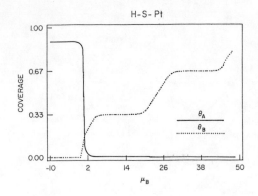

a. θ_A and θ_B vs. μ_B. The θ_A isotherm shows a very rapid change near $\mu_B = +0.8$, which does not, however, correspond to a phase transition. The θ_B isotherm shows rapid change at $\mu_B = +0.8$, followed by four second-order transitions at approximately $\mu_B = 4, 20, 28$, and 44, as expected from the ground-state diagram and asymptotic critical surfaces.

Scan is at $\mu_A = -5$

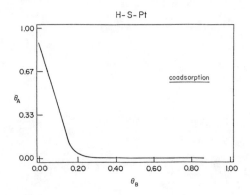

b. θ_A vs. θ_B.

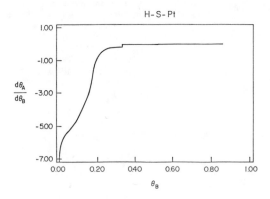

c. Derivative of the curve shown in b, $\dfrac{d\theta_A}{d\theta_B}$, corresponding to the number of A particles desorbed for each B particle adsorbed. The initial slope is approximately -7, corresponding to a hard-hexagon configuration of "holes" with one B particle at the center.

The poisoning observed in our simple model is not quite as strong as found experimentally in Ref. [7] but is the strongest possible with this model. This may be the result of longer-range interactions in the physical system. However, our result demonstrates that it is possible to obtain strong poisoning effects in models with only nearest-neighbor interactions without invoking the physically unreasonable infinite A-B repulsion inherent in the hard-hexagon model [6]. As previously remarked, the interaction constants in our model are consistent with results from thermodynamic and scattering experiments.

To summarize, we have demonstrated the usefulness of the three-state Ising model in obtaining theoretical adsorption isotherms for a complex multi-component adsorption problem with a surprisingly modest computational effort. In particular we can reproduce quite nicely the qualitative features of a system, H-S on Pt(111), that exhibits enhanced desorption, otherwise known as poisoning, with a suitable choice of interaction constants. Finally, we show how this particular set of interaction constants satisfies a general criterion necessary for such phenomena as enhanced adsorption and desorption.

ACKNOWLEDGEMENTS

Supported by the Commonwealth of Pennsylvania's Ben Franklin Partnership through the Advanced Technology Center of Southeastern Pennsylvania, a program of the University City Science Center. Also J.D.G. wishes to acknowledge the support of an Office of Naval Research Grant #N00014-83-K-0382.

REFERENCES

[1] G. Ertl, Langmuir 3, 4 (1987).

[2] L. Freedman, in "Proceedings of the 44th International Water Conference", Pittsburgh, Pennsylvania, October, 1983, p. 17. (Western Pennsylvania Engineering Society, Pittsburgh, 1984.); "Corrosion Control Through a Better Understanding of the Metallic Substrate / Organic Coating / Interface", edited by H. Leidheiser. (Center for Surface Coatings Research, Lehigh University, Bethlehem, Pennsylvania, 1984.)

[3] H. P. Bonzel, J. Vac. Sci. Technol. A 2, 866 (1984)

[4] K. J. Uram, L. Ng, and J. T. Yates, Jr., Surf. Sci. 177, 253 (1986).

[5] F. M. Hoffmann, J. Hrbeek, and R. A. dePaola, Chem. Phys. Lett. 106, 83 (1984); R. A. dePaola, J. Hrbeek, and F. M. Hoffmann, J. Chem. Phys. 82, 2484 (1985); J. J. Weimer, E. Umbach, and D. Menzel, Surf. Sci. 155, 132 (1985).

[6] A. E. Reynolds, J. S. Foord, and D. J. Tildesley, Surf. Sci. 166, 19 (1986); J. S. Foord and A. E. Reynolds, Surf. Sci. 152/153, 426 (1985).

[7] E. Protopopoff and P. Marcus, Surf. Sci. 169, L237 (1986).

[8] C. M. Pradier, Y. Berthier, and J. Oudar, Surf. Sci. 130, 229 (1983).

[9] e.g., P. A. Rikvold, K. Kaski, J. D. Gunton, and M. C. Yalabik, Phys. Rev. B. 29, 6285 (1984), and references cited therein.

[10] P. D. Beale, Phys Rev. B 33, 1717 (1986)

[11] W. Kinzel and M. Schick, Phys. Rev. B 23, 3435 (1981)

[12] N. C. Bartelt, T. L. Einstein, and L. D. Roelofs, Phys. Rev. B 34, 1616 (1986); L. D. Roelofs, T. L. Einstein, N. C. Bartelt, and J. D. Shore, Surf. Sci. 176, 295 (1986)

[13] J. B. Collins, P. A. Rikvold, and E. T. Gawlinski, submitted to Phys. Rev. B.; P. A. Rikvold, J. B. Collins, G. D. Hansen, J. D. Gunton, submitted to Surface Science.

[14] B. C. Schardt, J. L. Stickney, D. A. Stern, A. Wieckowski, D. C. Zapien, and A. T. Hubbard, Surf. Sci. 175, 520 (1986).

[15] J.-P. Muscat, Phys. Rev. B 33, 8136 (1986).

[16] E. Domany, M. Schick, J. S. Walker, and R. B. Griffiths, Phys. Rev. B 18, 2209 (1978); M. Schick, Prog. Surf. Sci. 11, 245 (1981)

[17] Y. Saito, J. Chem. Phys. 74, 713 (1981)

[18] R. J. Baxter, J. Phys. A 13, L61 (1980)

[19] H. Dreyssee, D. Tomanek, K. H. Bennemann, Ber. Buns.-Ges. Phys. Chem. 90, 245 (1986).

[20] J. Lee, J. P. Cowin, and L. Wharton, Surf. Sci. 130, 1 (1983).

[21] I. P. Batra, Surf. Sci. 137, L97 (1984).

[22] K. Hayek, H. Glassl, A. Gutmann, H. Leonhard, M. Prutton, S. P. Tear, and M. R. Welton-Cook, Surf. Sci. 175, 535 (1986).

[23] C. Domb, Adv. Phys. 9, 149 (1960)

[24] M. P. Nightingale, Physica A 83, 561 (1976); Phys. Lett. 59a, 486 (1977)

PILLARED CLAYS AND MICAS

JACK W. JOHNSON and JOHN F. BRODY
Corporate Research, Exxon Research and Engineering,
Annandale, NJ 08801

ABSTRACT

The interest in the petroleum industry in converting heavier feeds to liquid fuels has led
to a search for microporous materials with pore sizes larger than those found in the faujasitic
zeolites which form the basis of many petroleum processing catalysts. Materials with zeolite-like
pores in the 10 Å range can be synthesized by intercalating large polyoxocations between the layers
of smectite clays. Subsequent calcination dehydrates the cations and converts them into oxide
pillars that prop the clay layers apart, resulting in permanent microporosity in the interlayer region.
Pillared clays have been studied extensively during the last decade due to their potential use in
petroleum processing as cracking and hydrocracking catalysts. Previous workers have primarily
utilized smectite clays such as montmorillonite and hectorite as the starting layered material for
pillared clay. We now report that synthetic fluoromicas, clay-like materials of layer charge density
higher than that of smectites, can also be pillared with polyoxoaluminum cations to form alumina-
pillared fluoromicas that are thermally stable up to 700°C.

Introduction

2:1 Layered Clay Minerals. Clays are fine-grained silicate minerals that are
ubiquitously found in rocks and soils. Their structures are composed of layers that are strongly
bound in two dimensions, but more weakly in the third, leading to particles with platelet-like
morphologies. A major class of clay minerals exhibit structures composed of 2:1 layers[1]. These
layers contain two tetrahedral sublayers and one octahedral sublayer. The tetrahedral sublayers are
of composition T_2O_5 where T is a tetrahedral cation, most commonly silicon. Each tetrahedron is
linked to three other tetrahedra through T-O-T bonds to form a hexagonal pattern like that shown in
Figure 1a. The fourth oxygen atom of each tetrahedron projects in a direction perpendicular to the
sublayer forming part of the immediately adjacent octahedral sublayer. The octahedral sublayer,
sandwiched between two tetrahedral sublayers, is formed of octahedrally coordinated cations that
share edges within the sublayer (Figure 1b). The oxygen atoms of the octahedral sublayers are
provided by the fourth oxygen atoms of the tetrahedra composing the upper and lower tetrahedral
sublayers and by hydroxyl groups that lie at the same level as the apical tetrahedral oxygens.
These hydroxyls are located at the center of the sixfold rings of the tetrahedral sublayers.

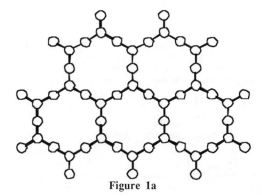

Figure 1a

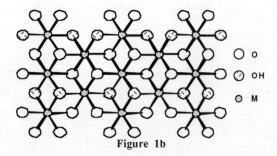

Figure 1b

The stoichiometry of a 2:1 layer is represented by the formula $A_x[M_{2-3}T_4O_{10}(OH)_2]$, shown in side view in Figure 2. The exchangeable interlayer cations are represented by A, the octahedral cations by M, and the tetrahedral cations by T. There are two main types of octahedral sublayer. When all the positions in the octahedral sublayer are filled, its structure is analogous to that of brucite ($Mg(OH)_2$) and the structure is designated *trioctahedral*. When only two-thirds of the octahedral positions are occupied and one-third are vacant, the structure of the octahedral sublayer is similar to that of gibbsite ($Al(OH)_3$), and the structure is designated *dioctahedral*. Most of the clay minerals found in nature have either a dioctahedral or a trioctahedral structure, but exceptions are known. One of the most interesting compounds we shall discuss in this report has an intermediate structure, with 2.5 of the 3 octahedral positions occupied. The cations in the octahedral sublayer are most commonly Al^{3+} and Mg^{2+}, but Fe^{2+}, Fe^{3+}, Li^+ and a wide range of transition metal ions are found. The cations in the tetrahedral sublayer are most often Si^{4+} or Al^{3+}, but may also be Fe^{3+}.

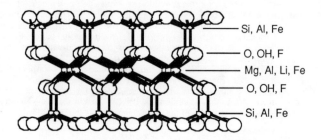

Si, Al, Fe

O, OH, F

Mg, Al, Li, Fe

O, OH, F

Si, Al, Fe

Exchangeable cations, $n\,H_2O$

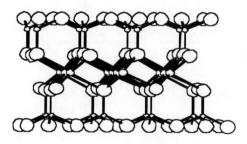

Figure 2

The key feature of a 2:1 clay mineral that determines its chemical behavior is the layer charge, represented by x in the formula $A_x[M_{2\text{-}3}T_4O_{10}(OH)_2]$. This layer charge arises from aliovalent substitution of the cations in the octahedral or tetrahedral sublayers. For example, talc is a 2:1 layered claymineral that has all Mg^{2+} in the octahedral sublayer and all Si^{4+} in the tetrahedral sublayer. Its formula is $Mg_3Si_4O_{10}(OH)_2$. There is no overall layer charge, so no interlayer cations are present. If some of the Mg^{2+} is replaced by Li^+, a negative charge equal to the degree of substitution is induced, forming the series of compounds $A_x[Mg_{3\text{-}x}Li_xSi_4O_{10}(OH)_2]$. If x is from ~0.2 to 0.6 the mineral would be called hectorite, while for x=1 the mineral is taeniolite. Similarly, the dioctahedral analogue of talc is pyrophyllite, $Al_2Si_4O_{10}(OH)_2$. Substitution of Mg^{2+} for Al^{3+} results in $A_x[Al_{2\text{-}x}Mg_xSi_4O_{10}(OH)_2]$ which for x from ~0.2 to 0.6 is called montmorillonite, a very common clay mineral. Substitutions can also be made in the tetrahedral sublayer. The most common substitution is Al^{3+} for Si^{4+}, giving in the dioctahedral case $A_x[Al_2Si_{4\text{-}x}Al_xO_{10}(OH)_2]$, which is beidellite for x from ~0.2 to 0.6 and muscovite for x=1. Clay minerals are divided into groups according to their layer charge. The talc-pyrophyllite group has layer charge of zero. If the layer charge is ~0.2 to 0.6 the mineral is in the smectite group. Vermiculites have layer charges from ~0.6 to 0.9, micas have layer charge ~1, and for brittle micas x is ~2.

The swelling behavior of 2:1 layered minerals is related to the layer charge. Talc and pyrophyllite, with no layer charge, do not readily take water or organic molecules into the interlayer space. Smectites, with a relatively small degree of layer charge and readily hydratable cations between the layers, swell freely in water or polar organic solvents, with the amount of water taken up depending on the identity of the interlayer cations. Vermiculites, with interlayer cations most often Mg^{+2}, are found with a double layer of water molecules in the interlayer space. As the charge density gets higher, the coulombic forces between the interlayer cations and the negatively charged layers become too strong to be overcome by intercalating molecules. Thus micas and brittle micas do not generally accept interlayer water or solvent molecules, with some important exceptions as will be discussed below.

Pillared Clays. In the mid 1970's, after the escalation in the price of crude oil, the petroleum industry was faced with the problem of processing heavy crudes rich in metals and large refractory molecules. Fluid catalytic cracking had the potential to handle such feeds, but the inherent problem with the zeolite catalysts used is the relatively small size of the micropores. This requires a large amount of relatively nonselective precracking on the catalyst matrix material before the large molecules present in heavier crudes can enter the zeolite pores, where they are cracked with high activity and selectivity. Very large pore zeolites with 16- or 20-ring pores, compared to the 12-ring pores of the faujasite catalysts used, are theoretically possible[2], but have thus far eluded synthetic chemists.

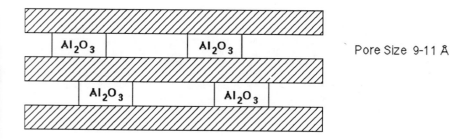

Pore Size 9-11 Å

Figure 3

Using smectite clays as the primary building blocks, an alternate approach to microporous materials was recognized. It had been known earlier that smectites could be made microporous by exchanging the interlayer alkali or alkaline earth cations with large organic or organometallic cations that could prop the layers apart and create microporosity between the pillars. Organoammonium cations[3] and transition metal chelates[4] were used as the pillaring cations, but the thermal stability of the resulting pillared clays did not approach that necessary for catalytic applications. The most significant advance in this area came from workers at W. R. Grace who discovered that partially hydrolyzed cation solutions of metals such as aluminum and zirconium contained polyoxocations that could exchange into smectites, forming pillared materials[5]. When calcined to temperatures well above 400°C, these polyoxocation pillars dehydrate to oxides, maintaining the layer separation and microporosity in a pillared clay that shows considerable thermal stability. These materials are represented schematically in Figure 3. Since that time, many different groups, both in the petroleum industry and in universities, have studied pillared clays. Different pillaring polycations have been utilized, including those of aluminum[6], zirconium[7], titanium[8], iron[9], and chromium[10]. More recently, oligimers of silica[11] and alumina-coated silica[12] have also been used to pillar smectites. The best materials have surface areas of 200-500 m²/g, pore volumes of 0.15-0.20 ml/g, and layer spacings of 16-20 Å which result from a 6-10 Å expansion of the 10 Å clay layers.

Water-Swelling Fluoromicas. Micas are defined as clay minerals having a structure composed of negatively charged 2:1 layers with an ideal layer charge of -1.0 per $O_{10}(OH)_2$ unit. In most naturally occuring micas this layer charge is balanced by potassium in the interlayer sites, although examples of other alkali cations are known[13]. Micas can also be synthesized in the laboratory. Since the layers contain hydroxyl groups as an integral part of the structure, micas must be synthesized under hydrothermal conditions. Pressures in the kilobar range are usually employed, making synthesis of large quantities impractical. However, the hydroxyl groups in mica can be replaced with similarly sized fluoride ions giving the class of compounds called fluoromicas. While fluoromicas are found in nature with partial or, more rarely, almost complete substitution of hydroxyl by fluoride, in this report we are concerned with synthetic fluoromicas.

Synthetic fluoromicas are prepared by high temperature reactions in the solid state or more efficiently by cooling a melt from ~1400° C. These materials were extensively studied by the Bureau of Mines, United States Department of the Interior, at the Norris Metallurgy Research Laboratory in Norris, Tennessee, during the 1950's. At least as early as 1952 it was recognized that some of the synthetic fluoromicas would swell in water[14]. Results from a thorough investigation of water-swelling fluoromicas is described in a series of reports[15] and in an early patent[16]. It was shown that in order to produce water-swelling mica phases the interlayer cation must be lithium or sodium, not potassium as is generally found in natural micas, and that the fluorine content had to correspond to 2 F per Si_4 unit, allowing for complete occupation of the hydroxyl positions by fluoride.

More recent studies of water-swelling synthetic fluoromicas have taken place in Japan. Sodium taeniolite, $Na[LiMg_2Si_4O_{10}F_2]$, was found to undergo limited water-swelling to form a two-layer hydrate[17], while lithium taeniolite--the same compound with Li^+ replacing Na^+ in the interlayer space, swells freely in water[18]. Further studies on taeniolites showed that germanium could be substituted for silicon in the tetrahedral sublayer, with some decrease in the tendency of the resulting mica to swell in water. It was also pointed out that in addition to the requirement of fluorine substitutition and the dependence on interlayer cation, water-swelling micas tend to be tetrasilicic; that is, the origin of the negative charge is substitution in the central octahedral sublayer of the 2:1 layer, rather than in the tetrahedral sublayer as is most common in micas[19]. In attempting to synthesize solid solutions of sodium and potassium taeniolite $Na_xK_{1-x}[LiMg_2Si_4O_{10}F_2]$, phase separation into K-rich(0≤x<0.5) and Na-rich(0.8<x≤1) phases occurs. Only the Na-rich taeniolite swells in water[20]. Further work demonstrated that a solid solution does exist between sodium fluorohectorite and sodium taeniolite. A continuous series of formula $Na_x[Li_xMg_{3-x}Si_4O_{10}F_2]$ exists for 0.32≤x≤1. The hectorite end member swelled freely in water while the Na taeniolite formed only a bilayer hydrate, with a smooth variation of the swelling behavior for intermediate compositions[21].

Another tetrasilicic fluoromica exists that has no natural analogue. The charge on the 2:1 layers results not from substitution of magnesium in the octahedral layer by lithium as in taeniolite, but rather by the formation of magnesium vacancies. This mica, $Na[Mg_{2.5}Si_4O_{10}F_2]$, is called simply tetrasilicic mica or TSM. It swells freely in water[22]. The crystal structures of the non-swelling potassium analogues of taeniolite[23] and TSM[24] and the corresponding phases[25] obtained upon substition of germanium for silicon in the tetrahedral layer have been solved by single crystal X-ray diffraction. The water-swelling fluoromicas and their colloidal solutions are covered by a series of U. S. patents[26].

Alumina-Pillared Fluoromicas

Properties of Commercial Materials. Sodium tetrasilicic mica is available from Topy Industries. Elemental analysis confirms the formula $NaMg_{2.5}Si_4O_{10}F_2$ claimed for the material. The manufacturer claims an cation exchange capacity of 67 meq/100g, measured by Ca^{2+} uptake. The cation exchange capacity clearly depends on the method used to measure it. By a different method[27] involving the uptake and then release of Ca^{2+}, we measured a cation exchange capacity of 84 meq/100g. However, by cation exchange with ammonium chloride, we measured a cation exchange capacity of only 2 meq/100g. The theoretical cation exchange capacity is ~250 meq/100g, so all the cations are not being exchanged. Clearly, the amount of cations exchanged depends on the swelling conditions during the exchange. A cation that does not allow the layers to swell, such as ammonium, will cause the sodium on the interior of the crystallites to become inaccessible. It would be interesting to examine the cation exchange capacity with lithium, as it generally causes swelling in water equal to or greater than that of sodium. We would predict that the measured ion exchange capacity would be higher, approaching the theoretical value.

Pillaring with Chlorhydrol™. There are a number of variations in the technique of exchanging polyoxoaluminum cations into smectites to form alumina pillared clays. In order to make reproducible, well ordered pillared fluoromicas, we investigated a number of different variations in the pillaring technique before settling on the way described below. One of the variables in producing alumina pillared clays is the nature of the alumina solutions. Some workers prefer to use alumina solutions prepared by hydrolyzing aluminum chloride solutions with sodium hydroxide. The reaction must be done slowly to avoid precipitation of $Al(OH)_3$. Various OH/Al

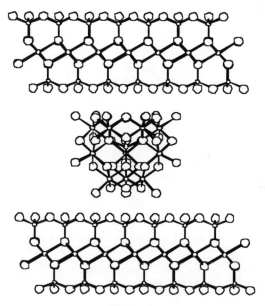

Figure 4

ratios and aging treatments were investigated, with one group[28] settling on solutions with OH/Al ratio of 1.85 aged at 95°C for 6 to 48 hours. Other groups have found commercially available Chlorhydrol™ solutions to produce better samples of pillared clays with respect to both maximum pore volume and enhanced thermal stability[5,6d]. Chlorhydrol™ is produced by the reaction of AlCl$_3$ solution with aluminum metal and is used in the manufacture of antiperspirants. Investigation of the actual pillaring solutions by ^{27}Al NMR indicates that both in hydrolyzed Al solutions and in Chlorhydrol™ there exists a variety of Al oligimers[29,30], including the Al$_{13}$O$_4$(OH)$_{24}$(H$_2$O)$_{12}^{7+}$ cation that has been structurally characterized in the solid state[31]. The relative concentrations of the various species depend on hydrolysis ratio, aging, dilution and other factors. The data of different investigators are not always in agreement. Generally, starting with Vaughan[5], it has been assumed that Al$_{13}$O$_4$(OH)$_{24}$(H$_2$O)$_{12}^{7+}$ is the pillaring cation, because the observed layer expansion is consistent with the known cation structure, as is the ^{27}Al MASNMR[32]. The structure of an alumina pillared clay or mica is represented in Figure 4. The details of how the pillar binds to the layer are not understood at this time.

Another variable in preparing pillared smectites is the method of exchanging the alkali ions initially present in the clay interlayer with the polyoxoaluminum cations. Most workers perform the exhange with a dilute (<1 wt %) clay suspension and heat the mixture at 60-90°C for a few hours, followed by thorough washing to remove all chloride and Al^{3+} monomer. However, the original patents[5] report successful pillaring of Ca-montmorillonite at much higher clay concentration, leading to efficient synthesis of larger amounts of material. After the addition of the alumina solution to the well-dispersed clay suspension, the ion exchange reaction can is completed during a period of heating. Some reports omit this heating and carry out the ion exchange at room temperature, but our initial experiments showed higher stability for pillared micas that were ion exchanged for an hour at 90° C. A recent study reports the variation in pillared clay microstructure and cracking activity as a function of the dilution and aging of the Chlorhydrol™ solution.[33]

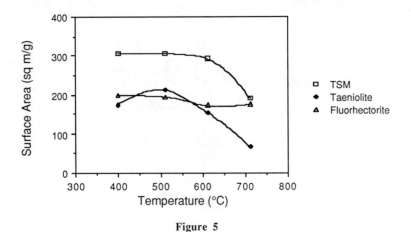

Figure 5

Our experiments also showed that in order to get well ordered pillared clays, the details of the calcination treatment were also important. The ion exchanged material must be washed free of excess chloride, then thoroughly dried in an oven at ~110° C. Once dry, the material is calcined in a muffle furnace at 200° C for two hours. Then the temperature is increased to 400° C at 50° C/hr and held for four hours. The resulting pillared micas have optimum surface area, X-ray crystallinity, and steam stability.

Characterization of Pillared Fluoromicas. We have prepared samples of alumina-pillared fluorohectorite, taeniolite, and tetrasilicic mica[34]. A strong, relatively narrow line is observed in the X-ray powder diffraction pattern at 18-19 Å (see examples following), indicating that the layer repeat distance is regular. The micas incorporate more aluminum than fluorohectorite. Pillared TSM contained 0.96 Al_{13} ions per Si_4O_{10} while fluorohectorite had only 0.62. Analysis of the nitrogen adsorption isotherms by the BET method gave surface areas of ca. 175-225 m²/g for pillared fluorohectorite and taeniolite, while pillared TSM consistently had a surface area of 300-350 m²/g. The hexane adsorption capacity of pillared TSM was slightly over 6 wt%.

Alumina-pillared TSM has higher surface area and is more stable in steam than the corresponding pillared fluorohectorite and taeniolite. In Figure 5, the effect on the surface area of treating the pillared micas with 25 Torr of steam in helium flow for two hours at various temperatures is displayed. The results of the surface area degradation with temperature are also reflected in the X-ray diffraction data, in which the 18Å line broadens and eventually disappears as the surface areas go below 100 m²/g. Some of the interlayer sodium remains after TSM is pillared by polyoxoaluminum cations. Elemental analyses indicate that about 20% of the sodium originally present in the cation exchange positions is left after extensive washing of the pillared material. After calcination, some of this residual sodium is labilized. By washing the pillared TSM with water after it has been calcined at 400°C, 60% of the residual sodium is removed. Removal of this residual sodium has a dramatic effect on both the steam stability and the acidity of the pillared TSM. It is widely known that the presence of sodium in zeolites and cracking catalysts adversely affects their stability in steam[35], and pillared micas are no exception. Removing the residual sodium by water washing after calcining results in increased steam stability, as is shown in Figure 6. Further treatment with ammonium ion removes a small additional amount of sodium, but does not result in a more stable product. The X-ray powder diffraction patterns of washed TSM after different temperature steam treatments are shown in Figure 7. Crystallinity is maintained to temperatures in excess of 700°C.

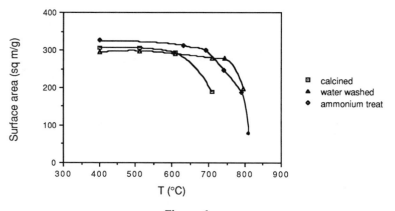

Figure 6

The acidity of pillared TSM is also affected by residual sodium. Acidity measurements based on the catalytic activity for 2-methylpent-2-ene isomerization[36] have shown that washing alumina pillared TSM results in a catalyst with an acidity slightly greater than that of γ-alumina, while calcined, but unwashed, Al-TSM has lower acidity[37]. Infrared spectroscopy of washed alumina pillared TSM films that have been exposed to pyridine indicate that the acidity is due primarily to Lewis-type sites, ie., the bands for pyridinium ion are not present[38].

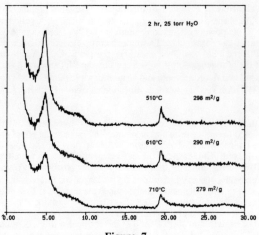

Figure 7

Zirconia-Pillared Tetrasilicic Mica

Sodium tetrasilicic mica can also be pillared with zirconia polyoxocations. By treating a dilute aqueous suspension of size-fractionated NaTSM with an aqueous solution of $ZrOCl_2$, followed by washing and calcination at 400°C, zirconia pillared TSM is formed. It has a layer spacing of 21Å, although it is not as well ordered as its aluminum analog, as demonstrated by the

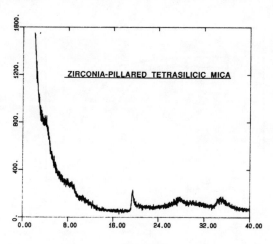

Figure 8

powder diffraction pattern shown in Figure 8. The peak representing the layer spacing is only a shoulder on the low angle background. Zr-TSM does have significant microporosity as can be seen by its BET surface area of 231 m^2/g and its hexane isotherm, shown in Figure 9. The initial portion of the isotherm shows rapid adsorption of about 5 wt% hexane at low partial pressure due to the micropores. The hysteresis on the desorption branch shows the presence of some mesopores as well.

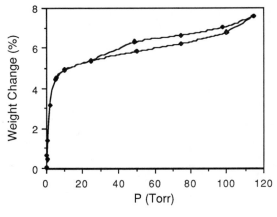

Figure 9

Summary

We have demonstrated that micas can be pillared by large cations in spite of their high layer charge density, as long as suitable swelling fluoromicas are chosen. The resulting pillared micas, particularly in the case of alumina pillared tetrasilicic mica, show good stability upon exposure to high temperature steam and have acidity and microporosity that make them interesting as potential cataytic materials.

Acknowledgements

We would like to thank J. R. Robbins, G. B. McVicker, J. J. Ziemiak, and L. N. Yacullo for aid in characterization and D. E. W. Vaughan for helpful discussions.

References

1. Brindley, G. W.; Brown, G., Eds.; "Crystal Structures of Clay Minerals and Their X-Ray Identification"; Mineralogical Society: London, 1980. Grim, R. E. "Clay Mineralogy"; McGraw-Hill: New York, 1968.
2. Smith, J. V.; Dytrych, W. J. *Nature*, **1984**, *309*, 607-608.
3. Barrer, R. M.; McCleod, D. M. *Trans. Faraday Soc.*, **1955**, *51*, 1290. Barrer, R. M. "Zeolites and Clay Minerals as Adsorbents and Molecular Sieves"; Academic Press: New York, 1978, pp. 407-483.
4. Knudson, M. I.; McAtee, J. L. *Clays and Clay Miner.*, **1973**, *21*, 19. Traynor, M. F.; Mortland, M. M.; Pinnavaia, T. J. *Clays and Clay Miner.*, **1978**, *26*, 319. Loeppert, R. H.; Mortland, M. M.; Pinnavaia, T. J. *Clays and Clay Miner.*, **1979**, *27*, 201.
5. Vaughan, D. E. W.; Lussier, R. J.; Magee, J. S. *U. S. Patents #4,176,090*, **1979**; *#4,248,739*, **1981**; *#4,271,043*, **1981**.
6. (a) Brindley, G. W.; Samples, R. E. Clay *Miner.*, **1977**, *12*, 229-237. (b) Lahav, N.; Shani, U.; Shabtai, J. *Clays and Clay Miner.*, **1978**, *26*, 107-115. (c) Lussier, R. J.; Magee, J. S.; Vaughan, D. E. W. *Prepr., 7th Canad. Symp. Catal.* Edmonton, Alberta **1980**, pp. 88-95. (d) Occelli, M. L.; Tindwa, R. M. *Clays and Clay Miner.*, **1983**, *31*, 22-28.

7. Yamanaka, S.; Brindley, G. W. *Clays and Clay Miner.,* **1979,** *27,* 119-124. Bartley, G. J. J.; Burch, R. *Appl. Catal.,* **1985,** *19,* 175-185. Burch, R.; Warburton, C. I. *J. Catal.,* **1986,** *97,* 503-510.
8. Sterte, J. *Clays and Clay Miner.,* **1986,** *34,* 658-664. Yamanaka, S.; Nishihara, T.; Hattori, M. *Mater. Chem. Phys.,* **1987,** *17,* 87-101.
9. Oades, J. M. *Clays and Clay Miner.,* **1984,** *32,* 49-57. Yamanaka, S.; Doi, T.; Sako, S.; Hattori, M. *Mater. Res. Bull.,* **1984,** *19,* 161-168.
10. Brindley, G. W.; Yamanaka, S. *Amer. Miner.,* **1979,** *64,* 830-835. Pinnavaia, T. J.; Tzou, M.-S.; Landau, S. D. *J. Amer. Chem. Soc.,* **1985,** *107,* 4783-4785.
11. Lewis, R. M.; Ott, K. C.; Van Santen, R. A. *U. S. Patent #4,510,257,* **1985.**
12. Ocelli, M. L. *J. Mol. Catal.,* 1986, 35, 377-389; Ocelli, M. L.; in Schultz, L. G.; van Olphen, H.; Mumpton, F. A., Eds.; "Proceedings of the International Clay Conference, Denver, 1985"; Clay Minerals Society: Bloomington, Indiana, 1987, pp.319-323.
13. Bailey, S. W. *Reviews in Mineralogy,* **1984,** *13,* 1-12.
14. Eitel, W. *Proc. Int. Symp. React. Solids,* **1952,** Gothenburg, Sweden, pp. 335-347.
15. Shell, H. R. *Bur. Mines Report 6076,* **1963.** Johnson, R. C.; Shell, H. R. *Bur. Mines Report 6235,* **1963.**
16. Hatch, R. A. *U. S. Patent #3,001,571,* **1961**
17. Kitajima, K.; Daimon, N. *Chem. Lett.,* **1972,** 953-956.
18. Kitajima, K.; Daimon, N. *Chem. Lett.,* **1973,** 241-244. Kitajima, K.; Daimon, N.; Kondo, R. *Nippon Kagaku Kaishi,* **1976,** 597-603.
19. Kitajima, K.; Daimon, N.; Kondo, R. *Clay Minerals,* **1978,** *13,* 167-175.
20. Kondo, R.; Daimon, M.; Asaga, K.; Nishikawa, T.; Kitajima, K.; Daimon, N. *J. Amer. Cer. Soc.,* **1980,** *63,* 41-43.
21. Kitajima, K.; Koyama, F.; Takusagawa, N. *Bull. Chem. Soc. Japan,* **1985,** *58,* 1325-1326.
22. Kitajima, K.; Daimon, N. *Nippon Kagaku Kaishi,* **1975,** 991-995.
23. Toraya, H.; Iwai, S.; Marumo, F.; Hirao, M. *Z. Kristallogr.,* **1977,** *146,* 73-83.
24. Toraya, H.; Iwai, S.; Marumo, F.; Daimon, N.; Kondo, R. *Z. Kristallogr.,* **1976,** *144,* 42-52.
25. Toraya, H.; Iwai, S.; Marumo, F.; Hirao, M. *Z. Kristallogr.,* **1978,** *148,* 65-81.
26. Daimon, N.; Izawa,T. *U. S. Patent #4,045,342,* **1977.** Daimon, N.; Kitajima, K. *U. S. Patent #4,067,819,* **1978.** Daimon, N.; Izawa,T.; Imai, M. *U. S. Patent #4,077,938,* **1978.**
27. Jackson, M. L. "Soil Chemical Analysis--An Advanced Course"; Published by the author: Madison, Wisconsin, 1979, pp.256-257.
28. Tokarz, M.; Shabtai, J. *Clays and Clay Miner.,* **1985,** *33,* 89-98.
29. Pinnavaia, T. J.; Tzou, M.-T.; Landau, S. D.; Raythatha, R. H. *J. Molec. Catal.,* **1984,** *27,* 195-212.
30. Akitt, J. W.; Farthing, A. *J. Chem. Soc. Dalton Trans.,* **1981,** 1606-1628. Teagarden, D. L.; Kozlowski, J. F.; White, J. L.; Hem, S. L. *J. Pharm. Sci.,* **1981,** *70,* 758-761. Schonherr, S.; Gorz, H.; Bertram, R.; Muller, D.; Gessner, G. *Z. anorg. allg. Chem.,* **1983,** *502,* 113-122. Bertram, R.; Gessner, W.; Muller, D.; Gorz, H.; Schonherr, S. *Z. anorg. allg. Chem.,* **1985,** *525,* 14-22. Bertram, R.; Gessner, W.; Muller, D. *Z. Chem.,* **1986,** *26,* 340-342.
31. Pauling, L. *Z. Kristallogr.,* **1933,** *84,* 442-452. Johansson, G. *Acta Chem. Scand,* **1960,** *14,* 771-773; *Ark. Kemi,* **1962,** *20,* 321-342.
32. Diddams, P.; Thomas, J. M.; Jones, W.; Ballantine, J. A.; Purnell, J. H. *J. Chem. Soc., Chem. Comm.,* **1984,** 1340-1342. Plee, D.; Borg, F.; Gatineau, L.; Fripiat, J. J. *J. Amer. Chem. Soc.,* **1985,** *107,* 2362-2369. Pinnavaia, T. J.; Landau, S. D.; Tzou, M.-T.; Johnson, I. D.; Lipsicas, M. *J. Amer. Chem. Soc.,* **1985,** *107,* 7222-7224.
33. Harris, J. R. *Prepr., Div. Petr. Chem. Amer. Chem. Soc.,* **1987,** *32,* 652-657.
34. Johnson, J. W.; Brody, J. F., to be submitted, **1988**
35. McDaniel, C. V.; Maher, P. K. in "Zeolite Chemistry and Catalysis", Rabo, J. A., ed. *ACS Monograph 171,* **1976,** pp.285-331
36. Kramer, G. M.; McVicker, G. B.; Ziemiak, J. J. *J. Catal.,* **1985,** *92,* 355-363
37. McVicker, G. B.; Ziemiak, J. J.; Johnson, J. W.; Brody, J. F., unpublished results.
38. Robbins, J. R.; Johnson, J. W., unpublished results.

VPI-5: A NOVEL LARGE PORE MOLECULAR SIEVE

MARK E. DAVIS*, CARLOS SALDARRIAGA*, CONSUELO MONTES*, JUAN GARCES** AND
CYRUS CROWDER**
*Department of Chemical Engineering, Virginia Polytechnic Institute and
State University, Blacksburg, VA, 24061
**The Dow Chemical Company, Midland, MI, 48640

ABSTRACT

This paper reports a novel family of aluminophosphate based
molecular sieves denoted as VPI-5. These molecular sieves are the first
to contain pores larger than 10 Å. The large pores of the VPI-5 sieves
consist of long channels circumscribed by eighteen membered rings and
are capable of adsorbing molecules excluded from other molecular sieves.

INTRODUCTION

The first report of a zeolite molecular sieve was that of stilbite
in 1756 [1]. Since that time, many natural and synthetic zeolites have
been discovered. Recently, Union Carbide has synthesized
aluminophosphate ($AlPO_4$) [2] and element substituted aluminophosphate
[3] molecular sieves, e.g., SAPO (silicon substitution), MeAPO (metal
substitution).

The largest rings or pores in natural zeolites [1] and synthetic
molecular sieves [1,3] contain 12 T-atoms. Thus, the free diameter
available for adsorption is bounded by approximately 10 Å. The
existence of the 12 membered ring has been known for quite some time.
In fact, the natural zeolites gmelinite and faujasite discovered in 1807
and 1842, respectively both contain 12 membered rings [1].

We report here the first molecular sieves which contain rings
consisting of greater than 12 T-atoms. Virginia Polytechnic Institute
number 5 (VPI-5) is a family of aluminophosphate based molecular sieves
with the same three dimensional topology and all which contain 18
membered rings.

RESULTS AND DISCUSSION

The structure of the VPI-5 molecular sieves has been solved by
x-ray powder diffraction methods and the complete details of the
refinement will be given elsewhere [4]. The symmetry of the structure
is hexagonal with unit cell dimensions of a = 18.989 Å, c = 8.112 Å and

the space group for the aluminophosphate member is P6$_3$ cm (P6/mmm). The topology is described by the net labelled 81(1) and was predicted by Smith and Dytrych [5] in 1984. The main channel system which is unidimensional consists of pores circumscribed by eighteen membered rings. Thus, VPI-5 is the first molecular sieve to exceed the 12 T-atom barrier which has existed since the 1800's. The framework density is defined to be the number of T-atoms per 1000 $Å^3$. VPI-5 possesses a framework density of 14.2. For comparison, the faujasite structure contains a framework density of 12.7 [6]. Thus, VPI-5 is slightly more dense than faujasite.

The existence of the large pores is confirmed experimentally by the adsorption of triisopropylbenzene and perflurotributylamine which possess kinetic diameters of 8.5 and 10.5 $\overset{o}{A}$, respectively. These molecules do not adsorb in known molecular sieves except for the case of triisopropylbenzene which adsorbs very slowly into faujasite type molecular sieves (equilibrium reached in approximately one day compared to approximately 5 minutes for VPI-5). In Figure 1, we illustrate the adsorption capabilities for several zeolites and VPI-5. We do not show an upper bound for VPI-5 since at this time we do not know what it is. However, the diameter of the large pore in VPI-5 is approximately 14 $\overset{o}{A}$ as determined from the crystal structure.

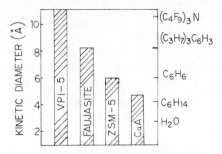

Figure 1. Kinetic Diameter of Adsorbates for Several Molecular Sieves

The VPI-5 family of molecular sieves include element substituted aluminophosphates, e.g., silicon and cobalt, as members. These molecular sieves show similar x-ray diffraction patterns to the original aluminophosphate member. Complete disclosure of the physicochemical properties of the VPI-5 materials is forthcoming [7].

ACKNOWLEDGEMENTS

We thank the National Science Foundation and the Dow Chemical Company for support of this work through the Presidential Young Investigator Award to M.E.D.

REFERENCES

[1] D. W. Breck, Zeolite Molecular Sieves, (Wiley, New York, 1974).

[2] S. T. Wilson, B. M. Lok, C. A. Messina, T. R. Cannan, and E. M. Flanigen, J. Am. Chem. Soc. 104, 1146-1147 (1982).

[3] E. M. Flanigen, B. M. Lok, R. L. Patton, and S. T. Wilson in New Developmnts in Zeolite Science and Technology, edited by X. Murakawi, A. Iijima, and J. W. Ward (Elsevier, Amsterdam, 1986), p. 103.

[4] C. Crowder, in preparation.

[5] J. V. Smith and W. J. Dytrych, Nature 306, 607-608 (1984).

[6] W. M. Meier in New Developments in Zeolite Science and Technology, edited by X. Murakawi, A. Iijima, and J. W. Ward (Elsevier, Amsterdam, 1986), p. 13.

[7] M. E. Davis, C. Saldarriaga, C. Montes, P. E. Hathaway, J. Garces, C. Crowder, and D. Hasha, in preparation.

COMPUTER SIMULATIONS OF DIFFUSION, ADSORPTION AND REACTION OF ORGANIC MOLECULES IN PILLARED CLAYS

MUHAMMAD SAHIMI*, THEODORE T. TSOTSIS* AND MARIO L. OCCELLI**
*Department of Chemical Engineering, University of Southern California, Los Angeles, California 90089-1211
**Unocal Science and and Technology Division, Unocal Oil Company of California, P. O. Box 76, Brea, California 92621

ABSTRACT

Diffusion and reaction of large organic molecules in pillared clays, a new class of catalysts capable of converting gas oil into transportation fluids, is investigated. We first discuss some recent experimental data and point out the possible difficulties for obtaining accurate data. We, then, discuss a new model for describing diffusion and reaction of large organic molecules in pillared clays. The model employs stochastic and random walk concepts to model the diffusion process, and a dynamic Monte Carlo method for predicting various properties of interest, such as the effective diffusivity of the molecules.

Introduction

Diffusion and reaction in catalysts with regular porous structures have been the subject of considerable research activity in the last few years [1-4]. Such catalysts, beyond their great industrial importance, also represent ideal porous model systems well suited for theoretical and experimental studies of hindered (configurational) diffusion and reaction phenomena. Of all such catalysts, zeolites have received by far the greatest attention. The zeolitic structure can be successfully modelled as a regular three-dimensional network of interconnected bonds and throats and several studies so far have utilized Monte Carlo numerical simulations and stochastic modeling techniques [2-4] to describe diffusion in such systems.

Considerably less attention has been focused on modeling diffusion and reaction phenomena in another class of catalysts, also characterized by very regular porous structures, namely pillared clays. The preparation and molecular sieving properties of pillared montmorillonites was first reported by Barrer and coworkers [5]. Montmorillonites are 2:1 dioctahedral clay minerals consisting of layers of silica in tetrahedral coordination, holding in between them a layer of alumina in octahedral coordination (see Fig. 1).

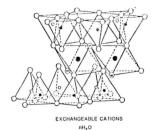

Fig. 1: Diagrammatic Sketch of the Montmorillonite structure.

EXCHANGEABLE CATIONS
nH_2O

O OXYGENS   ⊙ HYDROXYLS   ● ALUMINUM, IRON, MAGNESIUM
O and ● SILICON, OCCASIONALLY ALUMINUM

Mat. Res. Soc. Symp. Proc. Vol. 111. ©1988 Materials Research Society

Substituting Si^{4+} with Al^{3+} or Fe^{3+} or Al^{3+} with Mg^{2+} gives the silicate layers a negative net charge, which is normally compensated by Na^+, Ca^{2+}, and Mg^{2+} ions [6]. By exchanging the charge-compensating cations with large cationic oxyaluminum polymers one can synthesize molecular sieve type materials [7,8]. These inorganic polymers, when heated, form pillars which prop open the clay layer structure and form permanent pillared clays (see Fig. 2).

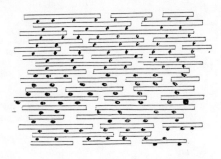

Fig. 2: Schematic representation of a pillared clay. Its structure is characterized by base to base connections. Oxide clusters (dark dots) prop the clay silicate layers.

Pillared clays behave like two-dimensional molecular sieves since molecules are forced to move in between the silicate layers and cannot diffuse from one layer to the other.

Access to the interior pore volume of pillared clays is controlled by the distance between the silicate layers (~8Å, determined by the dimensions of the interlayering cation) and by the distance between the pillars (~14-16Å, controlled by the charge density distribution on the silicate layers, which is nonuniform; one therefore expects the pillar distribution also to be nonuniform).

Pillared clays have pore volumes of 0.16-0.2 cm^3/g., surface areas 250-350 m^2/g [9-11], and have been shown to allow sorption of C_6-C_{10} normal paraffins, 1, 3, 5 trimethylbenzene (d=7.6Å) but not of 1, 2, 3, 5 tetramethyl benzene (d=8.0Å) and perfluorotributyl amine (d=10.4Å). They have both Brönsted and Lewis acidity. Pillared clays have shown high catalytic activities for gas oil cracking (similar to zeolite based FCC catalysts) but the Na and Ca bentonites interlayered with cationic oxyaluminum pillars do not have the hydrothermal stability of zeolite based commercial FCC catalysts [12, 13]. They have also shown high initial activities towards methanol conversion to olefins and toluene ethylation but they are substantially deactivated by coke deposition [11]. They have also been studied for cumene and isopropylnaphthalene dealkylation [8] and in hydrotreating applications [7].

Experimental Studies

Diffusion studies of various organic molecules in pillared clays have been performed by Occelli and coworkers [9-11]. Their experimental system has been described in detail elsewhere [9-10]. It consists basically of a Perkin-Elmer TGS-2 UHV thermogravitometric system. Small amounts of finely ground catalyst (5-6mg) were placed in the PE quartz microbalance, gas was introduced in the all quartz apparatus (0.1-0.12 Torr) and the uptake was measured. At small times a plot of (M_t/M_∞), (where M_t is the weight of the sample at time t, and M_∞ the weight when adsorption/desorption equilibrium is attained) versus $t^{\frac{1}{2}}$ is a straight line with a slope proportional to D_e'.

the effective diffusivity of the organic molecules in the pillared clay. Fig. 3 shows a plot of such diffusivities (C_6-C_{10} straight chain alkanes) as a function of temperature.

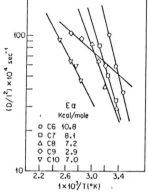

Fig. 3: Diffusivities of straight chain alkanes in pillared clays as a function of temperature.

A model of diffusion of organic molecules in pillared clays as the one described below should in principle, without resorting to adjustable parameters, predict the measured diffusivities. Unfortunately the available experimental data does not allow for such direct comparisons. There are several factors responsible for this problem. First the uncertainty in the particle size distribution (a 100x325 mesh portion of the ground sample was used) and shape result in an order of magnitude uncertainty in the measured diffusivities. This is a problem, which can of course be remedied by utilizing a narrower particle size distribution. The most serious problem with the available experimental data, however, is the direct consequence of the nature of the diffusion phenomenon itself. Since the size of most organic molecules (like straight chain alkanes) is comparable to the size of the pore opening, adsorption of such molecules on the pillars (and possibly on the silicate layers) results in a reduced pore space available for further diffusion and therefore in pore diffusivities, which are decreasing functions of experimental time. It would appear that this problem would be of importance only in cases with significant sorptive uptakes of the diffusing molecules, and therefore the existing data collected at small experimental times [9,10] and low pressures (equilibrium uptakes of the order of 15-30% of "saturation" coverage) will be hardly affected. This however might not necessarily be the case. The effect of molecular sorption on diffusivity strongly depends on the adsorption/desorption rates relative to the bulk diffusive fluxes. For high adsorption rates, sorption will primarily occur at and around the external perimeter of the particle and phenomena similar to pore mouth blocking will occur. Further detailed discussion can be found elsewhere [14].

Similar problems arise in the measurement of adsorption and desorption rate and adsorption/desorption equilibrium constants, since pore blocking due to sorption tends to falsify both the rates as well as saturation coverages. It would appear that the problem in correctly measuring adsorption/desorption rate constants and saturation coverages is unrelated to that of the measurement of diffusivities in the pores. Note, however, that the diffusivity one measures by uptake experiments, D_e, is related to the true diffusivity in the pore by the relationship,

$$D_e = \frac{D_c}{\epsilon p + K_e}$$

(1)

where K_e is the local adsorption/desorption equilibrium constant (assuming a linear isotherm). It is clear that some of the significant differences reported in the D_e values for normal alkanes in pillared clays (over 2 orders of magnitude, see Fig. 3) can only be explained in terms of differences

of adsorptive properties for these compounds. Furthermore the high
activation energies for pore diffusivities reported (see Fig. 3) and the
existence of a minimum activation energy for nonane can also be explained
in terms of differences in the sorption behavior. Similar observations were
recently made with zeolites by Garcia and Weisz [15] who showed that, when
one properly accounts for the effect of sorption, the pore diffusion process
is unactivated.

It should be clear from the discussion so far that a significant amount of
work still remains to be done before our understanding of diffusion and
reaction phenomena in pillared clays is complete. There is in particular
great need for carefully designed sorption/diffusion experiments, which can
account for the effect of sorption phenomena on the diffusion fluxes and vice
versa. It is, furthermore, imperative that in the design of such experiments
one is guided by a model of diffusion and sorption, which properly accounts
for the morphology of the porous medium, and which provides a detailed
description of the microscopic events occurring during diffusion and sorption.
Work in this direction is currently in progress in our group. A brief
description of the model of sorption and diffusion which is currently being
developed by our research group is given below.

THE MODEL

As discussed above, pillared clays have a very regular structure and,
therefore, it is relatively straightforward to simulate their pore structure
with the aid of a computer. Thus, after generating the pore structure of the
clays and identifying the locations of the pillars, we inject N organic
molecules into the system (this represents the time t=o). We assume that the
molecules can be represented by orthogonal parallelepipeds with three
effective dimensions b, t and l (see Table II). Each molecule performs a
random walk in the pore space of the catalyst particle; this random walk
represents the diffusion process. No same points of the pore space can be
occupied by two molecules, so that there is an effective hard core repulsion
operating between the molecules. To move the molecules in the pore space,
we assume that they take their steps through principle directions of a simple
cubic lattice or one of its main diagonals. Thus, at every step of the
simulations, a molecule can select one of the 12 available directions, with
an equal probability of 1/12, and make a transition to another point in the
pore space. The assumption of a simple cubic lattice structure for the pore
space is only a matter of convenience, and any other scheme for moving the
molecules in the pore space can be used. The fact that a molecule can also
move along the main diagonals of the lattice, allows one to take into account
the possible rotation of the molecules around a point.

If a molecule hits one of the two silicate layers, it is reflected back
onto its last position before the collision. This is based on the
experimental fact that there is practically no adsorption on the silicate
layer. However, if a molecule collides with a pillar, it can be adsorbed
with the probability p which depends on the temperature of the system and the
particular organic molecule adsorbed. Clearly, the limit of very small p
represents a system under kinetic control, in which diffusion does not play
an important role, whereas the limit p→1 represents a diffusion-controlled
system. If the molecule is adsorbed on the pillar, it stays there for the
rest of simulation (we assume here negligible desorption; the model can also
account for finite rates of desorption). In this case, part of the pillar is
covered by the molecule. However, if the molecule is not adsorbed on the
pillar, it is reflected back onto its last position before the collision. If
a molecule crosses the boundaries of the system, it is removed from it and the
simulations continue. When all of the initial N molecules have been adsorbed,
or have left the system, the simulation time is increased by one unit of time,
another N molecules are injected into the system and the simulations continue.
When enough number of molecules have been adsorbed, such that they can

effectively block macroscopic motion of unadsorbed molecules, the simulations are terminated. In this situation, the system is saturated and a volume fraction ϕ of the pore space is occupied by the adsorbed molecules. The simulation procedure just described is a dynamic Monte Carlo method, by which one can calculate all of the dynamic properties of the system. For example, to calculate the effective diffusivity of the molecules, we determine the time-dependence of the mean-squared displacement of the center of mass of the molecules, R^2 (t)

$$R^2(t) \equiv <[R_{cm}(o) - R_{cm}(t)]^2>$$ (2)

where $< \; >$ denotes an averaging over all molecules in the system. In the limit of long times, the ratio

$$D \equiv R^2(t)/t$$ (3)

is proportional to the diffusivity of the molecules. This method not only enables one to determine the long-time diffusivity of the molecules, it can also determine the time evolution of the system and, thus, it is a practical tool for analyzing the experimental data. The above problem is repeated for many randomly-selected initial positions of the molecules and the average values of the quantities of interest are computed.

Results and Discussion

Due to the problems associated with the currently-available experimental data only order of magnitude comparisons can be made at this point between the model and the experiments. The experimental $\acute{D}e$ values are in the range of $10^{-6} - 10^{-7} cm^2/s$ and the experimental De values in the range of $10^{-3} cm^2/s$. The model predicts same order of magnitude diffusivities with the experimental De values in the absence of sorption. One can also predict similar order of magnitude diffusivities with the De reported by using the adsorption probability (sticking coefficient) p as an adjustable parameter, and using the experimental $\acute{Ke}$ values.

An view of the uncertainty in the measured of $\acute{Ke}$ values and the lack of experimental information concerning the sticking coefficients such an agreement can only be considered at this point as parameter fitting.

An issue of interest, closely related to the phenomena of pore blocking due to sorption, previously discussed, is shown in Table I. This table shows the volume fraction ϕ at the high pressure sorption saturation limit for straight chain alkanes $(C_5 - C_{10})$ and its relationship to the overall catalyst pore volume. Note, that all occupied saturation values are less than 66%, which happens to be the percolation threshold [16] for a three dimensional continuum system, i.e. the maximum occupied volume that is allowed, if blocking of the pores occurs due to sorption. Note that as the molecular size increases the maximum saturation volume decreases. Of particular interest is the fact that a sudden drop in saturation ϕ occurs for heptane and that the saturation volumes for C_7 to C_{10} straight chain alkanes are almost indistinguishable. This latter behavior appears very peculiar until one examines the size of these organic molecules (see Table II). Note that for the $C_7 - C_{10}$ straight chain alkanes (b+t)>16 Å, which is the average distance between pillars for the pillared clays resulting from pillaring of Ca - bentonites.

Acknowledgements: This work was supported in part by the donors of the Petroleum Research Fund, administered by the American Chemical Society (grants 18950-67 and 18168-AC5,7) and United States Department of Energy.

Table I

Alkane	Liquid Saturation volume cm^3/gr	% Total pore volume
C$_5$	0.101	63
C$_6$	0.095	59
C$_7$	0.082	51
C$_8$	0.082	51
C$_9$	0.085	52
C$_{10}$	0.084	52.5

Liquid Sorption Saturation volume from (10).

Table II

Sorbate	b	t	l
C$_6$	4.9	4.0	10.30
C$_7$	4.9	4.0	11.51
C$_8$	4.9	4.0	12.30
C$_{10}$	4.9	4.0	15.24

Sorbate size in Å, b = breadth, t = thickness
l = length.

References

1. M. Sahimi, B. D. Hughes, L. E. Scriven and H. T. Davis, J. Chem. Phys. 78, 6849 (1983).

2. M. Sahimi, Chem. Eng. Sci. in press (1988); R. Mojaradi and M. Sahimi, ibid, in press (1988).

3. W. T. Mo and J. Wei, Chem. Eng. Sci. 41, 703 (1986).

4. S. K. Bhattia, J. Catal. 93, 192 (1986).

5. R. M. Barrer, Zeolites and Clay Minerals as Sorbents and Molecular Sieves (Academic Press, New York, 1978).

6. R. E. Grim, Clay Mineralogy (McGraw-Hill, New York, 1968).

7. D. E. W. Vaughan, R. J. Lussier and J. S. Magee, U. S. Patent No. 4 175 090 (1979).

8. N. Lahav, V. Shani and J. Shabtai, Clays and Clay Minerals, 26, 107 (1978).

9. M. L. Occelli, F. Hwu and J. W. Hightower, Preprints 182nd ACS Meeting, NY, 1981 (unpublished).

10. M. L. Occelli, V. N. Parulekar and J. W. Hightower, Proc. 8th Inter. Congress Catalysis, Volume IV, (Berlin, 1984) p. 725.

11. M. L. Occelli, R. A. Innes, F. S. S. Hwu and J. W. Hightower, Applied Catalysis 14, 69 (1985).

12. M. L. Occelli and R. W. Tindwa, Clays and Clay Minerals 31, 22 (1980).

13. M. L. Occelli, I & EC Prod. Res. Dev. Journal, 22(4), 553 (1983).

14. M. Sahimi, T. T. Tsotsis and M. L. Occelli, submitted for publication.

15. S. Garcia and P. B. Weisz, paper 16b presented at 1987 AIChE Annual Meeting, NY, Nov. 1987.

16. D. Stauffer, Introduction to Percolation Theory (Taylor and Francis, London, 1985).

STRUCTURAL STUDIES OF PILLARED CLAYS AND MODIFIED PILLARED CLAYS

KATHLEEN A. CARRADO, ARTHUR R. THOMPSON, RANDALL E. WINANS, AND
ROBERT E. BOTTO
Chemistry Division, Argonne National Laboratory, 9700 South Cass Avenue,
Argonne, Illinois 60439

ABSTRACT

The long-range order of pillared interlayered clays (PILCs) after acid activation with 0.05N HCl has been investigated by X-ray diffraction (XRD) methods. The data show that long-range order in PILCs decreases as AlCH-PB = ZrCH-PB >> Zr/AlCH-PB = Cr/AlCH-PB (PB = pillared bentonite; MCH = metal chlorohydroxy pillaring agent, where M = Al, Zr or Cr). Apparently, pure oxide clusters are more stable than mixed oxide clusters. Treatment of PILCs with dilute HCl at 25°C is less damaging than at reflux temperature, and calcined PILCs are more stable than air-dried materials. More structural damage occurs with 3M sulfuric acid treatment than with dilute HCl. Treatment with a weak base also causes some degradation of the pillars. ^{27}Al-MAS NMR has been used to study pillared hectorite (PH), as well as other clay systems. The large increase of the observable octahedral aluminum (Al(VI)) resonance seen after pillaring is explained by loss of water from the $[Al_{13}O_4(OH)_{24}(H_2O)_{12}]^{7+}$ (Al$_{13}$) cation. ^{27}Al spectra of PILCs derived from different pillaring agents and exposed to various heat and acid treatments are remarkably similar.

INTRODUCTION

The stucture elucidation of clay layers and pillars in pillared interlayered clays (PILCs) [1] is of fundamental importance in understanding their thermal stability and catalytic properties. Earlier studies have focused on X-ray diffraction methods to assess the degree of structural damage of PILCs with various chemical treatments [2].

Clays and pillared clays are converted into more acidic, hence more active catalysts by mineral acid treatment [3] by making H$^+$-forms [4], and by transition metal cation-exchange [5]. However, the discrete chemical changes associated with clay activation is not well understood. X-ray powder diffraction (XRD) has been used to follow the long-range structural order of PILCs during various chemical treatments. Changes in short-range order have been monitored by ^{27}Al-MAS NMR.

EXPERIMENTAL

Clays used are Bentolite L, a Ca^{2+}-bentonite from Southern Clay Products, and SHCa-1, a California hectorite from the Source Clays Repository. SHCa-1 was purified using wet sedimentation techniques to remove calcite and quartz.

AlCH-PB: Conditions for pillaring with the Al$_{13}$ cation were similar to those reported previously [4] using Chlorhydrol (AlCH) from the Reheis Chem. Co. These conditions called for pH control at 5.5 using dilute NH$_4$OH.

Zr/AlCH-PB: Pillaring with $[ZrOCl_2Al_8(OH)_{20}]^{4+}$ derived from Rezal-67 (Reheis) followed the procedure of Occelli and co-workers [6a].

ZrCH-PB: The procedure of Bartley and Burch [7] was followed to pillar with ZrO$_2$ clusters dervied from ZrOCl$_2 \cdot$8H$_2$O (Alfa).

Cr/AlCH-PB: Syntheses of PILCs containing Cr(III) ions were carried out under conditions described previously [8], with Cr(NO$_3$)$\cdot$9H$_2$O (Alfa).

AlCH-PH: Pillaring hectorite with AlCH was carried out under standard conditions [8]; however, a mechanical stirrer was needed to stir the thick (gelled) solution.

Materials were either air-dried or calcined in porcelain dishes in a muffle furnace at 400°C for 4 hours. Chromium-containing samples were heated at 300°C for 3 hours under N_2 atmosphere (to minimize oxidation to Cr(VI)) using quartz tubes in a tube furnace.

Acid treatments were carried out by stirring 0.1 g clay/10ml 0.05N HCl for 24 hours at either room temperature or at reflux. Calcined AlCH-PB samples were prepared with 3M H_2SO_4 following the procedure of Gregory and Westlake [4]. KOH saturation was accomplished by immersion in 1M solution for 4.5 hours.

XRD measurements were obtained on a Scintag PAD Model V, a theta-theta instrument with a solid-state detector and Cu/K-alpha radiation. Data was collected and stored on a DG Desktop Series computer. Samples were either mounted as powders or prepared as films on glass slides from aqueous suspension.

The ^{27}Al NMR spectra were recorded on a modified Bruker AM-300 spectrometer operating at 78.2 MHz (7.04T). A Doty Scientific MAS probe capable of attaining spinning speeds in excess of 10 kHz was interfaced to the NMR. A 3 microsecond pulse width (pulse angle = 15° for aq $AlCl_3$) was employed and pulse repetition times were varied from 0.1 to 5 seconds. Chemical shifts are reported relative to 1M $Al(H_2O)_6Cl_3$ in H_2O.

RESULTS

X-ray Diffraction

The d_{001}-spacings in XRD experiments were used to monitor long-range structural order of acid-treated PILCs. Several things can be pointed out for AlCH-PB in Table I.

TABLE I. XRD Results for PILCs; d_{001}, Å [a]

	Air-Dried	Calcined
AlCH-PB	19.5	18.7
HCl, 25°C	16.4	18.4
HCl, 100°C	13.6	16.8
H_2SO_4	--	17.3
KOH/H_2SO_4	--	16.2
ZrCH-PB	22.1	--
HCl, 25°C	22.1	--
Zr/AlCH-PB	19.2	18.8
HCl, 25°C	15.8	16.7
HCl, 100°C	--	15.0
Cr/AlCH-PB	19.2	19.2
HCl, 25°C	15.5	15.2
HCl, 100°C	15.4	13.6

[a]Estimated precision ±0.2Å

The d-spacing decreases by <1Å upon calcination as reported previously [1,6]. The calcined PILC retains higher d-values than the air-dried samples upon acid treatment, indicating an increase in structural stability with calcination. Room temperature HCl treatment is less destructive than refluxing HCl; the calcined PILC is hardly affected and decreases only 0.3Å. Use of H_2SO_4 at room temperature decreases the d-spacing more than HCl, and when used after KOH saturation the d_{001} reflection decreases to 16.2Å. The mixed Zr/AlCH system is more susceptible to acid attack than AlCH-PB, although calcining Zr/AlCH-PB makes the pillars more stable toward acid. On the other hand, pure zirconia pillars (ZrCH-PB) display excellent stability to acid treatment. Even an air-dried sample exhibits no change in d_{001} spacing, 22.1Å.

Association of Cr(III) ions with the aluminum pillar creates a PILC having d-spacings that are very stable toward calcination at 300°C for 3 hours. However, even mild acid treatments had devasting effects on the pillar structure. In both cases d_{001} collapsed to about 15Å.

^{27}Al-MAS NMR

Although SHCa-1 contains only 0.69 wt% Al_2O_3 [9], an [27]Al spectrum of a purified sample having good signal-to-noise ratio could be obtained in about 3 hrs., see Figure 1a. Two resonances of approximately equal intensity are evident. The resonance assigned to tetrahedrally coordinated aluminum [Al(IV)] occurs at approximately 65 ppm. The octahedral aluminum resonance [Al(VI)] occurs between 0 and 5 ppm. Fyfe and co-workers [10] have also observed resonances at 65.1 and 2.7 ppm for hectorite.

The increase in the signal-to-noise ratio observed for the air-dried AlCH-PH samples reflects the increase in aluminum content due to incorporation of the pillars, as shown in Figure 1b. Moreover, the increase in intensity is more pronounced for the octahedral Al(VI) resonance. A similar [27]Al spectrum was obtained for calcined AlCH-PH. This spectrum has largely the same features as reported previously by others [11,12].

Figure 1c displays the [27]Al spectrum of Bentolite L. An Al(VI) resonance near 0 ppm and two Al(IV) resonances at 55 and 65 ppm are observed. Calcined AlCH-PB displays only one broad Al(IV) resonance and no obvious differences in the spectrum are apparent with acid treatment. Zr/AlCH-PB samples also display a broad Al(IV) resonance (Figure 1d); however, acid treatment (25°C) of the air-dried sample reintroduces the two Al(IV) resonances at 55 and 65 ppm. The calcined Zr/AlCH-PB sample treated with refluxing HCl displays these two Al(IV) resonances, but of much lower intensity. The spectra of Cr/AlCH-PB, ZrCH-PB and acid-treated ZrCH-PB also exhibit the same two Al(IV) peaks.

DISCUSSION

It is generally thought that protons are generated during pillar formation as a result of dehydration of the interlayering polymeric cation [6b]. On heating, oxide clusters having Lewis acid character form and the protons are retained to neutralize the charge on the clay layer [6c]. These clusters include both pure aluminum and zirconium oxides (AlCH-PB, ZrCH-PB) and mixed zirconium-aluminum and chromium-aluminum oxides (Zr/AlCH-PB, Cr/AlCH-PB) in this study.

Table I shows XRD results from mineral acid treatments on various pillared clays. The fact that air-dried samples are less resistant than calcined samples to acid attack suggests the pillars are chemically altered after thermal treatment. After HCl treatment, d_{001} for air-dried AlCH-PB decreases by 16% versus only 2% for the corresponding calcined material.

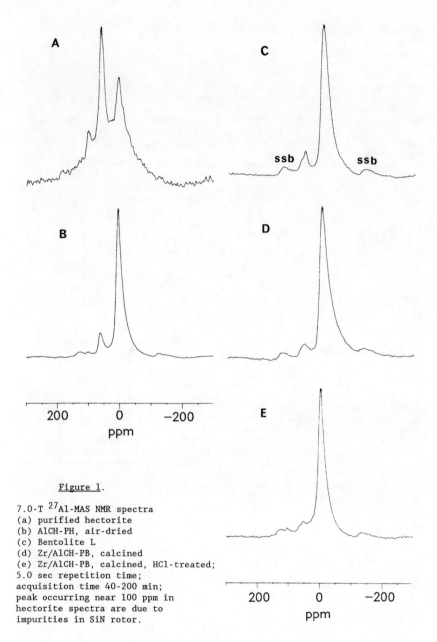

Figure 1.

7.0-T ^{27}Al-MAS NMR spectra
(a) purified hectorite
(b) AlCH-PH, air-dried
(c) Bentolite L
(d) Zr/AlCH-PB, calcined
(e) Zr/AlCH-PB, calcined, HCl-treated;
5.0 sec repetition time;
acquisition time 40-200 min;
peak occurring near 100 ppm in
hectorite spectra are due to
impurities in SiN rotor.

Even after harsher treatment with refluxing HCl, d_{001} for the calcined AlCH-PB decreases by only 10% while the air-dried sample collapsed to 13.6Å (30% loss). This is to be expected because aluminum oxide clusters should be more stable than aluminum hydroxy oligomers to mineral acids.

The cation exchange capacity (c.e.c.) of calcined PILCs is less than that of parent clays, probably due to proton migration into octahedral layers [12]. Treatment with a weak base has been shown to increase the c.e.c.[8], possibly because these protons react with the basic reagent. The method of KOH saturation followed by 3M H_2SO_4 exchange to make a H^+-PILC [4] is more severe than 0.05N HCl treatment. The severity of the KOH/H_2SO_4 treatment is reflected in their smaller d-spacings compared to HCl treated materials, although H_2SO_4-PILCs still have larger d-spacings than non-pillared clays. Furthermore, comparing d_{001} spacings in clays with (16.2Å) and without (17.3Å) KOH saturation suggests that the base is responsible for degrading the PILCs in some way.

The long-range structural order and stability for Zr/AlCH-PB and ZrCH-PB also can be assessed from the data presented in Table I. While Zr/AlCH-PB is quite stable to thermal activation, it is very susceptible to mild acid attack. On the other hand, ZrCH-PB is not affected at all by mild HCl. Data is also reported for Cr/AlCH-PB, where Cr(III) ions are associated with AlCH pillars [8]. This sample behaves very similar to the Zr/AlCH-PB material. In summary, the trend of chemical stability decreases as: AlCH-PB = ZrCH-PB >> Zr/AlCH-PB = Cr/AlCH-PB. The reason why pure oxide clusters are more stable than mixed oxide clusters is not fully understood at this time.

The occurence of two Al(IV) peaks in the ^{27}Al-MAS NMR spectra of montmorillonite may be explained by invoking a structure proposed long ago by Edelman and Favejee [14]. Recently, considerable support for this model has been offered to describe grafting of a pillar to the surface of beidellite [11,15]. Inversion of some tetrahedra in the tetrahedral layers, especially where Si is isomorphically substituted by Al, is suggested [3] and could lead to the two different Al(IV) sites observed.

While AlCH-PB and Zr/AlCH-PB both display an unresolved, broad Al(IV) resonance (air-dried and calcined), acid-treating these samples restores the 55 and 65 ppm Al(IV) resonances to at least some degree. This phenomenon is not yet well understood because the Al(IV) resonances from the pillars overlap these signals.

For hectorite samples, XRD data shows a d_{001} equal to 18.4Å for air-dried AlCH-PH and 18.0Å for the calcined and mildly acid-treated samples. After ion-exchange and air-drying of hectorite, ^{27}Al-MAS NMR reveals a substantial increase in the relatively narrow Al(VI) resonance, as shown in Figure 1b. This is not to be expected if the Al_{13} cation remains unaffected, since this cation has been shown previously to have an exceptionally broad Al(VI) resonance [12,16]. Plee and co-workers have proposed partial hydrolysis of Al_{13} cations in the layers to account for these spectral changes [11], but any further explanation by these authors was not given. We believe the water molecule on Al(VI) sites may have been replaced by a clay oxygen, as depicted below. Substitution of a clay oxygen would decrease the electric field gradient about Al(VI) in the Al_{13} cation [12,16] and would account for the observed narrow resonances:

$$O^{2-}-Al^{VI}-O^0H_2 \longrightarrow O^{2-}-Al^{VI}-O^{2-}-M \quad (M = Al, Si)$$

This would imply at least some interaction between the clay and Al_{13} cation prior to calcination.

282

Slight changes can be observed in the NMR spectrum after AlCH-PH has been calcined and treated with mild acid. This spectrum looks no different from a calcined AClH-PH before acid treatment. The octahedral ^{27}Al resonance broadens somewhat, which shows that heating has created new environments. The same result has been obtained for all pillared bentonite clays in this study, see Figure 1d. The tetrahedral resonance also changes and now seems to be composed of the two Al(IV) resonances similar to those observed for Bentolite L. These two resonances are observed for purified hectorite at longer pulse recycle times (5 sec) Fig. 1a. Acid treatments do not significantly alter the spectra for any of the PILCs examined here. In fact, ^{27}Al spectra of PILCs derived from different pillaring agents and exposed to various heat and acid treatments are remarkably similar (compare Figure 1d. with 1e.).

In summary, XRD results indicate that HCl treatment can drastically affect the d-spacings of pillared clays under certain conditions. While the observed changes imply acid hydrolysis of the pillars, only minor changes in ^{27}Al-MAS NMR spectra are observed. Further research to resolve these discrepancies is currently in progress.

ACKNOWLEDGEMENTS

This work was performed under the auspices of the Office of Basic Energy Sciences, Division of Chemical Sciences, U. S. Department of Energy, under contract number W-31-109-ENG-38.

REFERENCES

1. D. E. W. Vaughan, R. Lussier, J. Magee, Jr., U.S. Patent No. 4 176 090 (1979).
2. K. A. Carrado, A. Kostapapas, S. L. Suib, R. W. Coughlin, Solid State Ionics 22, 117 (1986).
3. C. Breen, J. M. Adams, C. Riekel, Clays and Clay Minerals 33, 275 (1985).
4. R. Gregory, D. J. Westlake, Eur. Patent No. 0 110 628 (1984).
5. J. Shabtai, F. E. Massoth, M. Tokarz, G. M. Tsai, J. McCauley, Proc. 8th Int. Cong. Catalysis (Vol. IV, Berlin, 1984), p. 735.
6. (a) M. L. Occelli, D. H. Finseth, J. Catal. 99, 316 (1986).
   (b) M. L. Occelli, R. M. Tindwa, Clays and Clay Minerals 31, 22 (1983).
   (c) M. L. Occelli, J. Molec. Catal. 35, 377 (1986).
7. G. J. J. Bartley, R. Burch, Appl. Catal. 28, 209 (1986).
8. K. A. Carrado, S. L. Suib, A. Kostapapas, R. W. Coughlin, N. D. Skoularikis, Inorg. Chem. 25, 4217 (1986).
9. Data Handbook for Clay Materials, edited by H. van Olphen and J. J. Fripiat (Pergamon Press, 1979).
10. S. Komarneni, C. A. Fyfe, G. J. Kennedy, H. Strobl, J. Am. Ceram. Soc. 69, C45 (1986).
11. D. Plee, F. Borg, L. Gatineau, J. J. Fripiat, J. Am. Chem. Soc. 107, 2362 (1985).
12. B. T. B. Tennakoon, W. Jones, J. M. Thomas, J. Chem. Soc. Farad. Trans. I 82, 3081 (1986).
13. D. E. W. Vaughan, R. J. Lussier, J. S. Magee, Jr., U.S. Patent No. 4 271 043 (1981).
14. C. H. Edelman, J. C. L. Favejee, Z. Krist. 102, 417 (1940).
15. T. J. Pinnavaia, S. D. Landau, M. S. Tzou, I. D. Johnson, M. Lipsicas, J. Am. Chem. Soc. 107, 7222 (1985).
16. A. C. Kunwar, A. R. Thompson, H. S. Gutowsky, E. Oldfield, J. Magn. Reson. 60, 467 (1984).

ADSORPTION AND ACIDIC PROPERTIES OF CLAYS PILLARED WITH OXIDE SOLS

S. YAMANAKA, T. NISHIHARA, AND M. HATTORI
Department of Applied Chemistry, Faculty of Engineering, Hiroshima University, Higachi-Hiroshima 724, Japan

ABSTRACT

Interlayers of montmorillonite were pillared with TiO_2, SiO_2-TiO_2 and SiO_2-Fe_2O_3 sols. The pillared structures were thermally stable at least up to $500°C$ and retained unusually large basal spacings in the range of 24-45Å and surface areas as high as 300-500 m^2/g. The TiO_2 pillared clay showed Type IV adsorption isotherm for nitrogen. Although SiO_2-TiO_2 and SiO_2-Fe_2O_3 pillared clays had basal spacings much larger than that of TiO_2 pillared clay, these mixed oxide pillared clays had small pores and exhibited Type I isotherm. The acidic strength distributions were determined by a titration method using n-butylamine and Hammett indicators. All of the pillared clays had large acidities, but the acidic strength decreased in the following order: $TiO_2 \approx SiO_2 >> SiO_2$-$Fe_2O_3$ pillared clays. Temperature-programmed desorption (TPD) spectra of ammonia were measured and interpreted in relation to the acidity distribution. Infrared spectra of pyridine adsorbed on TiO_2 pillared clay indicated that the acidity predominantly arises from Lewis acid sites.

INTRODUCTION

Clays pillared with oxides are considered to be a new type of two-dimensional zeolites. In these structures the two-dimensional silicate sheets of clay are pillared with positively charged small metal oxide particles such as alumina [1] and zirconia [2]. These pillared clays are thermally stable and retain high surface areas of 300-500 m^2/g at least up to $500°C$. Extensive studies have been conducted on their formation mechanism, structure [3,4] and catalytic properties [5,6]. From the catalytic viewpoint, the main interest in pillared clays is to generate tailor-made pore structures by controlling the size and the population of pillars between the silicate layers. The nature of metal oxides can also be varied. Although the pillared clays obtained thus far have small pore sizes in the range of 6-8Å, it appears to be possible to synthesize pillared clays with pore sizes larger than those of faujasite zeolites [7].

Recently, we have found that positively charged oxide sol particles such as TiO_2, [8] SiO_2-TiO_2 [9] and SiO_2-Fe_2O_3 can also be used as pillaring agents. These pillared clays with oxide sols show remarkably large basal spacings of 24-42Å. In the present paper, the acidity distributions of these pillared clays have been studied because these are important characteristics to evaluate their catalytic activities.

EXPERIMENTAL

Clay

A sodium montmorillonite supplied by Kunimine Industrial Company, Japan was used as the host clay. Its structural formula was determined to be $(Na_{0.35}K_{0.01}Ca_{0.02})$ $(Si_{3.89}Al_{0.11})$ $(Al_{1.60}Fe_{0.08}Mg_{0.32})$ O_{10} $(OH)_2 \cdot nH_2O$; its cation exchange capacity (CEC) was measured to be 100 meq/100g.

Sol Solutions

Titania Sol: Titanium isopropoxide was hydrolyzed by adding to 1 N HCl solution, and the resulting slurry was peptized to a clear solution by stirring for 3 hrs at room temperature. The molar ratio of HCl to the alkoxides was about 4 [8].

Silica Sol: Silica sol solution was prepared by mixing silicon ethoxide, 2 N HCl and ethanol in a ratio of 41.6g/10ml/12ml. Since the negatively charged silica sol cannot be used as pillaring agents, it was used in the form of mixed sols with titanium (IV) and iron (III) hydroxides deposited on the silica sols.

Mat. Res. Soc. Symp. Proc. Vol. 111. ©1988 Materials Research Society

Titania-Silica Mixed Sol: The silica and the titania sol solutions prepared were mixed in such a way that the molar ratio of TiO_2/SiO_2 = 1/10 and stirred for 1 hr at room temperature.

Iron Oxide-Silica Mixed Sol: About 0.05M trinuclear acetate iron nitrate $[Fe_3OH(OCOCH_3)_7]NO_3$, was added to the silica sol solution to obtain a molar ratio of Fe_2O_3/SiO_2 = 0.5/10, followed by titration with 0.2 N NaOH to a pH of 2.1. The titrated solution was allowed to stand for 1 hr at room temperature under stirring.

Pillaring

About 1% suspension of montmorillonite in water was mixed with each sol solution. The ratio of the sols to the CEC equivalent of the clay were 40, 30, and 30 for TiO_2, SiO_2-TiO_2 and SiO_2-Fe_2O_3, respectively. After the mixtures were allowed to stand for 3 hrs under stirring at 50^oC, the products were separated by centrifugation, washed with water several times, and then dried under a stream of dry air at room temperature.

Analyses

For powder X-ray diffraction (XRD) analysis, a small portion of the wet sample was spread on a glass slide before air-drying to achieve preferred orientation of the basal planes. The basal spacing was measured by using a diffractometer with Ni-filtered CuKa radiation.

Surface areas of the samples were measured by nitrogen adsorption at liquid nitrogen temperature. The samples were degassed for 3 hrs by evacuation at 200^oC prior to the measurement. The porosity of the sample was calculated as the liquid volume of nitrogen adsorbed at the relative pressure of 0.95.

Acid Strength

The acid strength and the acid amount were measured by the amine titration method [10]. The pillared clays were heated at 500^oC and dispersed in iso-octane. The dispersion was titrated with n-butylamine using various indicators with different pKa. The acid strength is expressed by the Hammet acidity function Ho corresponding to pKa of the indicator. The following indicators were used: benzalacetophenone (pKa = -5.6), dicinnamalacetone (pKa = -3.0), 4-benzenazo-diphenylamine (pKa = +1.5), p-dimethylaminoazo-benzene (pKa = +3.3), and methyl red (pKa = +4.8).

TPD Measurement

TPD spectra were recorded by a gas chromatograph with a thermal conductivity detector. About 0.6g of the sample was reheated at 500^oC in a stream of helium for 1 hr and cooled down to room temeprature. The sample was exposed to 1 atm of ammonia for 20 min and the excess ammonia was removed in a flow of helium at room temperature for about 2 hrs. The measurement was carried out up to 650^oC at a heating rate of $5^oC/min$ and a helium flow rate of 100 ml/min. The detector response was normalized for 1g of the sample.

Infrared Spectroscopy

Infrared spectra of the pyridine adsorbed on a self-supporting film of TiO_2 pillared clay were measured in an evacuable glass cell with NaCl windows. The films heated at 300° and 500^oC were degassed in the glass cell and exposed to pyridine vapor. The spectra were recorded after reducing the vapor pressure to about 1 torr.

RESULTS

Adsorption Properties and Pore Structures

The montmorillonites pillared with various oxide sols were found to be thermally stable at least up to 500^oC, and retained remarkably large basal spacings and high surface areas. Basal spacings and nitrogen adsorption data of the three samples heated at 500^oC are given in Table I.

TABLE I. Basal Spacings and Nitrogen Adsorption Data

Sol	Basal Spacing, Å	Δd,* Å	Surface Area, m^2/g	Type of Isotherm	Porosity, ml/g
TiO$_2$	23.2	13.6	284	IV	0.27
SiO$_2$-TiO$_2$	38	28.4	463	I	0.25
SiO$_2$-Fe$_2$O$_3$	42	32.4	425	I	0.22

*Pillar height (basal spacing - silicate layer thickness 9.6Å)

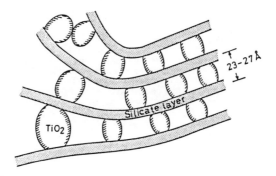

Fig. 1. Schematic structural model of TiO$_2$ pillared clay.

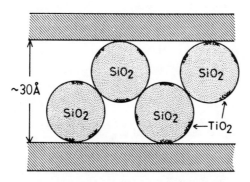

Fig. 2. Schematic structural model of SiO$_2$-TiO$_2$ pillared clay.

Most of the pillared clays thus far obtained, such as alumina and zirconia pillared clays, had basal spacings in the range of 16-18Å, i.e., pillar heights (basal spacing - silicate layer thickness 9.6Å) in the range of 6.5-8.5Å. In those pillared clays with small pillar heights, the nitrogen adsorption isotherms were usually of Type I, since the pore size is so small that the multilayer adsorption of nitrogen molecules in pores is physically impossible. However, the TiO_2 pillared clay prepared here showed Type IV isotherm (Table I). This is due to its large pillar height. A schematic structural model for TiO_2 pillared clay was proposed in a previous paper [8] and is shown in Fig. 1. It is interesting to note that the mixed oxide sol pillared clays showed Type I isotherm in spite of much larger basal spacings than that of the TiO_2 pillared clay. This indicates that the pore sizes in the mixed sol pillared clays are much smaller than the height of the pillars. It appears that the silica sol particles were intercalated in multilayers and thus small pores resulted between silica particles or between silica sol particles and silicate layer as shown in Fig. 2 [9]. Negatively charged silica sol particles cannot be used alone as pillaring agents. However, when titanium (IV) and iron (III) hydroxy cations were deposited onto the surface of the sol particles, the total charge of the sol particle must have been positive. Therefore, the silica sols coated with Ti or Fe hydroxides were taken up by the clay. Detailed studies on the preparation and structure of the pillared clays with mixed oxide sols will be published elsewhere.

Acidity Distribution

The acidity distributions of the three kinds of pillared clays were determined on the samples heated at 500°C and the results are shown in Fig. 3. Although all of the pillared clays showed large acid amounts, the distributions were very different. TiO_2 pillared clay was characterized by a very strong acidic solid and furthermore, a large portion of its acidity resulted from strong acidic sites (Ho ≤ -3.0); the portion of the weak acidic sites was very small. The SiO_2-TiO_2 pillared clay was found to be a strong acidic solid and had a similar acidity distribution to that of TiO_2 pillared clay, but it contained similar amount of weak acidic sites (Ho ≥ 4.8) as well. The SiO_2-Fe_2O_3 pillared clay also had large acidic amount, but most of the acid sites were very weak (Ho ≥ 4.8). The acid strength of the pillared clays decreased in the following order, TiO_2 ≈ SiO_2-TiO_2 >> SiO_2-Fe_2O_3 pillared clays.

TPD

TPD spectra of ammonia are often used to determine the amount and distribution of acid sites. The peaks at higher temperature range is considered to be due to the presence of stronger acid sites, and peak intensity is proportional to the amount of the acidic sites. TPD spectra of ammonia for the three pillared clays are shown in Fig. 4. The spectra consist of two peaks, low temperature and high temperature peaks. Part of the peak appearing from 500°C may be assigned to water, since the dehydroxylation of the silicate layers begins around 600°C. However, in the spectrum of SiO_2-Fe_2O_3 which has only weak acid sites, the intensity of the high temperature peak is very weak. It appears that the high temperature peak is also mainly due to ammonia desorption. Although it is very difficult to assign programmed temperature to a certain Hammett parameter, 500°C appears to be the dividing line. If we assume that the peak above 500°C corresponds to the very strong acid sites (Ho ≤ -3.0), the acidity distribution measured by titration is in good agreement with the TPD spectra.

Infrared Spectroscopy

Infrared spectra of pyridine adsorbed on solid surfaces are useful to distinguish the B or L acid sites [11]. The IR spectra measured on the self-supporting film of TiO_2 pillared clay are shown in Fig. 5. The strong absorption bands at 1445 and 1605 cm^{-1} can be assigned to pyridine molecules adsorbed on Lewis acid sites; whereas the band at 1545 cm^{-1} is the characteristic absorption due to pyridinium. The absorption at 1492 cm^{-1} can be assigned to both Bronsted and Lewis acid sites. In the spectrum measured after heating at 300°C, there is a weak absorption at 1545 cm^{-1} which is characteristic of Bronsted acid site. For the sample heated at 500°C, the intensity of the 1545 cm^{-1} band was much reduced. This indicates that Lewis acid sites are predominant in TiO_2 pillared clay.

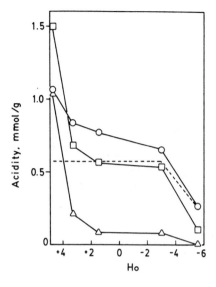

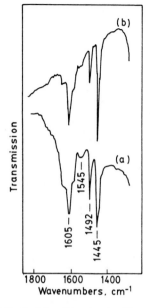

Fig. 3. Acid strength distribution for the clays pillared with oxide sols heated at 500°C: O, TiO₂;□, SiO₂-TiO₂;△, SiO₂-Fe₂O₃ pillared clays. A dashed line shows a distribution for a binary oxide TiO₂-SiO₂ (ref. 12).

Fig. 5. Infrared spectra of pyridine adsorbed on TiO₂ pillared clay heated at (a) 300°C and (b) 500°C.

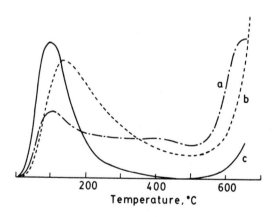

Fig. 4. TPD spectra for the pillared clays heated at 500°C: (a) TiO₂; (b) SiO₂-TiO₂ and (c) SiO₂-Fe₂O₃ pillared clays.

DISCUSSION

Itoh et al. [12] studied the acidic properties of TiO_2-SiO_2 binary oxide, and showed that single component oxides such as SiO_2 and TiO_2 had little acidity, but mixing the two oxides resulted in a remarkable increase in acidity. The acidity for TiO_2-SiO_2 (1:1) at Ho ≤ 1.5 is larger than those of SiO_2-Al_2O_3. The acidity distribution of TiO_2-SiO_2 (1:1) reported by Itoh et al. [12] is compared with those of sol oxide pillared clays in Fig. 3. Thus it is not surprising that SiO_2-TiO_2 mixed sol pillared clay had large acidity, but it should be noted that the clay pillared with single component TiO_2 sol also had very strong and large acidity. The acidic strength distributions of TiO_2 pillared clay and SiO_2-TiO_2 (1:1) binary oxide are similar.

Tanabe et al. [13] proposed the mechanism of the generation of acid sites of binary oxides. According to the mechanism, coordination numbers of the positive elements should be maintained even when mixed. When TiO_2 is dissolved in SiO_2 matrix, it generates excess valence unit of -2 and this position acts as Bronsted acid site, whereas when SiO_2 is dissolved in TiO_2 matrix, it generates excess valence unit of +4/3, which becomes a Lewis acid site. The IR spectra indicated that the predominant acidic sites were Lewis type. The strong acidity of TiO_2 pillared clay probably is due to the strong interaction between the TiO_2 pillars and interlayer surfaces of silicate layers.

Yamaguchi et al. [14] recently reported that Fe_2O_3 supported on SiO_2 can be promoted to super strong acid by treatment with SO_3 and can be used as powerful catalyst for isomerization of cyclopropane to iso-butane. Since SiO_2-Fe_2O_3 pillared clay has large acidity, it can be converted into an interesting catalyst if it is possible to improve it as above.

According to the analyses of nitrogen adsorption data, the pore size of TiO_2 pillared clay was about 18Å [8]. SiO_2-TiO_2 and SiO_2-Fe_2O_3 pillared clays had pore sizes of 8-10Å [9]. These pillared clays with a variety of pore sizes coupled with their strong acidity will open up a new family of heterogeneous catalysts.

ACKNOWLEDGEMENT   This work was supported in part by the Mazda Foundation. S. Y. acknowledges the support of this research by Nippon Sheet Glass Foundation for Materials Science.

REFERENCES

1.   G.W. Brindly and R.E. Sempels, Clay Miner. 12, 229 (1977).
2.   S. Yamanaka and G.W. Brindly, Clays Clay Miner. 27, 119 (1979).
3.   T.J. Pinnavaia, M.-S. Tzou, S.D. Landau, and R.H. Raythatha, J. Mol. Catal. 27, 195 (1984).
4.   A. Schutz, W.E. Stone, G. Poncelet, and J.J. Fripiat, Clays Clay Miner. 35, 251 (1987).
5.   M.L. Occelli, J. Mol. Catal. 35, 377 (1986).
6.   A. Schutz, D. Plee, F. Borg, P. Jacobs, G. Poncelet, and J.J. Fripiat, in Proc. Int. Clay Conf.,Denver, 1985, edited by L.G. Schultz, H. van Olphen, and F.A. Mumpton (The Clay Minerals Society) pp. 305.
7.   T.J. Pinnavaia, Science 220, 365 (1983).
8.   S. Yamanaka, T. Nishihara, and M. Hattori, Mat. Chem. Phys. 17, 87 (1987).
9.   S. Yamanaka, F. Okumura, and M. Hattori, Extended Abstracts of the 1986 Annual Meeting of the Ceramic Society of Japan, 1F07.
10.   H. Benesi, J. Phys. Chem. 61, 970 (1957).
11.   E.P. Parry, J. Catal. 2, 371 (1963).
12.   M. Itoh, H. Hattori, and K. Tanabe, J. Catal. 35, 225 (1974).
13.   T. Tanabe, T. Sumiyoshi, K. Shibata, T. Kiyoura, and J. Kitagama, Bull. Chem. Soc. Jpn. 47, 1064 (1974).
14.   T. Yamaguchi, T. Jin, T. Ishida, and K. Tanabe, Mat. Chem. Phys. 17, 3 (1987).

HIGH DISPERSION METAL OXIDE/MOLECULAR SIEVES: NEW BIFUNCTIONAL
SHAPE SELECTIVE CATALYSTS

R. SZOSTAK, V. NAIR, D.C. SHIEH AND T.L. THOMAS
Zeolite Research Program, Energy and Material Sciences Laboratory,
Georgia Tech Research Institute, Georgia Institute of Technology, Atlanta,
Georgia 30332, USA

ABSTRACT

Highly dispersed metal oxide/molecular sieve catalysts can be prepared
through modification of metallosilicate molecular sieves. The resulting
materials have been shown to exhibit enhanced catalytic activities for
selected reactions. The metallosilicates examined include the gallosilicates, iron (ferri-) silicates and cobalt silicates. The size of
the metal oxide particles as well as their location (in the pores or on
the surface) can be controlled through post-synthesis methods. Thermal
or mild hydrothermal treatment of the metallosilicate produces a highly
dispersed metal oxide phase, while higher temperatures or longer treatment
causes the metal oxide to migrate and form larger agglomerates of the
metal oxide phase. The size of these agglomerates are strongly dependent
on the conditions of hydrothermal treatment. Dispersion, location and
agglomeration of the metal oxide phase have been characterized for the
iron silicates using both physical and catalytic techniques. The magnetic
properties of the iron silicates are described.

METALLOSILICATE MOLECULAR SIEVES

The substitution of other elements for framework aluminum in the

molecular sieve zeolites has provided a means of modifying its acid

properties (1-3). However, framework substitution of elements with

catalytically active oxide forms followed by post-synthesis modification

provides a unique method for generating high dispersions of that

catalytically active metal oxide within the pores and channels of the

molecular sieve structure (4-9). Such catalysts should exhibit characteristic

properties of the active metal oxide phase as well as the shape selective

properties characteristic of the molecular sieve structure. Three metallo-

silicate molecular sieves have been synthesized and modified through post-

synthesis thermal and hydrothermal treatment to produce metal oxide/

molecular sieve catalysts. These include the gallosilicate, ferrisilicate

and cobalt silicate molecular sieves (4,8,9). The ferrisilicates are used

as a model system as the iron can be easily examined through a variety of

techniques. The ZSM-5 structure was the molecular sieve framework used

in this study as the aluminosilicate form is known to exhibit shape

selective properties for a variety of metal promoted processes.

Hydrothermal treatment of the zeolite molecular sieves is known to

encourage loss of structural aluminum. Characterization of these modified

zeolites, which includes both physical and catalytic techniques can

identify the amount of aluminum remaining in the structural framework sites

Mat. Res. Soc. Symp. Proc. Vol. 111. ©1988 Materials Research Society

but produces little information about the nature, size and location of the nonframework aluminum species. Similarly in the other metallosilicates, the loss of framework metal also occurs upon hydrothermal treatment resulting in increased hydrophobicity and a loss in ion exchange capacity (2,4,7,9). In the iron silicate systems, for example, nonframework iron oxide can be directly examined using magnetic measurements, EPR and Mossbauer spectroscopy. From such studies, the relative amounts of framework and nonframework iron can be determined and more importantly, variation in the size of the catalytically active iron oxide particles generated under different conditions can also be determined.

EXPERIMENTAL METHODS

The synthesis procedures used for the preparation of the ferrisilicate analogs of zeolite ZSM-5 as well as the thermal and hydrothermal treatment of these materials have been described in detail elsewhere (7,8). The calcination temperature was $550^{\circ}C$. All samples were treated at the designated temperatures in 100% steam. The notation used in this paper to represent the ferrisilicate ZSM-5 materials with varying SiO_2/Fe_2O_3 ratios is: Fe(xx) where (xx) is the bulk SiO_2/Fe_2O_3 ratio of that sample obtained from atomic absorption. Thus Fe(53) is the ferrisilicate ZSM-5 with a bulk SiO_2/Fe_2O_3 ratio of 53.

The magnetic moments of the ferrisilicates were evaluated using a Faraday microbalance. This method of measuring the magnetic properties of the materials is a very useful technique because it uses very small quantities of a sample (approximately 10-15 mg). The system was operated between room temperature and $-196^{\circ}C$. Complete details of the calibration and operation of the Faraday microbalance are described elsewhere (7).

IRON OXIDE DISPERSION AND AGGLOMERATION

Highly Dispersed Iron in the As-Synthesized and Calcined Forms

Table 1 shows that all of the as-synthesized and calcined materials have a magnetic moment between 5.5 and 5.8 Bohr magneton (independent of SiO_2/Fe_2O_3) which represents the lack of contribution of a magnetic moment from an external iron oxide phase. This is indicative (though not surprising for the as-synthesized form which cannot contain Fe-O-Fe linkages) of a high dispersion of iron present in these materials. The dependency of the magnetic moment on temperature was   examined and, for both the as-synthesized and calcined forms, the magnetic moment appears independent of temperature further verification that the Fe^{+3} ions are relatively far apart and exert a minimum influence on each other (i.e. the materials are magnetically dilute). This has been confirmed by Mossbauer spectroscopy (6,7). The as-synthesized samples contain nearly exclusively framework iron as determined by Mossbauer and EPR methods while the calcined material contains both tetrahedral framework and octahedral nonframework iron as determined from physical as well as catalytic techniques (5-8).

Table 1: Room Temperature Magnetic Moments for the Modified Ferrisilicates

sample	As-synthesized	Calcined	Hydrothermal treatment at: 550°C		700°C
			4 hrs	x** hrs	4 hrs
Fe(175)	5.5	5.8	6.4		13.2
Fe(92)	5.6	5.8	6.5		15.2
Fe(54)	5.7	5.7	6.5	10.9(12**)	31.7
Fe(24)	5.6	5.6	6.3	9.4(8**)	

** x hrs denotes that the ferrisilicate molecular sieves were hydrothermally treated for long periods of time. The actual number of hours treated appears within theparentheses next to the value in the table.

Agglomeration of Iron Oxide After Hydrothermal Treatment

When the ferrisilicates were hydrothermally treated for four hours (at 550°C) the magnetic moments increased to between 6.3 and 6.5 Bohr magneton. Since the expected value of the magnetic moment for a Fe^{+3} is under 6.0, any value over that is an indication that the iron oxide phase has begun to agglomerate to an extent which causes a significant contribution to the net magnetic moment of the material. The presence of agglomerated iron oxide in these materials has also been further verified using Mossbauer spectroscopy, EPR and catalytic reaction studies (5-7).

Hydrothermal treatment of the iron silicates at higher temperatures (at 700°C) causes the sharp increase in the magnetic moment from 6.5 BM to 31.7 BM) and is attributed to significantly large particles of iron oxide which are formed at these very high temperatures (greater than 550°C). The presence of these large particles has been confirmed by Mossbauer and EPR spectroscopy as well as high resolution transmission electron microscopy (5-7,10).

The individual magnetic moments of these materials treated at 700°C appears to change to a different extent unlike the 550°C treated materials. In addition, (not shown) the reduction of the magnetic moments with decreasing temperature was greater in the Fe(54) material than in the Fe(175) and Fe(92) samples, all of which were treated hydrothermally at 700°C. These differences are attributed to a dependence of the magnetic properties on the particle size. Cskooie-Tabrizi (11) notes that a particle which consists of more than a few atoms which are coupled together magnetically can possess a large magnetic moment. Thus it appears that similar high temperature treatment of ferrisilicates with different concentrations of iron in the framework can result in the formation of iron oxide particles of differing sizes which may contribute to the overall catalytic activity.

When the ferrisilicates were subjected to prolonged hydrothermal treatment (greater than 4 hours at 550°C) the magnetic moments of the materials increased by a large factor (6.5 BM to 10.9 BM). However, the increase is not as great as that observed for the samples treated hydrothermally at higher temperature for shorter periods of time. Prolonged hydrothermal treatment increases the magnetic moment by a factor of 1.5 whereas, the hydrothermal treatment at higher temperature causes the magnetic moment to be increased by a factor of 2.5 to 5.0. In addition, the temperature dependency of the magnetic moments differ between the 550°C hydrothermal treatments and those done at 700°C. No change in magnetic moment with temperature is observed when the samples are hydrothermally treated at 550°C unlike the sharp dependence observed for the 700°C treated materials. These results indicate that significantly larger particles of iron oxide are formed at higher temperatures of hydrothermal treatment as compared to prolonged treatment at lower temperatures. Thus it appears that the temperature of hydrothermal treatment has much more of an effect in determining the particle size than does the time of treatment. These results are in agreement with those obtained using Mossbauer spectroscopy, EPR and catalytic reaction studies (5-7).

Another factor which appears to contribute to the size of the iron oxide phase is the mode of preparation of the sample. It has been observed that the preparation of the ferrisilicate in either a stirred or unstirred crystallizer also effects the magnetic properties. The results are shown in table 2.

Table 2: Room Temperature Magnetic Moment for Stirred vs. Unstirred Ferrisilicate Molecular Sieves.

sample	As-syn.	Calc.	Magnetic Moment (BM) Hydrothermal treatment at:	
			550oC 4 hrs	650oC 1 hr
Fe(98) unstirred	5.4	5.9	6.7	30.4
Fe(92) stirred	5.6	5.8	6.5	12.0

The magnetic moments of both stirred and unstirred materials are almost the same for their as-synthesized, calcined and steamed 4 hours (at 550oC) forms.

However, when the unstirred material was hydrothermally treated at $600°C$ and beyond, its magnetic moment increases significantly indicating the formation of larger particles in the unstirred system than in the stirred system.

The magnetic properties alone do not provide the complete differentiation between the stirred and unstirred materials. Though the calcined samples show little variation in magnetic properties, the amount of framework iron in the two samples do differ. When Fe(92) is calcined only 12% of the framework iron is lost as compared to 29% in the Fe(98) material. Thus more iron is present as iron oxide in the unstirred sample prior to hydrothermal treatment. Examination of these samples for differences in morphology and homogeneity in composition among the crystals using electron microscopic techniques is described in a separate paper in these proceedings (10).

CONCLUSIONS

Examination of the ferrisilicates, as a model system for studying the generation of metal oxide/molecular sieve catalysts, provides significant insight into the role of the temperature and time of hydrothermal treatment, initial amount of M^{+3} (M=Fe) in the framework sites and the synthesis conditions (stirred vs. unstirred), in preparing these potentially useful catalyst materials. All of these factors contribute to the location and dispersion of the metal oxide phase within the molecular sieve pore system and must be taken into account in optimizing such bifunctional active catalysts.

References

1. R.Szostak, D.K.Simmons and T.L.Thomas, presented at the American Chemical Society Meeting, Philadelphia, Pa., Aug. 26-31,1984 (unpublished)

2. D.K.Simmons, Masters Thesis, School of Chemical Engineering, Georgia Institute of Technology, Atlanta, Ga., USA , 1986

3. C.T-w.Chu, C.D.Chang, J.Phys.Chem., 89, 1569 (1985)

4. R.Szostak, V.Nair, D.K.Simmons, D.C.Shieh, T.L.Thomas, R.Kuvadia, B. Dunson, in Studies in Surface Science and Catalysis, Elsevier, Amsterdam, 1988, (in press)

5. V.Nair, R.Szostak, T.L.Thomas, P.Agrawal, Proceedings of the 10th North American Meeting of the Catalysis Society, May 17-22, 1987.

6. A.Meagher, V.Nair and R.Szostak, Zeolites, 1987 (in press)

7. V.Nair, PhD Thesis, School of Chemical Engineering, Georgia Institute of Technology, Atlanta, Ga., USA , 1987

8. R.Szostak, V.Nair, T.L.Thomas, Journal Chemical Society, Faraday Trans., 87, 487, (1987)

9. D.K. Simmons, R.Szostak, P.K.Agrawal and T.L.Thomas, Journal of Catalysis, 106, 287, (1987)

10. R.Csencsits, R.Gronsky, V.Nair and R.Szostak, in Microstructures and Properties of Catalysts, MRS Symposium Series, 1988 (in press)

11. M.M.Oskooie-Tabrizi, PhD Thesis, Pennsylvania State University, Pa., 1983

Acknowledgement

The authors wish to thank Prof. A. Bertrand of the School of Chemistry, Georgia Institute of Technology for his assistance in obtaining the magnetic measurements. This work was supported, in part, under DOE contract # DE-AC-22-PC90007.

A NEW FAMILY OF PHOTOACTIVE CATALYSTS BASED UPON BISMUTH OXIDE

ANTHONY HARRIMAN[*], JOHN M. THOMAS[*], WUZONG ZHOU[**], AND DAVID A. JEFFERSON[**]

[*]Davy Faraday Research Laboratory, The Royal Institution, 21 Albemarle Street, London W1X 4BS, England

[**]Department of Physical Chemistry, University Chemical Laboratory, University of Cambridge, Lensfield Road, Cambridge CB2 1EP, England

ABSTRACT

Bismuth(III) oxide, which is pale yellow, functions as an amphoteric semiconductor, although it is relatively unstable upon illumination with UV light. The energy level positioning, corrosion limits and photoactivity of samples of n-type bismuth oxide have been determined by electrochemical techniques. It was found that this material is able to oxidise water to molecular oxygen upon illumination with UV light.

Bismuth oxide forms a wide range of well-defined solid solutions with many other oxides and, in particular, niobium oxide can be used to form several distinct structures, each based upon defective fluorite. Each of these structures will function as an n-type semiconductor, retaining the basic properties of undoped bismuth oxide. However, the exact composition of the solid solution determines the band-gap energy of the semiconductor. Low concentrations of niobium in the lattice induce a substantial lowering in the band-gap energy and give rise to orange coloured materials that can collect a higher fraction of the solar spectrum. Photoelectrochemical studies showed that the presence of niobium caused the valence band to move to lower potentials but had little effect upon the energy of the conduction band. The photochemical properties of these solid solutions were studied using the photooxidation of alcohols as a test system. The results are discussed in terms of the structure of the oxide.

INTRODUCTION

Bismuth oxide is polymorphic (1). The most stable modification at ambient temperatures, which is termed alpha-bismuth oxide, has a complex monoclinic structure with layers of bismuth atoms and oxygen atoms parallel to y and z axes in the model shown in Figure 1a. There are two types of bismuth atom in the model; one type (referred to as I in the figure) has five oxygen neighbours at the corners of a distorted octahedron with the bismuth lone-pair occupying the sixth corner as shown below:

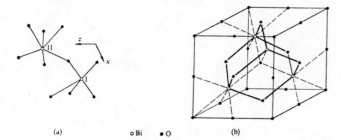

The other type of bismuth atom (referred to as II in the model) has six oxygen neighbours at the corners of an octahedron, distorted to suggest that here also the lone-pair affects the stereochemistry. The coordination polyhedra are linked to give tunnels parallel to the z axis into which the bismuth lone-pairs project.

Figure 1. (a) The coordination polyhedra present in alpha-bismuth oxide. (b) The fluorite structure of delta-bismuth oxide (one-quarter of the oxygen atoms are missing).

The cubic delta-form of bismuth oxide, which is stable at temperatures above $700^{\circ}C$, has a defective fluorite structure in which oxygen vacancies are randomly distributed throughout the lattice. Each bismuth atom has eight neighbours, as shown in Figure 1b. There is also a beta-modification of bismuth oxide, and several oxygen rich forms (2), related to the high temperature delta-structure. Some of the oxygen-rich forms are stabilised by small amounts of elements such as Ge or P and, consequently, they may be double-oxides. Other double-oxides can be obtained by fusing bismuth oxide with oxides of Ca, Sr, Ba, Cd or Pb. These materials have a layered structure.

Recently it has been shown (3) that there exists a vast compositional range of mixed oxides formed between bismuth oxide and oxides of Nb, Ta, Mo, W, V, and Si. Many members of this family crystallize into structures which are related to defective fluorite (anion vacancy). The $Bi_2O_3:Nb_2O_5$ system, for example, takes up four distinct, but related, phases, in the range 61:1 to 5:3 all of these being superstructural variants of the high temperature defect fluorite structure taken up by delta-Bi_2O_3. The structures of these solids have been elucidated (3) so that it is possible to study the relationship between crystal structure and photoelectrochemical properties. Such effects have not been well studied for a single, well-defined system and, here, we focus attention on the photoelectrochemical properties of the $Bi_2O_3.Nb_2O_5$ system.

EXPERIMENTAL

Bismuth oxide electrodes were prepared by anodic and thermal oxidation of pure bismuth. For the electrochemical preparation, a rod of pure bismuth was sealed into a glass tube and the open end was polished with successively finer grades of alumina to obtain a mirror finish. This electrode was anodised at 0.6 V vs SCE in Ar-purged aqueous solution at pH 9.2 (borate buffer) containing sodium sulphate (0.2 M). After 20 minutes, some 25 coulombs had passed which, with a surface area of 1.5 sq. cm, corresponds to a 16 mm thick layer of oxide. A similar rod of pure bismuth was heated at 400°C in air to form a surface coating of alpha-bismuth oxide.

Samples of bismuth oxide were obtained from Aldrich Chem. Co.. They were stored at 80°C to remove any surface moisture and pressed into discs (0.8 sq. cm surface area and 2 mm thick) at 80°C under high pressure. The discs were sintered in air at 800°C for 8 hours. After cooling, the discs were powdered and resintered as above. They were sealed into teflon holders using epoxy resin and the back of the disc was coated with a thick film (ca. 1 mm) of bismuth, which was applied by rf sputtering. A contact wire was attached to the bismuth film and the whole electrode was sealed into a glass tube with epoxy resin. Electrodes were made from the various solid solutions in the same manner and at least three electrodes were made for each sample.

Microcrystalline samples of the ternary oxides were prepared by direct combination of the binary oxides in pure oxygen at temperatures in the range 820-1000°C. Details relating to the preparation and analysis of these materials can be found elsewhere (3). The structure of the ternary oxide depends markedly upon the composition of the material and a general classification is given in Table 1.

TABLE 1. LIST OF SPECIMEN PREPARATIONS USED, WITH STARTING COMPOSITION, HEATING TEMPERATURE, INITIAL HEATING TIME, AND PHASES PRESENT

(The latter were derived solely from electron diffraction observations.)

starting composition $(Bi_2O_3:Nb_2O_5)$	heating temperature °C	initial time/h	main phases present (type)
61:3	820	113	I
19:1	820	162	I
15:1	820	169	I
9:1	820	168	II
17:3	820	168	II
4:1	820	24	II
3:1	820	24	III, II
7:3	820	24	III
17:8	820	24	III, IV
2:1	820	24	IV, III
5:3	1000	48	IV
1:1	1000	48	α-form

For the photochemical studies, the oxide (70 mg/25 ml) was suspended in nitrogen-saturated 1/1 water/2-propanol at pH 9.2 in a glass tube of total volume 35 ml equipped with a septum cap and a glass optical window. The solution was sonicated for 20 minutes and stirred constantly during the experiment. Irradiation was made at 30°C using a 950 W Xe lamp, filtered to remove IR and UV radiation. Samples of the vapour phase were removed periodically and analysed by gas chromatography.

RESULTS AND DISCUSSION

Photoelectrochemical properties of n-type Bi_2O_3

Bi_2O_3 possesses amphoteric character, functioning as both n- and p-type semiconductors under specific conditions (4). For samples of alpha-bismuth oxide prepared by both anodic and thermal oxidation of Bi, the optical band-gaps (E_{bg}) measured by reflectance spectroscopy and by photoconductivity were 2.78 eV, a value which compares well with previous measurements (4) which place the optical band gap in the range 2.8 to 3.0 eV. Flat-band potentials were determined by current-potential curves as shown by Figure 2 and by capacitance measurements in aqueous solution at pH 9.2 (borate buffer).

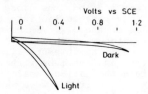

Figure 2. Current-potential curves observed for n-type Bi_2O_3 in the dark and under illumination.

Under white light illumination, n-type flat-band potentials (E_{fb}) of -0.33 and -0.12 V vs SCE respectively were derived for the anodically and thermally prepared samples. Again, these are consistent with literature values (4). The flat-band potential was found to decrease by 54 mV per unit increase in pH, consistent with surface protonation. The n-type decomposition level of Bi_2O_3 at pH 9.2 lies at -0.37 V vs SCE, giving rise to the energy level diagram shown in Figure 3.

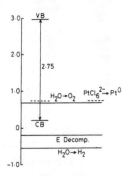

Figure 3. Energy level diagram for n-type Bi_2O_3 in aqueous solution at pH 9.2.

According to the data expressed in Figure 3, it is clear that n-type bismuth oxide can function as a powerful oxidising agent. Positive holes residing in the valence band should be able to perform many oxidation processes, including the oxidation of water to molecular oxygen. However, the conduction band is situated at quite positive potentials so that electrons promoted into this level are not strongly reducing. Indeed, this semiconducting material does not possess the capability of reducing water to molecular hydrogen. It will photoreduce chloroplatinate to atomic platinum, as shown in Figure 3. In steady-state photochemical studies, it was confirmed that irradiation of Bi_2O_3 suspended in water at pH 9.2 containing hexachloroplatinate ions resulted in formation of molecular oxygen and platinum metal. This, in itself, is an interesting photoreaction although it does not store energy.

A series of photoelectrochemical studies were performed to measure the electrical output of the semiconductor. For these experiments, the counter electrode was a Pt foil in contact with an aqueous solution of potassium ferricyanide containing sodium sulphate as background electrolyte. In the photolysis chamber, the electrolyte was aqueous sodium sulphate (0.2 M) at pH 9.2 and electrical contact was maintained with a salt bridge. For white light illumination of n-type Bi_2O_3 electrodes, the observed photocurrent was unstable and it decreased rapidly with continued irradiation time. Applying a positive potential to the electrode gave a marked increase in the stability of the photocurrent. At 0.6 V vs SCE applied potential, the photocurrent was stable over many minutes irradiation. Typically, the photocurrent density observed in such experiments was about 2 mA per sq cm and bubbles of oxygen were formed at the photoelectrode.

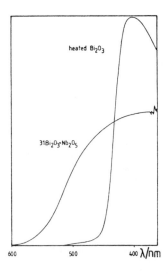

Figure 4. Reflectance spectra recorded for n-type Bi_2O_3 and one of the solid solutions formed with Nb_2O_5.

Photoelectrochemical properties of the mixed oxides

Incorporation of Nb_2O_5 into Bi_2O_3 has a marked influence upon the band-gap of the material, as shown by Figure 4. Surprisingly, low levels of Nb_2O_5 lower E_{bg} from 2.78 eV to 2.26 eV but further increases in the doping level raises E_{bg} (Figure 5). This effect is related to the detailed structure of the ternary oxide, and each particular phase appears to give a characteristic E_{bg}. Thus, subtle structural changes occur at Nb_2O_5 levels of <10, 10-20, 30 and >33%, and the corresponding phases show E_{bg} values of 2.26, 2.52, 2.67 and ca. 3.0 eV respectively. Samples having 2-6% Nb_2O_5 are clearly the most attractive as regards their ability to harvest sunlight, especially since such materials absorb appreciably in the visible region (Figure 4). Current-potential plots showed that the flat-band potentials of the mixed oxides decreased systematically upon increasing the level of Nb_2O_5. However, the decrease was slight (i.e. 1:1 $Bi_2O_3 \cdot Nb_2O_5$ gives E_{fb} = -0.22 V vs SCE), signifying that the valence band position changes for the various compositions of the solid solutions.

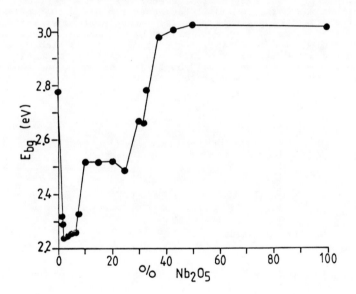

Figure 5. Band-gaps measured by photoconduction for the various $Bi_2O_3 \cdot Nb_2O_5$ solid solutions.

With pure Bi_2O_3, photocurrents can be detected upon
illumination with light of wavelength less than 450 nm (Figure
6). This corresponds to band-gap excitation. The photocurrent
action spectrum recorded for $31Bi_2O_3.Nb_2O_5$ is given also in
Figure 6. It is seen clearly that the mixed oxide is sensitive
to light of much lower energy and, for this mixed oxide, the
band-gap is very much lower.

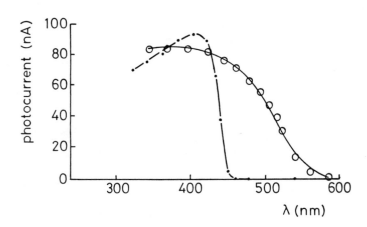

Figure 6. Photocurrent versus excitation wavelength plots for n-
type Bi_2O_3 and for $31Bi_2O_3.Nb_2O_5$.

The above studies show that incorporation of very low levels
of Nb into the lattice of Bi_2O_3 produces a material that is able
to harvest a higher fraction of the solar spectrum. All incident
light of energy greater than the band-gap can be used for
photoelectrochemical reactions and, for band-gaps of about 2.2
eV, the semiconductor has genuine applications for solar energy
storage. The Nb dopant affects the band-gap of the bulk
semiconductor by modifying its structure and there is a clear
relationship between the photoelectrochemical properties of the
mixed oxides and their known crystal stuctures (3). Preliminary
studies describing this effect have been published (5) and it is
clear that the most useful photochemical material possesses the
beta-bismuth oxide structure. This structure relates to the high
temperature delta structure and it is obtained by adding very low
concentrations of foreign ions into the crystal lattice.

302

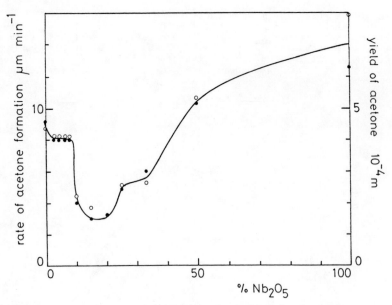

Figure 7. Initial rate (o) and total yield (•) of acetone formation upon photolysis of the oxides in 1/1 water/2-propanol solution.

Because of the unfavourable position of the flat-band potential, the $Bi_2O_3.Nb_2O_5$ system lacks the ability to photodissociate water. However, the ternary oxides showed more activity upon irradiation in 50% aqueous propan-2-ol solution. Upon illumination, acetone and H_2 were detected as reaction products. For naked Nb_2O_5, the ratio of products was consistent with dehydrogenation of the alcohol (at least over the early stages of reaction).

$$(CH_3)_2CHOH = (CH_3)_2CO + H_2 \quad (1)$$

With the $Bi_2O_3.Nb_2O_5$ solid solutions, acetone but not H_2 was detected during the early stages of irradiation (Figure 7). After an induction period, H_2 began to evolve but at a slow rate and the molar ratio of acetone to H_2 always exceeded ca. 5. This suggests that the major photoreaction is the oxidation of propan-2-ol by valence band holes giving rise to 2-hydroxy-2-methylethyl radicals and forming Bi at the surface of the semiconductor. The C-centred organic radicals disproportionate in the bulk solution to form acetone.

$$Bi_2O_3 + 3(CH_3)_2CHOH = 2Bi + 3(CH_3)_2CO + 3H_2O \quad (2)$$

The 2-hydroxy-2-methylethyl radicals evidently reduce water to H_2 upon the surface of a suitable catalyst, and it appears that Bi functions in that role since further irradiation results in formation of detectable yields of H_2. In this manner, turnover numbers with respect to Bi_2O_3 of 3 have been obtained for the acetone production reaction.

The rate of acetone production does show some dependence upon the composition of the material (Figure 7). Interestingly, the rate does not simply reflect the depth of the valence band but depends upon the precise structure of the ternary oxide, the materials with low levels of Nb_2O_5 doping again exhibiting relatively high photoactivity.

We thank the S.E.R.C. for support.

REFERENCES

1. G. Gattow and H. Schroder, Z. Anorg. Allgem. Chem. 318, 176 (1962).

2. S.C. Abrahams, P.B. Jamieson and J.L. Bernstein, J. Chem. Phys. 47, 4034 (1967).

3. W. Zhou, D.A. Jefferson and J.M. Thomas, Proc. Roy. Soc. London 406, 173 (1986).

4. K.L. Hardee and A.J. Bard, J. Electrochem. Soc. 26, 989 (1981).

5. A. Harriman, J.M. Thomas, W. Zhou and D.A. Jefferson, J. Solid State Chem. in press.

THE ROLE OF [M$^+$O$^-$] CENTERS (M$^+$= GROUP IA ION) IN THE ACTIVATION OF
METHANE ON METAL OXIDES

J. H. Lunsford, C.-H. Lin, J.-X. Wang and K. D. Campbell

Department of Chemistry
Texas A & M University
College Station, Texas 77843

ABSTRACT

If the ionic radii are compatible, alkali metal ions will substitute
for the divalent metal ions in magnesium oxide, calcium oxide and zinc
oxide. At high temperatures in the presence of molecular oxygen, centers
of the type [M$^+$O$^-$] are formed, where M$^+$= Group IA ion. The O$^-$ ions in
equilibrium with these centers are effective in H· abstraction from CH$_4$,
which is the first step in the oxidative dimerization reaction. At
reaction temperatures >720°C alkali metal oxides formed on the catalytic
surface may themselves become active centers. Alkali metal carbonates
also may inhibit the activity of the host oxide and sinter the oxide,
thereby eliminating corner sites which result in complete oxidation.

INTRODUCTION

The oxidative dimerization of methane to ethane and its sequential
conversion to ethylene have been studied over a variety of metal oxide
catalysts which are modified by the addition of alkali metal ions [1-7].
Maximum C$_2$ (ethane plus ethylene) yields of 15-20% have been achieved as
summarized in Table I. Without the alkali metal ion, the catalyst may
either be more active and less selective, as is the case with ZnO[5], or
less active and less selective, as is the case with MgO[1]. The purpose
of this paper is to explore the several roles which are played by the
alkali metal ion to improve the C$_2$ yield for this important reaction.

Table I. Oxidative Coupling of Methane Over
Promoted Metal Oxide Catalysts

Catalyst	Temp, °C	CH$_4$ Conv, %	C$_2$ Sel, %	C$_2$ Yield, %	Ref.
Li$_2$CO$_3$/MgO	720	43	45	19.4	1
Li$_2$CO$_3$/Sm$_2$O$_3$	750	38	54	20.7	2
NaNO$_3$/MgO	800	39	57	22.4	3
Li$_2$CO$_3$/ZnO	740	36	67	23.9	4
Na$_2$CO$_3$/CaO	725	33	45	14.8	6
NaMnO$_4$/MgO	925	22	70	15.4	7

The $[M^+O^-]$ centers have ESR spectra which are characteristic of the Group IA ion and the host oxide. As an example, the spectrum of the $[Li^+O^-]$ center in ZnO is shown in Figure 1 [5].

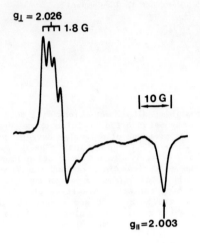

$g_\perp = 2.026$

$\overline{\text{⊓⊓}}$ 1.8 G

|10 G|

$g_\parallel = 2.003$

Figure 1. ESR Spectrum of the $[Li^+O^-]$ center in ZnO [Ref. 5].

The values of $g_\perp=2.003$ and $g_\parallel=2.048$ are in excellent agreement with those previously reported for the $[Li^+O^-]$ center produced in this material via irradiation [12]. The four-line hyperfine structure results from the interaction of the $^7Li^+$ nucleus (I=3/2, natural abundance=92.6%) with the unpaired electron localized mainly on the O^- ion.

In support of the experimental results describing the reactivity of O^- ions on MgO with CH_4, Anderson and co-workers [13] have demonstrated using molecular orbital calculations that the hydrogen atom abstraction occurs with a zero activation energy barrier and that the resulting methyl radicals can easily desorb. We have used a matrix isolation electron spin resonance (MIESR) system to detect these surface-generated radicals and to show that their formation rate correlates with the presence of the $[M^+O^-]$ centers in the oxide [14].

As shown in Figure 2, the correlation holds up to a rather large loading of Na^+ on CaO; however, at greater loadings Na_2CO_3 covers most of the surface and the CH_4 is no longer accessible to the $[Na^+O^-]$ centers. For this same set of catalysts, the $(C_2 \text{ yield})^{1/2}$ tracks the $CH_3\cdot$ radical formation rate remarkably well (Here the $(C_2 \text{ yield})^{1/2}$ is used because the coupling reaction is 2nd order in $[CH_3\cdot]$).

FORMATION OF ACTIVE CENTERS

[M$^+$O$^-$] Centers

As one might expect, the mechanism for the oxidative coupling of CH$_4$ involves the coupling of methyl radicals. These radicals are generated at active sites on the surface which are capable of abstracting a hydrogen atom from CH$_4$. In the gas phase O$^-$ ions react with CH$_4$ to form CH$_3$· and OH$^-$ ions with an 8% probability (i.e., 8% of all collisions result in reaction) at 25°C[8]. Moreover, we have demonstrated that O$^-$ centers on MgO surfaces react stoichiometrically with CH$_4$ at temperatures as low as -150°C[9]. These centers were formed by reacting trapped electrons with N$_2$O; however, of more importance in catalysis, O$^-$ centers may be formed thermally at elevated temperatures by substituting suitable Group IA ions in oxide catalyst.

Provided the size match is acceptable, Group IA ions substitute to a limited extent for the divalent metal ions in MgO, CaO and ZnO. At elevated temperatures in the presence of molecular oxygen an oxide ion adjacent to a monovalent ion in the lattice loses an electron (traps a hole), forming a center of the type [M$^+$O$^-$], where M$^+$ is an alkali metal ion [10,11]. The mechanisms for the formation of [Li$^+$O$^-$] centers in MgO, either thermally or as a result of UV irradiation, are depicted in scheme I. It is interesting to note that these centers were first observed in relatively large single crystals of MgO, provided the crystals were heated in the presence of air or oxygen [10]. Therefore, the bulk and the surface of the crystal are in communication via hole transport. As the holes hop from one oxide ion to another, transient O$^-$ centers are formed, both in the bulk and on the surface.

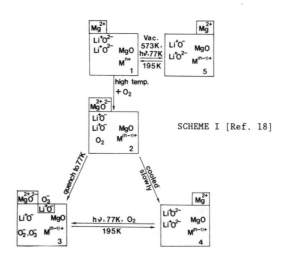

SCHEME I [Ref. 18]

It is particularly significant to note that up to a loading of 15% Na^+, the $(C_2 \text{ yield})^{1/2}$, the rate of $CH_3\cdot$ formation, and the $[Na^+O^-]$ concentration all increased in a parallel manner, whereas the surface area decreased [6, 15]. In addition, pure Na_2O_3 was ineffective as a catalyst for C_2 formation. These results suggest that at this reaction temperature, specific sites (O^- centers) are required for the selective reaction and the Na_2CO_3-covered surface is largely inactive for the generation of $CH_3\cdot$ radicals.

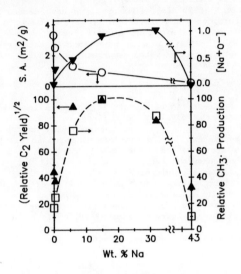

Figure 2. Effect of Na^+ concentration on $[Na^+O^-]$ concentration ($\blacktriangledown$), surface area (o), $(C_2 \text{ yield})^{\frac{1}{2}}$ ($\blacktriangle$), and methyl radical formation ($\square$). 1g of catalyst, 300 torr of CH_4, 20 torr of O_2, flow rate= 0.92 mL sec^{-1}, T= 675°C.

Because of the mismatch in ionic radii, it is not possible to substitute Na^+ or K^+ into MgO or K^+ into CaO at the temperatures of these experiments. This limitation provides yet another confirmation that $[M^+O^-]$ centers may be important in the activation of methane. The data of Table II give a comparison of catalytic data and the presence or absence of $[M^+O^-]$ centers in the host oxide [16]. Ionic radii for the relevant ions are $Li^+(0.68\text{Å})$, $Na^+(0.97\text{Å})$, $K^+(1.33\text{Å})$, $Mg^{2+}(0.66\text{Å})$, and $Ca^{2+}(0.99\text{Å})$. Clearly, with MgO only Li^+ is sufficiently small to substitute for Mg^{2+} and Li^+/MgO was the only catalyst in this set which gave a good C_2 yield. Over the CaO catalysts Li^+ and Na^+ promoted C_2 formation as expected, but K^+ was anomalous, for reasons which will be discussed below.

Table II. Comparison of MgO and CaO Catalysts Promoted
with Group IA Ions

| | MgO | | | CaO | | |
	pure	Li	Na	K	pure	Li	Na	K
wt%[a]	0	7	20	30	0	5	15	23
S.A.,m^2g^{-1}	12	4	6	7	6	2	2	2
Conv., %[b]	10.2	22.6	11.9	11.5	23.4	10.8[c]	22.0	21.9
(CH_4)								
C_2 sel,%[d]	6.3	56.7	17.8	0.7	8.1	67.2[c]	51.1	41.5
	(0.3)[e]	(1.0)	(0.5)	(0.0)	(0.1)	(0.7)	(0.9)	(0.7)
C_2 yield%	0.6	12.8	2.1	0.1	1.9	7.3	11.2	9.1
[M^+O^-]	No	Yes	No	No	No	Yes	Yes	No[f]

[a] Atomic ratio of Group IA ion = 0.43. There was not significant loss of Li^+ or Na^+ during the period of reaction.
[b] 1g MgO, 0.8 g of other catalysts (based on Group IA carbonate, Group IIA oxide); reaction run at 700°C, 76 Torr CH_4, 37 Torr O_2, 647 Torr He, 0.92 mLs^{-1}.
[c] At 750°C the conversion and selectivity were 22.9% and 52.8%, respectively.
[d] Other major products were CO and CO_2; the CO_2:CO ratio was >10 for all catalysts except MgO and K/MgO. Small amounts (<1%) of CH_3OH, HCHO, C_3H_6 and C_3H_8 were observed.
[f] A small concentration of [Na^+O^-] was observed, but no [K^+O^-] was detected.

Thus, there is considerable evidence which indicates that M^+O^- or O^- centers are involved in the oxidative dimerization of CH_4 over promoted MgO, CaO and ZnO. We do not distinguish between M^+O^- and O^- centers because the two are in thermal equilibrium, both at the surface and between the surface and the bulk. As the surface species react with CH_4, the equilibrium will immediately be reestablished. In this manner the bulk centers play an essential role in the surface reactions, which is in contrast to most conventional catalytic systems where the surface is only weakly coupled to the bulk.

Other Active Forms of Oxygen

As more systems are studied it becomes evident that centers other than [M^+O^-] are effective for the activation of CH_4, particularly at temperatures in excess of 720°C. For example, Aika and co-workers [3] have reported that a Na^+/MgO catalyst, prepared by heating a sample to 800°C, was active and selective for C_2 formation at 750°C. We demonstrated that such a catalyst did not contain [Na^+O^-] centers, and the activity pattern, shown in Figure 3, was substantially different from that of a Li^+/MgO catalyst [16]. Since no [Na^+O^-] centers were detected in this material, it appears reasonable that another active site, such as $Na^+O_2^-$ or $Na_2^{2+}O_2^{2-}$, were formed by the decomposition of Na_2CO_3 at the elevated temperatures.

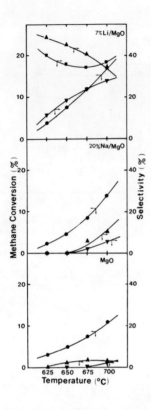

Figure 3. The effect of temperature on activities and selectivities over 1.0 g of 7% Li/MgO, 20% Na/MgO, and low surface are a MgO: •, CH_4 conversion; ▼, C_2H_2 selectivity; ▲, C_2H_6 selectivity; ■, C_1 selevitity. CH_4 (76 torr) and O_2 (38 torr) were used at a flow rate of 0.92 mLs^{-1} [Ref. 16].

Similarly, La_2O_3 is a very active catalyst for CH_4 conversion and at low conversions the C_2 selectivities are high [17]. Quenching experiments reveal the presence of superoxide ions, O_2^-, on the surface rather than O^- ions [18]. Based on the work of Loginov et al. [19] it has been suggested that O_2^- is formed via the reaction

$$O_2 + O_2^{2-} \rightleftharpoons 2O_2^- \tag{1}$$

thus, the presence of paramagnetic superoxide ions implies the existence of the diamagnetic peroxide ions.

Otsuka et al. [20] have studied stoichiometric reactions between CH_4 and peroxide ions on Na_2O_2, BaO_2 and SrO_2 at 400°C. They observed that the yield of C_2 products reached 11% over Na_2O_2 and concluded that the peroxide ion, rather than the superoxide ion, was responsible for the activation of methane over these oxides. Under catalytic conditions, where CO_2 is a reaction product, the carbonate phase is the favored sodium compound. Nevertheless, at elevated temperatures it is expected that the oxide phase would also exist on the surface.

The most stable oxide of potassium is KO_2, and the O_2^- ion may be capable of abstracting hydrogen from CH_4 at ca. 700°C. As an alternative, it is possible that under the catalytic conditions K_2O_2 may exist on the surface. Clearly, there is a need for more experimental evidence to identify these diamagnetic forms of oxygen on the selective catalysts.

OTHER ROLES OF ALKALI METAL IONS

Modification of Surface Activity

At temperatures ≥700°C most metal oxides have considerable activity for the oxidation of CH_4, although generally the C_2 selectivity is rather low. Oxides such as ZnO and Mg_6MnO_8 are effective for the complete oxidation of CH_4, yet when modified by an alkali metal ion they become good dimerization catalysts. Here the role of the alkali metal seems to be that of modifying the general (and nonselective) activity by physically blocking much of the surface with an inactive carbonate phase.

This phenomenon is illustrated by comparing in Table III the catalytic data obtained for a washed and unwashed sample of ZnO which had been modified with Li_2CO_3 [5]. One catalyst after pretreatment in flowing O_2 at 750°C was thoroughly washed with water to remove any residual Li_2CO_3 from the surface. The surface Li concentration of the two samples was determined by particle desorption mass spectrometry (PDMS), [21] which is a surface sensitive technique capable of analyzing only the top 1-3 monolayers of the sample. The analysis showed that the surface Li concentration fell from 10 wt% on the unwashed sample to 3 wt% on the washed material. The washed sample exhibited a C_2 yield which was comparable to the unwashed catalyst; however, the rate of C_1 product formation was considerably greater for the washed sample. The surface Li_2CO_3 may moderate the non-selective activity by prohibiting access of the methyl radicals to the ZnO surface (see below).

TABLE III: Effect of Washing on Activity[a,b]

Sample		A	B
Pretreatment		original	washed
Selectivity(%)	CO_2	41.7	57.4
	CO	5.6	3.0
	C_4H_4	24.5	20.5
	C_2H_6	28.1	19.0
	Total C_2	52.6	39.6
Conversion(%)	CH_4	21.4	28.0
	O_2	32.4	70.3
C_2 Yield(%)		11.3	11.1

[a]Original catalyst, (A)-0.18 wt% Li/ZnO; (B) -A washed 6 times with 50 ml cold H_2O. Over 4 g of catalyst at 720°C and flow rates of 42.5 mL min^{-1} He, 5.0 mL min^{-1} CH_4, 2.5 mL min^{-1} O_2.
[b]Ref. 5

Promotion of Sintering

In addition to modifying nonselective activity by partially covering an active oxide, alkali metals promote extensive sintering which decreases the surface available for reaction [3]. As described in Figure 2 this decrease in surface area is accompanied by an increase in C_2 yield [6]. The origin of this inverse relationship between yield and surface area is complex because of the heterogeneous-homogeneous nature of the reaction [14,15] and modeling studies are needed to identify the several factors which are most important. Qualitatively, however, the results may be understood in view of competitive reactions involving selective coupling of gas phase $CH_3\cdot$ radicals

$$2(CH_3\cdot)_g \rightarrow (C_2H_6)_g \tag{2}$$

and the reductive addition of methyl radicals to surfaces

$$(CH_3\cdot)_g + M^{n+1}O^{2-} \rightarrow M^n(OCH_3)^-. \tag{3}$$

The latter reaction produces methoxide ions which, under the conditions of oxidative coupling, result in the formation of CO_2. In order to maximize yield one would hope to couple the very reactive methyl radicals before they have an opportunity to react with the oxide surface.

Anderson and co-workers [13] have calculated for MgO the energetics for the reaction

$$O^-(s) + O^{2-}(s) + CH_4(g) \rightarrow OH^-(s) + OCH_3^-(s) + e^- \tag{4}$$

where the electron is promoted to the lowest surface state orbital belonging to a Mg^{2+} cation on a corner, edge or face site. Their results show that the energetics for this reaction are favorable only when the Mg^{2+} cation is at a corner site. Sintering would reduce the number of these corner sites and thereby decrease the nonselective oxidation of CH_4.

CONCLUSIONS

Alkali metal compounds influence the oxidative dimerization of methane both by generating active sites and by preventing undesirable secondary reactions on surfaces. On MgO, CaO and ZnO alkali metal ions of the appropriate size give rise to $[M^+O^-]$ centers which are effective in the formation of $CH_3\cdot$ radicals. At temperatures >720°C there is evidence that Na and K peroxides may be formed under catalytic conditions and these oxides are capable of activating CH_4. Alkali metal carbonates also inhibit secondary reactions between the $CH_3\cdot$ intermediate and the metal oxide, thus providing an opportunity for the radical coupling reaction to occur. This inhibition involves site coverage and sintering of the metal oxide which reduces the number of reactive corner sites.

ACKNOWLEDGMENTS

The author wishes to acknowledge the experimental and conceptual contributions of Daniel Driscoll, and Hong-Sheng Zhang. The research was supported by the National Science Foundation under Grant No. CHE-8617436.

REFERENCES

1. T. Ito and J.H. Lunsford, Nature (London) 314, 721 (1985); T. Ito, J.-X. Wang, C.-H. Lin and J.H. Lunsford, J. Am. Chem. Soc. 107, 5062 (1985).

2. K. Otsuka, Q. Liu, M. Hatano, A. Morikawa, Chem. Lett. 1986, 467.

3. E. Iwamatsu, T. Moriyama, N. Takasaki and K. Aika, J. Chem. Soc. Chem. Commun. 1987, 19; T. Moriyama, N. Takasaki, E. Iwamatsu and K. Aika, Chem. Lett. 1986, 1165.

4. I. Matsuura, Y. Utsumi, M. Nakai and T. Doi, Chem. Lett. 1986, 1981.

5. H.-S. Zhang, J.-X. Wang, D.J. Driscoll and J.H. Lunsford, submitted to J. Catal.

6. C.-H. Lin, J.-X. Wang and J.H. Lunsford, J. Catal. (in press).

7. J.A. Sofranko, J.J. Leonard, C.A. Jones, A.M. Gaffney and H.P. Withers, Petr. Div. Preprints, 32, 763 (1987).

8. D.K. Bohme and F.C. Fehsenfeld, Can. J. Chem. 47, 2717 (1969).

9. K. Aika and J.H. Lunsford, J. Phys. Chem. 81, 1393 (1977).

10. M.M. Abraham, W.P. Unruh and Y. Chen, Phy. Rev. B 10, 3540 (1974); D.N. Olson, V.M. Orera, Y. Chen and M.M. Abraham, ibid. 21, 1258 (1980).

11. J.-X. Wang and J.H. Lunsford, J. Phys. Chem. 90 5883 (1986).

12. J. Haber, K. Kosinski, M. Rusiecka, Disc. Faraday Soc. 58, 151 (1974).

13. S.P. Mehandru, A.B. Anderson and J.F. Brazdil, submitted to J. Am. Chem. Soc.

14. D.J. Driscoll, W. Martir, J.-X. Wang and J.H. Lunsford, J. Am. Chem. Soc. 107, 58 (1985).

15. K.D. Campbell, Ph.D. Dissertation, Texas A&M University, 1987.

16. C.-H. Lin, T. Ito, J.-X. Wang and J.H. Lunsford, J. Am. Chem. Soc. 109, 4808 (1987).

314

17. C.-H. Lin, K.D. Campbell, J.-X. Wang and J.H. Lunsford, J. Phys. Chem. 90, 534 (1986).

18. J.-X. Wang and J.H. Lunsford, J. Phys. Chem. 90, 3890 (1986).

19. A.Y. Loginov, K.V. Topchieva, S.V. Kostikov, N.S. Krush, Dokl. Akad. Nauk. SSSR 232, 1351 (1977).

20. K. Otsuka, A.A. Said, K. Jinno and T. Komatsu, Chem. Lett. 1987, 77.

21. W.R. Summers and E.A. Schweikert, Rev. Sci. Instrum. 57, 692 (1986).

MOLYBDATE CATALYSTS

WATARU UEDA
Research Laboratory of Resources Utilization, Tokyo Institute of Technology,
Nagatsuta-cho 4259, Midori-ku, Yokohama, 227 Japan

ABSTRACT

It is possible to classify the solid state catalyst systems into two categories; structure static catalyst and structure dynamic catalyst. The former is the catalyst in which the original structural circumstances of surface active center, such as coordinatively unsaturated state of active element, length and order of metal-ligand bond or of a bond between structural component, and their bond angle, are maintained unchanged during catalysis. The example is molybdenum sulfid which is a good catalyst for the isomerization and hydrogenation of olefins. On the contrary, the catalysts, in which the above circumstances of surface active center are changing dynamically during catalysis, are classified into the latter category. Molybdate catalysts for olefin oxidation by molecular oxygen at high temperature belong to this category. Main difference between two categories is whether the originally coordinated ligand on metal cation, such as sulfur anion of molybdenum sulfid and oxygen anion of metal molybdate, can participate in the catalysis directly or not. When the coordinated ligand is involved in the reaction, resulting abnormal coordinatively unsaturated state of surface active center must be sustained by the dynamic structure changes of both surface and bulk for the steady state catalysis. The bulk structure, therefore, affects strongly the properties of structure dynamic catalysts, whereas the properties of structure static catalysts depend on the surface structure like coner and edge mainly. The relationship between catalytic properties and structural stability of molybdate catalysts in the selective oxidation of propylene was discussed on the basis of ^{18}O tracer study.

INTRODUCTION

In past 20 years, remarkable developments in the heterogeneous oxidation catalysts have been achieved by the blending of several kinds of metal oxides to the catalytically active components. Typical examples are the so-called multicomponent bismuth molybdate catalysts, which are known as the most active and selective catalysts for the allylic oxidations(eq 1) of

$$CH_3CH=CH_2 + O_2 \xrightarrow{300-460°C} CH_2=CHCHO + H_2O \qquad (1)$$

lower olefins and have been widely used in the industrial catalytic oxidation process. Inspite of the importance of these catalyst systems, only a little work has been reported for the working mechanism and the role of each component owing to their complicated compositions and structures[1].

A simple question arising from the multicomponent catalyst systems is how the various kinds of metal components in these systems are functioning simultaneously in catalysis. Grasselli et al. proposed the multifunction catalyst concept[2], in which an allylic hydrogen abstraction component and the olefin chemisorbing/oxygen insertion component, which together perform the catalyst reduction cycle, are combined with a reoxidation component to complete the catalysis cycle. This concept has been extended to a number of complicated catalyst systems to explain the role of each component. They concluded that the α-hydrogen abstractors are partially reduced species, which have partially filled valence orbitals, while the chemisorbing/0-inserting species are fully oxidized. The redox element needs to be in easily interconverted multivalence state. It seems to be, however, still obscure how the redox element can take part in the catalysis.

An interesting thing is that as long as molybdate catalysts are concerned, the lattice oxygen can migrate in the lattice and its diffusion rate is much higher than the reaction rate. As revealed by the pioneer works by Keulks and Wraggs et al. using $^{18}O_2$ tracer(eq 2 and 3)[3],[4], the

$$CH_3CH=CH_2 + {}^{18}O_2 \xrightarrow[\text{BI-Mo-}{}^{16}O]{} CH_2=CHCH^{16}O \qquad (2)$$

$$CH_3CH=CH_2 + {}^{16}O_2 \xrightarrow[\text{Bi-Mo-}{}^{18}O]{} CH_2=CHCH^{18}O \qquad (3)$$

lattice oxygen is active oxygen species for allylic oxidation and can be supplied from the lattice of bismuth molybdate catalyst through the diffusion. The author also measured the mobility of lattice oxygen for many kinds of molybdate catalysts using $^{18}O_2$ tracer(Table I)[5] and concluded that although the diffusion phenomenon is common property for molybdate catalysts, there is a strong relationship between the mobility of lattice oxygen and the catalytic activity because the mobility tends to become higher when the molybdate catalysts are constructed with many kinds of metal components and showed high activity.

On the basis of above tracer results, the author proposed a concept of pseudo homogeneous phase to explain the simultaneous collaboration of each metal component in the multicomponent molybdate catalysts during catalysis. The pseudo homogeneous phase is where the lattice oxygen can migrate quite rapidly, like a reactant in solution in the case of homogeneous catalysis, between the active components for hydrogen abstraction, adsorption, O-insertion, and molecular oxygen uptake. The activity of reactive oxygen species is equalized in this way between the active components, so that many kinds of metal component can collaborate each other intimately at the same time during catalysis.

The main theme of this work are to discuss further the role of rapid diffusion of lattice oxygen during the oxidation of propylene over molybdate catalysts and to demonstrate the application of the concept for the design of effective oxidation catalysts.

CATALYTIC PROPERTIES OF SCHEELITE TYPE MOLYBDATE CATALYSTS

Catalytic activity and mobility of lattice oxygen

Scheelite oxide system, $Bi_{1-x/3}V_{1-x}Mo_xO_4$, was chosen for the present work because the catalyst composition can be changed without changing the fundamental structure of the catalyst. Table II summarised the catalytic activity and selectivity to acrolein in the oxidation of propylene over the scheelite catalysts having different compositions. Catalytic oxidation was carried out using a conventional flow microreactor under an atmospheric pressure(propylene, 16.7%; oxygen, 16.7%; nitrogen, balance; reaction temperature, 450 °C; catalyst weight, 1g). As already reported in the systematic study of scheelite oxide catalysts by Sleight et al.[6], catalytic activity increased drastically with the substitution of V^{5+} ion by Mo^{6+} ion in $BiVO_4$. $Bi_2(MoO_4)_3$ catalyst which belongs to this series of catalyst showed a low ability for propylene oxidation. The selectivity of the catalysts tested was slightly affected by the substitution.

Similar to the kinetic results reported previously for the bismuth molybdate catalyst system, propylene oxidation is first order with respect to propylene and independent of the oxygen pressure under the present conditions. The apparent activation energy is 19 ±0.5 kcal/mol for every catalyst. Inspite of the drastic change in the specific activity, the reaction kinetics were independent of the Mo content of the catalysts.

Table I Catalytic Properties and Mobility of Lattice Oxygen of Molybdate Catalysts and Other Related Catalysts in the Oxidation of Propylene[1]

Catalyst	Structure	Activity for propylene oxidation $/10^{-5}mol \cdot min^{-1} \cdot \bar{m}^2$	Selectivity to acrolein (%)	Mobility of lattice oxygen (%)[2]
Bi_2MoO_6	Keochlinite	4.9	90	100
$Bi_2Mo_2O_9$	layer structure	6.9	90	98
$Bi_2Mo_3O_{12}$	defect scheelite	5.9	84	10
$PbMoO_4$	scheelite	0.06	26	1-10
$CaMoO_4$	scheelite	0.02	0	1-10
$BiVO_4$	scheelite	0.8	76	37
$Pb_{1-3a}Bi_{2a}MoO_4$[3]	scheelite	8.3-11.8	72	50-100
$Bi_{1-b/3}V_{1-b}Mo_bO_4$[4]	scheelite	8.7-16.9	91	57-100
$CoMoO_4$	$\alpha\text{-}MnMoO_4$	0.7	46	1
$MnMoO_4$	$\alpha\text{-}MnMoO_4$	1.2	60	1
$NiMoO_4$	$\alpha\text{-}CoMoO_4$	3.2	53	1
$Fe_2Mo_3O_{12}$	$Fe_2(MoO_4)_3$	0.9	59	9
$Co_{11}Bi_cMo_{12}O_x$[5]	$\alpha\text{-}MnMoO_4$[6]	1.2-2.2	97	1-10
$Co_8Ni_3Bi_1Mo_{12}O_x$	$\alpha\text{-}MnMoO_4$[6]	0.7	97	5
$Co_8Al_3Bi_1Mo_{12}O_x$	$\alpha\text{-}MnMoO_4$[6]	5.1	95	59
$Co_8Cr_3Bi_1Mo_{12}O_x$	$\alpha\text{-}MnMoO_4$[6]	6.4	96	60
$Co_8Fe_3Bi_dMo_{12}O_x$[7]	$\alpha\text{-}MnMoO_4$[6]	12.2	96	45-60
$CoTeMoO_6$	$CoTeMoO_6$	0.26	89	13
$MnTeMoO_6$	$MnTeMoO_6$	0.69	77	10
$Fe_2Te_3Mo_3O_x$	$Fe_2(MoO_4)_3$[6]	0.46	97	16
$Fe_2Te_1Mo_1O_x$	-	0.74	97	26
$CuO\text{-}TeO_3\text{-}6MoO_3$	-	0.41	96	29
$TeO_3\text{-}6MoO_3$	Te_2MoO_7[6]	2.0	97	9
$Fe_2O_3\text{-}4Sb_2O_4$	$FeSbO_4$[6]	0.6	94	2
$Fe_2O_3\text{-}2Sb_2O_4$	$FeSbO_4$[6]	1.2	93	1
$ZnO\text{-}4Sb_2O_4$	$ZnSb_2O_6$[6]	0.2	95	2
$Cr_2O_3\text{-}4Sb_2O_4$	$CrSbO_4$[6]	0.2	93	2
$MoO_3\text{-}Sb_2O_4$	-	0.5	92	10
$Cu_2O/Celite$[8]	Cu_2O	0.2	59	0

1) All runs were carried out at 450 °C using a circulating system under a reduced pressure($P_{C_3H_6}=P_{O_2}=$ 70 Torr). 2) Percentage fraction of the complete mixing volume involved in the reaction to the total amount of lattice oxygen in the catalyst. 3) a=0.02-0.06. 4) b=0.09-0.45. 5) c=0.05-1. 6) Main phase. 7) d=0.05-1. 8) Reaction temperature, 260°C.

Table II   Comparison of the Catalytic Activity and Selectivity
to Acrolein with the Fraction of Lattice oxygen Involved
in the Oxidation of Propylene over
$Bi_{1-x/3}V_{1-x}Mo_xO_4$ Catalysts

X	Specific activity $/10^{-4}mol \cdot min^{-1} \cdot m^{-2}$	Selectivity to acrolein (%)	Fraction of lattice oxygen involved in the reaction (%)
0.00	0.08	75.6	37.0
0.09	0.87	85.1	57.4
0.15	1.35	91.1	63.0
0.21	1.43	93.7	75.1
0.27	1.31	90.8	72.2
0.45	1.69	90.8	100
1.00	0.59	84.4	10.2

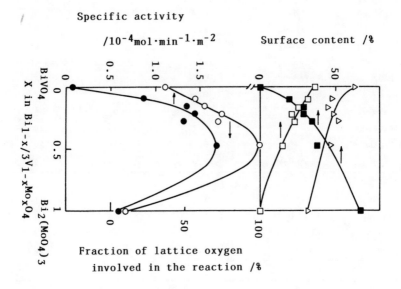

Figure 1.   Comparison of the specific activity of $Bi_{1-x/3}V_{1-x}Mo_xO_4$ catalysts
with the fraction of lattice oxygen involved in the oxidation of propylene to
the total lattice oxygen in the catalyst particles and with surface content
(M/Bi+V+Mo) of the constituent determined by XPS. △:Bi, □ :V, ■ :Mo.

Thus, the most plausible reason for the increase in the catalytic activity by the introduction of molybdenum to $BiVO_4$ catalyst is the increase of the number of active site on the surface. The mobility of lattice oxygen in the scheelite oxide catalysts during the catalytic oxidation of propylene was estimated from the $^{18}O_2$ tracer measurements. Propylene oxidation using $^{18}O_2$ gas was carried out in a circulating glass reactor system under the following standard conditions: initial pressure, P(propylene)=P(oxygen)=70 Torr; reaction temperature, 450 °C, catalyst weight, 0.1g. As mentioned above, molecular oxygen participates in the reaction after being captured in the oxide catalyst as lattice oxygen and scrambling with inner lattice oxygen. The increase of ^{18}O concentration in oxidized product with reaction time, therefore, depends upon the extent of the scrambling of the lattice oxygen during the reaction, which may reflect the mobility of lattice oxygen. However, it seems quite difficult to determine the quantitative correlation between the bulk diffusion rate of lattice oxygen and the results obtained in the tracer experiment. For the purpose of assuming the mobility of lattice oxygen, the amount of lattice oxygen with which the incoming $^{18}O_2$ equilibrates instantly was measured. This quantity can be regarded as the dilution volume or comlete mixing volume of lattice oxygen, ^{16}O, in the catalyst for the ^{18}O flux through the catalyst. The detail procedure was reported previously[5].

The fraction of the complete mixing volume to the total amount of lattice oxygen in the catalyst was listed in Table II for every catalyst with the activity data and the comparison between the specific activity of each catalyst tested and the fraction of the lattice oxygen involved in the reaction is shown in Fig. 1. As revealed from the comparison, there should be a correlation between the specific activity and the mobility of lattice oxygen. The kinetic results clearly showed that the increase of the catalytic activity by the introduction of molybdenum to the scheelite catalyst is attributable to the increase in the number of active site. Since no clear correlation was observed between the surface concentration of each component determined by XPS analysis and the specific activity(Fig. 1), the result suggests that there should be some direct effects of the bulk diffusion of lattice oxygen on the formation of surface active sites which concern with the rate determining step.

Structural stability

One of the serious problems regarding the practical use of molybdate catalysts is the deactivation of the catalysts due to the decomposition of the active composite oxides. In this part, the relationship between the structural stability and the mobility of lattice oxygen is dealt with.

The structural stability of two typical scheelite catalysts, $BiVO_4$ and $Bi_{0.85}V_{0.55}Mo_{0.45}O_4$, were examined by XRD measurements after reduction-oxidation cycle(Fig. 2). After the catalyst was reduced by 6% of the lattice oxygen of catalyst under the hydrogen at 450°C, free bismuth metal was detected in $BiVO_4$, whereas no liberated bismuth metal was detected in $Bi_{0.85}V_{0.55}Mo_{0.45}O_4$. XRD peak width of the former catalyst was unchanged but that of the latter catalyst was found to broaden prominently. After the reoxidation under the oxygen upto 450°C, the original XRD peaks of both catalysts reappeared but several additional unidentified peaks were detected in the case of $BiVO_4$. The unidentified peaks became more prominent after four reduction-oxidation cycles.

These observations can be explained as follows. Since the lattice oxygen in $BiVO_4$ could not diffuse faster than the rate of reduction on surface, the catalyst was reduced mainly in the vicinity of the surface oxide layer and was destroyed completely at the surface region after the repeated redox cycle. On the other hand, $Bi_{0.85}V_{0.55}Mo_{0.45}O_4$ has the highest mobility of lattice oxygen, so that the reduction was not limited to the vicinity of the surface oxide layer but could spread into the bulk of

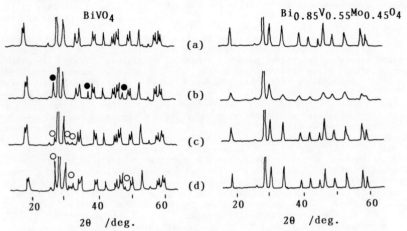

Figure 2. XRD patterns. (a), fresh catalyst, (b) after reduction by hydrogen at 450°C(6%), (c) after reoxidation at 450°C, and (d) after four reduction-oxidation cycles. ●: Peaks of metallic Bi, ○: unidentified.

the catalyst. The structural deformation caused by the reduction would be mild compared to that of $BiVO_4$ and the surface structure seems to be unimpaired during the reduction and reoxidation. In fact, no unidentified peaks or peaks assigned to the metallic bismuth were detected at any time during the redox cycle. The original XRD pattern was completely recovered and this catalyst showed high structural stability.

The structure of the composite oxide catalysts for the oxidation of hydrocarbons changes dynamically under the reaction conditions according to the Mars and van Krevelen mechanism. The structural changes of the catalysts result in changes of the catalytic properties. Formation of different metal oxide phase, growth of large crystals which decrease the surface area, transformation of amorphous states to crystals, liberation of active elements, and losses of a particular element due to vaporization, all these changes result in the deactivation of the catalyst. In order to diminish the change in catalytic properties, it seems to be important to suppress the structural deformation. The present study reveals clearly that the structural deformation during catalysis over molybdate catalysts would be minimized by the rapid diffusion of the lattice oxygen.

ROLE OF THE BULK DIFFUSION OF LATTICE OXYGEN

The surface structure of solid-state catalyst is very important, which strongly affects the catalytic properties. Although the surface structure is not always the same as the bulk structure, the bulk structure is a very important factor to decide what type of surface structure can be formed. If a corner or a edge is important for catalysis, crystal habit of solid catalyst will be another factor. Crystal habit also depends on the crystallographic structure of solid. Many investigators, therefore, have tried to explain the catalytic properties from the standpoints of bulk structure and surface layer structure determined by a static method. One must, however, look at the solid catalyst by taking it into account whether the surface structure is changeable during catalysis or not.

Tanaka et al[7]. studied the isomerization and hydrogenation of olefin over MoS_2 catalyst of which structure is layer type. They made the corner or edge enriched MoS_2 sample by cutting artificially and showed that the corner or edge part is active site for isomerization or hydrogenation. Their work is obviously based on that the surface structure is stable under the reaction conditions. In contrast to MoS_2 work, Haber et al.[8] demonstrated the dynamics of surface reconstruction using molybdate catalyst for propylene oxidation to acrolein. They observed the structural change of surface layer by varing the reaction conditions. Propylene conversion and selctivity to acrolein were also changed. They concluded that the surface of the catalyst has a dynamic character and reconstracts depending on the properties of reactants of the catalytic reaction until it attains the structure and composition corresponding to the given steady-state conditions of the reaction.

The main difference between these two catalysts is whether the originally coordinated ligand on metal cation(structural component), such as sulfur anion of molybdenum sulfid and oxygen anion of molybdate, can participate in the reaction directly or not. As already mentioned, oxygen anion in the lattice of molybdate catalysts needs to take part in the catalysis, so that the resulting abnormal coordinatively unsaturated state must be sustained by the dynamic character of the catalyst for the steady-state reaction. It was found that in the molybdate catalysts for allylic oxidation there was a good relation between the catalytic activity (or the stability of catalyst) and the mobility of lattice oxygen in the lattice. Since, in general, the phenomenon of lattice oxygen mobility does not reflect the property of the catalyst surface but of the catalyst bulk, the surface dynamic character results from the dynamic property of bulk like the mobility. On the other hand, no substantial structural change on surface needs to occur for the molybdenum sulfid catalyst because the catalytic cycle can be completed only by using coordinatively unsaturated active site which originally exists at fresh state. Although the catalyst structure in this system is also a important parameter, the static factors, such as crystallographic structure and defect structure, are anticipated to be important in determining the catalytic activity.

On the basis of above discussion, the solid-state catalyst systems were classified into two categories, the structural static catalyst and structure dynamic catalyst. The former is the catalyst in which the original structural circumstances of surface active center, such as coordinatively unsaturated state of active element, bond length, and so on, are maintained unchanged during catalysis. Example is the molybdenum sulfid catalyst. The catalysts, in which the circumstances of surface active center are changing dynamically during catalysis, are classified into the latter category. Molybdate catalysts belong to this category.

The homogeneous catalyst system was compared with the heterogeneous system. The same classification may be possible for the homogeneous transition metal complex catalyst systems. The Wilkinson complex, which is a active catalyst for hydrogenation of olefins, may be representative of the structural static catalysts in homogeneous system. Cobalt carbonyl catalyst for hydroformylation may be representative of the structural dynamic catalysts. Similar to the heterogeneous catalyst, the main difference between two types of homogeneous catalyst is whether the originally coordinated ligand can participate in the catalysis directly or not. In the case of the Wilkinson complex, active catalyst is in the state of coordinatively unsaturated state formed by the elimination of ligand(phosphine). The elimination of ligand is controlled by equilibrium. Once the coordinatively unsaturated state is formed, the catalytic cycle can proceed smoothly whithout changing the coordination structure of complex. The elimination to form coordinatively unsaturated state just corresponds to the formation of edge or corner of MoS_2 catalyst. In the case of hydroformylation catalyst, CO is not only reactant but also the supporting ligand of metal complex. Therefore, if CO can not be supplied sufficiently to the complex for coordination, catalytic cycle will not be attained with the decomposition of complex to form metallic particles. Further important things are that coordinated CO must be eliminated to make the complex in uncoordinated state for allowing the coordination of olefin and that CO has to coordinate immediately to accelerate the CO insertion into alkyl-metal bond(or the alkyl migration) or to prevent the hydrogen adsorption when the complex is uncoordinated state. Since the steps of coordination and elimination of CO are usually very rapid and in equilibrium, the hydroformylation catalyst can be anticipated to be dynamic, like as molybdate catalysts.

The above characteristic features of hydroformylation catalyst may be applicable to the molybdate catalysts. In the case of homogeneous system, CO is supplied from the gaseous phase through solvent. On the other hand, the reactant, lattice oxygen, in the case of molybdate catalysts can be supplied by diffusion through the lattice of catalyst, so that the rate of diffusion should be a key factor. As already shown, a good relationship was obtained between the mobility of lattice oxygen in molybdate catalyst and their catalytic activity and it is suggested on the basis of kinetic experiment that the mobile property of lattice oxygen is strongly associated with the number of active site on the surface. From the simple analogy to the hydroformylation catalyst where the performance of CO is very significant in not only maintining the active complex by the coordination but also providing the coordination site for olefin by the elimination, it is speculated that an oxygen vacant site, probably on molybdenum, of which concentration may depend strongly on the mobility of lattice oxygen, would be an adsorption site for allylic intermediate in the rate determining step.

APPLICATIONS

On the basis of above results and discussion, the following two applications have been made to demonstrate the importance of bulk diffusion of lattice oxygen for the design of active and stable catalysts. First, in order to get a active catalyst, Bi^{3+} ion, which is known as one of key element for allylic oxidation, was supported on cobalt and iron containing molybdate where the lattice oxygen can diffuse rapidly. For the comparison, Bi^{3+} ion was also supported on cobalt molybdate where on the other hand the mobility of lattice oxygen is very low(Table I). It was confirmed by XPS analysis that all Bi elements were located in the vicinity of surface for both samples. Catalytic activity data and fraction of lattice oxygen involved in the reaction are summarized in Table III. Very active and selective catalyst was attained by supporting Bi on molybdate support which has a mobile property of lattice oxygen. Molybdate supports were non-selective as catalyst. It is noteworthy that the catalyst which shows high mobility of lattice oxygen has a high ability for propylene

Table III   Catalytic activities of Co-Fe-Bi-Mo-O system[1]

Catalyst	Specifc activity $/10^{-5} mol \cdot min^{-1} \cdot m^{-2}$	Selectivity to acrolein (%)	Fraction of lattice oxygen involved in the reaction (%)
$CoMoO_4$	0.7	46	1
$Bi_{0.1}Co_{11}Mo_{12}O_x$	0.9	47	2
$Co_{11}Fe_4Mo_{12}O_x$	4.5	25	59
$Bi_{0.1}Co_{11}Fe_3Mo_{12}O_x$	12.2	97	45

1) Catalytic oxidation was carried out using a conventional flow microreactor under atmospheric pressure(propylene, 16.7%; oxygen, 16.7%; nitrogen, balance; reaction temerature, 450 °C; catalyst weight, 1g).

Table IV Participation of Lattice Oxygen in the Oxidation
of Propylene and Vaporization of Tellurium from
Catalyst during Catalysis[1]

Catalyst	Specific activity $/10^{-5} mol \cdot min^{-1} \cdot m^{-2}$	Selectivity to acrolein (%)	Fraction of lattice oxygen invovled in the reaction (%)	Amounts of vaporized tellurium[2] Weight(mg)	(%)
MoO_3	0.2	75	10	-	-
TeO_2-$6MoO_3$	0.3	95	14	0.75	1.2
$Fe_2Mo_3O_{12}$	0.9	59	9	-	-
$Te_{0.1}Fe_2Mo_3O_x$	2.5	96	30	0	0

1) Reaction conditions are the same as shown in Table I.
2) Total amount of vaporized tellurium element in 1h reaction.

oxidation without depending on Bi element. It is no doubt that Bi element
constructs the active site on the molybdate surface for selective
oxidation and its high performance is sustained by the bulk diffusion of
lattice oxygen.
    The second application was done using Te as a key element. The purpose
of this application is to show the effect of the bulk diffusion of lattice
oxygen on suppressing the vaporization of Te element from active site as
well as on the catalytic activity. Two catalysts were tested for propylene
oxidation; one is a catalyst prepared by combining TeO_2 with MoO_3 which
shows very low mobility of lattice oxygen and the other is prepared by
supporting Te element on ferric molybdate which shows relatively high
mobility. Table IV shows the catalytic activity, selectivity to acrolein,
the fraction of lattice oxygen involved in the reaction and the amount of
liberated Te from the catalyst during catalysis. TeO_2-MoO_3 catalyst was not
only very poor but also very unstable under the reaction condition with
liberating Te element. On the other hand, Te-$Fe(MoO_4)_3$ catalyst showed
superior catalytic properties.

REFERENCES

[1] M. W. J. Wolfs and Ph. A. Batist, J. Catal. **32**, 25(1974); I.
    Matsuura and Ph. A. Batist, J. Catal. **37**, 174(1975); R. K. Grasselli,
    J. D. Burrington, and J. F. Brazdil, Faraday Discuss. Chem. Soc.
    72, 204(1981); T. S. R. Prasada Rao and K. R. Krishnamurthy, J. Catal.
    95, 209(1985).
[2] R. K. Grasselli and J. D. Burrington, Adv. Catal. **30**, 133(1981);
    J. D. Burrington, C. T. Kartisek, and R. K. Burrington, J. Catal.
    81, 489(1983); R. K. Grasselli and J. D. Burrington, Ind. Eng. Chem.
    Prod. Res. Dev. **23**, 393(1984).
[3] G. W. Keulks, J. Catal. **19**, 232(1970).
[4] R. D. Wragg, P. G. Ashmore, and J. A. Hockey, J. Catal. **22**, 49(1971),
    28, 337(1973).
[5] Y. Moro-oka, W. Ueda, S. Tanaka, and T. Ikawa, in "Proceedings,
    International Congress on Catalysis," 7th, (Tokyo, 1980)(T. Seiyama and
    K. Tanabe, Eds.), Part B, p. 1086. Kodansha, New York/Elsevier,
    Amsterdam, 1981; W. Ueda, Y. Moro-oka, and T. Ikawa, J. Catal. **70**,
    409(1981); W. Ueda, Y. Moro-oka, and T. Ikawa, J. Catal. **88**, 214(1984);

W. Ueda, Y. Moro-oka, and T. Ikawa, J. Chem. Soc. Faraday Trans. I **78**, 495(1982); W. Ueda, K. Asakawa, C-L. Chen, Y. Moro-oka, and T. Ikawa, J. Catal. **101**, 360(1986); W. Ueda, C-L. Chen, K. Asakawa, Y. Moro-oka, and T. Ikawa, J. Catal. **101**, 369(1986).

[6] A. W. Sleight, K. Aykan, and D. B. Rogers, J. Solid State Chem. **13** 231(1975); A. W. Sleight, "Advanced Materials in Catalysis" p. 181, Academic Press, New York, 1977.

[7] K. Tanaka and T. Okuhara, Cat. Rev. **15**, 249(1977).

[8] J. Haber, Kinet. Katal. **21**, 1(1980).

CHARACTERIZATION BY XRD HIGH RESOLUTION ELECTRON MICROSCOPY
OF HYDRODESULFURATION CATALYSTS

J. Cruz, M. Avalos-Borja, S. Fuentes* and G. Diaz*
Instituto de Física, A. Postal· 2681, Ensenada, B.C.N. 22800
México. *Instituto de Física, A. Postal 20-364, México, D.F.
01000 MEXICO.

ABSTRACT

Recently a lot of work has attempted to correlate the activity of Co and Mo sulfide catalysts with observations of high
resolution electron microscopy. Hydrodesulfuration (HDS) bulk
sulfide catalysts are specially suited for this type of study.
We prepared unsupported molybdenum sulfide by three methods: the
homogeneous sulfide precipitation (HSP) (1), co-maceration (CM)
(2), and impregnated thiosalt decomposition (ITD) (3). Pure cobalt sulfide was prepared only by HSP, and two Co-Mo sulfides in
the ratios of 0.3 and 0.5 at. % were prepared by HSP and ITD.
The samples were treated with a mixture of 20% H2S/H2 for 4 hrs.
at 400 C. In all Co/Co+Mo mixtures the presence of Co9S8 and
MoS2 was identified in both, HSP and ITD treatments. Characterization was done by high resolution electron microscopy. The
ITD method can be considered a better method, as opposed to HSP
for the preparation of mixed catalysts. This method produces
poorly crystalline specimens that are usually associated to
higher chemical activity.

INTRODUCTION

Hydroprocessing catalysts based upon the transition metal
sulfides have been widely used for over 60 years. Application
have included among others dehydrogenation, hydrodenitrogenation
and hydrodesulfuration (HDS).
The understanding of the microstructure of HDS catalysts
has progressed remarkably in recent years. Nevertheless, we are
still far from a general agreement about the nature of the highly active sites created by the combination of group VIII and
group VI metal sulfides.
The objective of this work is to try to characterize by
HREM and XRD, the various catalysts prepared by HSP, ITD and CM
methods in order to elucidate the influence of preparation on
the catalyst morphology.

EXPERIMENTAL PROCEDURE

Samples of molybdenum sulfide, cobalt sulfide and mixtures
in the ratios of 0.3 and 0.5 at % were prepared according to the
usual HSP methods (1). Similar pure and mixed sulfides were prepared by impregnated thiosalt decomposition (3), that involves
the reaction of the intermediate (NH4)2MoS4 with the apropiate
salt. The precursors are subsequently sulfide in a mixture of
H2S/H2 for 4 hrs. at 400 C.
The catalysts were ultrasonically dispersed in distilled
water and mounted on carbon coated microscope grids for their
observation in a Jeol 100 C electron microscope.

Mat. Res. Soc. Symp. Proc. Vol. 111. ©1988 Materials Research Society

RESULTS AND DISCUSSION

Fig. la shows a typical electron diffraction pattern from
the pure molybdenum sulfide prepared by HSP. Rings can be inde_
xed as corresponding to the MoS2-2H phase according to data from
the X-ray Powder Data File (4). It is interesting to note that
some of the expected rings are missing, a possible explanation
for this effect is given below. As is evident from the full
rings, the specimen is, nevertheless, polycrystalline. MoS2
prepared by ITD and CM methods also give similar electron diffraction patterns. Fig. lb shows an electron diffraction pattern from a pure cobalt sulfide specimen prepared by HSP. In
this pattern we see that this method produces a less polycrystalline sample. All spots can be indexed as corresponding to
the Co9S8 phase.

Fig. lc shows a diffraction pattern from a 50 at. % mixture by the HSP method. A combination of planes is found, some
of them correspond to Co9S8 and the others to the MoS2 phase. A
different effect is observed for the same mixture prepared by
ITD, as illustrated in Fig. ld.

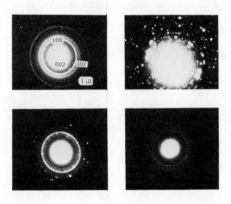

Figure 1. Electron diffraction pattern from a)MoS_2-2H b) Co_9S_8,
c) 50 at. % Co-Mo HSP, d) 50 at. % Co-Mo ITD.

X-ray diffraction pattern from MoS2 prepared by HSP produces (Fig. 2a) few broad peaks that are typical of poorly crystallized specimens. This last point is also supported by the
absence of some reflections (and the broadness of others)obser-
ved in the electron diffraction pattern (Fig. la). Spots corres_
ponding to 002, 101 and 103 planes are identified. The X-ray
diffraction pattern from pure cobalt sulfide (Fig. 2b) contains
well defined peaks. Planes 111, 222, 400, 333 and 440 are identified. Co-Mo sulfide mixtures prepared by ITD with 30 at. %

exhibit a X-ray diffraction pattern (Fig. 2c) with not well defined peaks, nevertheless, some planes can be associated with
the Co9S8 phase and others with MoS2-2H. An X-ray diffraction
pattern from a 50 at. % Co shows (Fig. 2d) peaks with more symme
trical form, better crystallinity is expected in this case.

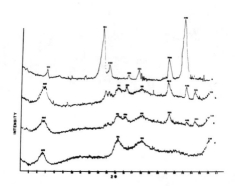

Figure 2.- X-ray diffraction pattern from a) MoS2-2H, b) Co9S8,
c) 30 at. % Co-Mo, ITD, d) 50 at. % Co-Mo, ITD.

It is interesting to note that mixtures of Co-Mo sulfide
catalysts prepared by ITD present 002, 101, 103 and 110 planes.
These MoS2-2H reflexions are usually observed in specimens prepared by the thermal decomposition of ammonium thiomolybdate (5)
Others planes are associated with the Co9S8 phase.
    High resolution electron microscopy shows the presence of
002 planes with spacing of 0.6 nm for the case of MoS2 (Fig.5),
this planes are found in HSP, CM and ITD methods. Recently Liang et. al. studied by X-rays some parameters associated with
the crystallinity of the MoS2 (6) like number of layers, stacking sequence and others. This study reveals the importance of
defects in the samples. Fig. 3 and 4 show Co-Mo sulfides prepa
red by ITD with 30 and 50 at % respectively. These pictures
show clearly two types of planes. One of them in well defined
parallel lines, with a spacing of 0.57 nm, which corresponds to
111 planes (See 'A' in Figs. 3 and 4). The other type assigned
to molybdenum sulfide (0.62 nm spacings) sometimes shows distor
tions, faults, kinks, and in some cases curvature of planes, as
illustrated by arrows in Fig. 3. This type of poorly crystallized specimens is usually considered a good catalyst. This is
most likely due to the formation of new surface sites. The
identification of these sites is an active field of research in
this and other laboratories.

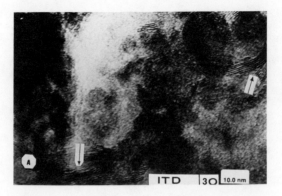

Figure 3.- High resolution electron image from Co-Mo sulfide
ITD with 30 at. % Co.

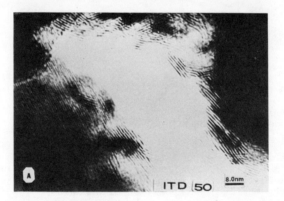

Figure 4.- High resolution electron image from Co-Mo sulfide
ITD with 50 at. % Co.

CONCLUSIONS

X-ray shows how cobalt addition in mixed catalysts improves
the sharpness of diffraction peaks. We can most likely assign
this change to an improvement of crystallinity, as Co9S8.

ITD preparation method can be considered as a good method
since it produced poorly crystallized specimens which are common
ly regarded as having the greatest chemical activity.

Defects, as the ones observed here by TEM, play an important role in the creation of new active sites in poorly crystallized specimens. These sites are considered to be responsible
for the higher chemical activity of these catalysts.

Figure 5.- High resolution electron image from MoS2 prepared by HSP.

ACKNOWLEDGEMENTS

Authors are very grateful to Mr. F. Ruiz for patient work with the TEM. J.C. acknowledges CONACyT and UABC for financial support for his graduate studies.

REFERENCES

1.- R. Candia, B.X. Clausen and H. Topsoe, Bull. Soc. Chim. Belg., 90 (1981) 1225.
2.- G. Hagenbach, Ph. Courty and B. Delmon, J. Catal., 23 (1971) 295 and 31 (1973) 264.
3.- S. Fuentes and G. Diaz, F. Pedraza H. Rojas, N. Rosas. Accepted for publication on J. Catal.
4.- J.H. Fang and F. D. Bolss. "X-ray Diffraction Tables". Sothern Illinois University Press, 1966.
5.- R.R. Chianelli and M.B. Dines, Inorg. Chem. 17 (1978) 2758.
6.- K.S. Liang, R.R. Chianelli, F.Z. Chien and S.C. Moss J. non-Cryst. Solids 79 (1986) 251.

CHARACTERIZATION OF HDS AND SYNGAS CATALYSTS DERIVED FROM HETEROBIMETALLIC CLUSTERS

M. D. CURTIS,[*] J. SCHWANK,[**] J. PENNER-HAHN,[*] L. THOMPSON,[**] O. BARALT[*] AND G. WALDO[*]

Departments of Chemistry[*] and Chemical Engineering,[**] The University of Michigan, Ann Arbor, MI 48109-1055

ABSTRACT

Discrete, organometallic clusters containing S, Mo, and Fe or Co have been deposited intact on metal oxide supports. These supported clusters are subjected to temperature programmed decomposition (TPDE) which causes loss of the organic ligands and produces catalysts active for CO hydrogenation and thiophene hydrodesulfurization (HDS). In-situ Mossbauer spectra reveal that the surface is remarkably uniform and that high spin ferrous iron is the predominate species (ca. 90%) under catalytic reaction conditions. EXAFS and XANES spectra show that the clusters are irreversibly oxidized by the surface at ca. 400 K under H_2.

INTRODUCTION

Hydrotreating of coal liquids and petroleum to remove sulfur (HDS) and nitrogen (HDN) is an exceedingly important step in the refining process. In spite of extensive work [1], the nature of the active site of the most commonly used catalyst, viz. "sulfided cobalt molybdate" supported on Al_2O_3, is still a matter of debate.

Conventional preparation of HDS Catalysts produces a variety of phases on the sufrace, making a precise identification of the active site difficult. XPS and Mossbauer emission spectroscopy (MES) have shown the presence of Co^{+2} (interstitial), Co_9S_8, MoS_2, and an ill-defined "CoMoS" phase on the surface [2]. More recently, EXAFS has been used to characterize these catalysts [3]. In the sulfided form, only Mo-S or Co-S vectors could be isolated in the EXAFS Fourier transforms. No Co-Mo vectors are evident in spite of the fact that the Co must be intimately associated with the active site [4]. The model which has grown out of these studies assumes that in the active site the Mo is present as small crystallites (ca. 10-15Å) of MoS_2 with the Co coordinated around the edges of the MoS_2 basal planes.

This paper reports the characterization by in situ Mossbauer and EXAFS of catalysts derived from the clusters, "MoFeS" and "MoCoS", whose structures are shown below. These clusters were chosen since the cluster cores contain Mo, S, and Fe and Co and might therefore be good models of the active site in conventionally prepared catalysts. The catalytic activity of these species has been compared to commercial catalysts in thiophene HDS. The Co hydrogenation activity of the supported clusters was also investigated since it was suspected that these catalysts would show high tolerance for sulfur in the feed stream.

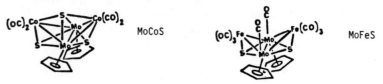

EXPERIMENTAL

The cluster precursors were prepared as previously reported [5]. The alumina (Strem or Catapal) was calcined at 770 K in oxygen, and handled thereafter in a dry box. The clusters were deposited by contacting a methylene chloride solution of the cluster with sufficient quantity of calcined alumina to give a 1% (wt/wt) loading of total metal (the clusters are quantitatively adsorbed from solution under these conditions). The solid is separated by filtration and dried under vacuum. All subsequent manipulations are under dry nitrogen.

The supported clusters were subjected to TPDE under flowing H_2 to a maximum temperature of 670 K.

Hydrogenation of CO and thiophene HDS was performed in a glass-lined, SS differential flow reactor at conversion of 1-5%. Products were analyzed by GC using FID for hydrocarbons, while permanent gases were detected by TCD.

Mossbauer spectra of MoFeS were obtained on samples enriched in ^{57}Fe under catalytic working conditions.

Xray absorption were carried out at the Stanford Synchrotron Radiation Laboratory on samples which had been subjected to various treatments (see below) and subsequently handled under inert atmosphere.

RESULTS

During TPDE, a burst of CO evolution is observed near 370 K, followed by the onset of loss of the cyclopentadienyl ligand near 470 K. Traces of methane, CO_2, and dimethylsulfide are released between 470 K and 670 K. The surface compositions after heating to the maximum temperature (670 K) are $C_5S_{1.8}Mo_2Fe_2O_mH_n$ and $C_{1.7}S_{1.8}Mo_2Fe_2O_xH_y$ for MoFeS/Al_2O_3 and MoCoS/Al_2O_3, respectively.

The Mossbauer parameters are summarized in Table I. The pure crystalline cluster has IS = -0.01 and QS = 0.80 mm/s. Following air oxidation, the supported cluster shows IS = 0.35 mm/s, QS = 0.98 mm/s which corresponds to Fe(+3) as the only detectable species. Under H_2, H_2+CO, or H_2+thiophene at 670 K, ca. 90% of the signal corresponds to Fe(+2) and 10% to a reduced iron species (Feδ+). The cluster could be cycled through numerous oxidation-reduction-catalysis cycles to give the same parameters shown in Table I.

Table I. In-situ Mossbauer Parameters for MoFeS/Al_2O_3.

	Fe(+2)			Fe(δ+)		
Treatment	IS	QS	A%	IS	QS	A%
670K/H_2	1.0	1.9	91	0.08	0.43	9
670K/H_2+CO	0.9	1.7	89	0.08	0.32	11
590K/H_2+C_4H_4S	1.1	2.0	86	0.14	0.60	14

Xray absorption spectra (XAS) were obtained on pure MoFeS cluster, MoFeS/Al_2O_2 (no heat treatment), and MoFeS/Al_2O_3 after heating under H_2 to 390 K and 570 K at both the Fe and Mo edges. The Fourier transforms of the EXAFS data are shown in Figures 1a and 1b for the Fe and Mo data, respectively.

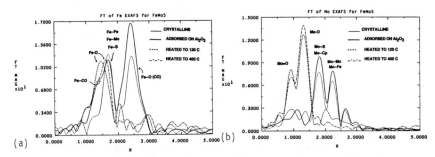

Figure I. (a) FT of Fe EXAFS; (b) FT of Mo EXAFS.

After heating to 370 K, both Fe and Mo EXAFS show only oxide ligands (for Fe, ca. 6 oxygens at 1.95 Å). The Mo XANES also shows a feature attributed to one or more Mo=O bonds. Heating the samples to 670 K causes no further changes in the EXAFS. The Fe EXAFS shows that 1-2 CO groups are lost when the cluster is initially absorbed on the support, whereas the Mo environment remains unchanged until the sample is heated.

Catalytic studies show these clusters are active for CO hydrogenation. Mostly methane (and CO_2) are produced with turn-over frequencies (TOF) of ca. 10^{-4}/s at 670 K. The clusters also show nearly identical behavior toward catalysis of HDS of thiophene as a commercial HDS catalyst (Catalco). Butenes (70%) and propene (18%) are the primary products with TOF's of ca. 10^{-4}/s at 570 K with no presulfiding. With presulfiding, butenes make up ca. 90% of the products and the TOF increases to 10^{-3}/s.

DISCUSSION

On alumina support, the MoFeS and MoCoS clusters form catalysts for CO hydrogenation and thiophene HDS which are very similar to those prepared by conventional methods. However, unlike the conventionally prepared catalysts, the surfaces of the cluster-derived catalysts are remarkably uniform as judged from the Mossbauer spectra.

The Mossbauer spectra further show that the Fe in MoFeS is oxidized to predominately Fe(+2) upon heating under H_2. About 10% of a further reduced iron is also present. Upon exposure to air, the heat treated samples are oxidized, and only Fe(+3) is present. Reduction with H_2 of the oxidized samples restores the signals due to Fe(+2) and Fe(δ+). Under H_2/CO or H_2/thiophene, the signals show little change from H_2 alone. Hence, it is concluded that the resting state of the MoFeS catalyst is an oxo-surface ensemble.

This conclusion is buttressed by the XAS data. For both Fe and Mo, the absorption edge is shifted ca. 5eV to higher energy upon heat treatment, consistent with oxidation of the metals. Furthermore, the XANES and EXAFS show predominately oxygen coordination for Fe and Mo after heating to 370 K under H_2.

Since the Mossbauer data show that the resting state is the same under H_2 or H_2/thiophene, we conclude that the oxo ensembles contribute substantially to the HDS activity, i.e. MoS_2-like domains are not necessary for the HDS activity. Further studies on the "sulfided" forms are in progress.

334

REFERENCES

1. (a) E. Furminsky, Cat. Rev. 22, 371 (1980); (b) P. Ratnasamy and S. Sivisanker, ibid. 22, 401 (1980); (c) P. Grange, ibid. 21, 135 (1980).

2. (a) I. Alstrup, I. Chockendorff, R. Candia, B. S. Clausen, and H. Topsoe, J. Catal. 77, 397 (1982); (b) C. Wivel, R. Candia, B. S. Calusen, S. Morup, and H. Topsoe, ibid. 68, 453 (1981); (c) H. Topsoe, B. S. Clausen, R. Candia, C. Wivel, and S. Morup, ibid. 68, 443 (1981).

3. (a) M. Boudart, R. A. DallaBetta, K. Foger, and D. G. Loffler, Springer Proc. Phys. 2, 187 (1984); (b) B. Clausen, H. Topsoe, R. Candia, B. Langeler, ibid 2, 181 (1984); (c) B. S. Clausen, H. Topsoe, R. Candia, J. Villadsen, B. Langeler, J. Als-Nielsen, and F. Christensen, J. Phys. Chem. 85, 3868 (1981).

4. R. R. Chianelli, T. Pecoraro, T. R. Halbert, W.-H. Pan, and E. I. Stiefel, J. Catal. 86, 226 (1984).

5. (a) M. D. Curtis and P. D. Williams, Inorg. Chem. 22, 2661 (1983); (b) P. D. Williams, M. D. Curtis, D. N. Duffy, and W. M. Butler, Organometallics 2, 165 (1983).

ADSORPTION AND REACTION MECHANISMS OF THIOPHENE OVER
SULFIDED RUTHENIUM CATALYSTS

RAY A. COCCO AND BRUCE J. TATARCHUK
Department of Chemical Engineering, Auburn University, AL 36849

ABSTRACT

Adjustable (hydrogenolysis/hydrogenation) ratios (i.e., selectivity) of
thiophene over sulfided ruthenium catalysts have been observed to depend on
the method of presulfidization. Sulfidization in 10% H_2S/H_2 at 300-800K
yielded a selectivity greater than 100, whereas a ratio of ca. 1 was noted
after presulfidization in 100% H_2S. The transition from one selectivity
regime to another is reversible and can be changed via subsequent annealing
or resulfidization procedures. X-Ray photoelectron spectroscopy shows
similar results where sulfidization in 100% H_2S yields: (i) a higher sulfur
uptake with a corresponding 0.5 eV shift of the S $2p_{1/2-3/2}$ binding
binding energy; (ii) multilayer sulfur incorporation at lower temperatures;
and (iii) similar reversible behavior following subsequent annealing or
resulfidization. Secondary ion mass spectroscopy and temperature programmed
desorption studies also reveal that sulfur coverages affect thiophene
adsorption, orientation and cracking on ruthenium surfaces.

INTRODUCTION

Previous studies [1-3] have shown that sulfided ruthenium catalysts have
a higher hydrodesulfurization (HDS) activity than traditional HDS catalysts,
including commercial CoMo catalysts. Furthermore, Kuo et al. [2,3]
demonstrated that sulfided ruthenium catalysts possess adjustable
hydrogenolysis to hydrogenation ratios (i.e., selectivity) which were
dependent on the sulfidization pretreatment and can be divided into regimes
separated by a "selectivity line".
        For one regime, sulfidization in 100% H_2S at temperatures above 573K
produced a catalyst with a RuS_2-like surface and provided equal rates of
hydrogenolysis and hydrogenation (i.e., hydrogenolysis/hydrogenation ratio or
selectivity of ca. 1). In contrast, sulfidization in 10% H_2S/H_2 at high
temperatures produced a regime, indicative of an amorphous skin covering the
ruthenium, with a selectivity of greater than 100 (i.e., a dominating direct
hydrogenolysis pathway). Moreover, Kuo et al. [3] also demonstrated that it
was possible to move from one selectivity regime to another in a reversible
fashion by appropriate annealing and/or resulfidization procedures.
        Since adjustable selectivities are not generally provided by commercial
hydrotreating catalysts, studies involving X-Ray Photoelectron Spectroscopy
(XPS) of unsupported ruthenium catalysts pretreated on both sides of the
"selectivity line" were performed in order to better examine the surface
compositions corresponding to the various selectivity regimes noted above.
Studies involving Temperature Programmed Desorption (TPD), Secondary Ion Mass
Spectrometry (SIMS) and Temperature Programmed Static SIMS (TPSS) were also
employed to determine the selectivity mechanism using thiophene over sulfur
precovered and clean Ru(0001) single crystal surfaces.

EXPERIMENTAL

XPS studies were performed on a ruthenium sponge pressed into a 1 cm
diameter pellet and reduced in pure hydrogen at 900K. Sulfidization was
performed in an ex situ reactor and subsequently transferred into the XPS
analysis chamber under high vacuum.

TPD, SIMS and TPSS studies on Ru(0001) were performed in an ultra-high vacuum chamber equipped with Leybold-Hereaus SIMS and mass spectrometer. TPD and TPSS were accomplished by resistively heating the single crystal at a heating rate of 10K/s. SIMS experiments were performed at a current density which did not exceed 1 nA/cm^2. At these rates, SIMS is considered relatively nondestructive (i.e., Static SIMS) since the probability of sampling a perturbed surface is small during the course of an experiment [4]. All SIMS experiments were performed in the positive ion mode using 3 KeV argon ions at incidence and take-off angles of 45°.

RESULTS AND DISCUSSION

XPS Studies of Unsupported Ruthenium Catalysts

Examination of the sulfur $2p_{1/2-3/2}$ peaks after sulfidization in 10% H_2S/H_2 are illustrated in Figure 1 and compared with elemental sulfur, laurite and metal sulfate standards. Using the known oxidation states of the zero valance sulfur in S_8, the S^{-1} state in laurite as well as the S^{+6} of the sulfate, a calibration of the valance state with respect to the peak position was obtained. The sulfur doublet indicative of multilayer uptake at 773K compared well with the laurite standard which corresponds to a S^{-1} state. The sulfur peak representative of submonolayer coverage at 573K was estimated to have an oxidation state labeled $<S^{-1}$. Similar behavior was observed for the peak intensity and chemical shift for sulfidization in pure H_2S.

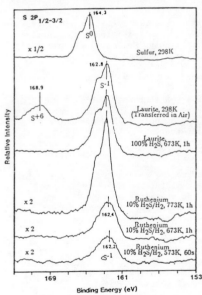

Figure 1: Peak position for sulfur, laurite, and sulfided ruthenium

A better illustration of the transition from submonolayer to multilayer uptake is shown in Figure 2 where the S/Ru atomic ratio and peak position are compared to the sulfidization conditions. The S/Ru ratio was only corrected for cross section and spectrometer efficiency differences. Escape depth corrections were not considered since the sample morphology at low sulfur coverages was unknown. Similarly, to account for sample charging or changes in the specimen's work function, the sulfur peak position is also displayed with respect to the observed binding energy difference between the Ru $3d_{5/2}$ and S $2p_{3/2}$ maxima.

As shown in Figure 2, both sulfidization conditions demonstrated similar trends for the S/Ru ratio and peak position. Sulfidization in pure H_2S resulted in a chemical shift to higher binding energies and a 10-fold increase in the S/Ru ratio at lower temperatures than for 10% H_2S/H_2 sulfidization. Furthermore, at high temperatures where the S/Ru ratio has stabilized, a 10% increase in this ratio was observed after sulfidization in pure H_2S.

Vacuum or hydrogen annealing (Figure 2) of ruthenium pellets following sulfidization in pure or 10% H_2S/H_2 at high temperatures resulted in a substantial decrease in the S/Ru ratio and the return of the S $2p_{3/2}$ peak position to 162.2 eV. Subsequent resulfidization in pure or 10% H_2S/H_2 provided evidence for reversible behavior reminiscent of the above noted

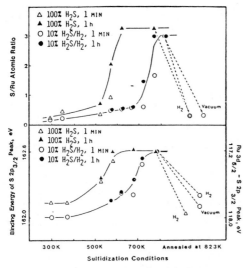

Figure 2: Sulfidization behavior
of polycrystalline ruthenium

reversible trends in catalytic
selectivity [3].

By comparing the chemical
state or peak position of the S
$2p_{3/2}$ peak with respect to the
sulfur uptake or S/Ru ratio, it
was found that the chemical
state of the sulfur doublet was
mostly dependent on the level
of sulfur incorporation into
the specimen. The peak
position was not influenced by
the sulfidization condition
employed, thereby indicating
that XPS cannot easily
distinguish the surface species
responsible for the above noted
selectivity behavior.

Further examination of
these subleties provided from
pure and 10% H_2S/H_2
sulfidization can be seen by
deconvolution of the S $2p_{1/2}$-
$3/2$ envelope as shown in Table
I. For each doublet used to
fit the sulfur envelopes,
constraints were used to
maintain the physical significance of deconvoluted spectra. Specifically,
these constraints involved: (i) area $2p_{1/2}$ / area $2p_{3/2} = 0.508$, based on
available cross-section data for Mg $K\alpha_{1,2}$ X-Rays [5], (ii) doublet peak
splitting of 1.15 ev, in agreement with data obtained from the standards
shown in Figure 1 and available literature values of 1.0 - 1.3 eV [6,7], and
(iii) equivalent Gaussian peak shapes and FWHM's for the $2p_{1/2}$ and $2p_{3/2}$
peaks equal to 1.0 eV, based on experimentally determined spectrometer
resolution factors. The number of doublets and corresponding peak positions
for each sulfur envelopes observed after pure or 10% H_2S/H_2 sulfidization
were determined from only physical evidence, a method specifically outlined
in Cocco, et al. [8]. Values obtained for pure H_2S sulfidization have been
multiplied by a factor of 1.12 corresponding to the 10% increase in the S/Ru
ratio.

Table I. Deconvolution of Sulfur $2p_{1/2\text{-}3/2}$ Spectral Region

		Sulfidization Conditions	
S $2p_{3/2}$ Peak Position (eV)	Sulfur Assignment	10% H_2S/H_2, 823K 1h, 101 kPa	100% H_2S, 773K 1h, 101 kPa
162.2	$<S^{-1}$	22.0%	22.5%
162.75	S^{-1}	78.0%	85.0%
164.25	S^o	0.0%	4.5%
	Total =	100.0%	112.0%

As illustrated in Table I, comparison of the deconvoluted spectra
following each sulfidization treatment shows that sulfidization in pure H_2S
results in an additional species corresponding to elemental sulfur and ca.
10% increase in the S^{-1} species representative of lattice sulfur in RuS_2. This
agrees with the results of Kuo et al. [3], where sulfidization in pure H_2S
yielded a RuS_2-like surface. Sulfidization in 10% H_2S/H_2 at high

temperatures produced a relaxed or reconstructed overlayer on a RuS_2 bulk. The presence of elemental sulfur following sulfidization in pure H_2S may be the result of the fact that specimens were cooled to room temperature under the flow of H_2S prior to anaylsis.

TPD, SIMS and TPSS of Sulfur Precovered and Clean Ru(0001)

TPD results (Figure 3), in conjunction with earlier studies involving high resolution electron energy loss spectroscopy and temperature programmed reaction spectroscopy demonstrate that thiophene adsorbs with a high sticking coefficient on a clean Ru(0001) and that a portion of the thiophene is cracked to a metallocycle species. Thiophene adsorption onto sulfur precovered Ru(0001), however, lowers the sticking coefficient and passiviates the surface toward the cracking reaction provided the sulfur coverage exceeds 0.25 monolayers.

Desorption features from these two surfaces are shown in Figure 3 where thiophene on clean ruthenium displays a multilayer peak at 145K, a σ-bonded compressed state at 165K and a Π-bonded state at 260K. For sulfur precovered surfaces, the multilayer peak was observed at 145K with a sharp feature (ca. 10K FWHM) at 161K. Typically, desorption peaks with sharp features similar to this one have been attributed to autocatalytic or explosive desorption mechanisms [9-11].

SIMS studies performed on a clean Ru(0001) surface showed that the AMU 186 (ruthenium-thiophene)$^+$ adduct increased to a maximum at approximately 2 Langmuirs thiophene exposure. The 102 AMU ruthenium$^+$ ion species, however, exhibited an exponential decrease with increasing thiophene exposure.

These results agree well with many available cluster-formation models such as: recombination, direct emission, cationization [12], precursor [13], and many thermal models [14,15]. The mechanism involves an increase consumption of the AMU 102$^+$ species to produce the more stable (ruthenium-thiophene)$^+$ adduct as the thiophene overlayer increases. The (ruthenium-thiophene)$^+$ cluster can be formed at the ruthenium surface as modeled by direct emission and precursor models or can be formed in the dense gas phase or selvedge produced by sputtering.

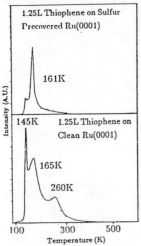

Figure 3: Thiophene TPD from clean and sulfur precovered Ru(0001)

For thiophene adsorption onto ruthenium surfaces precovered with one-half monolayer of sulfur, the behavior of the AMU 102$^+$ and 186$^+$ species was different. First, the addition of the sulfur overlayer attenuated the ruthenium$^+$ ion intensity. This was unexpected since typically, the addition of an anionic adsorbate increases metal yields, as often observed with oxygen adsorption [4]. The addition of a thiophene overlayer resulted in an increase in the AMU 102$^+$ species, a behavior opposite to that observed over clean Ru(0001). Second, a substantially larger AMU 186$^+$ intensity was observed over the sulfur precovered surface compared to the clean surface with the maximum intensity attained after a 2 Langmuir exposure.

Due to the sulfur overlayer, models which assume cluster formation at the surface (i.e., direct emission and precursor) no longer apply, since prior to sputtering, thiophene is separated from ruthenium by the sulfur layer. Thus, cluster formation could only be formed in the selvedge developed during a primary ion impact. The reduction in the ruthenium ion yield due to the sulfur overlayer may be the result of the formation of RuS$^+$

and RuS_2^+ species which consume Ru^+ ions. Furthermore, these species may also serve as intermediate transport species responsible for the increase in the AMU 186^+ and 102^+ species with increasing thiophene exposures. During a sputtering event, the RuS^+ and/or RuS_2^+ species bring ruthenium above the sulfur layer into the thiophene overlayer. The thiophene reacts with the species to produce the more stable $(ruthenium\text{-}thiophene)^+$ adduct and also $ruthenium^+$ ions.

To further examine the differences observed between these two surfaces, TPSS experiments were performed where the intensity of the AMU 186^+ peak was monitored during the course of a TPD experiment. As shown in Figure 4, the 186^+ intensity observed over clean ruthenium correlated well with decreasing thiophene coverages. For the sulfur precovered surface, however, the 186^+ AMU intensity did not exhibit a substantial decrease until after 165K. At this point, TPD demonstrated that more than half the thiophene had left the surface which suggests that the 186^+ AMU intensity is not completely dependent on coverage.

By dividing the AMU 186^+ intensity by the coverage at a specific temperature, a coverage corrected ion yield can be obtained as illustrated in Figure 5. For the clean ruthenium surface, this yield appeared to be relatively insensitive to coverage and temperature. However, for thiophene on sulfur-precovered surfaces, the yield exhibited a substantial increase at 165K which closely resembled the explosive desorption during TPD. This result suggests that the yield enhancement is tied to the explosive desorption process. Such an enhancement cannot be attributed to a direct thermal process on the yield since similar behavior was not observed on clean ruthenium surfaces.

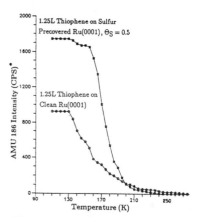

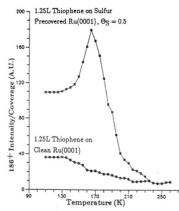

Figure 4: TPSS of thiophene on clean and sulfur precovered Ru(0001)

Figure 5: Coverage corrected ion yield during thermal desorption

A possible model for this desorption-assisted ion yield enhancement may be due to the above noted RuS^+ and RuS_2^+ species. By enhancing the dense gas phase provided by a thiophene explosive desorption event, higher concentration of ruthenium and ruthenium carrying species can "hitch-hike" with desorbing thiophene molecules. Furthermore, this enhancement appears to be relatively short range since similar behavior is not observed for multilayer desorption. Thus, this effect must be confined to a monolayer interaction.

This enhancement also provides further confirmation of recombination and cationization cluster models as applied to this system. Simply, by enhancing the dense gas phase or selvedge by an explosive desorption event, the cluster formation of the (ruthenium-thiophene)$^+$ adduct is substantially increased, thereby supporting mechanisms that assume cluster formation in the selvedge or dense gas phase.

SUMMARY

XPS studies show that the S/Ru ratio and S $2p_{3/2}$ peak positions indicate the transition from submonolayer to multilayer sulfur uptake occurs at lower temperatures for sulfidization in pure H_2S than 10% H_2S/H_2. Furthermore, annealing/resulfidization treatments exhibit a reversible behavior corresponding to the reversible behavior in selectivity observed by Kuo et al. [3]. At high temperatures, sulfidization in pure H_2S result in a 10% increase in the S/Ru ratio as well as a 10% increase in the lattice sulfur contribution.

TPD results show that thiophene desorbs from clean Ru(0001) at 145, 165 and 260K corresponding to multilayer, σ-bonded and Π-bonded states, respectively. Sulfur precovered surfaces provide a multilayer desorption feature at 145K and an explosive desorption feature at 161K.

TPSS studies demonstrate that the coverage corrected ion yield is enhanced and resembles the above noted explosive desorption feature. A possible origin for this enhancement has been described in terms of a "hitch-hiking" mechanism involving ruthenium attachment to already desorbed thiophene molecules.

REFERENCES

1. T.A. Pecoraro and R.R. Chiannelli, J. Catal., 67, 430 (1981).

2. Y.J. Kuo and B.J. Tatarchuk, Accepted in J. Catal.

3. Y.J. Kuo, R.A. Cocco and B.J. Tatarchuk, Accepted in J. Catal.

4. A. Benninghoven, Surf. Sci., 28, 541 (1971).

5. J.H. Scofield, J. Electron Spectrosp. Relat. Phen., 8, 129 (1976).

6. D.T. Clark and D.M.J. Lilley, Chem. Phys. Letters, 9, 234 (1971).

7. G. Panzner and B. Egert, Surf. Sci., 144, 651 (1984).

8. R.A. Cocco, Y.J. Kuo, C.R.F. Lund, R.T.K. Baker and B.J. Tatarchuk, in preparation.

9. Y.O. Park, W.F. Banholzer and R.I. Masel, Surf. Sci., 155, 341 (1981).

10. M.A. Barteau, E.I. Ko and R.J. Madix, Surf. Sci., 102, 99 (1981).

11. M.W. Lesley and L.D. Schmidt, Chem. Phys. Letters, 102, 459 (1983).

12. B.J. Garrison, Int. J. Mass Spectrom. Ion Phys., 53, 243 (1983).

13. A. Benninghoven, Int. J. Mass Spectrom. Ion Phys., 53, 85 (1983).

14. G.M. Lancaster, F. Honda, Y. Fukuda and J.W. Rabalais, J. Am. Chem. Soc., 101, 1951 (1979).

15. H.T. Jonkman and J. Michl, J. Am. Chem. Soc., 103, 773 (1981).

HIGH ACTIVITY DUAL COLLOID AMMONIA SYNTHESIS CATALYSTS[a]

THOMAS HENRY VANDERSPURT* AND MICHAEL A. RICHARD**
* Basic Chemicals Technology, Exxon Chemical Company, P.O. Box 4900, Baytown, TX 77522-4900
** Intermediates Technology, Exxon Chemical Company, P.O. Box 241, Baton Rouge LA 70821

Abstract
    A new synthesis technique produces a family of iron and iron alloy catalysts with a micro-morphology uniquely suited for ammonia synthesis catalysts.

Introduction
    Ammonia synthesis has intrigued researchers since the 1800's. Early in this century Haber and Oordt [1] reported the first catalytic synthesis   of ammonia from nitrogen and hydrogen. In the 75 years since the first commercialization of ammonia synthesis it has become one of the most important and most studied catalytic reactions.   Haber's success owed much to his understanding of the thermo-dynamics of ammonia synthesis. Since the reaction is exothermic and converts 4 moles of reactant gas into 2 moles of product at equilibrium he realized that high pressures would be needed to achieve significant ammonia concentrations at the high temperatures early catalysts required. The first practical catalyst, derived from magnetite, was discovered by Mittasch in 1910 [2]. This fused iron catalyst required temperatures around 550 °C for activity and was key to the 200-300 atm. Haber-Bosch process, which produced about 15-20% NH_3 . Later the lower pressure Mont Cenis process (100 atm.) was based on the Uhde catalyst which was sufficiently active in the 400-425 °C region to produce 5-12% NH_3. [2] Engineering improvements have made derivatives of the Haber-Bosch process dominant today.

Careful work by Mittasch showed that Uhde catalyst prepared through the decomposition of "$KAlFe(CN)_6$" catalyst was in fact   similar in composition to the catalyst he had discovered earlier: α-iron promoted with potassium and aluminum oxides [3].

a. This work was carried out at Corporate Research, Exxon Research and Engineering Co. under the sponsorship of the Agricultural Chemical Technology Division of Exxon Chemical Company.

Today's most used commercial catalysts [4, 5], like the earlier ones, are α-iron promoted primarily with potassium and aluminum oxides along with various other oxides: MgO, CaO etc. , and though more active and more stable than the earlier catalysts, high temperatures and high pressures are still needed. Depending on the design and the date of construction today's typical ammonia plants operate between about 150 and 300 atm. Taking 200 atm. and a 500 °C exit gas temperature as an example the equilibrium concentration of NH_3 in $3H_2$ to 1 N_2 is about 19 mole %, but it drops to about 5 mole % at 500 °C and 40 atm. Dropping the exit temperature to 400°C increases the equilibrium ammonia concentration to 13%. [6a] Lower exit temperatures without the need for overly large reactor volumes require new catalysts with much higher volumetric activities.

One limitation conventional catalysts have is simply their low active surface area. Nielsen [6b] estimates the exposed Fe surface for a commercial catalyst KM I, to be about 0.6 M^2/g or 5-6% of the BET surface area. The reason for this is fairly straight forward; typically, conventional catalysts consist of ~360Å Fe crystallites separated at the grain boundaries by "structural" promoters like aluminum oxide, and covered with clusters of "electronic" promoter species .

Not all of this Fe available surface is likely to be active because of the high structure sensitivity of ammonia synthesis over iron. A C7 site as suggested by Boudart and others [7] requires at least 3 surface atoms, three 2nd layer atoms and one third layer atom. Brill [8] has demonstrated that di-nitrogen preferentially adsorbs on the Fe(111) face, which is the more open of the low index faces since α-iron has a bcc structure. This agrees with Somorjai's [9] work where he has shown that the open Fe(111) face is 17 times more active than the next most open face (100) which is in turn 25 times more active than the close packed (110) face. Boszo, Ertl and co-workers [10] have shown that surface nitrides form on the Fe(111) surface which are stabilized by the reconstruction of Fe(111) to a form similar to the (111) plane of Fe_4N. This structure sensitivity may explain the reaction rate dependence on domain size. Topsoe et al [7b,11] have shown that Fe particles of 50-100 Å are more active than smaller particles.

The challenge then is to create a catalyst with the maximum effective and accessible density of Fe domains approaching the minimum effective size, ~50Å, separated and stabilized by the minimum of highly basic, thermally stable electronic promoter. This structure would be unlike either the conventional bulk metal catalysts already in use, or the traditional supported metal catalysts, where small metal crystallites are deposited on the surface of a highly porous support. Conceptually this structure consists of colloidal sized metal crystallites interdispersed with similar sized colloidal promoter particles thus creating an open highly active catalyst structure. Such a structure with α-iron could have several times the activity of conventional catalysts, and be useful at lower temperatures and pressures, even though there are indications that the kinetics or mechanism may change in this intermediate region. [7a,12]

Such a structure seemed inconsistent with the present technique of preparing a doped iron oxide and subjecting it to prolonged high temperature reduction. Consequently we re-investigated systems which did not require a high temperature iron-oxide reduction step.

A system of immediate interest to us was the Uhde aluminum ferrocyanide system. Direct reduction or decomposition of "$KAlFe(CN)_6$" prepared according to Uhde[13] doesn't yield the desired dual colloid structure [14]. However we have found that dual colloid structures can be produced by reacting a ferrocyanide salt with a soluble aluminum salt, and refluxing under conditions which allows the aluminum cation to undergo hydrolysis with the resultant formation of a ferro-ferrocyanide. This can be followed by the development of the characteristic 595 cm^{-1} Fe^{+2}-C stretching and 470 cm^{-1} Fe^{+2}-CN bending bands in its IR spectra. The small ferro-ferrocyanide domains are interdispersed with colloidal hydrous aluminum oxide domains, often containing KCN. Improved catalysts result when magnesium or other non-reducible nitrates are partially substituted for the aluminum nitrate. Some of these cations, which do not undergo hydrolysis under the conditions employed, remain as dopants in the the ferro-ferrocyanide structure . Their presence gives a characteristic orthorhombic distortion to the cubic ferro-ferrocyanide structure as shown in the powder X-ray diffraction pattern, and helps insure that the metal crystallites produced on reduction will be in the right size range.

After the hydrolysis reaction is complete the resulting solid dual colloid precursor phase is separated from the solution and thoroughly washed. The filter cake partially dried, then extruded into 1.5 mm diameter cylinders , and dried further in a vacuum oven at 50 °C. Reaction was carried out in $3H_2/1N_2$ mixture first at 1 atm. and 150 °C , then under 4.2 MPa with the temperature increased to 390 °C at 12 °C/hr [14].

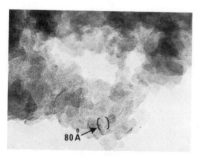
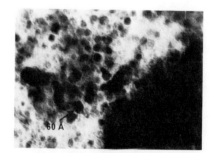

PRECURSOR   CATALYST

Figure 1. Dual colloid morphology in a type FeRu/KAlOx catalyst

This system is easily adapted to the production of a wide range of multi-metallic or alloy crystallites by substituting the appropriate Ru, Co and/or Ni nitrate for some of the aluminum nitrate [14]. Figure 1 shows both the precursor to and the activated form of a highly active RuFe catalyst. Powder X-ray crystallography confirms that these crystallites are primarily a Ru-Fe alloy , as no Fe only phase is detected.

Similar structures have been produced for an FeCo catalyst which approaches the predicted activity benefit for the dual colloid morphology. Since ammonia synthesis is dominated by equilibrium considerations the relative catalyst activity concept is used as shown on figure 2. This is the volumetric activity multiplier that a hypothetical bench mark catalyst would need to match the exit gas composition over a given catalyst under given conditions using Temkin-Pyzhev kinetics. The relative activity of a KM I type commercial catalyst under the same conditions of 4.2 MPa with $3H_2/1N_2$ synthesis gas is shown for comparison. The comparison is on an initial activity basis because like the early Uhde catalysts [2] these dual colloid catalysts may require much higher purity synthesis gas than the present commercial catalysts for stability. Likewise such activity cannot be achieved at all unless the starting materials are rigorously free of potential catalyst poisons such as sulfur.

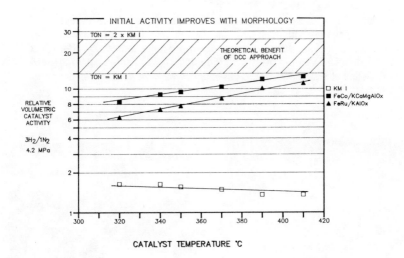

Figure 2. Relative Activity of KM I, FeRu/KAlO$_X$, and FeCo/KCaMgAlO$_X$ Types

Conclusion

Ammonia synthesis catalysts with initial activities at moderate temperature several times that of the present commercial catalysts can be prepared by the reduction of doped ferro-ferrocyanide/ alumina-aluminate precursors provided the precursors have been prepared in such a way as to insure the desired "dual colloid" micro-morphology in the finished catalyst.

Acknowledgements

The Authors would like to thank Gregory DeMartin for careful work in preparing and testing the catalysts, E.B. Prestridge for the electron microscopy and Dr. G. B. Ansell for the powder X-ray diffraction work. They would also like to thank Dr. A.A. Montagna for many interesting discussions.

References

1. F. Haber and G. Van Oordt, Z Anorg. Chem. , 43,111 (1905).

2. W.G. Frankenburg, in CATALYSIS Vol. III Hydrogenation and Dehydrogenation, edited by P.H. Emmett (Reinhold Publishing Corp., New York, 1955), pp171-263.

3a. A. Mittasch, and E. Kuss, Z. Elektrochem., 34, 159 (1928).
   b. A. Mittasch, E. Kuss and O. Emert, Z. Anorg. Allgem. Chem., 170, 193 (1928).

4a. A. Neilsen, Cat. Rev., 4, 1 (1970): Catal. Rev.-Sci. Eng., 23, 17 (1981).
   b. A. Nielsen, J. Kjaer, and B. Hansen, J. Catl., 3, 68 (1964).

5. A. Ozaki and K. Aika, in A Treatise on Dinitrogen Fixation, edited by R.W.F. Hardy, F. Bottomley and R.C. Burns, (John Wiley & Sons, New York, 1979), Ch.4.

6a. A. Nielsen, An Investigation on Promoted Iron Catalysts for the Synthesis of Ammonia, 3rd ed. (Jul. Gjellerups Forlag, 1968)pp. 14-20;
   b. ibid., pp202-227.

7a. M. Boudart, Catal. Rev.-Sci. Eng., 23, 1 (1981).
   b. J.A. Dumesic, H. Topsoe and M. Boudart, J. Catal., 37, 513 (1975).

8. R. Brill, E.L. Richter, and E. Ruch, Angew. Chem. Internat. Edit., 6, 882 (1967).

9. N.D. Spencer, R.C. Schoonmaker, and G.A. Somorjai, J. Catal., 74, 129 (1982).

10a. F. Bozso, G. Ertl, M. Grunze, and M. Weiss, J. Catal., 49, 18 (1977).
    b. G. Ertl, Catal. Rev.-Sci. Eng., 21, 201 (1980).

11. H. Topsoe, A. Topsoe, and H. Bohlboro, Proc. 7th Internatl. Cong. Catal. edited by T. Seiyama and K. Tanabe, (Kodansha, Tokyo, 1981), p.247

12. M.A. Richard and T.H. Vanderspurt, "New Observations During Steady State and Non-Steady Ammonia Synthesis", presented at the Advances in Catalytic Chemistry III Symposium, University of Utah, May 1985, Salt Lake City, Utah, (unpublished).

13. Uhde, U.K. Patents 253,122 (April, 1927) and 273,735 (October, 1928)

14. T.H. Vanderspurt and M.A. Richard, U.S. Patent No. 4,588,705 (May 13,1986).

THEORETICAL STUDIES of Pt CHEMISORPTION ON γ-Al$_2$O$_3$.

J. J. LOW* AND D. E. ELLIS**
*Allied-Signal Engineered Materials Research Center,50 East
Algonquin Rd., Box 5016,Des Plaines, IL   60017-5016
**Northwestern University, Evanston, Il 60201

ABSTRACT

Generalized Valence Bond (GVB) and Local Density Functional (LDF)
calculations were carried out to explore the influence of electronic
structure on the metal-support interaction in Pt/γ-Al$_2$O$_3$. On the basis of
GVB calculations it was found that ethylene can form either di-σ- or π-
complexes on Pt depending on the oxidation state of the metal. The LDF
calculations showed that chemisorbed Pt transfers charge to γ-Al$_2$O$_3$. These
results and experimental observations led us to propose a model of the
metal-support interaction in which the Pt exists in two oxidation states for
reduced Pt/Al$_2$O$_3$ catalysts.

INTRODUCTION

In spite of the importance of metal-support interactions in catalysis
little is known of the microscopic events which lead to these
interactions. A large fraction of industrially important catalysts contain
clusters of metal atoms deposited on a metal oxide support. A
representative catalyst of this type is the Pt/Al$_2$O$_3$ system. A realistic
model of these catalysts must therefore include both the metal cluster and
the oxide support. Only a model which includes both components will include
the interaction between the metal and the support which is important for
performance of the catalyst. We have initiated an effort to construct
theoretical models of supported catalysts in order to explore the electronic
structure of these interactions.

We started this effort with Generalized Valence Bond (GVB) [1]
calculations for Pt atoms interacting with ethylene. These single Pt atom
calculations are the simplest possible model for the chemistry of a Pt/γ-
Al$_2$O$_3$ catalyst. The calculations showed that ethylene chemisorbs in one of
two forms on Pt. The form of chemisorbed ethylene is important for
catalysis since it affects its performance.

The next level of complexity was to include the support in our model.
Density functional calculations for Pt atoms chemisorbed on γ-Al$_2$O$_3$ were
chosen as the best method for this model. This series of calculations is
most relevant for atoms in the Pt cluster which are directly interacting
with the support. Our results show that the surface of γ-Al$_2$O$_3$ oxidizes
chemisorbed Pt atoms. The insights gained with these calculations combined
with experimental observations led us to propose that two oxidation states
of Pt are present in reduced Pt/Al$_2$O$_3$ catalysts.

GENERALIZED VALENCE BOND CALCULATIONS: INTERACTION OF PT WITH ETHYLENE

The GVB calculations were carried out on Pt(C$_2$H$_4$) and PtH$_2$(C$_2$H$_4$) to
explore the nature of the bonding of C$_2$H$_4$ to Pt in different oxidation
states. These calculations showed that ethylene can form two different
types of complexes with Pt depending on the metal's oxidation state.
Ethylene interacting with Pt(0) in Pt(C$_2$H$_4$) forms a di-σ-complex. In this
complex the C-C π bond has been broken allowing the formation of two metal-
carbon sigma bonds between ethylene and Pt. The assignment of a di-σ-
complex was based on the combination of the calculated geometry [2] and
qualitative analysis of the GVB wavefunction. The calculated C-C bond

Mat. Res. Soc. Symp. Proc. Vol. 111. ©1988 Materials Research Society

length (1.43 Å) is between the observed C-C single bond length in ethane
(1.54 Å) and the C-C double bond length in ethylene (1.34 Å).[3]   The
calculated Pt-C bond length in Pt(C_2H_4) (2.07 Å) is close to the Pt-C length
(2.12 Å) [4] in Pt(CH_3)$_2$(PR_3)$_2$. The GVB orbitals for one of the symmetry
equivalent Pt-C bonds is shown in figure 1. This bond is composed of a Pt
5d6s hybrid orbital and a C 2s2p hybrid orbital, typical for Pt-C sigma
bonds.[5]   Ethylene interacting with Pt(II) in Pt(H)$_2$(C_2H_4) forms a
π-complex. In this complex the C-C π-bond is still intact and the Pt-C_2H_4
bonding is characterized by a Lewis acid/base interaction between Pt and the
pair of electrons in the π-bond. The calculated C-C distance in this
complex (1.35 Å) is close to the C-C distance observed in Zeiss's Salt
(1.37 Å) [6] which is a classical π-complex. The GVB orbitals for the C-C
π-bond in Pt(H_2)(C_2H_4) are shown in figure 1. The orbitals in this bond pair
are two C 2p orbitals each localized on different atoms very similar to the
C-C π-bond in ethylene.[7] These qualitative conclusions from the GVB
calculations suggest that ethylene can be used as a sensitive probe of the
oxidation state of Pt.

    Although this study was performed with only a single Pt atom, it is
consistent with assignment of the peaks in the Electron Energy Loss Spectra
of chemisorbed ethylene with and without coadsorbed oxygen on (111) surfaces
of Pt.[8] On the clean surface, ethylene was observed to form a di-σ-
complex at low temperatures. Although at higher temperatures the di-σ-
complex rearranges to form ethylidyne on Pt, this complex appears to be the
most stable form of undissociated chemisorbed ethylene on a clean Pt
surface. Ethylene was observed to form a π-complex when coadsorbed with
oxygen on a Pt surface. Experimentally, the π-complex is observed to desorb
before rearranging to form coke precursors. In spite of the differences one
would expect between free Pt atoms and atoms at the surface of bulk Pt,
their behavior follows the same general rule. Ethylene forms a di-σ-
complex with low valent Pt and a π-complex with high valent Pt.

    The different chemistry of the π- and di-σ- complexes of ethylene can
be explained by the relative energetics derived from GVB calculations. The
Pt-(C_2H_4) bond strength is calculated to be 11.8 kcal/mol in the di-σ-
complex and 6.5 kcal/mol in the π-complex. The difference in the bond
strengths explains the observed thermolysis behavior of chemisorbed
ethylene. The weaker interaction between Pt and ethylene in the π-complex
allows ethylene to desorb before rearranging. The stronger interaction in
the di-σ-complex holds the ethylene longer, allowing it to rearrange before
desorbing.

Pt-C Bond Pair in Pt(C_2H_4)

Pt 6s5d Hybrid Orbital   C 2p orbital

C-C π-Bond Pair in PtH$_2$(C_2H_4)

Left C 2p Orbital   Right C 2p Orbital

 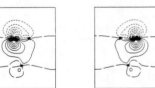

Figure 1.
GVB orbitals for a Pt-C bond in
Pt(C_2H_4) and the C-C π-bond in
Pt(H)$_2$(C_2H_4).

LOCAL DENSITY FUNCTIONAL CALCULATIONS: INTERACTION OF PT WITH THE SUPPORT

Local density functional calculations were used to model the interaction of a Pt atom with the support. The electron density for the cluster of Al_5O_7+ shown in figure 2, embedded in a semi-infinite lattice of frozen $Al+++$ and $O--$ ions was determined self consistently. The coulomb interactions of the lattice with the cluster were included through an Ewald summation. A defect spinel crystal structure with a lattice constant of 7.9 Å and a c/a ratio of 1.0 was chosen for the structure of γ-Al_2O_3.[9] The distribution of cations in this lattice is not well defined experimentally. NMR with magic angle spinning has determined that 25% of the Al cations in γ-Al_2O_3 are in tetrahedral sites.[10] This is most consistent with a model in which all the octahedral and two-thirds of the tetrahedral cation sites in the spinel lattice are occupied. The lowest energy distribution of tetrahedrally coordinated Al (Al(t)) was found by calculating the electrostatic energy of all possible distributions of Al(t) in an infinite lattice composed of supercells containing 9 unit cells of the spinel lattice. As our support surface in these calculations, we chose the (110) face of γ-Al_2O_3, which predominates on the surface of γ-Al_2O_3 crystallites.[9] The Pt was placed at a position which was equidistant from five oxygen atoms. Four of these oxygens were at the Al_2O_3 surface and the fifth was directly below the Pt atom in the first subsurface layer. The model described above includes the important interactions for a Pt atom chemisorbed on γ-Al_2O_3.

The calculated density for the $Pt(Al_5O_7)+$ model system can be analyzed to provide interesting qualitative conclusions. The analysis of the calculated electron densities through Mulliken populations (see table 1) for Pt chemisorbed on γ-Al_2O_3 with free Pt atom and clean γ-Al_2O_3 reveals the nature of interactions between the metal and the support. The Mulliken

Table I

Mulliken Populations	$Pt_{(adsorbed)}/\gamma$-Al_2O_3 [a]	$Pt_{(gas)}+\gamma$-Al_2O_3 [b]
Pt 6s+6p+6d	0.31	1.
5d	9.34	9.
Total	9.65 (0.35)[c]	10. (0.)[c]
O 2s (avg.)[d]	1.95	1.95
2p (avg.)[d]	5.57	5.45
Total (avg.)[d]	7.52 (-1.52)	7.40 (-1.40)[c]
Al 3s (avg.)[d]	0.45	0.47
3p (avg.)[d]	0.29	0.36
Total (avg.)[d]	0.74 (2.26)[c]	0.83 (2.17)[c]

a   Pt atom adsorbed on γ-Al_2O_3
b   Free Pt atom and clean γ-Al_2O_3
c   Numbers in parenthesis represents total charge on atom
d   Average over all atoms in $Al_5O_7^+$ cluster

populations show that the chemisorbed Pt atom has donated 0.35 electrons to the oxygens of Al_2O_3. Thus the chemisorbed Pt atom is partially oxidized by the support with 9.34 electrons in the 5d orbitals and 0.31 electrons in the 6s, 6p and 6d orbitals. The electronic structure of Pt/Al_2O_3 can be further characterized by examining the calculated density of states. The ground state Fermi level appears at -1.5 eV below vacuum, immediately above the large 5d peak in the total density of states (see figure 3a). The assignment of peaks in the density of states can be made with the partial density of states (PDOS) shown in figures 3b-e. The PDOS for O 2s and 2p (figure 3b) falls largely below the Fermi level, consistent with an O--

Figure 2.
The Al₅O₇⁺ Cluster. The large spheres are oxygen anions and the small spheres are aluminum cations.

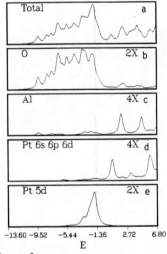

Figure 3.
Ground state density of states in
Pt(Al₅O₇)⁺. The dotted line desig-
nates the Fermi energy. The energy
is given in eV. The density of states
is given in arbitrary units.

ionic configuration. The largest peaks in the Al 3s and 3p PDOS (figure 3c) are above the Fermi level, corresponding to a nominal Al+++ state. This leads to almost fully occupied O valence orbitals and practically empty Al valence orbitals, which is consistent with ionic bonding between Al and O in γ-Al$_2$O$_3$. The Pt 5d PDOS (figure 3d) shows a band which is largely below the Fermi level while the 6s 6p and 6d bands lie above the Fermi level. This is consistent with a partially oxidized Pt atom which has donated its 6s valence electron to the support.

Much effort has been directed at understanding the interaction of C$_2$H$_4$ with Pt supported on Al2O3.[11,12] In particular Moshin, Trenary and Robota [12] have observed using infrared spectroscopy the presence of both di-σ- and π-complexes on reduced Pt/γ-Al$_2$O$_3$ catalysts. The theoretical work outlined above is consistent with this observation if we propose two states of Pt in this catalyst. The Pt in this catalyst is known to contain particles of about 250 atoms. Some of the Pt on the catalyst may become partially oxidized through interaction with γ-Al$_2$O$_3$. This Pt influenced by the support would form π-complexes with ethylene. Some of Pt atoms in the particles may be shielded from γ-Al$_2$O$_3$ by other Pt atoms which are closer to the support. These zero valent Pt atoms would form disigma complexes with ethylene. The presence of both zero valent Pt and partially oxidized Pt on γ-Al$_2$O$_3$ would lead to the observation of both di-σ- and π-complexes after exposure of the catalyst to ethylene.

SUMMARY

In this paper we modeled the chemistry of Pt supported on γ-Al$_2$O$_3$ using two different approaches. We used GVB wavefunctions to model the interaction of ethylene with Pt. We found that oxidized Pt forms π-complexes, while low valent Pt forms di-σ complexes with ethylene. The strengths of the GVB approach were the abilities to optimize geometries and to derive GVB orbitals. The optimized geometries and GVB orbitals aided in the understanding of the fundamental differences between di σ- and π complexes. LDF methods were used to model the interaction of metal with the support. The advantage of this technique was the ability to correctly include the Madelung potential of the support. These calculations showed that Pt is oxidized by the support. The concepts derived from the theoretical methods combined with the experimental observations led us to propose a model in which both low valent and oxidized Pt are present on the support.

REFERENCES

1. W.A. Goddard III, L.B. Harding, Annu. Rev. Phys. Chem. 29, 363 (1978).

2. The GVB calculations used a relativistic effective core potential double-ζ basis for Pt {J.O. Noel, P.J. Hay, Inorg. Chem. 21, 14 (1982)}, a valence double-ζ basis sets for C, and an unscaled triple-ζ basis for H {T.H. Dunning, Jr., P.J. Hay, In Modern Theoretical Chemistry: Methods of Electronic Structure, edited by H.F. Schaefer III (Plenum, New York, 1977) vol. 3, chapt. 4, pp. 79-127}. Geometry optimizations were carried out at the Hatree-Fock level with the Gaussian 82 package (J.S. Binkley, M.J. Frisch, D.J. Defrees, K. Raghavachari, R.A. Whiteside, H.B. Schelgel, E.M. Fuder, J.A. Pople, GAUSSIAN 82, Dept. Of Chem., Carnegie-Mellon University, Pittsburg Pa, 1983). The correlated wavefunctions used to determine bond energies were optimized using the GVB3 multiconfigurational self consistent field program {L.G. Yaffe, W.A. Goddard III, Phys. Rev. A, 13, 1682 (1976)}.

3. G. Herzberg, *Molecular Spectra and Molecular Structure, III. Electronic Spectra and Electronic Structure*, (Van Nostrand Reinhold Comp., New York, 1966) p. 646.

4. J.M. Wisner, T.J. Bartczak, J.A. Ibers, J.J. Low, W.A. Goddard III, J. Am. Chem. Soc. 108, 347 (1986).

5. J.J. Low, W.A. Goddard III, J. Am. Chem. Soc. 108, 6115 (1986).

6. J.A.J. Jarvis, B.T. Kilbourn, P.G. Owston, Acta Crystallogr. B27, 366 (1971).

7. P.J. Hay, W.J. Hunt, W.A. Goddard III, J. Am. Chem. Soc. 94, 8293 (1972).

8. H. Steininger, H. Ibach, S. Lehwald, Surf. Sci. 117, 685 (1982).

9. S.J. Wilson, J. Solid State Chem. 30, 247 (1979).

10. C.S. John, N.C.M. Alma, G.R. Hyas, Applied Catalysis 6, 341 (1983).

11. Y. Soma, J. Catalysis, 75, 267 (1982); T.P. Beebe, Jr., J.T. Yates, Jr., J. Phys. Chem. 91, 254 (1987); P.-K. Wang, C.P. Slichter, J.H. Sinfelt, J. Phys. Chem. 89, 3606 (1985)

12. S.H. Mohsin, M. Trenary, H.J. Robota, J. Phys. Chem. Submitted for Publication.

CHARACTERIZATION OF SUPPORTED METAL OXIDES BY LASER RAMAN SPECTROSCOPY: SUPPORTED VANADIUM OXIDE ON Al_2O_3 AND TiO_2

ISRAEL E. WACHS AND FRANKLIN D. HARDCASTLE
Zettlemoyer Center for Surface Studies, Departments of Chemical Engineering and Chemistry, Lehigh University, Bethlehem, PA 18015, U.S.A.

SHIRLEY S. CHAN
Photon Technology International, Princeton Corporate Plaza, Suite F, Deer Park Drive, South Brunswick, NJ 08852, U.S.A.

ABSTRACT

The interaction of supported vanadium oxide with Al_2O_3 and TiO_2 substrates is examined with Raman spectroscopy. The Raman spectra of the supported vanadium oxide reveal that the strong interaction of the vanadium oxide with the Al_2O_3 and TiO_2 supports results in the formation of an atomically dispersed surface vanadium oxide phase as well as supported crystalline V_2O_5. The relative concentrations of the atomically dispersed surface vanadium oxide and crystalline V_2O_5 depend on the vanadium oxide loading and the surface area of the oxide support.

INTRODUCTION

Supported metal oxides are formed when one metal oxide phase is dispersed on a second metal oxide substrate. The dispersed supported metal oxide phase can simultaneously possess several different molecular states. The multiple molecular states that can simultaneously be present in the supported metal oxide phase have acted as a source of confusion and hampered progress in the understanding of supported metal oxide catalysts. This confusion has resulted primarily because of the lack of applicable characterization techniques capable of discriminating between these different molecular states. Conventional catalyst characterization techniques provide general information concerning the physical characteristics of the supported metal oxide phase, but they are not adequate to discriminate between the different molecular states that are simultaneously present [1]. In the past few years, however, characterization studies of supported metal oxides have shown that the different molecular states in the supported metal oxide phase can be discriminated with the use of laser Raman spectroscopy [2]. This technique can discriminate between the different molecular states of the supported metal oxides because each state possesses a unique vibrational spectrum that is related to its structure. Therefore, Raman spectroscopy provides direct information about the structure of each state as well as a method of discriminating between the various states. In the present investigation we report on the Raman spectroscopy of supported vanadium oxide on Al_2O_3 and TiO_2 supports. The influence of the oxide supports upon the structural nature of the supported vanadium oxide phases and their catalytic properties will be discussed.

EXPERIMENTAL

The supported metal oxide catalysts were prepared by the incipient-wetness impregnation method. The oxide supports consist of γ-Al_2O_3 (Harshaw, 180 m^2/g) and TiO_2 (Degussa P-25, anatase/rutile$\approx$2, 55 m^2/g). The V_2O_5/TiO_2 samples were made with $VO(OC_2H_5)_3$ in ethanol, dried at room temperature for 16 hrs, dried at 110-120°C for 16 hrs, and calcined at 450°C for 2 hrs. The V_2O_5/Al_2O_3 samples were made with $VO(OC_3H_7)_3$ in methanol, dried at room temperature for 16 hrs, dried at 110-120°C for 16 hrs, and calcined at 450°C for 16 hrs. Several of the above samples were further calcined at higher temperatures in order to examine the influence of calcination temperature upon the supported vanadium oxide systems.

The Raman spectra were obtained with a Spectra-Physics Ar$^+$ laser delivering 1-100 mW of incident radiation measured at the sample, where the exciting line was typically 514.5 nm. The scattered radiation was then directed into a Spex Triplemate Spectrometer (Model 1877) coupled to a Princeton Applied Research OMA III optical multichannel analyzer (Model 1463) with an intensified photodiode array cooled thermo-electrically to -30°C. The Raman spectra were recorded using an OMA III dedicated computer and software. All Raman spectra were obtained at room temperature and under ambient conditions. Additional details concerning the optical arrangement of the laser Raman apparatus can be found elsewhere [2].

RESULTS AND DISCUSSION

The interaction of vanadium oxide with the Al_2O_3 support is reflected in the Raman spectra presented in Figure 1. The nature of the supported vanadium oxide phases is determined by comparison of the Raman spectra of the supported vanadium oxide catalysts with spectra of vanadium oxide reference compounds [3-8]. The 3-20 wt% V_2O_5/Al_2O_3 catalysts do not contain the Raman features of either crystalline V_2O_5 (major bands at 1000, 703, 525, and 150 cm^{-1}) or $AlVO_4$ (major bands at 1017, 988, 949, 899, 510, 391, 319, 279, and 133 cm^{-1}), but possess weak and broad Raman bands characteristic of an atomically dispersed surface vanadate species. The surface vanadate species possesses bands in the 900-1000 cm^{-1} region which are characteristic of V=O symmetric stretching modes and bands at $\approx$500 and 220 cm^{-1} which are charateristic of V-O-V linkages. The Raman spectra of the 22-30% V_2O_5/Al_2O_3 catalysts, however, are composed of the Raman features diagnostic of crystalline V_2O_5 which overshadow the weaker Raman features of the atomically dispersed surface vanadate phase. Thus, a monolayer of the surface vanadate species on alumina corresponds to $\approx$20% V_2O_5/Al_2O_3 for this system.

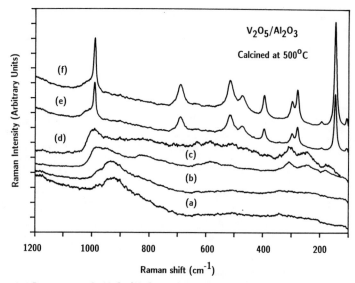

Figure 1. Raman spectra for V_2O_5/Al_2O_3 catalysts at sub-monolayer coverages: (a) 3% V_2O_5/Al_2O_3, (b) 5% V_2O_5/Al_2O_3, (c) 10% V_2O_5/Al_2O_3, (d) 20% V_2O_5/Al_2O_3; and at coverages exceeding the monolayer: (e) 22% V_2O_5/Al_2O_3, (f) 30% V_2O_5/Al_2O_3.

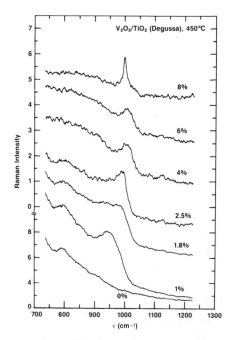

Figure 2. Raman spectra for V_2O_5/TiO_2 (Degussa) catalysts at sub-monolayer coverages: 1%, 1.8%, 2.5%, 4%, and 6% V_2O_5/TiO_2; and above monolayer coverage: 8% V_2O_5/TiO_2.

The Raman spectra for the supported vanadium oxide on TiO_2 catalysts are presented in Figure 2. In general, the very strong TiO_2 Raman features below 700 cm^{-1} prevent the acquisition of Raman spectra in this region for supported metal oxides on TiO_2. Furthermore, the TiO_2 support possesses a weak band at ≈ 796 cm^{-1} which arises from the first overtone of its very strong band at 398 cm^{-1}. The 1-6% V_2O_5/TiO_2 catalysts do not exhibit the Raman features of crystalline V_2O_5, but possess weak and broad Raman bands characteristic of an atomically dispersed surface vanadate species. The surface vanadate species possesses bands in the 900-1000 cm^{-1} region which are characteristic of V=O symmetric stretching modes. The intensity of the weak TiO_2 band at ≈ 796 cm^{-1} decreases as the vanadium oxide coverage is increased because of the dramatic increase in the absorptivity coefficient (sample color) of the sample in the visible region of the spectrum (i.e., a fraction of the Raman signal is absorbed by the sample matrix). The Raman spectrum of the 8% V_2O_5/TiO_2 catalyst is dominated by the band of crystalline V_2O_5 at 1000 cm^{-1} and overshadows the weaker Raman features of the atomically dispersed surface vanadate phase. Thus, a monolayer of the surface vanadate on titania corresponds to $\approx 7\%$ V_2O_5/TiO_2 for this system.

TABLE I

Influence of Calcination Temperature on BET Surface Area of Supported Vanadium Oxide on TiO_2 and Al_2O_3

Sample	Calcination Temperature*			
	450°C	700°C	850°C	950°C
TiO_2	56 m^2/g	31 m^2/g	12 m^2/g	—
5% V_2O_5/TiO_2	46 m^2/g	17 m^2/g	5 m^2/g	—
Al_2O_3	180 m^2/g	—	—	94 m^2/g
5% V_2O_5/Al_2O_3	178 m^2/g	—	—	3 m^2/g

* TiO_2 samples calcined for 2 hrs; Al_2O_3 samples calcined for 16 hrs.

The influence of elevated calcination temperatures upon the V_2O_5/Al_2O_3 and V_2O_5/TiO_2 systems was also investigated [8,9]. Both of these systems exhibited very dramatic decreases in surface area upon calcination between 700-950°C, and the corresponding surface areas are shown in Table I. Note that the presence of the supported vanadium oxide phase dramatically accelerates the loss in surface area of both the Al_2O_3 and TiO_2 supports. The significant loss in surface area of the oxide supports serves to increase the surface density of surface vanadium oxide and transforms the surface vanadate species to crystalline V_2O_5. The supported crystalline V_2O_5 is stable on the Al_2O_3 support and does not undergo any solid state reactions at elevated temperatures. The supported crystalline V_2O_5, however, is not stable on the TiO_2 support and does undergo a solid state reaction with the titania support to form crystalline $V_xTi_{1-x}O_2$ (rutile) after reduction of the vanadium from the +5 to the +4 oxidation state.

Many recent studies have shown supported V_2O_5 catalysts to be superior catalysts to unsupported V_2O_5 catalysts for the selective oxidation of many hydrocarbons [7,8,10-17]. Furthermore, the surface vanadate species were found to be the active sites for these partial oxidation reactions, and to possess a higher activity and selectivity than crystalline V_2O_5 for many hydrocarbon oxidation reactions. The oxide supports must be covered by a complete monolayer of the surface vanadate species since exposed support sites lead to combustion of the partial oxidation products. Moderate amounts of crystalline V_2O_5 do not significantly affect the catalytic performance of supported vanadium oxide catalysts because of the low effective surface area and poor catalytic activity of this phase. The unique properties of the surface vanadate monolayer on oxide supports are related to the specific nature of the vanadia/oxide support interactions.

CONCLUSION

The interaction of supported vanadium oxide with Al_2O_3 and TiO_2 supports was investigated with Raman spectroscopy. Both the alumina and titania supports exhibit very strong capacities for forming atomically dispersed surface vanadium oxide phases. The supported vanadium oxide phase is found to influence the surface area of the oxide support upon calcination.

REFERENCES

1. M.L. Deviney and J.L. Gland , eds. "Catalyst Characterization Science," ACS Symposium Series 288, American Chemical Society, Washington D.C., 1985.

2. I.E. Wachs, F.D. Hardcastle, and S.S. Chan, Spectrosc. 1(8) 30 (1986).

3. W.P. Griffith and T.D. Wickins, J. Chem. Soc. (A) 1087 (1966).

4. F. Roozeboom, J. Medema, and P.G. Gellings, Z. Phys. Chem. (Frankfurt am Main) 111 215 (1978).

5. F. Roozeboom, M.C. Mittlemeijer-Hazeleger, J.A. Moulijn, J. Medema, V.H.J. de Beer, and P.G. Gellings, J. Phys. Chem. 84 2783 (1980).

6. S.S. Chan, I.E. Wachs, L.L. Murrell, L. Wang, and W.K. Hall, J. Phys. Chem. 88 5831 (1984).

7. I.E. Wachs, R.Y. Saleh, S.S. Chan, and C.C. Chersich, Appl. Catal. 15 339 (1985).

8. R.Y. Saleh, I.E. Wachs, S.S. Chan, and C.C. Chersich, J. Catal. 98 102 (1986).

9. F.D. Hardcastle and I.E. Wachs, unpublished results.

10. D. Vanhove and M. Blanchard, Bull. Sci. Chim. Fr. 3291 (1971).

11. R. Grabowski, B. Grzybowska, J. Haber, and J. Slocgnski, React. Kinet. Catal. Lett. 2 81 (1975).

12. G.C. Bond, J. Sarkany, and G.D. Parfitt, J. Catal. 57 476 (1979).

13. Y. Murakami, M. Inomata, A. Miyamoto, and K. Mori, in Proceedings, 7th International Congress on Catalysis, Tokyo, 1980, p. 1344. Elsevier, Amsterdam (1981).

14. G.C. Bond and P.J. Konig, J. Catal. 77 309 (1982).

15. G.C. Bond and K. Bruckman, Faraday Discuss. 72 235 (1981).

16. A.J. Van Hengstum, J.G. Van Ommen, H. Bosch, and P.J. Gellings, Appl. Catal. 8 369 (1983).

17. M. Gasior, I. Gasior, and B. Grzybowska, Appl. Catal. 10 87 (1984).

INVESTIGATION OF NOVEL IRON OXIDE-CHROMIA ON ALUMINA OR MAGNESIA AEROGEL CATALYSTS FOR THE SELECTIVE REDUCTION OF NITRIC OXIDE BY AMMONIA

RONALD J. WILLEY*, J. OLMSTEAD*, V. DJUHADI*, AND S.J. TEICHNER**
*Northeastern University, Department of Chemical Engineering, Boston, MA 02115
**Universite Claude Bernard Lyon I, Laboratoire de Thermodynamique et Cinetique Chimiques, 69622 Villeurbanne Cedex, France

ABSTRACT

A series of iron oxide-chromia on alumina and iron oxide on magnesia materials were made by the aerogel process. These materials had high surface areas (300 to 700 m^2/g) and demonstrated good activity for the selective reduction of nitric oxide by ammonia. A combination of 4 parts Fe to 1 Part Cr gave the best catalyst for the reaction. Fe-Cr to alumina ratios were varied from 0.1 to 0.4. Activity per unit weight increased with increasing Fe-Cr/Al_2O_3 ratios, however, activity per unit Fe-Cr favored the lower Fe-Cr to alumina ratios.

The fresh material from the autoclave was found reduced (Fe_3O_4) and initially had poor activity for the selective reduction of nitric oxide by ammonia (525 ppm NO & 525 ppm NH_3 in air feed conditions). However, after several testing cycles, in which the catalyst was in continuous contact with oxygen (as air), activity improved dramatically. Upon removal from the reactor the material had a distinct color change (from brown to reddish brown) and magnetic properties were lost. This suggests that Fe_2O_3 is part of the active site for the selective reduction of nitric oxide by ammonia. At higher temperatures activity decreased because of a competing reaction, ammonia oxidation to nitric oxide.

INTRODUCTION

The selective catalytic reduction of nitric oxide by ammonia has been proposed as a method to control nitric oxide emissions from stationary sources for the past 20 years. Nitric oxide emissions are considered as a source for 30 to 40% of the acid rain [1]. The reaction of primary interest is:

$$4 \ NO + 4 \ NH_3 + O_2 \ \text{-----}> 4 \ N_2 + 6 \ H_2O \tag{1}$$

The use of iron-chromium based materials as catalysts for the above reaction has been investigated in the past. Notable work includes Wu and Nobe [2] who used iron oxide and chromia supported on alumina, and Naruse et al. [3] who studied iron oxide catalysts obtained from waste steel processes. More recently, Willey [4,5] reported that 430 SS (80% iron and 20% chromium) etched metal catalysts had good activity for Reaction 1.

On the other hand, the suggestion of aerogel materials as catalysts go back to their discovery by Kistler [6]. Advances in the aerogel process [7-9] has renewed interest in aerogel materials as catalysts. Past work includes the conversion by nitric oxide of paraffins [10], olefins [11], and alkylaromatics [12] into nitriles. With the combination of expertise in two areas, deNOx reactions and aerogel catalysts, several novel iron oxide-chromia supported aerogel catalysts were synthesized with the hopes that activity of these highly divided materials would surpass previous iron oxide-chromia materials. The ability to vary metallic oxide/support proportions is another advantage to the aerogel process.

EXPERIMENTAL

Preparation of Aerogel Catalysts

The procedure for preparation of iron oxide-chromia aerogel catalysts can be explained in four steps. For an example, consider an aerogel composed of the following metallic atomic ratios: 0.4:0.1:1.0 Fe:Cr:Al. The first step involves the preparation of two separate mixtures which are then combined. The first mixture consists of a solution of 12.5% aluminum sec-butoxide in sec-butanol. The second mixture consists of a 2.6% solution of ferric acetylacetonate and chromium acetylacetonate in methanol in which the ferric to chromium acetylacetonates ratio is adjusted to give an atomic ratio of iron to chromium of 0.8 to 0.2. Also added to the methanol solution is a small amount of water which will hydrolyze ferric and chromium acetylacetonates and aluminum sec-butoxide to their respective hydroxides. The two solutions are then combined at appropriate ratios to form an alcogel (hydrolyzed aluminum precursor which is first to precipitate).

For the second step, the combined solutions are placed into an autoclave and heated. As the autoclave contents heat up, pressure begins to increase because of the sealed environment, and the solvents follow a vapor-liquid equilibrium line up to the critical point. In the third step, pressure is slowly relieved after the temperature is about 10 K above the critical temperature of methanol (512.7 K). The fourth step involves purging the contents of the autoclave with nitrogen and cooling the material back to room temperature. The resultant aerogel material comes out in slightly reduced state as seen by its magnetic properties and has a brownish color. Table I is a listing of aerogel materials prepared and evaluated as catalysts for the selective reduction of NO by NH_3.

TABLE I.

Summary of Aerogel Materials Investigated for the Selective Catalytic Reduction of Nitric Oxide by Ammonia

Metallic Atomic Ratios				Initial Surface Area m^2/g (BET)	Maximum rate (1)	Maximum rate (2)	Temperature of peak activity K
Fe	Cr	Ni	Al				
0.088	0.022	-	1.0	700	3.03	29.57	680
0.40	0.10	-	1.0	300	5.55	18.20	640
0.40	-	-	1.0	265	1.62	6.04	640
-	0.10	-	1.0 -	-	0.20	2.32	518
-	1.00	-	- -	265	0.67	0.98	512
1.0	-	-	1.0(Mg)	600	0.90	1.93	605
0.20	-	0.80	1.0	395	1.73	3.77	645

(1) μmoles/g sec
(2) μmoles/[Fe+Cr+Ni] sec

Description of Reactor System

About 0.25 grams of the aerogel material were placed in a quartz tube (2.0 cm I.D.) along with a 3 mm pyrex bead packing to form a 4 cm long bed. The quartz tube was then placed vertically into a Lindberg 3 zone oven where temperature was controlled from 460 to 780 K. A gas mixture composed of 525 ppm nitric oxide and ammonia in air was passed across the material at a flow rate of 100 cc/sec (NTP)(weighted space velocity of 366 cc/[sec g] at STP). Nitric oxide analysis was done with a Thermo Electron Series 44 chemiluminescent nitric oxide analyzer. Catalyst bed temperature and the NOx analyzer signal were acquired by an IBM-PC via a Metrabyte Dash-16 analog to digital converter and a software package by Lotus Development Corp. called Measure. A typical experiment involved acquiring nitric oxide outlet concentration as oven temperature was increased. By knowing inlet nitric oxide concentrations, conversion as a function of catalyst bed temperature could be determined.

RESULTS

Figure 1 shows an example of performance for an aerogel material for the selective reduction of nitric oxide by ammonia. The first notable result is a peak in activity seen as temperature is increased from 460 to 780 K. For the 0.4:0.1:1.0 Fe/Cr/Al aerogel, the peak activity of 47.5% conversion of NO occurs at 640 K. The reason for the peak is because a secondary reaction begins to dominate at higher temperatures (<670 K). Specifically, ammonia oxidation to nitric oxide (shown in Equation (2)) creates nitric oxide and thus negative conversion is observed.

$$4 \ NH_3 \ + \ 5 \ O_2 \ -----> \ 4 \ NO + 6 \ H_2O \tag{2}$$

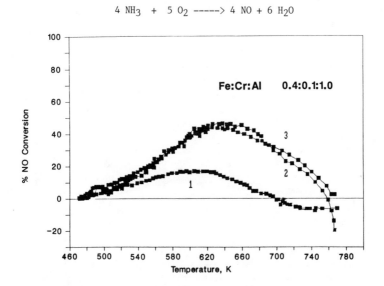

Figure 1. Performance of 0.4:0.1:1.0 Fe:Cr:Al Aerogel Material for the Selective Catalytic Reduction of Nitric Oxide by Ammonia at a Weighted Space Velocity of 366 cc/[sec g] (STP). Numbers represent testing sequence.

Figure 1 also shows that the catalyst required activation because Run 1's activity is below that of Runs 2 and 3. Upon inspection of material removed from the reactor, the color of the aerogel changes from a brown to a reddish brown indicative of a valence state change for iron related to iron oxide changing from Fe_3O_4 to Fe_2O_3. This is further confirmed because the aerogel loses magnetic properties after the testing. Besides oxidation, heating may improve the crystalline state of the catalyst and eventually lead to the formation of spinels. Crystalline formation has yet to be confirmed.

Figures 2 and 3 show interesting results when iron and chromium are used separately with the same support. In Figure 2, iron is present in the same atomic ratio to aluminum as is shown in Figure 1 (0.4:1.0 Fe:Al). Chromium is not present in this material. The results show lower overall conversion of nitric oxide with the peak conversion reaching 12% at 640 K. Again, the material required some activation because activity improved over the 3 runs. Experience showed that peak activity was generally seen by the third run with Fe containing aerogel materials. Previous work with Fe-Cr etched metal catalysts also found a more active catalyst with the Fe-Cr versus the pure Fe treated catalyst [14].

In Figure 3, chromium is present in the same atomic ratio to aluminum as is shown in Figure 1 (0.1:1.0 Cr:Al). Iron is not present in this material. Only on the third run was NO conversion observed. The peak activity observed was 2.5% conversion of NO at 540 K. The negative conversion represents NH_3 oxidation to NO as shown in Equation (2). Why, when chromium is added to iron enhances activity for NO conversion, is an intriguing question, especially in light of these results.

Table I summarizes the remaining results from this work. At a lower iron-chromium to aluminum ratio aerogel (0.088:0.022:1.0 Fe:Cr:Al) overall activity does drop. Performance curves are similar to those shown in Figure 1 with peak conversion reaching 30% conversion of NO. Although overall conversion is lower, conversion per unit Fe+Cr in this material is higher then the 0.4:0.1:1.0 Fe:Cr:Al material. The improve activity per unit Fe+Cr probably means better dispersion of the Fe-Cr active site.

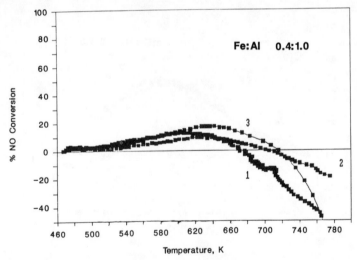

Figure 2. Performance of 0.4:1.0 Fe:Al Aerogel Material for the Selective Catalytic Reduction of Nitric Oxide by Ammonia at a Weighted Space Velocity of 366 cc/[sec g] (STP). Numbers represent testing sequence.

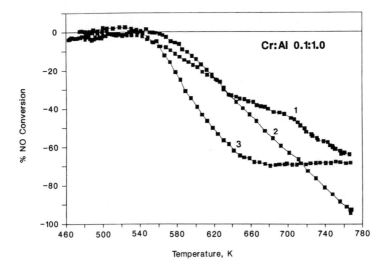

Figure 3. Performance of 0.1:1.0 Cr:Al Aerogel Material for the Selective Catalytic Reduction of Nitric Oxide by Ammonia at a Weighted Space Velocity of 366 cc/[sec g] (STP). Numbers represent testing sequence.

Table I also shows results for iron oxide on a different support (1:1 Fe:Mg). The performance of this material was not as good as iron oxide on alumina. Overall, NO conversion of about 30% slowly increased with temperature between 500 to 620 K (indicative of a low apparent activation energy). Above 630 K conversion quickly fell crossing 0% at about 700 K (where ammonia oxidation prevails).

Another material examined was a 0.2:0.8:1.0 Fe:Ni:Al aerogel. In Run 1 with this material no positive conversion of nitric oxide was observed. In fact, as temperature increased from 470 to 750 K, negative conversion continued to increase (the only fresh material investigated which showed activity for ammonia oxidation at temperatures below 500 K). However, Run 2 showed some positive conversion of nitric oxide for temperatures from 470 to 620 K, and, by Run 3, positive conversion of nitric oxide was observed from 470 to 710 K with the peak activity of 15% conversion at 640 K. Again, the oxidation state of the material changed from Run 1 to Run 3. Further, because the temperature exceeded 700 K the possibility exist that spinels formed which have activity for Reaction (1) and retard Reaction (2).

CONCLUSIONS

1. Iron oxide/chromia/alumina aerogel materials are good catalysts for the selective reduction of nitric oxide by ammonia at nitric oxide concentrations of 500 ppm in air.

2. Activity for the selective reduction of nitric oxide by ammonia of the iron containing aerogels is dependent on the oxidation state of iron from Fe_3O_4 to Fe_2O_3. In this work, higher the oxidation state, the higher nitric oxide conversion observed.

3. The combination of iron oxide and chromia on alumina shows better activity for NO conversion compared to materials composed of only iron oxide on alumina or chromia on alumina.

4. Stability of iron oxide-chromia-alumina aerogel materials for the selective reduction of nitric oxide by ammonia is excellent after an initial oxidation pretreatment.

5. Activity for the selective reduction of nitric oxide by ammonia reaches a peak with temperature for all materials investigated. Above about 720 K, activity for a competitive reaction, ammonia oxidation to nitric oxide, dominates.

ACKNOWLEDGEMENTS

The authors wish to acknowledge support from the IBM scholars fund which allowed Professor Teichner to visit Northeastern University for a period of 3 months. Also, a portion of this work was supported by the Cabot Foundation. Finally, the authors wish to acknowledge Lotus Development Corporation for computer software support.

REFERENCES

1. R.E. Wyzya, in Proceedings of the Joint Symposium on Stationary Combustion NOx Control 5, Oct 1980. p. 24.
2. S.C. Wu, K. Nobe, Ind. Eng. Chem. Prod. Res. Dev. 16, 136, (1977).
3. Y. Naruse, T. Ogasawara, T. Hata, and H. Kishitaka, Ind. Eng. Chem. Prod. Res. Dev. 19, 57 (1980).
4. R.J. Willey, W.C. Conner, and J.W. Eldridge, J. Catal. 92, 136 (1985).
5. R.J. Willey, J.W. Eldridge, J.R. Kittrell, Ind. Eng. Chem. Prod. Res. Dev. 24, 226 (1985).
6. S.S. Kistler, U.S. Pats 2,039,454; 2,188,007; 2,249,767, J. Phys. Chem., 36, 52 (1932).
7. S.J. Teichner, G.A. Nicolaon, M.A. Vicarini, and G.E.E. Gardes, Adv. Colloid. Interface Sci., 5, 245 (1976).
8. M. Astier, A. Bertrand, D. Bianchi, A. Chenard, G.E.E. Gardes, G.M. Pajonk, M.B. Taghavi, S.J. Teichner, and B.L. Villemin, in "Preparation of Catalysts", Delmon, B., Jacobs, P.A. and Poncelet, G., Eds., (Elsevier Sci. Pub. Co., Amsterdam 1976), p. 315.
9. Teichner, S.J., in "Aerogels", Fricke, J. Ed., (Springer Proc. in Physics, Vol. 6, Springer-Verlag, Berlin-Heidelberg 1986), p. 22.
10. H. Zarrouk, A. Ghorbel, G.M. Pajonk, and S.J. Teichner, in "Adsorption and Catalysis on Oxide Surfaces", Che, M. and Bond, G.C. Eds, (Elsevier Sci. Pub. Co., Amsterdam, 1985), p. 429.
11. F. Zidan, G. Pajonk, J.E. Germain, and S.J. Teichner, J. Catal. 52, 133 (1978).
12. A. Sayari, A. Ghorbel, G.M. Pajonk, and S.J. Teichner, S.J., Bull. Soc. Chim. France, 1, 16, 24, 220 1981; 39 1982.
13. Abouarnadasse, S., Pajonk, G.M., Germain, J.E. and Teichner, S.J., Appl. Catal. 9, 119 (1984).
14. R.J. Willey, Ph.D. Dissertation, University of Mass., 1984.

A NEW POLYMER-BASED HYDROGEN GETTER

LAURA R. GILLIOM
Sandia National Laboratories, Division 1811, Albuquerque, NM 87185-5800

ABSTRACT

Styrene-butadiene triblock copolymer PS-PB-PS was hydrogenated in the bulk using the Crabtree catalyst [Ir(COD)(py)(tcyp)]PF$_6$ (COD = 1,5-cyclooctadiene, py = pyridene, tcyp = tricyclohexylphosphine). Since this polymer/catalyst mixture reacts rapidly with hydrogen at ambient temperature and low hydrogen pressures, it should act as an effective hydrogen getter.

INTRODUCTION

The term "getter" was coined to describe substances capable of removing undesirable impurities from a gaseous environment through associative reactions [1]. One frequently targeted impurity, hydrogen (H_2), can present an explosion hazard, damage electrical and optical components, and embrittle metals. Hydrogen getter systems consisting of active metals such as titanium are used regularly for the removal of small amounts of hydrogen to maintain high vacuum conditions [1]. Many other applications could benefit from the development of a material effective at removing substantial quantities of hydrogen under a variety of conditions.

Previous approaches to a more adaptable hydrogen getter have used the hydrogenation of crystalline organic alkynes with heterogeneous catalysts to remove unwanted hydrogen gas. Molecules such as 1,6-diphenoxy-2,4-hexadiyne or 1,4-di(ethynylphenyl)benzene were combined with supported catalysts such as Pt/CaCO$_3$ or Pd/C [2,3]. The resulting mixture either was placed in small bags or was pressed into pellets. Both fabrications have proven successful at gettering hydrogen from closed systems. Mobility of the substrate to the active site is presumed to be provided by a thin surface layer of organic melted by the reaction exotherm. Problems encountered in the actual use of such getters have been due to volatility and diffusion of the constituents, encapsulation of the catalyst, or excessive production of water.

A new getter based on homogeneously-catalyzed hydrogenation of reducible organic polymers is described here. This scheme could eliminate some of the above-mentioned problems and could yield an efficient hydrogen getter in a form which might also be used as a structural material. We have previously reported the bulk hydrogenation of polyolefins catalyzed by a variety of homogeneous catalysts [4]. The work discussed below focuses on the single polymer/catalyst system identified as best suited to hydrogen getter applications.

EXPERIMENTAL

The getter material was cast from solution. Triblock copolymer (Kraton D1102 from Shell Chemical Co., 1.0g) was dissolved in 40 mL dichloromethane. After addition of 0.1g [Ir(COD)(py)(tcyp)]PF$_6$ [5], the solvent was removed under vacuum leaving the polymer/catalyst mixture(9.1 wt % catalyst).

Hydrogenations were performed in a pressure reaction vessel (total volume = 180mL) consisting of a glass sample container, appropriate pressure gauges, and gas inlet ports. One gram of the getter material was used in each experiment. The pressure vessel was evacuated, other gases (when used) were admitted, and then hydrogen was added to the chosen initial pressure. All reactions were run at 23°C.

RESULTS AND DISCUSSION

The system studied as a potential getter material consists of polystyrene-block-polybutadiene-block-polystyrene (PS-PB-PS), which is the reducible organic polymer, and [Ir(COD)(py)(tcyp)]PF_6, the homogeneous hydrogenation catalyst. The thermoplastic nature of PS-PB-PS provides structural integrity while its rubbery domains permit sufficient mobility for catalyst diffusion. Reduction of the polymer should consume hydrogen as shown below:

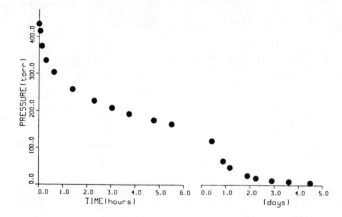

The iridium-based catalyst, first reported by Crabtree, is notable for its high activity in hydrogenation reactions and its resistance to poisoning by oxygen [6].

Test samples were prepared containing 9.1 wt % catalyst. They were somewhat porous due to the foaming action of the casting solvent. On reaction with a constant volume of hydrogen, the system pressure falls as the polymer is hydrogenated. Figure 1 illustrates the pressure drop

Figure 1. Room temperature hydrogenation of PS-PB-PS containing 9.1 wt % catalyst.

observed when twice the theoretically necessary amount of getter was used.
Incorporation of hydrogen into the polymer is proven both by the weight
increase of the getter sample and by NMR spectra of the reduced polymer.
Proton NMR spectra of the above getter materials before and after complete
reduction are shown in Figure 2.

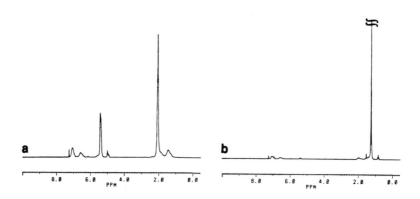

Figure 2. Proton NMR spectra of the getter sample (a) before
hydrogenation and (b) after complete hydrogenation.

On hydrogenation, the rubbery centerblock of PS-PB-PS is converted to a
semi-crystalline low density polyethylene-like polymer. This
transformation is indicated by the loss of the peak at $\delta5.3$ attributable to
the olefinic protons and the new peak at $\delta1.4$ assigned to chain methylene
groups. It is also observable in the increased stiffness of the
hydrogenated material. The theoretical capacity of the 9 wt % catalyst
getter is 274 cc hydrogen per gram getter. Based on the observed hydrogen
consumption in higher pressure (40 psig hydrogen) experiments, the actual
uptake is 253 cc/g. This uptake is also consistent with the sample's
weight gain and its NMR spectra [7].

Sensitivities of this material were assessed by introducing substantial
amounts of other gases into the hydrogenation reaction. The most
significant result is that the presence of air did not diminish the
hydrogen uptake rates or the capacity of the getter. Analysis of the fully
hydrogenated getter material by proton NMR spectroscopy showed that the
hydrogen consumed was fully incorporated into the polymer. No peak
attributable to water was present. Regardless, a dramatic color change in
the material from orange to green suggests that the catalyst was altered by
air in contrast to results reported for solution work [6]. The hydrogen
uptake was unaffected by the presence of water vapor. The reaction was
poisoned by carbon momoxide. Trimethylamine acted as an inhibitor,
reducing both the reaction rates and the capacity of the material.

For the experiments discussed above, a plug of the getter, slightly
foamed by solvent evaporation, was used without modification. Pellets of
the getter were prepared in a room temperature press and by 120°C molding

operations. Hydrogen uptake rates were significantly reduced when the getter was in pellet form. The getter material has also been applied to glass slides by spin-coating and by slow solvent evaporation leaving a uniform coating of the polymer/catalyst mixture. No gross structural changes (cracking, peeling, etc.) occurred on hydrogenation.

CONCLUSION

A polymer-based hydrogen getter was prepared and was shown to consume hydrogen at low temperature and pressure. One particularly interesting feature of the new getter is its efficacy in air-containing environments without detectable production of water. A variety of fabrications were demonstrated. Work on optimizing the specific formulation and on improving fabrication techniques is in progress.

ACKNOWLEDGEMENT

This work was performed at Sandia National Laboratories under DOE contract #DE-AC04-76P00789. J. Kawola provided technical assistance.

REFERENCES AND NOTES

1. T. A. Giorgi, B. Ferrario, B. Storey, J. Vac. Sci. Technol., A, 3 (2), 417 (1985).

2. R. L. Courtney and L. A. Harrah, J. Mater. Sci. 12, 175 (1977). D. R. Anderson, R. L. Courtney, L. A. Harrah, U. S. Patent No. 3 896 042 (22 July 1975) and U. S. Patent No. 3 963 826 (15 July 1976).

3. H. M. Smith, personal communication.

4. L. R. Gilliom, presented at the 1987 ACS Spring Meeting, Denver, CO, 1987 (manuscript in preparation).

5. The catalyst was prepared by literature methods: R. H. Crabtree, S. M. Morehouse, J. M. Quirk, Inorg. Synth. 24, 173 (1986).

6. R. Crabtree, Accts. Chem. Res. 12, 331 (1979) and references therein.

7. Recent evaluation of samples containing 1 wt % catalyst indicate that even faster initial hydrogen uptake rates can be obtained at lower loadings. The observed capacity, however, is reduced as compared to the 9 wt % material.

SURFACE CHEMICAL CHARACTERIZATION OF INTERNAL INTERFACES GENERATED WITHIN
THIN-FILM FE-TI HYDRIDES

JEFFREY H. SANDERS AND BRUCE J. TATARCHUK
Department of Chemical Engineering, 230 Ross Hall, Auburn University, AL
36849

ABSTRACT

FeTi is considered an excellent candidate for the reversible storage of
hydrogen and has been studied extensively in an attempt to understand the
bulk activation needed for this material before use. Segregation of TiO_2 to
the surface has been noted to occur during activation explaining a slight
loss of efficiency per hydride cycle, however, characterization by a host of
bulk and surface sensitive techniques has not revealed the cause of this
decomposition process.
10 nm FeTi samples were prepared in a UHV evaporator both with and
without palladium coatings. Post treatment characterization was performed
with backscatter conversion electron Mossbauer spectroscopy (CEMS), XPS and
SIMS. CEMS is a powerful tool for providing stoichiometric, electronic,
magnetic, chemical, and particle size information of iron at depths down to
100 nm. XPS and SIMS are useful to gain quantitative and chemical state
information from the topmost 2 nm and the topmost monolayer, respectively.
Activation treatments consisted of annealing at 573K and 623K followed by
reduction at 573K. Results indicate that ppm levels of H_2O in H_2 are
sufficient to decompose the FeTi alloy and produce TiO_2 and Fe metal domains
at the surface. Also, at 573K in vacuum, a solid-state reaction was found to
occur between Fe oxides and FeTi to produce Fe metal and TiO_2. The Pd-FeTi
interface was probed with CEMS and the results demonstrate hydrogen
dissociation and migration in the absence of alloy decomposition. Our
approach uses nondestructive-depth profiling of non-Pd coated FeTi samples
along with interfacial information from Pd-FeTi specimens to obtain unique
insight into the decomposition process.

INTRODUCTION

FeTi is an interesting alloy capable of storing 37% more hydrogen
per volume than liquid hydrogen [1]. To achieve this performance, the alloy
must first be subjected to an activation procedure which typically consists
of a heat treatment (673K) followed by exposure to ca. 65 atm of hydrogen
[2]. This activation procedure has been determined to occur from surface
decomposition of the FeTi alloy to produce hydrogen dissociators, e.g.
metallic Fe domains, and TiO_2. These products allow hydrogen incorporation
into the bulk, however, decrease the total amount of effective storage
material. The decomposition is triggered by a thermodynamic driving force
for FeTi to react with O_2/H_2O impurities in the hydrogen charging gas to form
an oxide [3-6].
This study was initiated to gain insight into the chemical phenomena
that arise at the extreme surface, i.e. topmost 10 nm, when FeTi is exposed
to hydriding conditions. Three surface science techniques were used to give
a non-destructive characterization of a 10 nm FeTi specimen prepared by
evaporation of successive 1.0 nm layers of iron and titanium onto quartz
substrates. These techniques were Backscatter Conversion Electron Mossbauer
Spectroscopy (CEMS), XPS and SIMS. CEMS is ideally suited for this study
because of its inherent capability to provide stoichiometric, electronic,
magnetic and chemical information of ^{57}Fe nuclei at depths to ca. 100 nm.
XPS and "static" SIMS are useful to gain quantitative and chemical
information from the topmost 2 nm and topmost monolayer, respectively. The
use of a 10 nm thin-film specimen with 93% isotopically enriched ^{57}Fe was
necessary to provide surface specific information in reduced collection

times. Also, hydriding conditions were simulated with reducing conditions
to investigate early stages of decomposition including precursor compounds.
The approach used verified the segregation phenomena proposed by Schlapbach
and Riesterer which consists of FeTi decomposition into TiO_2 at the surface
and Fe metal domains near the surface [3].
Our results obtained from the investigation discussed above demonstrate
the need for further investigation of the FeTi system which includes the use
of palladium as a surface coating and the determination of the mechanism
involved with a solid-state reaction whereby Fe_2O_3 and FeTi react to yield
TiO_2 and Fe metal domains. CEMS was ideally suited for both these endeavors
because it can probe below the specimen surface and relay local and
environmental information concerning ^{57}Fe nuclei at internal interfaces or
regions. A palladium coating was found to deter surface decomposition by
effectively prohibiting O_2/H_2O impurities from coming in contact with the
FeTi alloy while still allowing hydrogen to diffuse into the lattice. The
solid-state reaction noted above is currently under investigation. A
description of equipment and experimental results will now be given.

EXPERIMENTAL

EQUIPMENT

CEMS data were collected in an UHV chamber equipped with 7 spiraltron
electron multipliers operated at ca. 10^{-10} torr. Spectra were obtained
simultaneously in the constant acceleration mode and summed to increase
effective counting rates. A 200 mCi $^{57}Co/Pd$ source was used with positive
velocity defined as the source approaching the absorber and zero velocity
referenced to the centroid of a metallic iron spectrum. Further details of
the spectrometer and the data fitting routine are described elsewhere [7].
XPS was performed using a Leybold-Heraeus LHS-10 spectrometer operated
at base pressures of ca. 10^{-10} torr. An aluminum anode (1487 eV radiation)
was used for these studies. Lattice oxygen (@ 531.0 eV) within the specimen
was used to calibrate the work function of the analyzer.
SIMS was performed using a Leybold-Heraeus QMG-511 quadrupole mass
spectrometer with a sweep range of 0-250 amu. The ion gun rastered Ar^+ ions
over a 1 cm^2 area on the specimen surface at a base pressure of 2 X 10^{-10}
torr. An ion current of 2 X 10^{-9} A was used for 2 minutes to ensure a
"static" SIMS scan which minimizes chemical changes induced by the ion beam.
All layered samples were prepared in a high-vacuum evaporator equipped
with two-3 KW electron beam guns with three source crucibles. A calibrated
Inficon XTC crystal monitor and a pneumatic shutter assembly were used to
measure evaporation rates of ca. 0.1 nm/s and also to control the final film
thickness. The sample mount is equipped with heating capabilities and all
3.81 cm diameter quartz substrates were cleaned and annealed to 773K before
evaporation.
To ensure cleanliness of the samples, a high-vacuum sample transporter
was used to transfer the samples under vacuum (ca. 10^{-8}) among the various
analytical chambers. This device allowed multitechnique analyses without
exposing the specimen to significant amounts of O_2, H_2O or other background
gases.

Sample Fabrication and Treatments

Samples were prepared by evaporation of successive 1.0 nm layers of Ti
(99.9+%, AESAR) and ^{57}Fe (93%, Oak Ridge National Laboratory) at 0.1 nm/s
until a total of 10 nm were applied. Subsequent experiments consisted of an
additional application of a 5.0 nm Pd overlayer.
After fabrication, an FeTi sample was annealed at 573K for 1 hr and 623K

for 3 hrs. Subsequent treatments consisted of reduction in 1 atm of H_2 with ppm levels of O_2/H_2O impurities for 45 minutes at 573K, followed by a brief 3 minute exposure to air at room temperature. CEMS and XPS analyses followed the fabrication, annealing and reduction processes with an additional static SIMS being performed after reduction. The Pd-coated sample was subjected to similar treatments consisting of a UHV annealing at 623K for 3 hrs followed by reduction as above. CEMS and XPS analyses were again performed after each treatment.

RESULTS

Figure 1 presents CEMS spectra for the uncoated and Pd-coated samples after various treatments. Figure 2 presents Fe $2p_{1/2-3/2}$ and Ti $2p_{1/2-3/2}$ scans for the uncoated sample. Fe and Ti XPS scans for the Pd-coated sample could not be obtained because the elements were below the sampling depth; however, a Pd $3d_{3/2-5/2}$ scan was taken and revealed only zero valent metal.

Uncoated Sample - CEMS

The CEMS spectra reveal singlets at -0.12 mm/s for the uncoated sample after fabrication and annealing treatments, (a) and (b), respectively. This position is indicative of FeTi as found in the literature [8,9] and suggests that Fe and Ti mix readily upon evaporation. SIMS depth-profiling of identically prepared samples further substantiated complete mixing.

Figure 1(c) shows the CEMS spectrum of the same specimen after reduction. The domination of the spectrum by a sextuplet with a hyperfine field of 316.3 kOe suggests decomposition of the FeTi alloy to produce Fe metal domains. The small hyperfine field as compared to a bulk Fe metal standard (330 kOe) was used to calculate a domain volume consisting of 8000 Fe atoms according to the theory of collective magnetic excitations [10].

Oxidation in air followed reduction and the CEMS spectrum is shown in Figure 1(d). Comparison of the fit spectra after reduction and after oxidation reveals a 20% decrease in area associated with the sextuplet and a slight increase in center of mass from 0.046 to 0.158 mm/s. These characteristics signify slight oxidation of the Fe domains and suggest that these domains are partially exposed or very near the surface.

Uncoated Sample - XPS

Fe $2p_{1/2-3/2}$ and Ti $2p_{1/2-3/2}$ XPS scans for the uncoated sample are shown in Figure 2. Scans for the sample as fabricated are shown in Figure 2(a) indicating domination of the surface by zero valent Fe and Ti. There is some surface Fe oxide noted from the high binding energy structure on the Fe peaks. The Fe:Ti ratio was calculated by standard techniques to be 4.4 reflecting Fe enrichment at the surface. An additional O 1s scan was performed and used to calculate an Fe:O ratio of 3.26 which, if all the oxygen is assumed present as Fe_2O_3, correlates to a 0.2 nm thick surface layer.

Annealing the sample to 573K for 1 hr created no change in the XPS scans from those discussed above; however, subsequent anneal at 623K for 3 hrs did create significant differences, Figure 2(b). The Fe peak lost the oxide structure indicating the presence of only a pure metallic state while most of the Ti peak area shifted +5 eV suggesting that Ti is now in the form of TiO_2 with only a small amount still metallic, ca. 25%. The Fe:Ti ratio remained essentially the same at 4.6 which demonstrates that diffusion of the metals has not occurred.

Fe and Ti scans after reduction are shown in Figure 1(c) and demonstrate segregation of Ti to the surface has now occurred. The Fe:Ti ratio has been reduced to 0.9 with the Ti scan indicating only TiO$_2$ present within the XPS sampling depth while the Fe scan remains metallic. Static SIMS was performed on the sample at this point allowing a relative concentration of Fe to Ti of 0.1 to be calculated. This value is sufficiently lower than the 0.9 ratio determined from XPS suggesting that the domains exist mostly in near surface regions.

XPS scans of the sample after brief exposure to air are shown in Figure 2(d). The Fe peaks shifted toward higher binding energies associated with Fe$_2$O$_3$ yet a significant portion remained metallic. The Ti peaks remained unchanged as did the Fe:Ti ratio of 0.9.

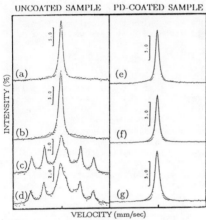

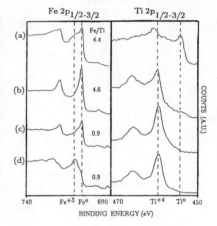

Figure 1: CEMS Spectra for Uncoated and 5.0 nm Pd-Coated FeTi Samples. Uncoated: (a) As fabricated, (b) Annealed at 623K, (c) Reduced, and (d) Exposed to air. Pd-Coated: (e) As fabricated, (f) Annealed at 623K, and (g) Reduced.

Figure 2: Fe 2p$_{1/2-3/2}$ and Ti 2p$_{1/2-3/2}$ XPS Scans for an Uncoated FeTi Sample. (a) As fabricated, (b) Annealed at 623K, (c) Reduced, and (d) Exposed to air.

Pd-Coated Sample - CEMS

Figure 1(e) shows spectra for the sample after fabrication and as for the uncoated sample discussed above, the peak at -0.12 mm/s is indicative of FeTi.

Annealing the sample to 623K for 3 hrs yielded no change as can be seen in Figure 1(f) in which the spectrum is still a singlet. It should be noted that the areas of all spectra for the Pd-coated sample are less than those of the uncoated sample and may be attributed to the inability of low energy electrons to penetrate the Pd overlayer.

The spectrum in Figure 1(g) parallels that of Figure 1(c) (both obtained after reduction) and as can be seen, the spectrum in this case remains a singlet. The Pd overlayer has effectively prevented impurity O$_2$/H$_2$O in the H$_2$ gas from coming in contact with FeTi so that the thermodynamic driving force to decompose the alloy is eliminated.

DISCUSSION

Figure 3 presents schematic profiles of the sample after various treatments based on combined information from CEMS, XPS and SIMS. Figure 3(a) depicts the uncoated sample after fabrication and shows the bulk to be FeTi with a 0.2 nm surface layer of Fe_2O_3. The Fe_2O_3 surface layer was easily detected with XPS, however, it was undetected with CEMS. This is due to the integral nature by which CEMS data is acquired over the entire sample depth so that the surface monolayer contribution is effectively overshadowed. The source of the oxygen causing the Fe_2O_3 layer was background impurities impinging on the sample during transport.

Figure 3(b) represents the sample after being annealed at 623K for 3 hrs and still shows the bulk to be FeTi as evidenced with CEMS; however, the surface has undergone an oxygen transfer from Fe to Ti as evident from the XPS scans. This solid-state reaction is a pathway for reactivation of bulk FeTi alloys used for hydrogen storage so that clean Fe metal can be produced to dissociate and diffuse hydrogen into the lattice below. Verification of this reaction as proceeding through a solid-state mechanism as opposed to gas phase intermediates is underway.

The reduction of the uncoated sample is depicted in Figure 3(c). CEMS gives positive evidence for Fe domain formation as a product of Ti segregation from bulk FeTi to react with O_2/H_2O impurities and form TiO_2 at the surface (evident from XPS scans). These results verify the activation procedure of bulk FeTi as proposed by Schlapbach and Riesterer whose investigation utilized evidence from XPS, H_2/D_2 exchange, H_2 chemisorption and magnetic susceptibility measurements on bulk samples [3]. Other researchers hold similar bulk activation schemes based on non-Fe metal particle mechanisms; however, under the reducing conditions we used, Fe domains were evident.

Figure 3(d) is a profile of the uncoated sample after exposure to air demonstrating oxidation of Fe in the surface and near surface regions. This treatment was performed as further evidence for locating the depth of the Fe domains and along with XPS and SIMS results we may conclude with certainty that the Fe domains are near the surface.

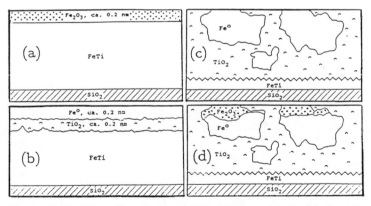

Figure 3: Profiles of an Uncoated FeTi Sample. (a) As fabricated, (b) Annealed at 623K, (c) Reduced, and (d) Exposed to air.

The addition of a 5.0 nm Pd coating was effective for preventing decomposition of FeTi when exposed to reducing conditions. CEMS proved to be unique for verifying this because it can probe beneath the overlayer to give information in a non-destructive fashion. Of course, hydriding conditions are much more severe than the conditions used, typically consisting of at least 60 atm of H_2 charging gas, so that performance of a Pd coating under hydriding conditions must still be evaluated.

CONCLUSIONS

The mechanism proposed for bulk activation of FeTi whereby TiO_2 and Fe metal domains are formed at the surface is substantiated by this investigation. The use of a thin-film system under controlled conditions and analyses with CEMS, XPS and SIMS highlighted the surface phenomena that occur. The use of a Pd overlayer was found to be effective for eliminating surface decomposition, however, requires further investigation under actual hydriding conditions. Also, a solid-state reaction involving the decomposition of FeTi by Fe_2O_3 was noted and further investigation into the mechanism behind this reaction undertaken.

REFERENCES

1. K. Ramakrishna, S.K. Singh, A.K. Singh and O.N. Scrivastava, in Progress in Hydrogen Energy, edited by R.P. Dahiya (D. Reidal Publishing Co., Holland, 1987), p. 84.

2. J.J. Reilly and R.H. Wiswall, Inorg. Chem. 13, 218 (1974).

3. L. Schlapbach and T. Riesterer, Appl. Phys. A 32, 169 (1983).

4. T. Schober and D.G. Westlake, Scr. Metall. 15, 913 (1981).

5. T. Schober, J. Less-Common Met. 89, 63 (1983).

6. D. Khatamian, G.C. Weatherly, F.D. Manchester and C.B. Alcock, J.Less-Common Met. 89, 71 (1983).

7. J.S. Zabinski and B.J. Tatarchuk, Nuclear Instr. Methods in Phys. Res. B (accepted for publication, 1987).

8. L.J. Swartzendruber, L.H. Bennett and R.E. Watson, J. Phys. F: Metal Phys. 6, L331 (1976).

9. D. Finkler, H.-G. Wagner, S.J. Campbell, N. Blaes and U. Gonser, Z. Phys. Chem. 145, 147 (1985).

10. S. Morup, J.A. Dumesic and H. Topsoe, in Applications of Mossbauer Spectroscopy, Vol. II, edited by R.L. Cohen (Academic Press Inc., New York, 1980) pp. 1-48.

AUGER ELECTRON SPECTROSCOPY OF AMORPHOUS CU-30 at.% ZR ACTIVATED FOR CATALYSIS

FABIANO VANINI, MEHMET ERBUDAK, GERNOT KOSTORZ AND ALFONS BAIKER*
Institut für Angewandte Physik, ETH Zürich, CH-8093 Zürich, Switzerland
*Technisch-Chemisches Laboratorium, ETH Zürich, CH-8092 Zürich, Switzerland

ABSTRACT

Amorphous $Cu_{70}Zr_{30}$ ribbons heat treated at 473 K in a hydrogen atmosphere yield a very active catalyst for the hydrogenation of ethene and butadiene. The catalytic activity is due to finely dispersed Cu particles formed on a surface containing numerous cracks. Auger depth profiles and scanning electron micrographs reveal details of the surface chemistry and morphology. The observations are related to the free-energy curve of bulk amorphous Cu-Zr.

INTRODUCTION

After suitable activation, some of the amorphous alloys prepared in ribbon form by rapid quenching (see e.g. [1]) show a high catalytic activity (see e.g. [2] for a summary). Catalytic activation of amorphous Cu-Zr alloys has been studied by Yamashita et al. [2] and by Shibata et al. [3]. Baris [4] has investigated the catalytic potential of several Fe-Zr and Cu-Zr ribbons. Some of these alloys were studied with surface analytical methods [5]. In the present paper, Auger depth profiles of as-received and activated amorphous $Cu_{70}Zr_{30}$ are reported, and the strong segregation of Cu after hydrogen treatment is related to the particular surface structure (cracking during hydrogenation) and the inherent thermodynamic instability of amorphous Cu-Zr.

EXPERIMENTAL

The samples were prepared in Ar by the melt spinning technique (see [6]). Auger electron spectroscopy (AES) measurements were performed in an UHV apparatus with a total pressure near 10^{-8} Pa. A hybrid gun [7] produced electrons for AES (beam size ~ 300 μm) and ions for cleaning the surface by sputtering. A cylindrical mirror analyzer (CMA) coaxial with the gun (resolution $\Delta E/E \sim 0.8\%$) and an electron multiplier were employed to detect the secondary electrons. Sputtering was performed with 2000 eV Ar^+ ions with a current density of 1 μA/mm^2. Auger electron spectra were measured in the first-derivative mode employing a phase-sensitive detector and modulating the pass energy of the CMA with 2.5 V (peak-to-peak). Depth profiles were obtained by subsequent cycles of sputtering and measurements of Auger electron spectra. Details of the apparatus have been described elsewhere [8]. The activated samples were hydrogen treated at 200°C for 16 hours in a special vacuum chamber, evacuated to 10^{-2} Pa and filled with 1 bar of hydrogen, passing through a liquid-nitrogen cooled trap. After cooling down to room temperature, they were quickly transferred into the UHV apparatus. Scanning electron micrographs of some of the samples were taken in a JEOL JSM 840.

RESULTS

Auger depth profiles were obtained recording the $Cu-M_{2,3}M_{4,5}M_{4,5}$, the $Zr-M_{4,5}N_1N_{2,3}$ and the $O-KL_{2,3}L_{2,3}$ transitions as a function of sputtering time. The sputtering time scale was transformed to a depth scale using

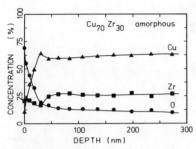

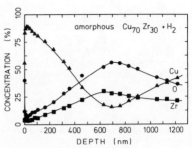

Fig. 1   AES depth profile of as-received amorphous $Cu_{70}Zr_{30}$.

Fig. 2   AES depth profile of amorphous $Cu_{70}Zr_{30}$ annealed for 16 hours at 473 K in H_2 at 1 bar.

sputtering yields of four for copper and one for zirconium. The sputtering rate was about 0.2 nm/s (see e.g. [9]).

In Fig.1 the concentrations of copper, zirconium and oxygen in as-received amorphous $Cu_{70}Zr_{30}$ are shown as a function of depth. In an initial layer about 15 nm deep, a strongly lowered copper concentration, a slight reduction of zirconium and a large oxygen contamination are measured. In this surface region, the ratio of Cu to Zr is clearly lowered indicating a zirconium segregation to the surface (see also [10]). The zirconium segregation is attributed to the large oxygen affinity of zirconium. Layers deeper than 70 nm show constant concentrations. The values were used to calibrate the atomic ratio of Cu and Zr to give 70/30. A large amount of oxygen (at least 12 at.% O) is always present in the films. The oxygen calibration is based on AES and XPS measurements on the same sample, see [5]. This oxygen pick-up is unlikely to have occurred during the preparation of the sample, hence it must have happened afterwards. The chemical states of Cu and Zr were studied by XPS and UPS on the same sample after prolonged Ar$^+$ sputtering [5]. From the relative intensity of the two pairs of 3d-peaks for metallic and oxidized Zr it is possible to estimate that up to one third of the zirconium present in the sample is oxidized. The $Cu-2p_{3/2}$ spectra do not yield any information on the chemical state of Cu because this level does not show a measurable chemical shift or a change in the line shape during oxidation. Attempts to reduce the oxygen content of the sample by different cycles of annealing and sputtering led either to no appreciable result or to crystallisation.

An AES depth profile of amorphous $Cu_{70}Zr_{30}$ treated with hydrogen is shown in Fig.2. A copper-rich layer about 400 nm thick is formed on the surface. The copper concentration reaches a maximum of 90% at a depth of ~ 20 nm and decreases afterwards to reach a minimum around 700 nm. At this depth, Zr and O contents, appearing at a constant ratio, both have a maximum. At depths exceeding 700 nm, the Zr and Cu concentrations tend to reach the bulk values of the untreated sample. Cu has thus diffused to the surface, leaving a zirconium-enriched region below. The estimated oxygen content implies that most of the Zr is present as ZrO_2. This is confirmed by XPS measurements [5]. Possibly, oxygen present as impurity in H_2 (≤ 1 ppm) was bound to zirconium during the hydrogen treatment. X-ray diffraction of hydrogen-treated amorphous $Cu_{70}Zr_{30}$ confirms that the segregated copper has crystallized. No Bragg peaks appeared after the copper layer had been etched off, i.e. the underlying material is amorphous. No copper segregation was observed for samples heated at 473 K for 16 hours in vacuum. The presence of hydrogen is essential for the process of copper enrichment at the surface. Crystalline $Cu_{70}Zr_{30}$ and amorphous and crystalline $Cu_{30}Zr_{70}$ were also

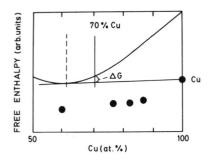

Fig. 3 Schematic of the free enthalpy of Cu-rich amorphous alloys in comparison with pure crystalline Cu and some intermetallic compounds (indicated by dots).

studied and showed no comparable effects (see [11]).

DISCUSSION

Among all the samples studied, only the amorphous $Cu_{70}Zr_{30}$ alloy after hydrogen treatment at 473 K showed an enhancement of catalytic activity [4]. This is due to the enrichment of Cu at the surface, which becomes possible when hydrogen is present. If it is assumed that at this relatively low temperature where the formation of intermetallic compounds is kinetically impossible, nucleation and growth of Cu is allowed (Cu is known to diffuse more rapidly than Zr, see e.g. [1]), a metastable equilibrium between Cu (crystalline) and amorphous Cu-Zr with a lower Cu concentration can be constructed as schematically illustrated in Fig.3, and precipitation of Cu from an amorphous Cu-Zr matrix becomes possible. It is presently unknown whether any precipitation occurs in the "bulk" of the samples, but it does occur at the surface at 473 K. The presence of oxygen reduces the effective metallic Zr content of the alloy and will thus enforce the basis of the present argument as long as some Zr remains unoxidized. The absence of Cu segregation on $Cu_{70}Zr_{30}$ annealed in vacuum can be attributed to the ZrO_2 layer on the surface of these as-received samples which hinders Cu diffusion and nucleation. As already pointed out by Shibata et al. [3], hydrogen treatment introduces fresh surfaces in cracks occurring because of the appreciable diffusion of

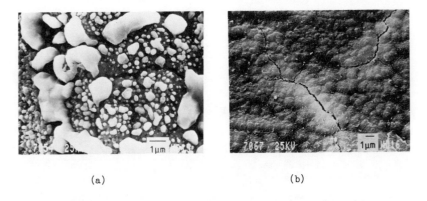

(a)   (b)

Fig 4 Scanning electron micrographs of a hydrogen treated sample (a) without etching, (b) after etching off most of the Cu surface layer.

hydrogen into the sample (see e.g. [1]) and the accompanying strains. Fig.4a shows a scanning electron micrograph of amorphous $Cu_{70}Zr_{30}$ after hydrogen treatment. Numerous Cu particles have formed on the freshly provided surfaces, grown and coalesced. The size distribution and arrangement of these particles is very irregular. The AES depth profile has a bad lateral resolution (~ 300 μm) and thus averages over these details. Fig.4b shows a similar micrograph after most of the Cu has been removed by etching in HNO_3 (13 mol.%). Cracks and pores are visible now, and these must be the sites where the decomposition reaction started.

CONCLUSION

The catalytic activation of amorphous $Cu_{70}Zr_{30}$ can be attributed to the chemical driving force for the metastable decomposition of an amorphous Cu rich alloy into Cu and amorphous Cu-Zr with less Cu. A subtle interplay of hydrogen embrittlement, oxidation and phase decomposition at fresh surfaces seems to be necessary for the observed activation of amorphous $Cu_{70}Zr_{30}$.

ACKNOWLEDGEMENTS

The authors are grateful to G. Heinzen (Lonza AG, Switzerland) for the samples used in this study, to R. Bormann (University of Göttingen) for useful discussions, and to Ph. Aebi, A. Meier and P. Wägli for their help with the experiments.

REFERENCES

1. R.W. Cahn, in Physical Metallurgy, 3rd ed., (edited by R.W. Cahn and P. Haasen), pp.1779-1852. Elsevier, Amsterdam (1983).

2. H. Yamashita, M. Yoshikawa, T. Kaminade, T. Funabiki and S. Yoshida, J. Chem. Soc., Faraday Trans. 1, 82, 707 (1986).

3. M. Shibata, Y. Ohbayashi, N. Kawata, T. Masumoto and K. Aoki, J. Catal. 96, 296 (1985).

4. H. Baris, Katalytische Studien an amorphen Fe-Zr- und Cu-Zr-Legierungen, Dissertation ETH Nr. 8227 (1987).

5. F. Vanini, St. Büchler, Xin-nan Yu, M. Erbudak, L. Schlapbach and A. Baiker, Surf. Sci. 189/190, 1117 (1987).

6. A. Baiker, H. Baris, and H.J. Güntherodt, Applied Catalysis 22, 389 (1986).

7. E.B. Bas, E. Gisler and F. Stucki, J. Phys. E17, 405 (1984).

8. E. Gisler and E.B. Bas, Vacuum 36, 715 (1986).

9. H.H. Andersen and H.L. Bay, in Sputtering by Particle Bombardment I, Topics of Applied Physics (edited by R. Behrisch), Vol.47, pp. 145-218. Springer, Heidelberg (1981).

10. P. Sen, D.D. Sarma, R.C. Budhani, K.L. Chopra, and C,N,R, Rao, J. Phys. F: Met. Phys. 14, 565 (1984)

11. F. Vanini, M. Erbudak, G. Kostorz and A. Baiker, to be published.

SUBSTANTIALLY TRANSPARENT Pt, Pd, Rh, AND Re FILMS:
PREPARATION AND PROPERTIES

D. E. ASPNES* AND A. HELLER**

*Bellcore, Red Bank, N.J. 07701-7020

**AT&T Bell Laboratories, Murray Hill, N.J. 07974

ABSTRACT

Films of Pt, Pd, Rh, and Re with metal volume fractions of 0.3 to 0.5 have been prepared by mass-transport-limited photoelectrodeposition onto (001) p-InP photocathodes from ~5 x 10^{-5} M solutions of the metal ions in 1 M $HClO_4$. These films exhibit their normal catalytic activities (e.g., in hydrogen evolution) and have normal crystal structures, yet are substantially more transparent than equivalent dense films of the same metal loading per unit area. Effective-medium analysis of the spectroellipsometrically measured dielectric functions of these films shows that the anomalous transparency is due to microstructure: depolarization factors and metal packing fractions obtained by best-fit model calculations indicate dendritic (Rh), particulate (Pt, Pd), or platelet (Re) forms that are poorly interconnected in directions parallel to the surface, and whose dimensions are all small compared to the wavelength of light. Transmission electron micrographs confirm these results and reveal that these films consist of primary building blocks of ca. 5 nm crystallites that are organized into relatively loosely packed secondary structures. Potential applications of these films include the formation of efficient metallic-catalyst-coated photoelectrodes on poor-quality semiconductors.

INTRODUCTION

Highly transparent metal films are of potential interest for a variety of reasons. They allow optical spectroscopies to be used to study kinetics and mechanisms of heterogeneous catalysis, properties of adsorbed or low-dimensional phases, and modifications of semiconductor surfaces or transparent oxide electrodes by metals. They also increase efficiencies of systems as diverse as catalyst-mediated photoelectrochemical conversion processes and optical detectors. We recently described methods of preparing highly transparent supported metal films of Pt [1,2] and of Pd, Rh, and Re [3] by mass-transport-limited photoelectrodeposition from extremely dilute [~10^{-4} - 10^{-5} M] solutions of the appropriate metal ions under diffusion-limited conditions. The resulting films were up to 60 nm thick, had metal volume fractions between 0.2 and 0.5, yet were substantially more transparent than dense films of equivalent metal loadings per unit area.

In this paper, we summarize and extend our previous results [1-3], describing methods of preparation, methods of analysis, and physicochemical properties of these anomalously transparent Pt, Pd, Rh, and Re films. The connection between microstructure, which gives rise to their high optical transmittances, and optical properties is discussed. With the aid of simple analytic effective medium models, the microstructure that optimizes these favorable optical properties can be deduced [4]. By least-squares fitting these effective medium models to spectroellipsometric (SE) data, we determine the macroscopic, or average, microstructural properties (packing frac-

tions, connectedness, and shapes of the metal particles) directly and nonde-
structively. Details of the microstructures themselves are obtained by
transmission electron microscopy (TEM). The TEM micrographs reveal that the
films consist of primary building blocks of ca. 5 nm metal crystallites that
are organized into relatively loosely packed secondary structures. Although
the shapes and sizes of these secondary structures are different for the
different metal films, the common characteristics of high optical transmittances and relatively low metal packing fractions suggest that control of
microstructure by mass-transport-limited deposition, accompanied by limiting
the kinetics of interfacial electron transfer between the solution and the
deposited metal film, may be a general way of producing the open structures
needed for high optical transmittance.

EXPERIMENTAL

Microstructured metal films were photoelectrodeposited onto 0.05-0.10
cm^2 (100) faces of p-doped (6 x 10^{16} cm $^{-3}$ Zn) single crystal InP samples of
approximate dimension 3 x 3 x 1 mm^3. Ohmic back contacts were formed by Au-
Zn diffusion at 420C. The samples were then encapsulated in an epoxy resin
except for their front surfaces. Immediately prior to an experiment, the
front surface of the selected sample was degreased, stripped in 1:1
$NH_4OH:H_2O$, swabbed with 0.05 vol. % Br_2 in methanol, rinsed with methanol,
rinsed with 1 M HF, rinsed with H_2O, and dried in N_2 [3].

Photoelectrodeposition was done from 1 M stock solutions of $HClO_4$ (Alfa
Ultrapure) to which appropriate amounts of $PtO_2 \cdot nH_2O$ (prepared as described
in ref. [2]), $ReCl_3$ (Alfa), $RhCl_3$ (Aldrich), or $PdCl_2$ (Engelhard) had been
added to reach molar concentrations of 1.2 x 10^{-5} (Pt) or 3-6 x 10^{-5} (Re,
Rh, Pd). Each solution was purged of O_2 by passing N_2 through it before
use. O_2 recontamination was avoided by keeping the solution surface in an N_2
ambient. Metal was deposited on the p-InP photocathodes at potentials (with
respect to the standard hydrogen electrode, SHE) of 0.236 V (Pt), 0.136 V
(Pd, Re), or 0.286 V (Rh) while maintaining a constant irradiance of no more
than 0.4 mW cm^{-2} from a tungsten-halogen lamp. Deposition was slow enough to
allow H_2 to dissolve in the solution and to diffuse to its surface. As
mass-transport-limited deposition was essential to the formation of these
open microstructures, the solutions were maintained stationary with respect
to the electrode unless stated otherwise. During one of the Pt depositions,
ellipsometric spectra were obtained at occasional intervals by holding the
photocathode in the "dark" at +0.436 V SHE for the 15 min necessary to
obtain each spectrum. Changes in transparency were periodically monitored
by determining the light-limited photocurrent with a 20 mV s^{-1} potential
sweep between -0.5 and +0.2 V SHE. Metal depositions during each of these
procedures were insignificant.

The average microstructural properties of these films were determined
from least-squares model analyses of SE data, and the detailed microstructural properties were obtained from TEM micrographs. We used a rotating-
analyzer spectroellipsometer that has been described in detail elsewhere
[5]. Transmission electron micrographs of lifted-off portions of the films
were obtained with JEOL (JEM 200A) and Philips (Model 400) transmission
electron microscopes operated at 120 kV. Details of the lift-off procedure
have also been described elsewhere [2].

THEORY AND PRINCIPLES

The near transparency of photoelectroplated metals was discovered in studies of light-assisted hydrogen evolution on platinized semiconducting photocathodes [6]. The quantum yields of these photocathodes did not decline significantly even for Pt films 50 nm thick. This was surprising, as dense Pt films this thick would have absorbed 99% of the incident visible and near-IR light. Consequently, these results indicated that absorption and reflection losses in moderately thick metal films were not inevitable. Further analysis, as discussed below, revealed that this anomalous transparency was due to microstructure, or spatial inhomogeneities on a scale small compared with λ, the wavelength of light. Analysis also showed that these favorable optical properties could be controlled within limits by controlling the microstructure.

Microstructure in the form of porosity clearly reduces light absorption simply because porous films contain less light-absorbing material than dense films of equivalent metric thicknesses. However, if the films could be fabricated of metal particles whose dimensions were small compared to λ, and if these particles could be electrically isolated from each other in the direction of the polarization of the light, then upon illumination a depolarization or screening charge would develop at the surface of each particle. This depolarization charge would act to screen, or exclude, the optical field, and thus any coupled power, from the particles themselves. This shielding effect would render the metal fraction far less important than that expected from volume-fraction arguments alone.

Assuming that the ambient medium permeates all interstices between metal particles, the relationships among the microstructure of a film, its measured or apparent dielectric function, $\epsilon = \epsilon_1 + \epsilon_2$, and the dielectric functions $\epsilon_m = \epsilon_{m1} + i\epsilon_{m2}$ and $\epsilon_a = \epsilon_{a1} + i0$ of its metal and ambient constituents, can be expressed quantitatively by effective medium theory. For a two-component composite [7]

$$\frac{\epsilon - \epsilon_H}{\epsilon + \kappa\epsilon_H} = f_m \frac{\epsilon_m - \epsilon_H}{\epsilon_m + \kappa\epsilon_H} + f_a \frac{\epsilon_a - \epsilon_H}{\epsilon_a + \kappa\epsilon_H}, \tag{1}$$

where f_m and f_a are the volume fractions of metal and ambient respectively ($f_m + f_a = 1$), ϵ_H is a "host" dielectric function, and κ is a screening parameter. For a cermet, or coated metal particle, microstructure $\epsilon_H = \epsilon_a$ (Maxwell Garnett effective medium theory.) For an aggregate, or random, metal particle - "ambient particle" mixture, $\epsilon_H = \epsilon$ (Bruggeman effective medium approximation.) If the metal phase is connected, then $\epsilon_H = \epsilon_m$ (a variant of the Maxwell Garnett effective medium theory.) The screening parameter κ assumes the values 0 for maximum screening (laminar microstructure, polarization perpendicular to the laminations), 1 for a two-dimensionally isotropic (cylindrical) microstructure, 2 for a three-dimensionally isotropic (spherical) microstructure, and ∞ for no screening (filamentary microstructure, polarization parallel to the filaments.)

In the limit of small metal packing fractions f_m and large metal dielectric functions $|\epsilon_m| \gg \epsilon_a$, Eq. (1) reduces to

$$\epsilon \cong \epsilon_a[1 + f_m(1 + \kappa)]. \tag{2}$$

Note that ϵ_m does not appear in Eq. (2), meaning that in the perfect-screening limit absorptive losses have vanished completely. Also, the effective dielectric function of the film is increased above ϵ_a because the photons have been "squeezed" into the interstices between the metal parti-cles, thereby increasing the apparent refractive index of the ambient fluid. In principle, f_m, κ, and the film thickness d can be chosen to perform an antireflective function and reduce optical losses to nil.

In practice, finite values of $|\epsilon_m|$ always result in some loss. For a constant metal loading thickness d_0 (thickness of the film with void frac-tion removed), and neglecting interference effects, the film transmittance increases with thickness (dilution.) In the optimum infinite-dilution limit it can be shown that [3]

$$\lim_{d \to \infty} (\alpha d) = \frac{2\pi n_a d_0}{\lambda} \frac{\epsilon_a \mathrm{Im}(\epsilon_m)(1 + \kappa)^2}{|\epsilon_m + \kappa \epsilon_a|^2},$$ (3)

where α is the absorption coefficient of the film and $n_a = (\epsilon_a)^{1/2}$ is the index of refraction of the ambient. For metals, for which $\mathrm{Re}(\epsilon_m) < 0$, Eq. (3) is minimized for $\kappa = 0$, the maximum-screening configuration consisting of parallel metal lines oriented perpendicular to the polarization vector of the incident radiation [4].

In principle, established procedures [8] can be used to determine the microstructural parameters f_m (f_a) and κ from data at any two wavelengths for any single film, but as a consistency check our microstructural parame-ters were determined by analysis of complete 250-point spectra from 1.5 to 5.5 eV. Our data were found to be consistent with the aggregate microstruc-ture described by the Bruggeman effective medium approximation ($\epsilon_H = \epsilon$ in Eq. (1)).

RESULTS AND DISCUSSION

1. Analysis of ellipsometric spectra.

The energy dependences of the real and imaginary parts of the pseudo-dielectric function, $<\epsilon>$, of the p-InP/Pt film/1 M $HClO_4$ system measured periodically during Pt deposition are shown in Fig. 1. The spectra labeled (a) correspond to the film-free InP sample in the 1 M $HClO_4$ plating solu-tion. Spectra labeled (b)...(s) represent the p-InP/Pt film/1 M $HClO_4$ system for film thicknesses from 0.27 to 27.7 nm at approximately equal intervals in log d. The E_1 and E_2 structures of the InP substrate, at 3.3 and 5.1 eV, respectively, appear in every $<\epsilon>$ spectrum. This can only happen if a substantial portion of the incident radiation returns through the film after being back-reflected from the InP/Pt film interface. Because a dense 27.7 nm Pt film would transmit only about 5% of the incident light, the anomalous transparency of these films is already evident.

The macroscopic parameters f_m, κ, and d that give rise to this trans-parency were investigated quantitatively by the standard approach [8]. The optical properties of the p-InP/metal-film/ambient system were expressed by the three-phase (substrate-film-ambient) Fresnel reflectance equations,

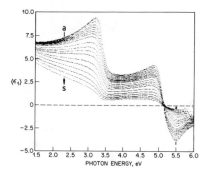 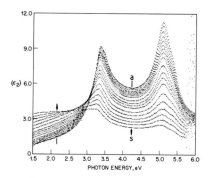

Fig. 1. Energy dependences of the real (left) and imaginary (right) parts of the pseudodielectric functions $\langle\epsilon\rangle$ of the p-InP/Pt-film/HClO$_4$ system measured periodically during deposition. The metal-free spectrum is (a). Mean Pt film thicknesses from (b) to (s) increase from 0.27 to 27.7 nm at approximately equal intervals of log d (after [1]).

which give an analytic representation of $\langle\epsilon\rangle$ in terms of d and the dielectric function spectra ϵ_a, ϵ, and ϵ_s of the ambient, film, and substrate, respectively. Here, ϵ is represented analytically in terms of the microstructural parameters f_m and κ by Eq. (1). For ϵ_s, we used previously reported values for InP [9]. Values of ϵ_m for Pt were obtained from measurements on a dense film prepared by electron beam evaporation [2]. Values of ϵ_a for 1 M HClO$_4$ were found to be given accurately by multiplying previously reported dielectric function data for pure H$_2$O [10] by a factor of 1.032 [2]. The choice of ϵ_H is determined according to whether ϵ_a, ϵ, or ϵ_m gives the best representation of $\langle\epsilon\rangle$. We found from trial calculations that $\epsilon_H = \epsilon$ (Bruggeman effective medium approximation, aggregate microstructure) yielded lowest values of the least-squares residual δ by factors of about 3.

The wavelength-independent quantities d, f_m, and κ were then used as free parameters in a least-squares fit of the combined Fresnel and effective medium equations to $\langle\epsilon\rangle$. To the extent that the three-phase p-InP/Pt-film/ambient model accurately represents the actual physical situation, i.e., that the films are homogeneous and the substrate/film and film/ambient interfaces are abrupt, and to the extent that the dielectric function spectra ϵ_s, ϵ_a, ϵ_m, and $\langle\epsilon\rangle$ are accurate, the free parameters d, f_m, and κ give an accurate __macroscopic,__ or average, summary of the actual microstructural details of the films.

The result of applying this procedure to one of the spectra of Fig. 1 is shown in Fig. 2. The dashed and solid curves show the $\langle\epsilon\rangle$ data for the p-InP/1 M HClO$_4$ (a) and ninth p-InP/Pt film/1 M HClO$_4$ (j) spectra, respectively. The dot-dashed curves, which overlap the solid curves over most of the accessible spectral range to the scale of Fig. 2, show the result of the best-fit model calculation with d = 4.9 $\pm$ 1.8 nm, f_m = 0.16 $\pm$ 0.09, and κ =

384

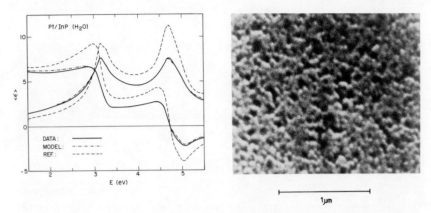

Fig. 2 (above, left.) Fit of the Bruggeman EMA to a typical $<\varepsilon>$ spectrum of Fig. 1. Dashed lines: p-InP/HClO$_4$ spectrum, (a); solid lines: p-InP/Pt-film/HClO$_4$ spectrum (j); dot-dashed line: best-fit model representation of (j) with d = 4.9 $\pm$ 1.8 nm, f$_m$ = 0.16 $\pm$ 0.09, and κ = 1.8 $\pm$ 0.2 (after [2]).

Fig. 3 (above, right.) Scanning electron micrograph at ~ 20 nm resolution of a Pt film photoelectrochemically deposited on p-InP as described in the text (after [2]).

1.8 $\pm$ 0.2. As the excellent agreement between calculation and experiment is obtained for both real and imaginary parts of $<\varepsilon>$ over an extended spectral range with only three parameters, we conclude that the three-phase aggregate-microstructure model accurately represents the physical situation and that the determined parameters provide an accurate macroscopic summary of the microscopic details of the films. This particular spectrum was chosen because it showed the largest residual δ (disagreement between model and data) of all spectra of Fig. 1.

2. Microstructural properties of deposited metal films.

In further experiments we deposited a number of Pt films under different conditions [2]. We discuss in detail the analysis of one such film, which was prepared by slowly stirring the plating solution to an extent equivalent to rotating the electrode at 25 rad s^{-1}. This had the effect of increasing mass transport during deposition, which, as shown below, increased the metal fraction in the film. This film was remarkable in that it was over 30 nm thick, yet its efficiency in light transport was 92% relative to that of the immersed InP electrode alone. A scanning electron micrograph, shown in Fig. 3, appeared to show that the platinum "particles" comprising the film were quite solid to the ~20 nm resolution of the scanning electron microscope, yet phase-contrast microscopy could not distinguish between bare and Pt-covered portions of the InP surface owing to the transparency of the Pt film.

SE measurements of the dielectric function of this film were consistent with the phase-contrast microscopy observations, revealing that the dielec-

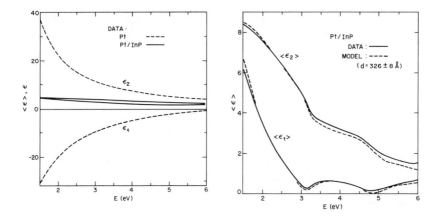

Fig. 4 (above, left.) Solid lines: ε spectrum of the Pt film of Fig. 3, measured in air. Dashed lines: reference ε spectrum of Pt obtained from an electron-beam-evaporated film (after [2]).

Fig. 5 (above, right.) Solid lines: <ε> spectrum of the p-InP/Pt film system of Fig. 3, measured in air. Dashed lines: best-fit model representation with d = 32.6 ± 0.8 nm, f_m = 0.50 ± 0.02, and κ = 0.80 ± 0.08 (after [2]).

tric function of the film was greatly reduced from that of dense Pt, as shown in Fig. 4. Here, the ε spectrum of the film was obtained by solving the three-phase model with d chosen to minimize substrate-related artifacts in ε [11]. The thickness obtained by this procedure was d = 33.7 ± 0.5 nm. A direct microstructural analysis of the <ε> spectrum by the standard approach [8] yielded d = 32.6 ± 0.8 nm, in good agreement with that obtained by direct solution of the three-phase model, and also yielded the microstructural parameters f_m = 0.50 ± 0.02, and κ = 0.80 ± 0.03. The excellent best-fit representation of the <ε> data is shown in Fig. 5. The microstructural analysis reveals that this film had a considerably greater metal packing fraction than that of Fig. 1, a result of higher mass transport rates during growth. The best-fit value of κ~1 in an aggregate microstructure indicates an essentially two-dimensional array of particles with some connectedness in the plane of the film.

The apparent contradiction between the optical measurements, all of which show that the film structure is relatively open, and the SEM micrograph, which appears to show that the film structure is relatively dense, was resolved by high-resolution transmission electron microscopy. A TEM micrograph of a lifted-off portion of the film of Fig. 3 is shown in Fig. 6. Dark-field micrographs from (001) and (002) oriented particles (not shown) revealed that the smallest resolvable features in Fig. 6 are single Pt grains from ~2 to ~10 nm in diameter. The majority of the grains have diameters of ~5 nm. Close examination of Fig. 6 shows that the Pt grains are arranged in cross-linked chains, forming a three dimensional network similar to that observed by Oleviera and Weitz and described by Lubensky and Pincus [12] in fractal aggregates of 5 nm Au particles. Examination of the

Table I. Microstructural parameters of photoelectrodeposited metal films as determined by least-squares analysis of $<\epsilon>$ data measured in dry N_2.

Metal	d (nm)	f_m	q	δ
Pt	32.6 $\pm$ 0.8	0.50 $\pm$ 0.02	0.56 $\pm$ 0.06	0.20
Rh	41.8 $\pm$ 0.8	0.49 $\pm$ 0.01	0.45 $\pm$ 0.01	0.13
Pd	48.3 $\pm$ 1.2	0.32 $\pm$ 0.02	0.32 $\pm$ 0.03	0.20
Re	59.8 $\pm$ 2.0	0.27 $\pm$ 0.02	0.23 $\pm$ 0.02	0.34

relative intensities of rings in electron diffraction patterns showed no preferred orientation of the grains. Thus the generally open microstructure of the deposit seen in Fig. 6 is compatible with the transparency deduced from phase-contrast microscopy and the photochemical and SE measurements.

Studies were also performed on similarly deposited films of Pd, Rh, and Re [3]. Microstructural analyses of the $<\epsilon>$ spectra of these films gave the microstructural parameters listed in Table I, along with those for the Pt film of Figs. 3-6. The screening parameters are represented in Table I as Lorentz depolarization factors $q = 1/(1 + \kappa)$. The optical data show that all these films have relatively open microstructures, which are poorly connected in the directions parallel to the surface. These conclusions are supported by the TEM micrographs of the films, which are given in Figs. 7, 8, and 9 for Rh, Pd, and Re, respectively. Electron diffraction patterns of the films all indicate that the basic building blocks of these films are single-phase metal particles of the order of 2-5 nm in diameter, having structures consistent with those of the normal dense metals. The TEM micrographs of Figs. 6-9 show that these basic building blocks are organized into relatively loosely packed secondary structures. These consistencies allow us to conclude that the development of open microstructure by mass-transport-limited deposition may be a fairly general phenomenon.

Nevertheless, examination of details shows distinct differences among the four metals studied here. The Lorentz depolarization factor for Pt and Rh approximates the value 0.5 expected for samples with a cylindrical (two-dimensionally symmetric, or dendritic) microstructure, while that for Pd suggests a spherical (three-dimensionally symmetric, or particulate) micro-structure, and that of Re indicates a flattened spherical (stacked oblate spheroidal, or platelet) microstructure. The detailed investigation of the microstructure of these films, as shown by the micrographs of Figs. 6-9, is consistent with the macroscopic analysis. The Pt and Rh films consist of 2-10 nm crystallites aggregated into 50-150 nm spheroids, the Pd films consist of 3-6 nm particles aggregated into 50-100 nm spheroids, and the Rh films consist of 6-10 nm platelets organized into 100-150 nm stacks. In particular, the agreement between the macroscopic and microscopic platelet analyses for the Re film is striking.

Transmittances of the four films were independently determined from the initial and final Faradaic photocurrents in the light-limited-potential domain, and are compared in Table II to values of selected optical functions calculated from the microstructural parameters of Table I. In every case the calculated relative transmittance based on microstructural parameters is less than that obtained from limiting Faradaic photocurrents. This discrepancy indicates that the actual conversion of normal incidence near-IR photons to hydrogen for immersed films is more efficient than that expected from the microstructural parameters deduced from the analysis of non-normal-

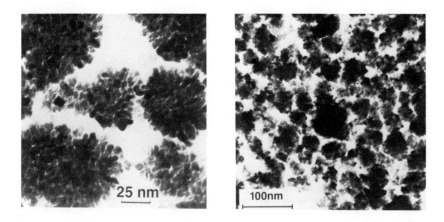

Fig. 6 (above, left.) Transmission electron micrograph of a lifted portion of the Pt film of Fig. 3 (after [2]).

Fig. 7 (above, right.) Transmission electron micrograph of a lifted portion of a Rh film photoelectrochemically deposited on p-InP (after [3]).

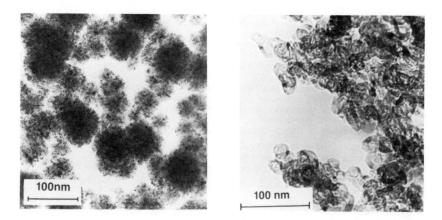

Fig. 8 (above, left.) Transmission electron micrograph of a lifted portion of a Pd film photoelectrochemically deposited on p-InP (after [3]).

Fig. 9 (above, right.) Transmission electron micrograph of a lifted portion of a Re film photoelectrochemically deposited on p-InP (after [3]).

Table II.   Selected optical functions of the p-InP/metal-film/water systems calculated at λ = 700 nm from the parameters of Table I.   The water ambient is assumed to permeate all voids in the films.

Metal	R	$e^{-\alpha d}$	T	Rel. T Calc.	Exp.
Pt	0.28	0.47	0.41	0.52	0.75
Rh	0.32	0.28	0.28	0.35	0.65
Pd	0.24	0.38	0.38	0.47	0.67
Re	0.15	0.48	0.47	0.59	0.85

Table III.   Comparison of attenuations $e^{-\alpha d}$ for dense metal films of the same metal loadings as the films of Table I (column a), for the films of Table I (column b), for infinitely dilute films with the metal loadings and depolarization factors of Table I (column c), and for infinitely dilute films with the same metal loadings but optimal microstructures (column d).

Metal	(a)	(b)	(c)	(d)
Pt	0.20	0.47	0.976	0.993
Rh	0.14	0.28	0.95	0.991
Pd	0.27	0.38	0.89	0.991
Re	0.45	0.48	0.75	0.982

incidence visible-near uv optical spectra of the same films in dry N_2.   As photocurrent-voltage curves of the p-InP photocathodes before and after photoelectrodeposition of the metal films all show shifts of the i-V curves to substantially more positive potentials with metal deposition, thereby indicating normal catalytic activity for hydrogen evolution, and because systematic chemical or microstructural changes of the films between water and dry N_2 ambients is hardly likely, the probable source of the discrepancy is an oversimplified description of the spatial inhomogeneities of the films in the optical analysis.   For example, pinholes or substantial thickness variations on a macroscopic scale (one comparable to or greater than λ) would substantially increase the apparent transmittance of normally incident radiation.   However, these same variations would either scatter light out of the field of the ellipsometer or add incoherent components to the measured flux, thereby not affecting the microstructural analysis.   Screening inhomogeneities in the plane of the film could also lead to greater local transmittances than expected from average parameters.

Additional insight can be gained by calculating attenuations $e^{-\alpha d}$ for dense metal films of equivalent metal loading per unit area, for the films with properties as listed in Table I, for infinitely diluted films of the same metal loading and microstructure, and for infinitely diluted films of the same metal loading and optimal microstructure.   The last two attenuations are calculated from Eq. (3) with κ values given in Table I and $\kappa = 0$, respectively.   Assuming water saturation at the representative wavelength λ = 700 nm (ϵ_a = 1.77156 [10]), we obtain the results shown in Table III.

Because depolarization effects are largest for the Pt film, it shows the greatest improvement in transparency relative to a dense film of equivalent metal loading even though its metal volume fraction is the highest. Conversely, the relatively inefficient oblate-spheroid screening in the Re film leads to little improvement even though its metal volume fraction is the lowest. The attenuation values of the films in the infinite-dilution limit also indicate the importance of screening, especially when compared to the optimum values achieved for metal lines oriented perpendicular to the polarization vector of the incident light ($\kappa = 0$).

We can identify three levels of microstructure determined by nucleation and growth processes. Primary microstructure, i.e., the size distribution of individual metal grains, is related to the overall rates of lattice growth. Secondary microstructure, i.e., chain length and branching, depends on the rate of formation of viable nuclei on the growing primary subunits. Tertiary microstructure is determined by the surface density and occupation rate of nucleation sites on the substrate, and on mass transport conditions in the ambient plating solution during film growth. The variables used to manipulate nucleation and growth processes, and hence microstructure, were surface preparation, surface chemistry (especially through solution composition), complex metal ion chemistry, mass transport rates, temperature, applied potential, and incident light intensity. We chose chemistries that would tend to maximize metal nucleation and growth rate constants and low to moderate rates of mass transport so that deposition took place under mass-transport control in concentration gradients normal to the substrate surface.

Highly transmitting catalysts are likely to allow poorer quality semiconductors to be used in photoelectrolytic cells. Until now, efficient hydrogen-evolving cells could only be made with good quality single-crystal semiconductors [6]. To maximize the collection efficiency, essentially all of the photogenerated minority carriers must reach a surface region covered by a hydrogen-evolution catalyst before being lost through recombination. Microstructured films offer the possibility of densely covering the semiconductor surface with catalyst without introducing excessive loss due to absorption in the film. As these films also enhance photon electric fields in the interstitial regions between metal particles, Raman and IR absorption spectra of adsorbates are also likely to be enhanced. Finally, the confinement of electrons to narrow channels of unusual dimensionality may result in unusual conductance properties at low temperatures. Indeed, we are hopeful that these results will open new and fertile areas of research in several fields.

REFERENCES

[1] J. D. Porter, A. Heller, and D. E. Aspnes, Nature (London) 313, 664 (1985).

[2] A. Heller, D. E. Aspnes, J. D. Porter, T. T. Sheng, and R. G. Vadimsky, J. Phys. Chem. 89, 4444 (1985).

[3] Y. Degani, T. T. Sheng, D. E. Aspnes, A. A. Studna, J. D. Porter, and A. Heller, Electroanal. Chem. 228, 167 (1987).

[4] D. E. Aspnes, A. Heller, and J. D. Porter, J. Appl. Phys. 60, 3028 (1986).

[5] D. E. Aspnes and A. A. Studna, Appl. Opt. 14, 220 (1975); Rev. Sci. Instrum. 49, 291 (1978).

[6] A. Heller, Science 223, 1141 (1984).

[7] D. E. Aspnes, Thin Solid Films 89, 249 (1982).

[8] D. E. Aspnes, J. B. Theeten, and F. Hottier, Phys. Rev. B20, 3292 (1979).

[9] D. E. Aspnes and A. A. Studna, Phys. Rev. B27, 985 (1983).

[10] G. M. Hale and M. R. Querry, Appl. Opt. 12, 555 (1973).

[11] H. Arwin and D. E. Aspnes, Thin Solid Films 113, 101 (1984).

[12] T. C. Lubensky and P. A. Pincus, Phys. Today 37, 50 (1984).

THE COMPOSITION AND STABILITY OF NiCu SURFACE ALLOYS
GROWN BY Ni(CO)$_4$ DECOMPOSITION ON Cu

PAUL F.A. ALKEMADE[*], H. FORTUIN, R. BALKENENDE, F.H.P.M. HABRAKEN AND W.F.
VAN DER WEG
Department of Atomic and Interface Physics, State University of Utrecht,
P.O. Box 80,000, 3508 TA Utrecht, The Netherlands.
[*]Present address: Department of Physics, The University of Western
Ontario, London, Ontario, Canada N6A 3K7

ABSTRACT

The composition of NiCu surface alloys, prepared by decomposition of
nickel carbonyl on a Cu(100) surface, has been studied before as well as
after annealing using MeV Rutherford Backscattering Spectroscopy.
NiCu films with a thickness of 2 nm and of 10 nm, respectively, having
a sharp interface with the copper substrate, were grown. The nickel
fraction is found to be 65% and 93% respectively. The outermost monolayer
is slightly enriched in copper. During annealing interdiffusion between
the film and the interface occurs and the enrichment of the surface
increases. In case of the 10 nm film the observed diffusion rate is much
higher than is expected on the basis of diffusion coefficients published in
the literature. This is ascribed to the presence of defects and/or grain
boundaries in the grown surface alloy.

INTRODUCTION

For the understanding of the selective catalytic properties of surfaces
of transition metal alloys [1] knowledge of the physical properties of the
alloy surfaces is of importance. This holds especially, since the surface
composition of an alloy often deviates much from the bulk composition
[2,4]. With regard to this much attention has been given to alloys of
copper and nickel and to thin layers of these elements grown on a substrate
of the other. Thin layers of nickel on copper surfaces are also studied
because of the assumed pseudomorphic character of the film growth [5-7].

Composition profiles of nickel, deposited on copper single crystal
surfaces by means of decomposition of nickel carbonyl at temperatures above
375 K, have been studied by Pietersen et al. [8] and by Mesters et al. [9]
using Auger Electron Spectroscopy (AES) in combination with Ar[+] sputtering.
Copper enrichment of the surface and alloying of copper and nickel in the
outermost 20-100 nm were observed. It was also observed that the
composition of the NiCu surface alloys did not change, even after prolonged
annealing. This effect was ascribed to a possible metastability of the
surface alloy. Gijzeman and Lekkerkerker showed [10] that metastability
can indeed theoretically be derived. The metastibility seems, however, to
be inconsistent with observed [11] interdiffusion of CuNi bimetals at
elevated temperatures. It is noted that the use of a stable, thin surface
alloy for a catalyst, may have advantages above a bulk alloy.

This paper describes the results of a study of the composition of thin
NiCu surface alloys prepared by means of decomposition of nickel carbonyl
at 575 K on a Cu(100) surface and its stability during annealing. The
composition is investigated using Rutherford Backscattering Spectroscopy
(RBS). The crystallographic structure of the surface alloys has been
investigated by ion-channeling. The results of the latter study are
published elsewhere [12].

Because of the relatively small difference in atomic mass of copper and
nickel, the copper and nickel signals in a RBS spectrum overlap. We have
shown [13,14], however, that even a single RBS spectrum, measured at one

beam energy and scattering geometry, is sufficient to determine unambiguously the composition of a sample containing two elements with almost equal atomic mass. Nevertheless, in order to reduce systematic uncertainties and to improve depth resolution, it is advantageous to choose more scattering geometries and/or beam energies.

EXPERIMENTAL METHOD

Apparatus

The experiments are performed in a stainless steel ultra-high vacuum (UHV) system which is coupled to the 3 MV Van de Graaff accelerator of the State University of Utrecht. Sample handling without disturbing the vacuum is possible using a vacuum interlock system. The UHV system is equipped with facilities for gas introduction, LEED optics, a cylindrical mirror analyzer for AES, a furnace for sample heating and an ion sputter gun.

During the RBS experiment, the sample is positioned in a three-axis target manipulator. Backscattered ions are energy analyzed by means of a surface barrier detector having an energy resolution of 14 keV (FWHM) for 1.5 MeV He^+ ions. The target manipulator and the detector can be positioned with an angular accuracy of better than $0.1°$. The circular aperture of the detector corresponds to $4°$ in scattering angle. More details about the apparatus are given in ref. [14].

Sample preparation

A Cu(100) crystal is cleaned by cycles of 600 eV Ar^+ sputtering and annealing at 750 K. The cleaning procedure is monitored by AES. Nickel is deposited onto the copper crystal by decomposition of nickel carbonyl, $Ni(CO)_4$, at a pressure of at most $4x10^{-3}$ Pa [15] and at a substrate temperature of 575 K. The method of, in situ, nickel carbonyl synthesis is described in detail in ref. [14] and by Pietersen et al. [8]. The gas mixture of CO and $Ni(CO)_4$ in the UHV system during the deposition is continuously renewed by pumping with a turbo-molecular pump.

Two different exposures are chosen: 14 Pa s ($1x10^5$ Langmuir) and 80 Pa s ($6x10^5$ Langmuir) $Ni(CO)_4$ plus CO. Deposition times are 1 and 7 hours, respectively. All other conditions are being kept the same. The sample exposed to 14 Pa s is annealed at 650 K during 1, 3 and 5 hours, subsequently. The sample exposed to 80 Pa s is annealed at 575 K during 16 hours and subsequently during 135 hours. AES measurements after deposition showed the presence of both nickel and copper at the surface without any carbon or oxygen contamination.

RBS measurements

The RBS measurements, using a 2 MeV He^+ beam, are performed after deposition and after each annealing treatment. In each measurement, spectra at nine different orientations of the sample with respect to the primary beam are recorded. The angle (θ) between the surface and the (centre of the) detector is varied between $65°$ and $1°$. The scattering angle is always $155°$. The depth resolution (FWHM) ranges from about 1 nm near the surface to 3 nm at a depth of 10 nm. Alignment of the beam with major crystallographic axes has been avoided by continuously rotating the sample around the surface normal. The beam dose per spectrum is 2 μC. The background pressure during the RBS measurements is about 10^{-7} Pa.

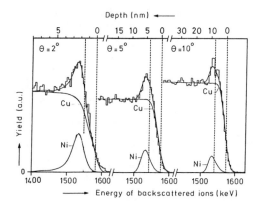

Depth (nm) ←——

θ = 2°   θ = 5°   θ = 10°

Energy of backscattered ions (keV)

Figure 1. RBS spectra of the Cu(100) surface exposed during 1 hour to Ni(CO)₄, measured at different glancing detection angles. Smooth curves are calculated spectra. The peak near the edge is caused by the overlap of the spectrum of nickel near the surface and that of the (slightly) heavier copper in slightly deeper layers.

EXPERIMENTAL RESULTS

RBS spectra

Figure 1 shows a series of RBS spectra of the sample exposed to 14 Pa s Ni(CO)₄ plus CO. These spectra are measured directly after exposure for different geometries. Figure 2 shows spectra measured after various annealing treatments at 2° detection angle. Clearly, nickel is present in the surface layers of the sample and it is seen that the nickel profile broadens during annealing. The sample exposed to 80 Pa s Ni(CO)₄ plus CO (the spectra are not shown here, see ref. [14]) contains appreciably more nickel than the former. The nickel composition profile broadens also during annealing. Quantitative results about the depth profiles are obtained by comparison of the measured spectra with theoretically calculated [14] RBS spectra.

The smooth curves in figs. 1 and 2 represent the calculated spectra which fit best the measured spectra. The lower curves indicate the separate contributions of copper and nickel, the upper curve indicates the total, theoretical spectrum. Figure 3 shows the nickel profiles for which the theoretical spectra are calculated. Within the indicated uncertainties.

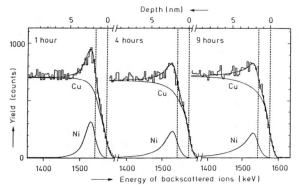

Depth (nm) ←——

1 hour   4 hours   9 hours

Energy of backscattered ions (keV)

Figure 2. Spectra measured at 2° detection angle after various annealing treatments.

394

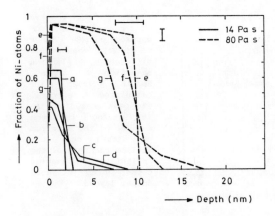

Figure 3. Composition profiles of the Cu(100) sample exposed during 1 hour (a: after exposure, b-d: after 1, 4 and 9 hours annealing at 650 K) and exposed during 7 hours (e: after exposure, f and g: after 16 and 150 hours annealing at 575 K). The bars indicate (systematical) uncertainties in the scales. Relative differences between the different profiles are more accurate.

these profiles represent the composition of the investigated samples. After an exposure to 14 Pa s Ni(CO)$_4$ plus CO, a 2.0 ± 0.5 nm thick layer, containing 65% ± 10% nickel and 35% ± 10% copper, is formed (see curve a). The interface between the NiCu layer and the copper substrate is sharp relative to the experimental depth resolution. After an exposure to 80 Pa s, a 10 ± 1.5 nm thick layer containing 93% ± 5% nickel is formed (curve e). The interface between the grown film and the copper substrate is, once again, sharp relative to the experimental depth resolution.

Figure 3 also shows the composition after the various annealing treatments (curves b,c,d and f,g respectively). The nickel composition profile broadens during annealing: nickel diffuses from the film into the copper substrate and/or copper diffuses into the film. The copper enrichment at the surface of the 2 nm film hardly changes, but the surface enrichment of the 10 nm film increases to about 85% (assuming enrichment in the first monolayer only).

In fig. 4 the diffusion distance Δz of the nickel composition profile is plotted against the square root of the annealing time [16]. The dependence is roughly linear. Assuming diffusion from a planar source [5], we derive a diffusion coefficient of 1.1x10^{-18} cm^2/s and 2x10^{-19} cm^2/s at 650 K and 575 K, respectively.

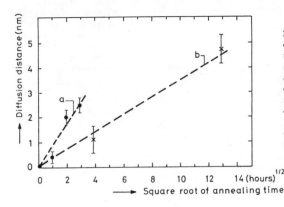

Figure 4. The diffusion distance of nickel against the square root of the annealing time. If diffusion from a planar source is assumed, the dependence must be linear. a: 2 nm film, 650 K; b: 10 nm film, 575 K.

SUMMARY AND DISCUSSION

Rutherford Backscattering Spectroscopy, performed at various glancing and non-glancing detection angles is used to study the near surface composition of Cu(100) on which nickel is deposited by decomposition of nickel carbonyl. After an exposure of less than 4×10^{-3} Pa during one hour, a 2 ± 0.5 nm thick film containing 65% ± 10% nickel, is formed. The Ni/Cu ratio in the film is, apart from some enrichment at the surface, independent of depth. When the Cu(100) surface is exposed to nickel carbonyl during seven hours, a 10 ± 1.5 nm film, containing 93% ± 7% nickel is formed. Thus, both the thickness of the film and the average nickel fraction have increased as compared to the former sample. The interfaces between the films and their substrates are sharper than the experimental depth resolution, which is 1.0 and 3.5 nm (FWHM) for the 2 nm and the 10 nm layer, respectively. The total amount of copper in the surface regions of these samples is about equal: $7.0 \times 10^{15}/cm^2$, equivalent to about 4 monolayers.

In studies by a number of authors [1-4], it has been found that nickel initially grows layer-by-layer on copper. Although these authors did not observe alloying at temperatures below 475 K, they also concluded that it could neither be excluded. For a thickness larger than about 4 monolayers Abu-Joudeh et al. [6] have found indications of island formation. Pietersen et al. [8] and Mesters et al. [9] have found by AES that, when nickel is deposited by decomposition of nickel carbonyl at substrate temperatures of 500 - 600 K, the nickel atoms are incorporated below the outermost copper surface layers.

By RBS alone the formation of a surface alloy cannot be distinguished from the formation of pure nickel islands on a pure copper substrate. However, ion-channeling measurements [12] suggest that the grown layers do not consist of separate, pure nickel islands, but of a homogeneous mixed surface alloy.

Because of the large difference in the surface free energy of copper and nickel, enrichment of the surface with copper is theoretically expected. In the present study, the copper enrichment at the surface of the thick NiCu film reaches a value of 85% of a monolayer after annealing during 150 hours at 575 K. This value has been found by many other authors (e.g. [4,17]).

We have concluded that during annealing the nickel composition profiles broaden by interdiffusion between the surface alloy and the substrate. This observation is not in agreement with the metastability of the nickel composition profiles, observed by Mesters and explained theoretically by Gijzeman [8-10]. Extrapolation of the diffusion coefficient of nickel in copper at high temperatures (above 1100 K) [18] results in a diffusion coefficient of 7×10^{-19} cm^2/s and 3×10^{-21} cm^2/s at 650 K and 575 K, respectively. (The coefficient for diffusion of copper in nickel is three orders of magnitude lower [19]). When our measurements are compared with diffusion measurements in CuNi bimetals by Van Dijk et al. [11], diffusion coefficients of 8×10^{-20} cm^2/s and 2×10^{-23} cm^2/s respectively, are expected. Our figures for annealing of the 2 nm surface alloy at 650 K are consistent, while those for annealing of the 10 nm surface alloy at 575 K are not consistent with the above mentioned figures. The ion channeling study [12] of these samples has revealed that (only) the latter surface alloy contains many defects and/or grain boundaries. It is therefore likely that the observed high interdiffusion rate is caused by the large number of defects and/or grain boundaries in the grown film.

In conclusion: decomposition of carbonyl is a suitable technique to produce thin surface alloys. The surface alloys, however, are not stable and therefore less useful as a catalyst with special, selective properties, than bulk alloys. A comprehensive discussion of the growth mechanism of thin NiCu surface alloys on a copper substrate will be presented in a forthcoming publication.

REFERENCES

1. J.L. Morán-López and K.H. Benneman, Surf. Sci. 75, 167 (1978).
2. J.R. Chelikowsky, Surf. Sci. 139, L197 (1984).
3. J.W. Matthews and J.L. Crawford, Thin Solid Films 5, 187 (1970).
4. H.H. Brongersma, M.J. Sparnaay and T.M. Buck, Surf. Sci. 71, 657 (1978).
5. S.P. Tear and K. Röll, J. Phys. C: Solid State Phys. 15, 5521 (1982).
6. M.A. Abu-Joudeh, B.M. Davies and P.A. Montano, Surf. Sci. 171, 331 (1986).
7. P.F.A. Alkemade, W.C. Turkenburg and W.F. van der Weg, Nucl. Instr. and Meth. B15, 129 (1986).
8. C.A. Pietersen, C.M.A.M. Mesters, F.H.P.M. Habraken, O.L.J. Gijzeman, J.W. Geus and G.A. Bootsma, Surf. Sci. 107, 353 (1981).
9. C.M.A.M. Mesters, G. Wermer, O.L.J. Gijzeman and J.W. Geus, Surf. Sci. 135, 396 (1983)
10. O.L.J. Gijzeman and H.N.M. Lekkerkerker, Chem. Phys. Lett. 95, 152 (1983).
11. T. van Dijk and E.J. Mittemeijer, Thin Solid Films 41, 173 (1977).
12. P.F.A. Alkemade, H. Fortuin, R. Balkenende, F.H.P.M. Habraken and W.F. van der Weg, in Proceedings of the 2nd Int. Conf. on the Structure of Surfaces, Amsterdam 1987, edited by J.F. van der Veen and M.A. Van Hove, (Springer-Verlag, Berlin).
13. P.F.A. Alkemade, to be published.
14. P.F.A. Alkemade, PhD thesis, Utrecht, 1987.
15. Since the exact composition of the $Ni(CO)_4$-CO gas mixture is not known - $Ni(CO)_4$ decomposes at the filament of a mass spectrometer - stated values of the exposures do not relate to the $Ni(CO)_4$ pressure alone, but to the total pressure of $Ni(CO)_4$ and CO.
16. The diffusion distance Δz has been defined as half the distance for which the nickel fraction decreases from 84% to 16% of its maximal value.
17. M. Shikata and R. Shimizu, Surf. Sci. 97, L363 (1980).
18. U. Gradmann, Ann. Phys. Lpz. 17, 91 (1966).
19. G. Brunel, C. Cizeron and P. Lacombe, C.R. Acad. Sci. Ser. C 270 (4), 393 (1970).

KINETICS AND MECHANISM OF THE
COPPER-CATALYZED ETCHING OF SILICON BY F_2

N. SELAMOGLU, J.A. MUCHA, D.L. FLAMM AND D.E. IBBOTSON
AT & T Bell Laboratories, Murray Hill, NJ 07974

ABSTRACT

The copper catalyzed fluorination of silicon is first-order in $[F_2]$ and in $[Cu]_s$ until the coverage reaches ~ 4 monolayers. Above ~ 4 monolayers the reaction rate is zero order in copper, suggesting a limited number of catalytically active Cu/Si sites. Surface diffusion of copper leads to decrease in the etch rate as a function of time as well as feature size-dependent etch depths. The copper compounds CuF_2, CuO, and copper silicides, Cu_5Si and Cu_3Si all catalyzed the F_2-Si reaction which suggests that they are all converted to the same active species. The results can be explained by mechanisms involving copper fluorides or copper silicides as active intermediates.

INTRODUCTION

We recently reported on the copper-catalyzed fluorination of silicon, [1] where trace amounts of copper increased the etching rate 100-fold. Copper deposits on the silicon surface were obtained by the selective removal of aluminum from deposited Al/0.5%Cu films. The temperature dependence of the F_2 reaction with Cu/Si was reported as well as the catalytic activity of other metals and gases.

In the present paper we report the dependence of the catalyzed etch rate on Cu and F_2 concentrations. We find that the selective removal of aluminum from deposited films of Al/Cu is a quantitative method for obtaining copper deposits on silicon. Because the finite supply of copper covers a surface area that increases with etching time, etch rates are dependent on time and feature size.

EXPERIMENTAL

The experimental setup and procedure were described earlier. [1] Briefly, 4000Å of thermal SiO_2 was grown on (100) silicon wafers and patterned by conventional resist-plasma processing. Al/0.5% Cu alloy was sputter-deposited onto the oxide-patterned wafers. Immediately before exposure to F_2, the aluminum film was etched in 20% aqueous HF, leaving the insoluble copper as a residue on the silicon surface. For most experiments a 10,000 Å film of Al/Cu

Mat. Res. Soc. Symp. Proc. Vol. 111. ©1988 Materials Research Society

was deposited, leaving ~14 monolayers of copper residue (see below). Auger and SEM analyses indicated uniform Cu deposition on a 0.5 μm scale for aluminum films <5000Å and island formation for thicker films.

Samples were bonded to a heated stage with gallium and exposed to fluorine in the flow system described in reference 1. Etch depths, measured by a stylus thickness monitor, were used to obtain average etch rates in most experiments. For all measurements on concentration and temperature dependence, [1] large Si/Cu areas (~2-3 mm size) on the sample were selected so that the final etch depths ($\leq$ 200 μm) were much smaller than the feature size. This ensured that surface diffusion had a negligible effect on the etch rate. For these measurements typical exposure times were about 20-30 minutes at 120° C and 1 Torr of F_2, during which time the etch rate remained constant. Samples used in experiments on feature size dependence had square windows in the SiO_2 mask, ranging from 2.5 μm to 3 mm. These were exposed to fluorine for up to 3 hours.

Quantitative determination of copper deposits was carried out by dissolving the surface copper residue of a blanket-coated wafer in aqueous HNO_3 and analyzing the solution by atomic absorption (AA) spectrometry. The surface copper concentration was obtained from the AA measurement and the known area of the wafer. The total amount of copper initially in the aluminum film was determined by AA analysis after the Al/Cu film was dissolved in HCl/HNO_3 solution. Figure 1 shows the surface copper concentration, $[Cu]_s$, and the total copper concentration, $[Cu]_{max}$, plotted as functions of the aluminum film thickness. Note that the two results are identical and indicate that copper is quantitatively transferred from the film to the silicon surface, thereby giving the linear dependence in Figure 1. Also shown are results for $[Cu]_r$, the concentration of copper that is transferred to the HF solution during initial cleaning. The latter results are small compared to $[Cu]_s$, and confirm that removal of copper by HF is negligible. Coverages in monolayers were estimated based on the surface atom density of Si(100), 1ML = 6.8 x10^{14}cm^{-2}.

Hence, by accurately depositing Al films 500-20,000 Å thick, uniform copper deposits between a fraction of a monolayer up to 30 monolayers can be obtained. Such deposits are difficult to obtain by conventional methods. Thus this selective removal of aluminum from deposited Al/Cu films is an easy and reliable method for depositing low, uniform, concentrations of copper on silicon.

RESULTS AND DISCUSSION

Dependence of Etch Rate on $[F_2]$ and $[Cu]_s$

We found that the etching rate was proportional to fluorine pressure in the range 0.1-10 Torr at T > 100° C and T < 60° C. The F_2 dependence could not be determined in the 60-80° C region because of data scatter and the high sensitivity to temperature.

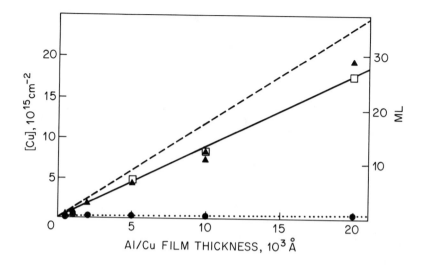

Figure 1. Surface copper concentration as a function of Al/Cu film thickness. $[Cu]_s$: Copper remaining on surface after HF treatment. $[Cu]_r$: Copper removed by HF. $[Cu]_{max}$: Total amount of copper in the Al/Cu film. The dashed line is the copper concentration computed from the composition of the aluminum sputter target (0.46 % Cu) and the thickness of deposited Al films.

The etching rate is shown as a function of surface copper concentration, $[Cu]_s$, in Figure 2. For small concentrations, the rate is nearly linear in $[Cu]_s$, but above $[Cu]_s \sim 4$ ML the etch rate is almost constant at about 4.0-4.5 x 10^4 Å/min. Although the reaction is first-order in copper up to ~ 4 ML, it is independent of $[Cu]_s$ for higher coverages.

This saturative dependence of etch rate on $[Cu]_s$ can be explained if there is a limited density of catalytically active sites. The latter is consistent with our observed formation of copper islands. [2] Copper deposits as a uniform layer for small coverages (~ 3 ML), however, island formation is observed at high coverages (≥ 14 ML). The first order reaction below 4 ML is probably associated with a intermixed copper-silicon phase limited to several monolayers, corresponding to the uniform coverage observed. Islands, which form with excess coverage, have little influence on the etch rate and result in the saturation observed in Figure 2. Although excess copper coverage does not affect the instantaneous etch rate, it does affect etch depths over long reaction times by compensating for loss of catalyst. [2] Slight masking by the copper islands probably accounts for the small falloff between $[Cu]_s \sim 5$ and 30 ML in Figure 2. With a large excess of copper, well beyond 30 ML, we observe a large drop in the etching rate, apparently caused by extensive masking.

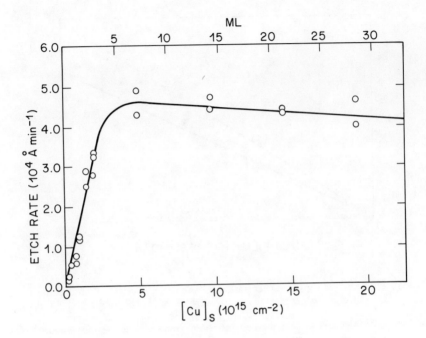

Figure 2. The etching rate of Si as a function of surface copper concentration, $[Cu]_s$, at 140 °C and 1 Torr F_2.

Etch Depth as a Function of Time and Feature Size

Our study of the kinetics, i.e., the temperature- and concentration-dependences of the reaction rate reported in reference 1 and in the previous section, were carried out under conditions where the rate was constant in time. However, we found that etch rates decreased over long exposure times and were feature size-dependent. Figure 3 shows measured etch depths as functions of time for Cu/Si feature sizes ranging from 2.5 μm to 3 mm, where the dashed lines represent smoothed curves through the data. (The solid lines represent the model, discussed below.) Note that for each feature, the etch rate decreases with time as indicated by the convex curvature of the etch depth versus time data. Furthermore, not all features are affected to the same extent so that the limiting etch depth is a function of feature size. Even the large 3 mm feature, which etches at a constant rate for 70-80 min, shows a decrease in the etching rate for longer times. We associate the declining etch rates with a decrease in surface copper concentration, $[Cu]_s$.

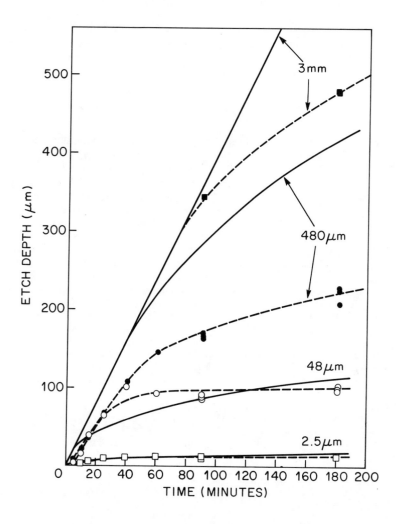

Figure 3. Etch depth as a function of time for feature sizes 2.5 μm to 3 mm. The dashed lines represent smoothed curves through each data set. The solid lines represent model results (see text).

Decrease in the surface concentration of copper could arise from irreversible chemical reaction, bulk diffusion in silicon or surface diffusion. Of these, surface diffusion will cause a size dependence along with a decrease in etch rate versus time. As a feature is etched its surface area increases and, if surface diffusion is fast enough, a dilution in $[Cu]_s$ occurs. This eventually leads to lower etch rates, following the curve in Figure 2. On a relative scale, the surface areas of small features increase more rapidly than those of large features, leading to the observed size dependence.

We carried out simple model calculations assuming that surface diffusion of copper on silicon is fast compared to etching and that the etching is isotropic. The input data were the feature size, the initial surface copper concentration, and the etch rate as a function of $[Cu]_s$ (taken from Figure 2). Etch depth versus time was calculated by obtaining an incremental etch depth for a small time interval Δt, computing the corresponding surface area, and adjusting $[Cu]_s$ and the etch rate for this area before iterating the procedure to obtain new areas and etch rates. The model results of etch rate versus time are shown as solid lines in Figure 3.

Note that, as expected, the model predicted decrease in the etch rate for each feature as a function of time as well as size dependent etch depths. For the smaller 2.5 and 48 μm features, the model reproduces the experimental curve within 30 % error. However, note that for the larger features, the model predicts etch depths that are much higher than the data. For example, for a 3 mm feature the model predicts that etching should continue at a constant rate until the thickness of the wafer, 508 μm, is etched through in about 130 minutes. Yet, even after 180 min, the sample etched only to 480 μm because of the marked decrease in the etch rate during this time. In other words, while dilution caused by surface diffusion can account for the size dependence and the falloff in etch rates for small features, it underestimates the rate falloff on large features. Thus, another process besides surface diffusion must affect the surface concentration of copper and etch rates. This is probably a chemical deactivation of copper into an uncatalytic form. We believe bulk diffusion is unimportant because, although metallic copper has a high diffusivity in Si ($D = 10^{-8}$ cm^2/s at 100°C), its solubility is too low (10^5 cm^{-3}) to support significant loss of surface copper (10^{16} cm^{-2}) over these time intervals. [3,4] (In fact, experimental etch profiles of features are anisotropic, contrary to the assumption used in the model. This is because of limited surface diffusion, discussed in reference 2).

Chemical Mechanism

We reconsider our earlier results [1] on the temperature dependence of the F_2 reaction with Cu/Si together with results of this study. The reaction exhibited three distinct temperature regimes. Below 60°C, the rate was virtually the same as the uncatalyzed reaction [5] with no apparent inhibition by copper at the surface. Between 60°C and 80°C a dramatic 100-fold rate increase was observed. Finally, above 80°C the large enhancement saturated - that is, the rate increase with temperature displayed a much smaller slope. In this high temperature regime (T>80°C), copper lowered the activation energy from the uncatalyzed value of 12.9 kcal/mol to 8.8 kcal/mol.

The experimental results were interpreted by mechanism (1):

$$2\,Cu + F_2 \xrightarrow{k_1} 2\,Cu-F \qquad (1a)$$

$$Cu-F + Si \xrightarrow{k_2} (Si-F) + Cu \qquad (1b)$$

Here (Si-F) undergoes further fluorination to form the final products. The break in the Arrhenius plot in the middle 60-80 °C region could be explained by a shift in the rate determining step between (1a) and (1b).

Reactions similar to (1) were originally invoked by Rochow [6,7] to interpret the catalytic effect of copper on the reaction of alkylchlorides with silicon. Although reactions (1) are consistent with the experimental kinetics we observed, conceivably other copper fluoride cycles involving species such as CuF, CuF_2 or $CuSiF_6$ would be equally satisfactory. Similarly, a mechanism involving silicide formation

$$(Cu)_s + (Si) \xrightarrow{k_2} (CuSi)_s \qquad (2a)$$

$$(CuSi)_s + F_2 \xrightarrow{k_1} (Cu)_s + (SiF) \qquad (2b)$$

could display identical kinetics. Here, the subscript "s" indicates that the species are bound to the silicon surface. A large body of kinetic and Auger data [8-13] on organosilane formation from alkylchlorides and silicon favor reactions analogous to (2a) and (2b), with $\eta-Cu_3Si$ as the active catalyst. However, it should be noted that copper concentration in those studies, $Cu/Si > 0.2$, is orders of magnitude higher than that in our study, $Cu/Si < 1.7 \times 10^{-5}$.

We tested several copper compounds for catalytic activity in order to gain insight into the mechanism. These were CuF_2, CuO, Cu_5Si, a mixture of Cu_3Si and Cu_5Si and metallic copper. If any of these were catalytically inactive, they might be candidates to account for the observed loss of catalyst. Each compound was pressed on a pure silicon surface using a spatula and excess powder was blown off, leaving particles in intimate contact with the silicon. All samples showed a dramatic catalytic effect upon exposure to fluorine at ~150 °C. The silicides formed metallic copper on the surfaces, demonstrating the ability of fluorine to extract Si from the silicide. Deep pits were etched in the treated surfaces, whereas pure silicon control samples etched slightly and were uniform in appearance. The results clearly rule out these compounds as inhibitors.

The effect of copper on reaction between Cl_2 and silicon was briefly examined in our laboratory. [14] With 14 ML Cu coverage, etching increased ~300-500 times over the uncatalyzed reaction at 450 °C. $CuCl$, $CuCl_2$ and pure copper powder all catalyzed etching by Cl_2, just as the copper fluorides catalyzed etching by F_2. Cl-atom etching was not enhanced by copper, which is also analogous to the negative result reported for F-atoms. [1]

Our experimental results on the kinetics of the Cu catalyzed etching of silicon cannot differentiate between a variety of possible mechanisms. In order to determine reactive intermediates and active catalytic species on the Cu/Si surface, techniques such as Auger electron spectroscopy or X-ray photoelectron spectroscopy need to be utilized *in situ*. These would be expected to yield information on Cu-F and Si-F bonding during the fluorination reaction, as well as on bonding between copper and silicon.

ACKNOWLEDGEMENT

We thank Kenneth J. Orlowsky and Jack Mayberry for sample preparation and Patricia O'Hara for the atomic absorption analyses.

REFERENCES

1. N. Selamoglu, J. A. Mucha, D. L. Flamm and D. E. Ibbotson, *J. Appl. Phys.*, **62**, 1049 (1987).

2. N. Selamoglu, J. A. Mucha, D. L. Flamm and D. E. Ibbotson, submitted to J. Appl. Phys.

3. R. N. Hall and J. H. Racette, *J. Appl. Phys.*, **35**, 379 (1964).

4. E. R. Weber, *Appl. Phys.*, **A 30**, 1 (1983).

5. J. A. Mucha, V. M. Donnelly, D. L. Flamm and L. M. Webb, *J. Phys. Chem.*, **85**, 3529 (1981).

6. E. G. Rochow, *J. Amer. Chem. Soc.*, **67**, 963 (1945).

7. D. T. Hurd and E. G. Rochow, *J. Amer. Chem. Soc.*, **67**, 1057 (1945).

8. R. J. H. Voorhoeve and J. C. Vlugter, *J. Catalysis*, **4**, 123 (1965).

9. R. J. H. Voorhoeve and J. C. Vlugter, *J. Catalysis*, **4**, 220 (1965).

10. T. C. Frank and J. L. Falconer, *Appl. Surface Sci.*, **14**, 359 (1982-83).

11. T. C. Frank, K. B. Kester and J. L. Falconer, *J. Catalysis*, **91**, 44 (1985).

12. A. K. Sharma and S. K. Gupta, *J. Catalysis*, **93**, 68 (1985).

13. T. M. Gentle and M. J. Owen, *J. Catalysis*, **103**, 232 (1987).

14. E. A. Ogryzlo, private communication.

STUDY OF MODEL COPPER-BASED CATALYSTS BY SIMULTANEOUS DIFFERENTIAL SCANNING CALORIMETRY, X-RAY DIFFRACTION AND MASS SPECTROMETRY.

ROBERT A. NEWMAN, JOSEPH A. BLAZY, TIMOTHY G. FAWCETT, LARRY F. WHITING, and ROBERT A. STOWE
Michigan Applied Science & Technology Laboratories, Dow Chemical USA, 845 Building, Midland, MI 48667

ABSTRACT

Recently, a new materials characterization instrument and technique employing simultaneous differential scanning calorimetry/X-ray diffraction/mass spectrometry (DSC/XRD/MS) have been developed at Dow. Use of this technology can be illlustrated by the study of various materials such as polymers, organics/pharmaceuticals, inorganics, and catalysts.

Presented here is the use of the DSC/XRD/MS instrument to study the thermostructural behavior of four model copper-based catalyst systems during activation and regeneration. The instrument allows simultaneous generation of thermal, structural and chemical data in real-time during temperature programmed analysis and provides useful insights into the chemical and physical processes occurring. In addition, the calorimetry data yield qualitative information on the magnitude and rate of heat flow, while the diffraction data provide structural dynamics of reduction, oxidation and crystallite growth.

The results of this study conclusively show cuprous oxide as an intermediate in the reduction of the copper oxide portion of each of the model catalysts. However, such features as the onset temperature and copper surface area varied widely among the four catalysts in response to the same chemical event conducted under similar experimental conditions. On the other hand, oxidation runs on the reduced catalysts were all similar to each other, first producing cuprous oxide from copper metal over a broad range of temperature, followed by the oxidation of cuprous oxide to copper oxide at even higher temperatures.

INTRODUCTION

Copper-based catalysts find use in diverse industrial applications, such as in the low temperature water gas shift reaction, methanol synthesis, selective hydrogenation of C4 acetylenes in the presence of butadiene, hydrogenation of edible fats and oils, and scavenging of oxygen from inert gases. In many cases, the catalyst must be activated by conversion of its copper oxide content to copper metal before use in the catalysed process. Often, a requirement for high catalytic activity is maximizing the copper metal surface area [1-3].

X-ray diffraction (XRD) has long been used as a tool for catalyst development and optimization research. Its advantages as a bulk structural analysis method are well-known. Thermal methods such as thermogravimetry (TG) and differential scanning calorimetry (DSC) have also found use in materials characterization and catalyst development. Reactive and/or inert atmospheres are frequently employed. Mass spectrometry (MS) has been coupled with both TG and DSC instruments to obtain "real time" analysis of the gaseous products.

Recently, a new instrumental technique for materials characterization has been developed at Dow [4,5]. It employs simultaneous DSC/XRD/MS and its use in the study of pharmaceutical materials and polymers has been discussed [5-7]. The instrument is shown schematically in Figure 1. It allows us to perform computer controlled DSC experiments in the temperature range from 25-600°C in a variety of atmospheres, with continuous real-time monitoring of the DSC thermogram. Throughout the experiment, XRD scans are continuously collected on a position sensitive detector (PSD) which observes a preselected 25 degree 2θ region of the diffraction spectrum. Meanwhile, the exhaust gases from the custom-designed DSC cell are monitored by the mass spectrometer, which is continuously cycling through the mass range from 8-58 amu.

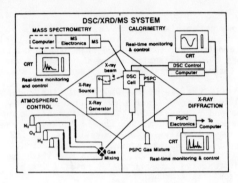

Fig 1. Schematic diagram of the fully-automated DSC/XRD/MS system.

This instrument was used to study the thermal/structural/reactivity behavior of several model copper catalysts during activation and regeneration (oxidation followed by reduction). The purpose of the study was to investigate differences in reduction behavior and to compare influences of chemical composition on the resulting copper crystallite sizes among the catalysts in order to better understand the factors which yield a well dispersed, high surface area, high efficiency copper catalyst.

EXPERIMENTAL

Oxidation and reduction experiments were carried out on four model catalyst systems. Initially, all four samples were bulk mixtures containing tenorite (CuO) as a major constituent. Similar conditions were used in all of the experiments. Gases were 5.6 H_2/He or 5% O_2/He at 50 cc/min. The heating rate was 2°C/min through the range of 100-500°C.

RESULTS AND DISCUSSION

Both the activation and regeneration of these model catalysts involves the reduction of cupric oxide to copper metal under a hydrogen atmosphere. The reaction of tenorite, or cupric oxide, with hydrogen at elevated temperatures is described by equations (1) and (2) listed below. The reaction is a two-step process, with the first reaction being slightly more exothermic than the second [8].

$$2CuO + H_2(g) \xrightarrow{\quad \Delta H_f^° \ -23.4 \ kcal \quad} Cu_2O + H_2O(g) \qquad (1)$$

$$Cu_2O + H_2(g) \xrightarrow{\quad \Delta H_f^° \ -18.0 \ kcal \quad} 2Cu + H_2O(g) \qquad (2)$$

Some typical results from a DSC/XRD/MS experiment performed on a model copper catalyst under the conditions cited above are shown in Figures 2-4. Figure 2 consists of a number of XRD spectra selected from the total set of 45 scans to illustrate the structural conversion taking place during the reduction of copper oxides. The solid line of Figure 3 illustrates the DSC thermogram which corresponds to the XRD spectra shown in Figure 2, while the corresponding MS data for mass 18 (H_2O) are plotted in Figure 4. (Unless otherwise noted, the ordinates of the Figures in this paper are scaled in arbitrary units.)

The XRD spectra in Figure 2 show the conversion of copper oxides to copper metal, first detected at about 250°C, followed by the rapid increase in the peak intensity from copper metal. Crystallite growth is discernable from the narrowing of the copper peaks, particularly obvious at 1.808 Å.

Sensitivity for Detecting Chemistry

Visual comparison of the DSC data (Figure 3) and the XRD data (Figure 2) illustrates the difference in sensitivity between these two techniques. While the onset of an exothermic reaction (reduction of CuO) is obvious as low as 150°C in Figure 3, no discernable change is seen in the XRD spectra until about 250°C, where the exotherm of Figure 3 is at maximum. Similar lags were observed in all the experiments. The explanation for the lag lies in the fact that X-ray diffraction is a bulk technique, and to observe a diffraction signal, sufficient lattice planes must be present in a position to diffract. By contrast, the DSC, which is measuring heat flow to and from the sample, is detecting the chemical event with much greater sensitivity than is possible for the XRD. The sensitivity of the MS for detecting the onset of reduction is similar to that of the DSC, as seen in Figure 4, where the three techniques are compared. Note also the good agreement between the MS and the DSC data which both maximize at about 250°C, then tail off to a plateau at approximately 280°C.

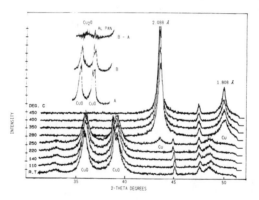

Fig. 2. Selected XRD spectra from experiment performed on sample A.

Presence of Cuprous Oxide

One of the chemical aspects we sought to elucidate in the comparison of the model catalysts was the question of the occurence of a two-step reduction, as shown in equations (1) and (2), above. If the reduction could be done stepwise, we expected that unwanted temperature excursions in commercial reactors would be easier to control. Referring again to Figure 2, the major peak of cuprous oxide, if present, would appear just to the right of the first CuO peak. In fact, a small shoulder can be detected growing in this region in the temperature range of about 140-250°C. A more satisfying verification of the existence of such a peak is illustrated in the inset of Figure 2. The XRD spectra for 210°C and 220°C are shown at A and B. The top spectrum is the result of subtracting A from B and clearly shows the Cu_2O peak which was almost obscured as the shoulder on the CuO peak on the left.

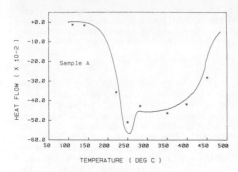

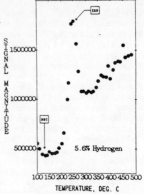

Fig. 3. DSC thermogram from DSC/XRD/MS experiment on Sample A. (*) denotes spectra of Fig. 2.

Fig. 4. Mass 18 spectra for experiment on Sample A.

In some experiments, H_2 concentration and/or heating rate were too high, making detection of the intermediate Cu_2O more difficult. In these cases, isothermal runs, or runs with lower heating rates, along with lower H_2 concentrations were made. Figure 5 shows selected XRD spectra for Catalyst B, taken with 1% H_2. Here, the presence of Cu_2O is clearly established. The presence of Cu_2O was verified for each model catalyst.

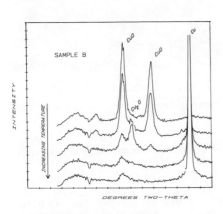

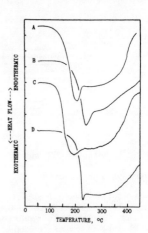

Fig. 5. Sample B XRD spectra showing presence of Cu_2O

Fig. 6. DSC data for Samples A-D, not normalized for sample weight

Comparison of Model Catalysts

Figure 6 shows a comparison of the DSC thermograms for the four samples. Clearly the reaction rates, as indicated by the slope of the exothermic event, vary considerably. These rates have important ramifications in industrial situations, where a reactor may contain thousands of pounds of catalyst, and there is limited capability to dissipate excess heat. The onset of reduction, as discussed above, is seen to vary over a broad range of temperature. At the point in each exotherm where 10% of the total heat has been evolved, this range is about 55°C.

Other parameters of interest in the comparison of the model catalysts are the rate of copper metal formation and crystallite growth, both of which could affect the surface area and efficiency of a catalyst. To a first approximation, the concentration of a component in a mixture is proportional to the area under a diffraction peak of that component. Thus, by obtaining the area under the major XRD peak for copper (2.088 Å) in selected XRD data scans, one can study the rate of formation of copper as a function of temperature. The results are shown in Figure 7. All the catalysts are similar, showing a rapid formation of copper metal, followed by a level region where presumably no further copper formation is occurring.

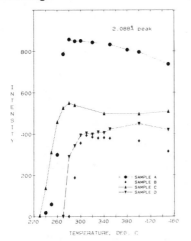

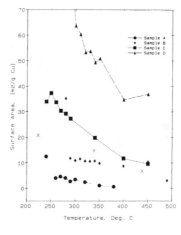

Fig. 7. Integrated peak intensities (areas) for samples A-D

Fig. 8. Surface areas for samples A-D

For this discussion, we have employed the conventional Scherrer method and an instrument correction to measure crystallite sizes. (The width of the 2.088 Å peak at half its maximum height was obtained from a computer-generated curve which had been regression fitted to the diffraction profile [9]. The fitted curve also provided a value of the area under the 2.088 Å peak, as discussed above.) The resulting crystallite diameters obtained from the Scherrer equation were converted to calculated values of surface area, assuming spherical crystallites. These values, in m^2/g Cu, are shown in Figure 8 and strikingly illustrate the different response in surface areas for the four catalysts, arising from the same treatment. Continuous annealing is evidenced for Samples C and D as the temperature is increased. Samples A and B seem to have sintered already to a relatively low surface area at the lower temperatures. The "X" points are surface areas obtained by O_2 chemisorption in other experiments on reduced Catalyst B, annealed at the indicated temperatures. The calculated values from the XRD data agree well with these data, showing the validity of the XRD technique.

Oxidation Runs

The oxidation of the reduced catalysts is illustrated in Figure 9 for the run with Catalyst B, which was quite typical for all the samples. The data show the intensity of the Cu_2O peak increasing while the intensities of the Cu^o peaks diminish until the Cu^o is nearly gone. Oxidation of the intermediate cuprous oxide does not occur until higher temperatures (about $270^{\circ}C$) are reached. Finally, at very high temperatures, the oxidation to CuO is complete. The two-step oxidation was plainly evidenced in the DSC traces for all runs.

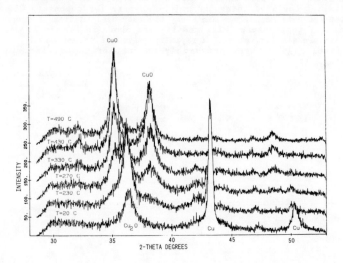

Fig. 9. XRD scans for oxidation of reduced Sample B.

CONCLUSIONS

In addition to illustrating some of the capabilities of the DSC/XRD/MS instrument, these experiments have yielded detailed information about the thermostructural behavior of the four model copper catalysts under well-defined and precisely controlled thermal and atmospheric conditions of the sample environment. The results show conclusively the two-step nature of the reduction of copper oxide which involves the formation of cuprous oxide as an intermediate. In response to this same chemical event, the behavior of the four catalysts varied widely with respect to such features as onset temperature and surface area production. On the other hand, oxidation runs of the reduced catalysts were quite similar to each other. These also involved step-wise formation of Cu_2O as an intermediate.

REFERENCES

1. D. K. Ghori, D. Datta, N. C. Ganguli and S. P. Sen, Fert. Technol. 18, 146 (1981).

2. N. Roy, S. K. Nath, A. Sarkar, D. P. Das, and S. P. Sen, J. Res. Inst. Catal., Hokkaido Univ. 21, 187 (1973).

3. G. E. Parris and K. Klier, J. Catal. 97, 374 (1986).

4. T. G. Fawcett, C. E. Crowder, L. F. Whiting, J. C. Tou, W. F. Scott, R. A. Newman, W. C. Harris, F. J. Knoll and V. J. Valdecourt, Adv. X-ray Anal. 28, 227 (1985).

5. T. G. Fawcett, Chemtech 1987, 565.

6. C. E. Crowder, S. Wood, B. G. Landes and R. A. Newman, Adv. X-ray Anal. 29, 315 (1986).

7. T. G. Fawcett, E. J. Martin, C. E. Crowder, P. J. Kincaid, A. J. Strandjord, J. A. Blazy, D. N. Armentrout and R. A. Newman, Adv. X-ray Anal. 29,323 (1986).

8. D. A. Johnson, Some Thermodynamic Aspects of Inorganic Chemistry, (Cambridge University Press, London, 1968), Appendix 5.

9. A. Brown and J. W. Edmonds, Adv. X-ray Anal. 23, 361 (1980).

Diffusion in High Occupancy Zeolites

James Wei

Department of Chemical Engineering, MIT

Abstract

Diffusion of small molecules in zeolites most often takes place in the configurational diffusion region, where the molecular diameter is approximately the same or slightly greater than the channel diameter. Since two molecules may not pass each other in a pore, the random walk based diffusion equation does not apply under high occupancy conditions in zeolites with one-dimensional pores.

For zeolites with multi-dimensional pores, such as ZSM-5 and A, one sometimes encounters the counter-intuitive result that diffusivity dramatically _rises_ with occupancy. There are two explanations for this behavior: one from irreversible thermodynamics and the Darken equation, which predicts that diffusivity will always rise with occupancy; the other from a Markov model of random activated jumps between low energy positions, such as pore crossings, which predicts that diffusivity will increase with occupancy if the activation energy of diffusion decreases with occupancy-- such as due to swelling.

1. Gaseous, Knudsen and Configurational Diffusions

We are all familiar with molecular diffusion in the kinetic theory of gases, where a molecule is in free flight until collision with another molecule. The mean free path describes the average distance of a molecule between collisions, and together with the average velocity, determines the diffusivity. This mean free path distance is given by the equation

$$\lambda = 1/(\sqrt{2}\pi d_M^2 \cdot n)$$

where d_M is the diameter of the molecule, and n is the molecular density per volume. Since the average velocity $\bar{u} = \sqrt{8kT/(\pi \cdot M)}$, thus the gaseous diffusivity is $D_G = \bar{u}\lambda/3$. See Fig. 1.

The Knudsen diffusion is based on the notion that n is very small so that collision between molecules is infrequent, but the channel walls are smaller than the mean free path, so that the dominant mechanism for limiting diffusion is the molecule-wall collision rather than the molecule-molecule collision. Thus the formula for Knudsen diffusivity is similar to the equation for gaseous diffusivity, where the channel diameter plays the role of mean free path in the distance travelled per free flight. $D_K \alpha d_c \sqrt{T/M}$, where d_c is channel diameter.

When the gas becomes more dense, a molecule spends less time in free flight and more time in collision, and eventually in a condensed state, a molecule is always under the influence of other molecules. In the liquid or solid diffusion regime, an often used theory is the Eyring theory where there are discrete spatial sites (or defects) with low energy where molecules normally reside, and diffusion takes place by molecular jumps between such sites when it has acquired sufficient energy to overcome the barrier. In that case, there is a very significant activation energy of diffusion, which may be in the region of 10 kcal/mol. In contrast, for gaseous and Knudsen diffusion near room temperature, the equivalent activation energy arising from the square root of T would amount to less than 1 kcal/mole. In all three cases, the rate of diffusion is determined by the product

$$D = L \cdot u \cdot \exp(-E/kT)$$

where L is the distance between jumps, u is the average velocity, E is the activation energy (if any) between jumps.

The configurational region is reached when the molecular diameter is about the same as the channel diameter, such as for benzene with a diameter of 6 Å in ZSM-5 with a channel diameter of 5.4 Å [1]. See Fig. 2. It is not easy to decide on the diameter of a molecule, especially if it is not spherical. Various definitions of molecular diameters, such as the van der Waals, the Lennard-Jones, and the hydrodynamic diameters would lead to different results. Furthermore, the molecules are not rigid, and all the bond lengths and angles are flexible to some extent, especially at high temperatures. Experimentally, it is easy to adsorb benzene into ZSM-5, so that these nominal diameters should not be taken as rigid and inflexible. For a molecule in the shape of a spheroid or ellipsoid, the smaller diameter seemed the limiting factor on whether the molecule can squeeze into a tight channel. For molecules that are available in various conformations, such as the chair-boat form of cyclohexane and long chain normal paraffins, the most compact conformation would seem to be the relevant diameter. For elliptical channels, we take the smaller diameter as rate controlling.

Fig. 3 give the diffusion regions, plotted according to the ratio λ/d_M and d_C/d_M. The 45 degree line, where $\lambda = d_C$, divides the gaseous from the Knudsen regions. When $\lambda/d_M = 1$, we have the configurational regime.

2. Experimental results on configurational diffusion

The mean free path of oxygen at room temperature and pressure is about 1000 Å. We generally regard that the diffusion of oxygen, with a molecular diameter of 3-4 Å, in an amorphous silica-alumina with a channel diameter of 100 Å, to be in the Knudsen diffusion regime. For oxygen diffusion in A zeolite, with a channel diameter of 3-4 Å, the molecule of oxygen is almost never out of the influence of the channel wall, so that we have arrived at the regime of configurational diffusion. Perhaps, the molecule has a lower state of energy in the crossings of the three-dimensional channel network, which forms cavities that are much larger than the channels. Thus to travel from crossing to crossing, the molecule needs to acquire sufficient activation energy to overcome the energy requirements of the tighter channel. For oxygen diffusion in the 5-6 Å channels of ZSM-5 and in the 7-8 Å channels of the faujasites X or Y, it is not obvious whether we will encounter the activated configurational diffusion or the freer Knudsen diffusion, or approach gaseous diffusion in high occupancy cases. There may be a transition region so that the activation energy may gradually (or abruptly) change from 1 to 10 kcal/mol.

We are led by these arguments to expect that there may be a qualitative difference in the laws of diffusion for the "tight-fit" cases where the ratio of channel diameter/molecule diameter is one or smaller, such as for oxygen in A and for benzene in ZSM-5, and for the "loose-fit" cases where the ratio is greater than one, such as for oxygen in ZSM-5. Two molecules cannot pass each other in a tight-fit case, which has been named single-file diffusion. For a zeolite with a one-dimensional channel, such as mordenite, one can still describe the passage of a single molecule in the channel by random walk, which obeys the diffusion equation. However in high occupation cases, since a molecule cannot move past another molecule, the diffusion equation does not apply. Monte Carlo simulations show that on a square 2-D grid, the mean square displacement is proportional to time; but for a linear 1-D grid, the mean square diplacement is roughly proportional to the square root of time [2].

3. The various ways to define diffusivity

We need to define a few notions here. There are a number of ways to measure and to define diffusivity. In theory and in experimentation, the value of diffusivity depend on the methods of measurement and its definition [3]. In catalysis, the most relevant measure is that of <u>steady-state counter diffusion</u>. Catalysis must concern itself with a multi-component system where in the simplest case, a molecule A diffuses inside the catalyst, is converted into molecule B, then diffuses out. Catalysis is normally concerned with steady-state behavior. Thus any other way of measuring diffusivity would be indirect and may not be conclusive. Diffusivities measured at any other temperature, pressure, or molecular composition, would also be indirect. Thus the only diffusivity that is authoritative is derived from simultaneous measurements of diffusion and catalysis [4].

The steady-state Wicke-Kallenbach method of diffusion across a barrier is very difficult with these micron sized zeolite crystals. There are a number of recent attempts to embed giant crystals in membranes, but the problem of tight seals in elevated temperatures and pressures is very difficult to solve. The transient uptake method with a gravimetric Cahn balance has been used most extensively. Other methods currently used include the chromatographic column method. Experimental results from transient nuclear magnetic resonance methods are often several orders of magnitude different from the other methods, and are difficult to rely on for predicting catalytic behavior [5].

Another point of confusion in defining diffusivity concerns the nature of the driving force. It is common to use concentration as the driving force. One may use the gaseous concentration differences across a barrier as the driving force, or one may use the adsorbed concentration differences. The difference between these two definitions can be enormous when occupancy is high. If the adsorbed concentration and the gaseous concentration are related by a Langmuir isotherm, these two diffusivities can be related by the formula

$$D_\theta = D_p (1+Kp)^2$$

so that at high pressure and high occupancy, even small differences in occupancy would mean large differences in pressure. If D_p were constant, then D_θ would be highly concentration dependent.

Advocates of irreversible thermodynamics would like to use the chemical potential as the driving force, and use the notion of mobility [6,7]. This would result in yet another definition of diffusivity, and a different dependence on occupancy.

Figure 1.

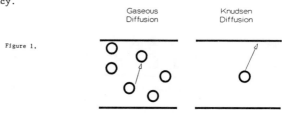

Gaseous Diffusion   Knudsen Diffusion

$D = L \ u \ exp(-E/kT)$

4. The influence of occupancy on configurational diffusivity

It is a common sense notion that if the streets are full of cars, traffic will slow down. Thus the usual expectation is that when occupancy goes up, diffusivity will go down. The simplest derivation of the occupancy effect assumes the Random Walk model, that a molecule has a tendency to move randomly to an adjacent site, provided that it has sufficient activation energy and that the adjacent site is not already occupied [8,9]. The probability that the adjacent site is occupied is simply the occupancy, which is defined as the number of molecules in the crystal divided by the maximum number of molecules that can be adsorbed. Therefore, we have the simple formula

$$D = D \ (\theta = 0) \ x \ (1 - \theta)$$

where θ is the occupancy [10,11]. There are other derivations which gives a faster decline of diffusivity with occupancy, such as $(1 - \theta)^2$ [2].

On the other hand, experimental values from uptake or chromatography of the effect of occupancy on diffusivity, not a very extensive list, may be divided into four classes:

(1) Diffusivity is independent of occupancy, such as methane in 4A zeolite [12].

(2) Diffusivity falling with occupancy, such as the loose-fit cases of oxygen in 5A [12].

(3) Diffusivity rising with occupancy, such as the tight-fit cases of benzene in ZSM-5 [2,13] and ethane in A [14].

(4) Diffusivity falling and then rising with occupancy, such as cyclohexane in 13X [15].
See Fig. 4.

The Markov theory teaches us that the self-diffusion diffusivity should be independent of occupancy, but the uptake-diffusivity should fall with occupancy, which agrees with class (2). It is not immediately obvious what would give rise to the rising behavior of class (3). Empirically, almost all the class (3) rising behavior derive from tight-fit cases, where the channel diameter is equal to or smaller than the molecular diameter.

There are two theories why this should happen: one based on irreversible thermodynamics, which suggests that rising class (3) behavior should always prevail; the other based on a Markov model, which suggests that rising class (3) behavior should prevail only if there is some swelling due to adsorption.

In the irreversible thermodynamics version, one uses the chemical potential as driving force for diffusion and derives the Darken equation [7]

$$D = D_0 \ \frac{\partial \ln a}{\partial \ln c}$$

Here the term D_0 is taken to be a "mobility", also termed the "corrected diffusivity", which is hopefully relatively independent of concentration effects. To use this Darken equation, one must have a relation between concentration and activity. If one identifies gaseous concentration with concentration, and adsorbed concentration with activity, we need the adsorption isotherm. If the adsorption isotherm is Langmuir in nature, then the Darken equation simplifies to:

Configurational Diffusion

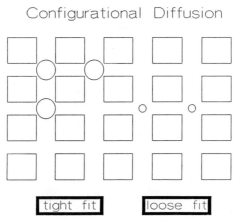

tight fit   loose fit

Figure 2.

Diffusion Regimes

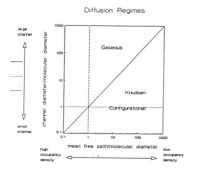

Figure 3.

Occupancy effects on diffusivity

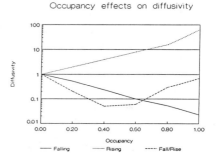

Figure 4.

$$D = D_0/(1-\theta)$$

If Do is constant, then D rises sharply when θ increases, and approaches infinity when θ approaches one. This behavior is approximated by the rising class (3) behavior of small molecules in zeolite A, and sometimes by benzene and toluene in ZSM-5. However, this equation runs into trouble when explaining falling class (2) behavior, as it would require that Do drops much faster than $1/(1-\theta)$ to compensate. There is no molecular mechanism in irreversible thermodynamics, and there is silence on why Do would behave differently in a tight-fit vs. a loose-fit situation.

The other approach is based on a Markov model of random walk diffusion, jumping from crossings to crossings when the molecule has sufficient activation energy to overcome the barrier between crossings. If the activation energy of diffusion is a linear and declining function of occupancy, rising class (3) behavior follows naturally but D does not approach infinity when θ approaches one [2].

But why does activation energy decline with occupancy? This phenomenon occurs mainly in tight-fit cases, where some squeezing or distortion is necessary to fit the larger molecules into the smaller channels. This may lead to a swelling of the channel, which means a lowering of activation energy of diffusion [16]. This type of behavior was observed in two other solid state situations, the diffusion of zinc in brass alloys and in polymer resins swelling by solvents. It may be possible to examine this hypothesis by x-ray diffraction of zeolite crystals before and after heavy loading. This theory avoids the dilemma of forcing all the data into one mode, it predicts rising class (3) behavior when there is swelling, and it predicts falling class (2) behavior when there is no swelling.

5. Conclusions

Occupancy effects become important in low temperature and high pressure regions, in cases of strong adsorption, and in liquid state catalysis. Even more bizarre behaviors are encountered in counter-diffusion situations. There is no theory today to deal with the behavior of long chain molecules, such as high carbon number normal paraffins, in zeolite channels. These are fruitful problems for the future.

References

[1] P.B. Weisz, Chemtech 3, 498 (1973).

[2] J. Tsikoyiannis, PhD Thesis, Massachusetts Institute of Technology, 1986.

[3] J. Karger and D. Ruthven, J. Chem. Soc., Faraday Trans. 77, 1485 (1981).

[4] W.O. Haag, R.M. Lago and P.B. Weisz, Faraday Discussion 72, 18 (1981).

[5] J. Karger and M. Bulow, Chem. Eng. Sci. 30, 893 (1975).

[6] L. Onsager, Phys. Rev. 37, 405 (1931).

[7] L.S. Darken, Trans. AIME 175, 184 (1948).

[8] D. Theodorous and J. Wei, J. Catalysis 83, 205 (1983).

[9] W.T. Mo and J. Wei, Chem. Eng. Sci. 41, 703 (1986).

[10] R.M. Barrer and W. Jost, Trans. Faraday Soc. 45, 928 (1949).

[11] L. Riekert, Advances in Catalysis 21, 281 (1970).

[12] D. Ruthven, Spec. Publ. Chem. Soc. 33, 43 (1980).

[13] A. Zikanova, M. Bulow, H. Schlodder, Zeolites 7, 115 (1987).

[14] M. Bulow, P. Struve, S. Pikens, Zeolites 2, 267 (1982).

[15] D. Ruthen, AIChE J. 22, 5 (1976).

[16] C.A. Fyfe, G.J. Kennedy, C.T. DeSchutter and G.T. Kokotailo, J. Chem. Soc. Chem. Commun. 1, 541 (1984).

MOLECULAR DIE CATALYSIS: HEXANE AROMATIZATION OVER Pt/KL

SAMUEL J. TAUSTER AND JOHN J. STEGER
Engelhard Corporation, Menlo Park, CN40, Edison, NJ 08818

ABSTRACT

It is proposed that the high aromatization selectivity of Pt supported in zeolite L is caused by the ability of the channels to collimate the flux of hexane molecules, leading to end-on adsorption at the Pt surface.

INTRODUCTION

Pt dispersed in the K^+-exchanged form of zeolite L (Pt/KL) is a recently discovered catalyst with exceptional selectivity for the aromatization of hexane. In tests at 520°C and 203 psig Bernard [1] found that Pt/KL made nearly five times as much benzene as dry gas (C_1 plus C_2), whereas a conventional reforming catalyst (Pt/Al$_2$O$_3$) gave a benzene/dry gas ration of only 1.2. This high aromatization selectivity has been confirmed in other laboratories [2]. However, an explanation for it has still not been established. It is important to understand such a striking change in Pt selectivity and, in particular, to establish what role the zeolitic nature of Pt/KL plays in this behavior. It should be noted in this regard that while metal particles within zeolitic channels are necessarily small (<15A), these dimensions are also found with well dispersed supported metals on alumina or silica. Thus, metal-particle-size does not provide a satisfactory explanation.

In this paper we present a hypothesis to account for Pt/KL's high aromatization selectivity. This hypothesis was prompted by a kinetic study in which several samples of Pt/KL were compared among themselves and compared, as a group, to a conventional Pt catalyst, Pt/SiO$_2$. Pt/SiO$_2$ was chosen for this purpose because it, like Pt/KL, is "monofunctional", meaning that all reactions occur at the Pt surface. Conventional Pt reforming catalysts, e.g., Pt/Al$_2$O$_3$, are termed "bifunctional" because the acidic nature of the support promotes carbonium ion mechanisms which contribute to activity for isomerization, cracking and aromatization. In the cases of Pt/KL and Pt/SiO$_2$, however, the supports are essentially nonacidic, so that carbonium ion processes make much smaller contributions to the overall catalysis. An important part of our studies concerned the cracking patterns of Pt/KL and Pt/SiO$_2$. It is well known that cracking catalyzed by solid acids favors mid-molecule cleavage, whereas this is not true for metal-promoted cracking. Therefore, it was important that Pt/KL, being essentially monofunctional, be compared only with other monofunctional catalysts in order to obtain useful information from their respective cracking patterns. It should also be noted that we compared Pt/KL and Pt/SiO$_2$ under conditions in which secondary cracking was minimal, so that their primary cracking patterns could be observed.

EXPERIMENTAL

Catalysts were tested for the aromatization of n-hexane in a stainless-steel downflow reactor at 510°C and a pressure of 100 psig. The space velocity was 50 w/w/hr and the H$_2$/n-C$_6$ ratio was six. The catalyst particles were 20/40 mesh and these were mixed with four times their weight of 20/40 mesh silica particles. Catalysts were reduced in situ in flowing H$_2$ at temperatures up to 525°C. The products were analyzed using on-line gas chromatography. The analysis for methane, however, was uncertain. Thus, we define a selectivity parameter for aromatization as the yield (wt %) of benzene divided by the combined yields of benzene plus C_2-through-C_5. Data was collected at 2 - 4 hours after introducing the n-C6 feed and, in some cases, after a further 16 -20 hours.

RESULTS AND DISCUSSION

Early in our investigations it became apparent that there were two important distinctions between Pt/SiO$_2$ and Pt/KL. As expected, we found Pt/KL to be much more selective for hexane aromatization. In addition, their cracking patterns were very different. Pt/KL showed a marked tendency to crack the terminal carbon-carbon bond in the hexane molecule, whereas Pt/SiO$_2$ showed no such tendency. As noted above, neither Pt/KL nor Pt/SiO$_2$ promote significant amounts of carbonium ion cracking. Thus, there was no ready explanation for their dissimilar cracking patterns.

A related observation involved the several Pt/KL preparations that were studied. The aromatization selectivities of these samples varied considerably, as well as their tendencies towards terminal cracking. It was noticed that these two features were, in fact, correlated.

This is seen in Figure 1, in which the aromatization selectivity parameter is plotted against the terminal cracking index (TCI). The latter is simply defined as the mole ration of pentanes to butanes in the cracked product. For Pt/SiO$_2$, the TCI is about 0.9, indicating a slight preference for cracking the next-to-terminal carbon-carbon bond versus the terminal one. For Pt/KL, however, most TCI values are much greater than one, and extend to more than double the Pt/SiO$_2$ value. Despite a considerable amount of scatter, there is an unmistakable trend for aromatization selectivity to increase along with the TCI. Thus, for TCI values below 1.2 the mean selectivity parameter is 0.58, whereas for TCI's greater than 1.7 the mean selectivity parameter is 0.79. It should be noted that while Pt/KL is essentially nonacidic, small amounts of residual acidity will have a significant (lowering) effect on the TCI, and this may well account for much of the scatter seen in Figure 1.

The wide diversity seen among the Pt/KL samples requires an explanation. It seems reasonable to relate this to differences in the distribution of Pt, i.e., between in-channel and out-of-channel locations. We assume that as the in-channel/out-of-channel ratio increases, the aromatization selectivity and TCI increase as well. Out-of-channel Pt is characterized by low selectivity and a low TCI. For Pt/SiO$_2$, all the Pt is, of course, "out-of-channel", and the selectivity and TCI are both very low.

Figure 1 shows two Pt/KL catalysts with about the same TCI as Pt/SiO$_2$, but with significantly greater selectivity. It is thus not clear that a single selectivity-versus-TCI curve applies to all monofunctional catalysts. Nevertheless, the data supports the assumption of an intrinsic connection between good aromatization selectivity and a high TCI. We seek now to account for such a relationship.

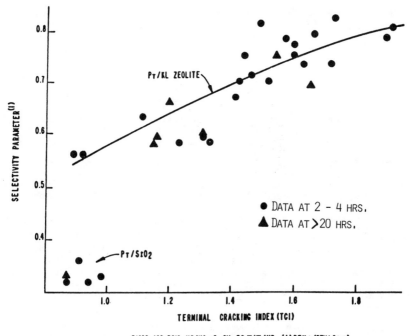

FIG. I

The Terminal-Adsorption Hypothesis

Clearly, the aromatization of hexane and the cracking of its terminal carbon-carbon bond cannot proceed from the same intermediate. Hexane aromatization over a monofunctional catalyst occurs via 1,6-closure [4]. This requires that the C1 and C6 carbon atoms coordinate in some way to the Pt surface. The latter mediates the introduction of electronic charge into the region between these atoms, leading to bond formation and ring closure. Hydrogenolysis of the C1-C2 bond, on the other hand, requires coordination either of the C1 and C2 atoms, or more probably, the C1 and C3 atoms, to the Pt metal surface [4]. Thus, the correlation between aromatization selectivity and TCI cannot reflect a common intermediate.

A possible explanation arises, however, if we focus on the initial adsorption of hexane at the Pt surface. Suppose this occurs, for a particular molecule, through the C1 atom. If the C6 atom subsequently attaches (this might involve a pi bond between the C5=C6 double bond and the metal) benzene may ensue. Of course, C1 coordination to the metal can have other outcomes. C1 attachment followed by C3 attachment will lead, as noted above, to cracking. However, if the initial attachment involves the C2 or C3 atom, benzene cannot form. Thus, if a catalyst could promote a higher probability of C1-initial-attachment, selectivity for hexane aromatization would increase. So, too, would the probability that cracking, if it occurred, would be terminal. We hypothesize that Pt/KL exhibits high hexane aromatization selectivity and a high TCI because it promotes a high degree of terminal adsorption and next consider why it should do so.

The 'Molecular Die' Hypothesis

We propose that the extraordinary hexane aromatization selectivity of Pt/KL is intrinsically related to its zeolitic nature. More specifically, we hypothesize that diffusing hexane molecules are oriented with their long axes parallel to the zeolite channel axis, leading to end-on approach to the Pt surface and terminal adsorption.

We believe there are two ways of justifying this 'molecular die' hypothesis. Attraction (i.e., van der Waals interaction) between a paraffin (or olefin) molecule and the channel wall would orient the molecule in the above-described manner. Alternatively, a simple geometrical description may suffice. These two rationales will be discussed in turn.

Weisz et al. [5] have shown that surprisingly large molecules can penetrate the intracrystalline volume of the zeolite ZSM-5 and undergo catalytic reaction there. An example is corn oil ($C_{57}H_{104}O_6$), the structure of which contains three n-C_{17} chains radiating from a central carbohydrate nucleus. Diffusion of such a molecule obviously requires each C_{17} chain to be oriented with its long axis parallel to the channel axis. It would seem that interactions between these chains and the channel wall are involved in such orientation.

Recent deuterium solid state NMR studies by Eckman et al. [6] have indicated that the para-xylene molecule in ZSM-5 is oriented with its long axis parallel to the channel axis. These measurements were carried out at temperatures up to 150°C but it is possible that the effect persists at higher temperatures as well.

Quite apart from molecule-zeolite attraction, it seems that simple geometrical effects may be capable of collimating the diffusive flux of hexane molecules. A butane molecule is about 7A long. The channels of zeolite L are one-dimensional, with a diameter that oscillates from about 7A to about 12A; the 7A "throats" recur every 7 1/2 A. Thus, molecules longer than n-butane will not be able to pass through the throat if oriented transverse to the channel axis. Of course, bent molecular conformations and zeolite "breathing" at high temperatures will relax these restrictions somewhat. Still, it seems plausible that some collimating of the diffusing hexane molecules will occur, leading to an increased likelihood of end-on approach to the Pt surface.

Finally, we wish to stress that the hypothesis described in this paper is quite distinct from the familiar concept of shape-selectivity. The classical example of the latter is the difference in reactivity between a linear and branched isomer, caused by the larger kinetic diameter of the branched molecule. No such distinction is involved in the proposed 'molecular die' effect, which pertains to the orientation of diffusing molecules, linear or branched, caused by the zeolite channels.

SUMMARY

The exceptionally high selectivity of Pt/KL for hexane aromatization correlates with its pronounced tendency to crack the terminal bond in the hexane molecule. The common cause of these effects is proposed to be the end-on approach of hexane molecules to the Pt surface, leading to attachment through the C1 atom. This end-on approach is ascribed to the collimating action of the zeolite channels.

REFERENCES

1. J.R. Bernard, Proc. 5th International Conference Zeolites, Naples,
    p. 686

2. T.R. Hughes, W.C. Buss, P.W. Tamm and R.L. Jacobson, presented at the
    1986 AICHE Spring Meeting, New Orleans, LA

3. F.M. Dautzenberg and J.C. Platteeuw, J. Catal. 19, 41 (1970)

4. J.R. Anderson and N.R. Avery, J. Catal. 5, 446 (1966)

5. P.B. Weisz, W.O. Haag, and P.G. Rodewald, Science 206, 57 (1979)

6. R.R. Eckman, and A.J. Vega, J. Phys. Chem. 90, 4679 (1986)

TEMPERATURE DEPENDENCE OF COPPER (II) CATALYZED OXIDATION OF PROPYLENE ON X ZEOLITE: RELATION TO COPPER (II) MIGRATION

JONG-SUNG YU AND LARRY KEVAN
Department of Chemistry, University of Houston, Houston, Texas   77004

ABSTRACT

The copper(II) catalyzed oxidation of propylene on CuK-X zeolite was studied over the temperature range of 150 to 400 °C in a flow system. Arrhenius activation energies could be obtained at less than 30 % conversion from 160 to 210 °C and were found to decrease with increasing copper(II) content. These results are correlated with electron spin resonance studies of the migration of copper(II) within the zeolite lattice at the reaction temperature. The change in activation energy with copper(II) content shows that a substantial part of the total activation energy is due to copper(II) migration.

INTRODUCTION

The nonacidic catalytic properties of zeolites are largely dependent on the type, amount and distribution of the exchangeable cations.[1,2] In such systems the location of a catalytically active transition metal ion must be a key aspect controlling the catalytic efficiency. At sufficiently low transition metal ion loadings, the transition metal ion will occupy only one type of site within the zeolite lattice and it should be possible to correlate the catalytic reactivity with site location.

We have recently studied[3,4] copper(II) catalyzed propylene oxidation over Cu(II)-exchanged X zeolites where the copper(II) location is monitored by its electron spin resonance (ESR) parameters which have been calibrated as indicative of location in the X zeolite structure by electron spin echo modulation spectroscopy.[5] It has been found that Cu(II) migrates during reaction from inaccessible site SI in the hexagonal prism of the X zeolite structure to accessible site SII' near the six-ring between the α- and β-cages at which site coordination with propylene occurs prior to reaction.

In the present work we report the temperature dependence of propylene oxidation over CuK-X zeolite as a function of Cu(II) content and find evidence that a substantial part of the total activation energy is associated with Cu(II) migration prior to reaction.

EXPERIMENTAL

Linde NaX (Si/Al = 1.4) zeolite from Alpha Chemicals was exchanged with 0.1 M sodium acetate solutions to decrease any excess Fe^{3+} impurities and was then exchanged with 0.05 M solutions of potassium acetate. To obtain maximum cation incorporation the exchange was carried out four times at 70 °C. These zeolites were exchanged with different volumes of 10 mM cupric nitrate solutions per 1 g of zeolite. The zeolite was further washed with triply distilled water and dried in air at room temperature. The zeolites are denoted by Cu(x)K-X where x gives the number of Cu(II) ions per unit cell.

Experiments were carried out in a fixed type reactor with continuous gas flow at atmospheric pressure. The catalyst (0.4 - 1.0 g) was placed in an electrically heated Pyrex 'U-tube' reactor, heated at 400 °C for 12 h in a stream of oxygen and then brought to the reaction temperature. The reac-

Mat. Res. Soc. Symp. Proc. Vol. 111. ©1988 Materials Research Society

426

tion temperature was monitored by a thermocouple in a thermowell located at
the center of the reactor tube. The reactant gas consisted of 10 mole %
propylene, 50 mole % oxygen, and 40 mole % helium and the total flow rate
was about 15 cm³/min.

Reactants and products (CO_2) were analyzed on-line with a sampling valve
to a Varian model 1400 gas chromatograph equipped with a 0.32 cm o.d. by 183
cm long stainless steel column with 80/100 mesh Porapak N maintained at 120
°C. The propylene conversion percentages were reproducible to better than
10 % average deviation.

RESULTS AND DISCUSSION

The cupric ion catalyzed propylene oxidation reaction is shown in expression (1).

$$C_3H_6 + 4.5\ O_2 \xrightarrow[\text{flow system}]{Cu^{2+}} 3\ CO_2 + 3\ H_2O \tag{1}$$

Under our conditions of excess oxygen the propylene is oxidized to comple-
tion as shown by the equation above. It was shown that the percent of propylene conversion was dependent upon the number of cupric ions per unit cell
in the zeolite demonstrating that this is indeed a cupric ion catalyzed
reaction.[3]

Figure 1 shows the temperature effect on the catalytic propylene oxida-
tion reaction at low conversion (< 30 %) in Cu(II)-exchanged X zeolite with
Cu(II) varying from 1 to 3.9 per unit cell. Note the monotonic effect on
Cu(II) content.

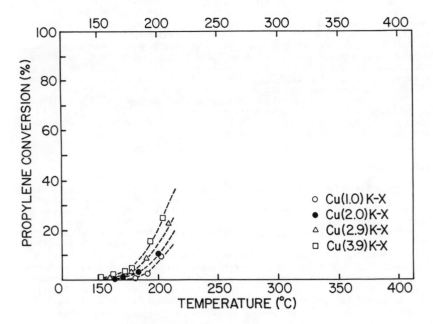

Figure 1. Effect of the reaction temperature on the oxidation of propylene
by CuK-X zeolite. Total reaction time 30 min.

Arrhenius-type plots of the temperature data are shown in Figure 2. These plots are linear in the temperature range of 160 - 210 °C corresponding to less than ~ 30 % conversion. The activation energies are determined by least squares linear fits; they are given in Table I and show a monotonic trend with Cu(II) content.

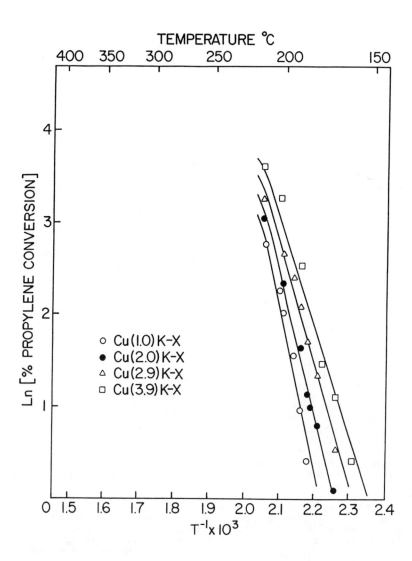

Figure 2. Arrhenius-type plots of propylene oxidation over CuK-X zeolites.

Table I. Activation Energies for Propylene Oxidation over CuK-X Zeolites
with Different Copper (II) Content

Zeolite	Activation Energy[a]
Cu(1.0)K-X	160 kJ mol^{-1}
Cu(2.0)K-X	128
Cu(2.9)K-X	112
Cu(3.9)K-X	106

[a] Estimated errors are ± 5 kJ mol^{-1}

At Cu(II) contents of four or less Cu(II) per unit cell in X zeolites,
ESR data have shown that the Cu(II) occupies mainly SI sites in the hexagonal
prism. There are 12 SI sites per unit cell. ESR and electron spin echo data
have also shown the Cu(II) migrates from site SI to site SII' where it coordinates with propylene to effect reaction. The decreasing activation energy
with increasing Cu(II) content suggests that the activation energy associated
with this Cu(II) migration depends on the number of Cu(II) per unit cell.
With such a small number of migrating or diffusing species per macroscopic
reactor, i.e. the unit cell, the probability of Cu(II) getting to site SII'
depends on the small number of Cu(II) per unit cell.

This interpretation suggests that the activation energy for reaction,
once Cu(II) has complexed with propylene, is less than 106 kJ mole^{-1}. The
only apparently comparable literature data is that of Gentry et al.[6] who
found an activation energy for propylene oxidation in excess oxygen over
Cu(8.6)Na-X of 76 kJ mole^{-1}. This was the same as their activation energy
of 77 kJ mole^{-1} for propylene oxidation over Na-X where the reaction is not
Cu(II) catalyzed but is acid catalyzed. The data of Gentry et al. is probably not an appropriate comparison because CuNa-X is about ten times less reactive than CuK-X for propylene oxidation.[3]

The other possibly comparable literature data is that of Mochida et al.[7]
who studied propylene oxidation in excess oxygen over CuNa-Y zeolite with
very high, but unspecified, Cu(II) loading. They report an activation energy
of only 18 kJ mole^{-1}. At high Cu(II) loading no Cu(II) migration is necessary prior to reaction so their low activation energy might be construed to
be associated with the reaction after Cu(II)-propylene coordination occurs.
However, their measured temperature range was in the 250 - 400 °C range which
is considerably higher than our temperature range.

We conclude that the activation energy of propylene oxidation catalyzed
by CuK-X zeolites consists of two parts. One part seems due to Cu(II) migration within the zeolite lattice before Cu(II)-propylene coordination and
this depends significantly on the Cu(II) content or loading per unit cell.
For one or less Cu(II) per unit cell this activation energy is at least 54
kJ mole^{-1} based on the decrease observed in Table I. The other part of the
activation energy is associated with the propylene oxidation process once
propylene is coordinated to Cu(II) and this part from the literature seems
to be between 20 to 75 kJ mole^{-1}. More comprehensive work is in progress

to clarify this activation energy interpretation.

ACKNOWLEDGEMENT

This research was supported by the National Science Foundation (CHM-8514108) and the Texas Advanced Technology Research Program.

REFERENCES

1.   I.E. Maxwell, Advances in Catalysis, Vol. 31. (Academic Press, New York, 1982), p. 1.

2.   J.H. Lunsford, Catal. Rev. 12, 137 (1975).

3.   H. Lee and L. Kevan, J. Phys. Chem. 90, 5781 (1986).

4.   J.S. Yu, H. Lee and L. Kevan, Proc. 10th North Amer. Catal. Soc. Meeting, (1987), in press.

5.   L. Kevan, Acct. Chem. Research 20, 1 (1987).

6.   S.J. Gentry, R. Rudham and M.K. Sanders, J. Catal. 35, 376 (1974).

7.   I. Mochida, S. Hayata, A. Kato and T. Seiyama, J. Catal. 15, 314 (1969).

THE CHARACTERIZATION OF HYDROCARBON INTERMEDIATES IN H-ZSM-5

*T. J. GRICUS KOFKE, R. J. GORTE AND **W. E. FARNETH
Department of Chemical Engineering, University of Pennsylvania, Philadelphia, PA 19104
**E. I. DuPont de Nemours and Comp., Central Research and Development, Wilmington, DE 19898

ABSTRACT

We have examined the adsorption of simple alcohols and 2-propanamine on H-ZSM-5 zeolites with Si/Al$_2$ ratios between 38 and 520. Thermogravimetric analysis (TGA) demonstrated that most of the molecules display a clearly defined adsorption state corresponding to a coverage of one molecule per Al site. Temperature programmed desorption (TPD) and transmission infrared spectroscopy results for each of the molecules in this 1:1 adsorption state are consistent with adsorption being due to the transfer of a proton from the zeolite to the adsorbed molecule. These results provide additional evidence that carefully prepared H-ZSM-5 is a Bronsted acid, with one acid site per framework Al atom, in which all of the acid sites are identical in strength.

INTRODUCTION

In a series of recent publications, our research group has reported the results of adsorption studies for simple alcohols and toluene on a H-ZSM-5 zeolite with a Si/Al$_2$ ratio of 70.[1-5] An important result in each of these papers was the finding that, except for 1-butanol, an adsorption state corresponding to a coverage of one molecule per Al site could be stabilized in the zeolite. The reactivities of the molecules in the 1:1 adsorption complex varied in a manner which was consistent with a picture of adsorption in which proton transfer plays a major role and in which all of the Al sites possess identical acid strengths.

Our previous results suggest that Al concentration in high silica H-ZSM-5 affects only the number and not the strength of the acid sites. To test this hypothesis, we have examined the adsorption of methanol, 2-propanol, and 2-methyl-2-propanol on H-ZSM-5 zeolites with Si/Al$_2$ ratios of 38 and 520. While minor differences are observed in TPD due to secondary reactions, the results reinforce the idea of identical acid sites in a concentration of one per framework Al atom. We have also examined the affect of constituent group on the 1:1 adsorption complex by comparing our previous TPD-TGA curves for simple alcohols to those observed for 2-propanamine. Again for 2-propanamine, we observe a 1:1 adsorption stoichiometry with the Al sites. This 1:1 complex reacts to form propene and ammonia between 600 and 650K, independent of Al concentration, further strengthening the picture of discrete acid sites associated with framework Al atoms.

EXPERIMENTAL

The experimental techniques have been described in previous papers.[1-5] The zeolites were exposed to the various reagents by admitting the vapors above the reagents to the zeolite samples for approximately 1-4 minutes. The samples were then evacuated for times ranging from one to 24 hours with a turbomolecular pump in a chamber which had a base pressure of approximately 10^{-7} Torr. TPD and TGA experiments were then carried out simultaneously using a Cahn 2000 microbalance mounted within the vacuum chamber and a mass spectrometer interfaced with a microcomputer. The three zeolite samples with Si/Al$_2$ ratios of 38, 70, and 520 were obtained from the Mobil Oil Company in the NH$_4$-ZSM-5 form and will be designated as ZSM-5(38), ZSM-5(70), and ZSM-5(520) in this paper.

The Al concentrations were checked using ^{27}Al NMR of the hydrated samples and were found to be in good agreement with the values reported to us by Mobil. All adsorption experiments reported in this paper were performed on H-ZSM-5 prepared by heating NH_4-ZSM-5 to 700K in vacuum.

RESULTS AND DISCUSSION

We chose to compare the adsorption of methanol, 2-propanol, and 2-methyl-2-propanol on different zeolites because our previous results had demonstrated that the reactivity of each of these three alcohols was considerably different.[3,4] For H-ZSM-5(70), we found that a clearly identifiable adsorption complex with a 1:1 stoichiometry with the Al atoms could be identified for each of the alcohols. For methanol, this complex decomposed upon heating to give only methanol and the empty site. For 2-propanol, the adsorption complex reacted completely above 360K and desorbed as propene and water. For 2-methyl-2-propanol, reaction occurred rapidly at 295K and the water in the reaction product could be evacuated at this temperature.

The adsorption of each of these alcohols on the two other H-ZSM-5 samples gave very similar results. We will discuss each of the alcohols individually. For methanol, TPD measurements on all three zeolites showed mostly methanol desorbing from the samples in a broad peak centered at approximately 400K with a small amount of ether desorbing at higher temperatures on H-ZSM-5(38); however, the TGA results did show a dependence on the Al concentration. On the H-ZSM-5(38), the coverage of methanol following several hours of evacuation was approximately 0.8 molecules/Al rather than the 1.0/Al coverage observed for H-ZSM-5(70). This seems to imply that a small fraction (approximately 20%) of the Al atoms in the zeolite are not associated with acid sites. This could be due to incomplete exchange of the zeolite cations in the H-ZSM-5(38); therefore, we cannot determine if this difference is significant.

On H-ZSM-5(520), we were unable to remove methanol below a coverage of approximately 1.5 molecules/Al by extended evacuation of 46 hours. At least part of this difficulty in evacuating to a one molecule/Al coverage is due to the fact that measurement of coverage based on units of molecules/Al is very sensitive to small quantities of physically adsorbed species. That evacuation to 1.0 molecule/Al is difficult, even for H-ZSM-5(70), has already been shown by the fact that this coverage could not be achieved at 295K without high pumping speeds.[1] However, while most of the methanol on H-ZSM-5(520) desorbed in a peak at 400K, just as we found for the zeolites with higher Al concentrations, a quantity of methanol corresponding to approximately 0.2 molecules/Al site would not desorb, even when the sample was heated to 700K. The zeolite sample weight could be returned to its original weight following the adsorption and subsequent desorption of methanol only by heating the sample for 30 minutes in 10 Torr O_2. We tentatively attribute this small amount of methanol which would not desorb thermally as being due to methoxylation of silanol groups.[6] Due to the fact that these could be removed only by high temperatures in an oxidizing environment, we do not believe they are important in catalysis.

Our results for 2-propanol were easier to interpret. On H-ZSM-5(520), we were again unable to evacuate to below a coverage of 2.3 molecules/Al; however, slightly more than half of the alcohol desorbed unreacted below 400K. The rest of the 2-propanol desorbed as water and propene between 400 and 450K. Unlike the case for methanol, the sample weight returned to its initial value following desorption of 2-propanol. We conclude, therefore, that, above 400K, an adsorption state exists with one molecule at each Al site and that the molecules associated with this state desorb as propene and water. These results are essentially identical to those reported previously for 2-propanol on H-ZSM-5(70)[2,3]. For 2-propanol on H-ZSM-5(38), however, secondary reactions began to be important. On this zeolite sample, we were able to lower the weight of the sample to a coverage of 0.9/Al by evacuation for several hours. Of this 2-propanol, roughly 10% desorbed as unreacted alcohol in TPD, implying, as did the results for methanol, that approximately 20% of the Al atoms in this zeolite sample are not associated with acid sites. During TPD, we observed propene and water desorbing just above 400K, just as we found for the other two zeolite samples. However, between 420 and 450K, significant quantities of higher molecular weight olefins were observed in the TPD spectrum. Since propene oligomerization has been shown to occur very rapidly on H-ZSM-5 at 295K[2], these olefin products are almost certainly due to bimolecular reactions at the Al sites. This bimolecular chemistry is apparently observed for 2-propanol only with the H-ZSM-5(38) sample due to the higher concentrations of propene during desorption and the higher concentration of active sites for carrying out the oligomerization reaction.

The results for 2-methyl-2-propanol on the three different zeolites confirm our previous work which showed that this alcohol reacts rapidly at 295K on the Al sites[3] and provides further insight into the bimolecular oligomerization chemistry which can occur at these acid sites. On both H-ZSM-5(38) and H-ZSM-5(70), the weight change in the samples following exposure to the alcohol and evacuation corresponded to slightly more than one molecule/Al. No unreacted alcohol and only a small amount of water were observed in TPD, indicating that dehydration of the adsorbed molecules occurs at 295K and that the water produced by the reaction can be removed during evacuation. Most of the desorbing product in TPD consisted of higher molecular weight olefins desorbing in a peak centered at approximately 425K, with only a very small amount of butene being observed at lower temperatures. ^{13}C NMR measurements of the intermediates formed by adsorption of $(CH_3)_3{}^{13}COH$ on H-ZSM-5(70) demonstrated that this migration of the intermediates to form oligomer products occurs to some extent already at 295K.[7]

The oligomerization chemistry could be observed and controlled more readily on H-ZSM-5(520) due to its lower concentration of acid sites. Following exposure to the alcohol and three hours evacuation, the weight change for the zeolite corresponded to a coverage just above one/Al. Very little water and no unreacted alcohol could be observed in TPD; however, we did observe two olefin desorption features, with butene desorbing at 360K and higher molecular weight products desorbing at approximately 425K. Following 20 hours of evacuation, the coverage could be decreased to approximately 0.9 molecules/Al and only higher molecular weight olefins desorbed during TPD. Apparently, the intermediates formed by the dehydration reaction can desorb from the Al sites as butene at reasonably high rates at 295K. With the lower Al concentration on H-ZSM-5(520), some of this butene left the zeolite channels before undergoing additional reactions. However, even with the dilute concentration of acid sites found on H-ZSM-5(520), a significant fraction of the desorbing butene reacted with carbenium ion-like intermediates at the acid sites to form larger olefins which cannot desorb readily at 295K.

This readily explains why only larger molecules are observed in TPD on the two zeolites with higher Al concentrations. On these materials, the probability of a molecule reacting before it can diffuse out of the zeolite channels is much higher due to the higher concentration of both acid sites and olefin products. The dehydration chemistry which occurs at the Al sites at 295K for 2-methyl-2-propanol, however, appears to be completely independent of the Al concentration.

The adsorption of 2-propanamine was examined on H-ZSM-5(70) and H-ZSM-5(520) to see if the chemistry for this molecule was analogous to that observed for 2-propanol and, again, to see if any changes could be observed as Al concentration of the zeolite varied. On both of the zeolites examined, we were unable to evacuate to a coverage of one/Al site in reasonable times. Following 7-13 hours of evacuation, the coverage on H-ZSM-5(70) corresponded to two molecules/Al site and the coverage on H-ZSM-5(520) corresponded to almost four molecules/Al. However, during TPD, molecular 2-propanamine desorbed from both samples in a broad feature below 500K, leaving a clearly defined adsorption state on the surface with a coverage corresponding to a 1:1 stoichiometry with the Al sites. We interpret this result as indicating that the molecules adsorbed in excess of the stoichiometric quantity are physically adsorbed and are not associated with acid sites. Physically adsorbed 2-propanamine is more strongly bound than physically adsorbed 2-propanol and requires higher temperatures to desorb, but the presence of these molecules does not imply weak acid sites. The 1:1 adsorption complex present after heating either sample to above 500K appears to be best described as an isopropyl ammonium ion. First, IR spectroscopy showed that the hydroxyl peak associated with the protons at the Al site at 3605 cm^{-1} was strongly affected by the adsorption of 2-propanamine at this coverage, indicating that the amine molecules in this adsorption state are located at the Al sites. Furthermore, on both zeolites examined, this 1:1 adsorption complex reacted in TPD to desorb as propene and ammonia between 600 and 650K. This reaction gives additional evidence for the formation of a protonated species in the 1:1 adsorption complex. The 1:1 adsorption stoichiometry on both zeolites again implies that there is one acid site for every Al atom in the structure and the fact that the reaction temperature was independent of the Al concentration provides further evidence that the strength of the proton donor sites is independent of Al concentration.

CONCLUSIONS

All of the adsorption results discussed in this paper agree with the picture of the acid sites in H-ZSM-5 which we had developed previously.[3] For these high-silica zeolites, the sites are present in a coverage of one/framework Al atom. They are Bronsted sites in which each acid site is identical in strength and is independent of Al concentration.

ACKNOWLEDGEMENTS

We are grateful to the Mobil Oil Corporation for supplying us with the ZSM-5 samples. This work was supported by the NSF, MRL Program, under Grant DMR82-16718. Equipment was purchased with support from the NSF, Grant #CBT-8604492 and the University of Pennsylvania Research Fund.

REFERENCES

1) A. Ison and R.J. Gorte, J. Catal. 89, 150 (1984).
2) M.C. Grady and R.J. Gorte, J. Phys. Chem. 89, 1305 (1984).
3) M.T. Aronson, R.J. Gorte, and W.E. Farneth, J. Catal. 98, 434 (1986).
4) M.T. Aronson, R.J. Gorte, and W.E. Farneth, J. Catal. 105, 455 (1987).
5) W.E. Farneth, D.C. Roe, T.J. Gricus Kofke, C.J. Tabak, and R.J. Gorte, Langmuir, in press.
6) B.A. Morrow, L.W. Thompson, and R.W. Wetmore, J. Catal. 23, 334 (1973).
7) M.T. Aronson, PhD Thesis, University of Pennsylvania (1987).

ACTIVE SITES IN OXIDE-PROMOTED METAL CATALYSTS -- EVIDENCE FOR THE
SIGNIFICANCE OF PERIMETER SITES IN AlO_x AND MnO_x/Ni(111) CATALYSTS

YONG-BO ZHAO, TZER-SHEN LIN AND YIP-WAH CHUNG
Northwestern University, Department of Materials Science and Engineering,
and Catalysis Center, Evanston, Illinois 60208

ABSTRACT

MnO_x (0.2 < x < 0.5) and AlO_x (0.5 < x < 1.1) /Ni(111) model catalysts
were prepared by depositing the oxides onto clean Ni(111) surfaces under
ultrahigh vacuum conditions, with the average oxide coverage varying from 0
to 1.3 monolayers (ML) for MnO_x and 0 to 2.8 ML for AlO_x. CO chemisorption
over these catalysts at 200 K was used to titrate different adsorption
sites. Thermal desorption showed the usual prominent CO desorption peak at
415 K from the clean Ni(111) surface. This 415 K peak decreases with
increasing oxide coverage in an almost linear fashion, indicating a site-
blocking effect by the oxide. At the same time, a new thermal desorption
peak appears at 305 K from MnO_x/Ni(111) and 300 K from AlO_x/Ni(111),
reaching maximum intensity at some intermediate oxide coverage. For
MnO_x/Ni(111), this coincides with the appearance of a new C-O stretching
mode at 1620 cm^{-1}, which disappears upon heating the surface to 325 K for a
few minutes, suggesting that both the 305 K thermal desorption peak and the
1620 cm^{-1} loss peak represent the same adsorbed CO state. This is
attributed to CO adsorbed on MnO_x/Ni perimeter sites. Results from Monte
Carlo simulation and scanning tunneling microscopy are presented and
compared with CO chemisorption data. The significance of perimeter sites
in strong metal-support interaction is discussed.

INTRODUCTION

Ever since the term of strong metal-support interactions(SMSI) was
introduced by Tauster et al.[1], there have been numerous studies exploring
the nature of such interactions. The suppression of CO and H_2
chemisorption [1-4], the enhanced activity of certain chemical reactions
such as methanation [5-8], and the migration of the reduced support species
onto the metal surface upon high temperature reduction [2, 8-12], have been
accepted as the basic features related to SMSI.
When a metal surface is partially covered with reduced oxide patches,
there are three types of surface sites: uncovered metal sites, oxide
sites, and perimeter sites between the two. Recently, suggestions have
been made that new active sites could be created at the metal-oxide
interface which are responsible for the enhanced CO hydrogenation activity
[13-14]. So far, however, there has been little experimental evidence
presented.
In this paper, we report our recent CO chemisorption study over MnO_x
and AlO_x-modified Ni(111) model catalysts using Auger electron spectroscopy
(AES), thermal desorption spectroscopy (TDS) and high-resolution electron
energy loss spectroscopy (HREELS). In order to determine the surface
mophology of these model catalysts, we also made some preliminary studies
using scanning tunneling microscopy (STM).

EXPERIMENTAL

Specimen preparation and surface characterization were carried out in
an ultrahigh vacuum (UHV) chamber equipped with an Auger electron
spectrometer, a high-resolution electron energy loss spectrometer, a
quadrupole mass spectrometer, an ion gun, and a metal/oxide evaporator.

Mat. Res. Soc. Symp. Proc. Vol. 111. ©1988 Materials Research Society

The specimen used in these experiments was a Ni(111) single crystal disk about 1 cm^2 in surface area and 1 mm in thickness, which can be cooled down to 150 K or heated at up to 7 K/sec to 900 K. A schematic diagram of the chamber and detailed description on CO chemisorption experiments over MnO_x/Ni(1111) model catalysts can be found elsewhere[4].

In the case of CO chemisorption over AlO_x/Ni(111), metallic Al was first evaporated onto a clean Ni(111) surface. The amount of Al deposited was determined with a quartz crystal thickness monitor which has a resolution of 1/4 monolayer (ML) for Al. The specimen was oxidized under 1×10^{-5} Torr oxygen at room temperature for 60 minutes, and then reduced under 2×10^{-6} Torr hydrogen at 773 K for 15 minutes. From our experience, such a reduction treatment will produce an unoxidized and smooth nickel surface. The oxygen-to-aluminum atomic ratio x was calibrated using 35 eV (Al) and 503 eV (O) AES peak intensities. AES data showed that the oxygen-to-aluminum ratio after reduction was between 0.5 to 1.1. Each time after a surface was prepared, it was cooled down to 200 K and 300 Langmuirs (L) CO was intoduced into the chamber.

The core of the STM is a single-tube piezoelectric scanner mounted on a sliding specimen platform to facilitate specimen transfer. The scanner is sufficiently tightly coupled to the specimen platform that a set of three stainless steel supporting plates stacked on viton is adequate for vibration isolation. Digital images composed of 128 × 128 pixels were acquired in real time (three seconds per frame) so that many images were taken from one sample in a reasonable time to obtain good statistics.

RESULTS

(A) CO Desorption from Clean Ni(111), MnO_x/Ni(111), and AlO_x/Ni(111)

Several surfaces were studied with different oxide coverages. For MnO_x/Ni(111), the manganese-to-nickel Auger ratio ranged from 0 to 0.3, the latter value corresponding to an average oxide coverage about 1.3 ML [4]. Results of CO thermal desorption from these surfaces are shown in Fig.1. For AlO_x/Ni(111) surfaces, our quartz crystal thickness monitor showed that the amount of aluminum deposited varied from 0 to 4.2×10^{15} atoms/cm^2, the latter corresponding to an average oxide coverage about 2.8 ML. Results of CO desorption from those surfaces are shown in Fig.2. From clean Ni(111), the CO desorption peaks at 415 K. This 415 K peak intensity and the overall CO desorption decrease with increasing oxide coverage. Meanwhile, a new feature (which peaks at 305 K for MnO_x/Ni(111) and 300 K for AlO_x/Ni(111)) grows with increasing oxide coverage at first, reaches a maximum at some intermediate oxide coverage, and then decreases with higher oxide coverage until it is undetectable. Areas under these high and low temperature peaks are plotted as a function of AlO_x coverage in Fig.3.

(B) Monte Carlo Simulation of a Partially Covered FCC(111) Surface

From Fig.1 and 2 it is seen that the low temperature peaks attain a maximum at some intermediate oxide coverage. Since these low temperature peaks vanish at zero or full oxide coverage, it is unlikely that these CO were adsorbed on clean Ni or on the reduced oxide sites alone. Our interpretation at this moment is that the low temperature TDS peaks come from CO adsorbed at or near the perimeter of oxide islands on the Ni surface.

This interpretation is supported by a series of Monte Carlo simulation with the Ni(111) surface. In such a simulation, a 50 × 50 hexagonal matrix was employed to represent the Ni(111) surface, and a given amount of oxide was deposited on the surface randomly. The oxide was deposited as clusters of different sizes. In each simulation, the cluster size n was fixed but the clusters were allowed to overlap. All the oxide species formed above the first layer were allowed to redisperse onto the surface to simulate surface diffusion during reduction. The redispersion was repeated r times.

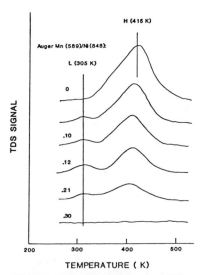

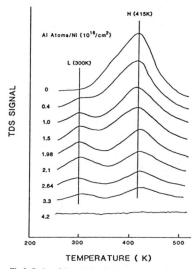

Fig.1. Series of thermal desorption spectra of CO fro the MnO$_x$-modified Ni(111) surface with different MnC coverage as indicated by the Auger Mn (589 eV)/Ni (8 eV) peak intensity ratio varying from 0 to 0.3. C chemisorption was made by introducing 30 L CO into t UHV chamber while the specimen was at 200 K. Duri the thermal desorption process, the surface was heatec about 7 K/sec.

Fig.2. Series of thermal desorption spectra of CO from the AlO$_x$-modified Ni(111) surface with different AlO$_x$ coverage as indicated by the number of Al atoms on Ni varying from 0 to 4.2×10^{15}/cm^2. CO chemisorption was made by introducing 300 L CO into the UHV chamber while the specimen was at 200 K. During the thermal desorption process, the surface was heated at about 7 K/sec.

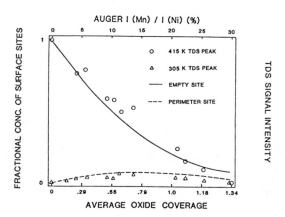

Fig.3. Superimposed CO thermal desorption from MnO$_x$/Ni(111) and Monte Carlo simulation results. For the TDS results, areas under both the 415 K TDS peak and the 305 K TDS peak are plotted as a function of average MnOx coverage. The circles stand for the 415 K peak and the triangles for the 305 K peak. For the simulation results, fractional surface concentration of a partially covered FCC(111) surface is plotted as a function of average oxide coverage. The oxide cluster size is fixed such that it blocks 169 surface sites. Oxide surface diffusion during reduction is simulated by allowing those adatoms beyond the first oxide layer to redisperse randomly three times onto the (111) surface with the same cluster size of 169 atoms. The solid curve stands for the uncovered sites and the broken one for perimeter sites. The top scale shows the Auger Mn (589 eV)/Ni (848 eV) peak intensity ratio (%). The bottom one gives the corresponding approximate average MnO$_x$ coverage, as inferred from seperate Auger calibration.

The perimeter site is defined as uncovered Ni next to oxide sites [4]. The concentration of clean Ni sites, oxide-covered sites, and perimeter sites was then calculated as a function of the oxide coverage, the cluster size n, and the number of times r the overlapping oxide is allowed to redisperse. Assuming a one-to-one correspondence between the number of surface adsorption sites (clean Ni and perimeter sites) and the amount of CO desorbed, a good match to the experimental data was obtained when n = 169 and r = 3. The fractional concentration of clean Ni and oxide-covered sites, using n = 169 and r = 3, are superimposed onto the TDS data for all oxide coverage in Fig.3. The agreement between the experimental results and the simulation is quite remarkable.

(C) HREELS Results of CO/Ni(111) and CO/MnO$_x$/Ni(111)

Fig.4(a) shows the HREELS spectrum of CO adsorbed on a clean Ni(111) surface at 200 K. The major peak at 1943 cm^{-1} is from C-O stretching of bridge bonded CO on Ni [15-17]. Fig.4(b) shows the HREELS spectrum of CO adsorbed on a Ni(111) surface partially covered with MnO$_x$ at 200 K. Compared with Fig.4(a), two new features are seen here: a high frequency shoulder at 2050 cm^{-1} and a weak peak at 1620 cm^{-1}. The 2050-cm^{-1} shoulder can be characterized as CO adsorbed on one-fold sites or on cationic sites [15-17]. A similar observation was made previously from CO on TiO$_x$/Ni [2]. For the 1620-cm^{-1} peak, we interpret it as C-O stretching due to CO adsorbed at perimeter sites with its oxygen end activated by oxygen-deficient MnO$_x$ species (see discussion below). A similar low stretching frequency of CO and

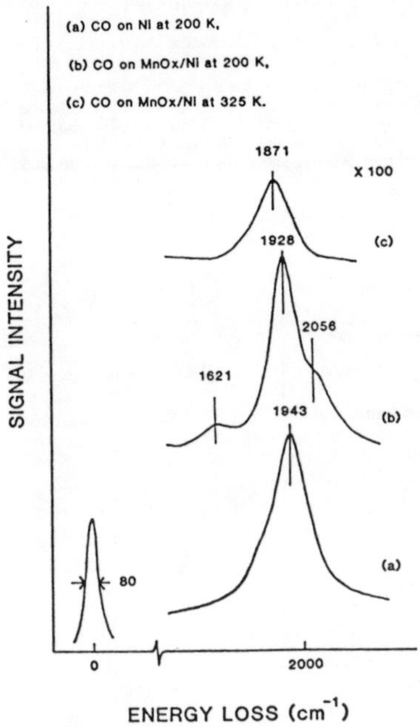

(a) CO on Ni at 200 K,

(b) CO on MnOx/Ni at 200 K,

(c) CO on MnOx/Ni at 325 K.

1871

X 100

1928 (c)

2056

1621

1943 (b)

(a)

→ ← 80

SIGNAL INTENSITY

0   2000

ENERGY LOSS (cm^{-1})

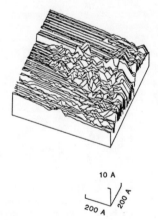

10 A

200 A

Fig.4(a) HREELS spectrum of 30 L CO adsorbed on Ni(111) at 200 K;
(b) HREELS spectrum of 30 L CO adsorbed on a MnO$_x$-modified Ni(111) surface at 200 K, Auger Mn (589 eV)/Ni (848 eV) = 0.11;
(c) Surface of Fig.4(b) after being heated at 325 K for 5 min.

Fig.5. STM image from an MnO$_x$-covered Ni(111) surface. The area is 800 A x 800 A. The oxide coverage is about 1 monolayer.

interpretation were reported previously with a Mn-promoted Rh/SiO_2 catalyst [18-19]. Fig.4(c) was taken at room temperature after the surface shown in Fig.4(b) was heated to 325 K for 5 minutes. The 2050 cm^{-1} shoulder and 1620-cm^{-1} peak disappeared immediately. This suggests that the 305 K TDS peak and the 1620-cm^{-1} HREELS peak represent the same adsorbed CO species.

(D). *Preliminary STM Observation of $MnO_x/Ni(111)$*

Fig.5 shows a Y-modulation image of Ni(111) covered by about 1 ML of MnO_x, obtained in the constant current mode. One must be cautious in interpreting the image due to the dependence of the tunneling current on both the topography and the electronic properties of the surface. Assuming topography to be the major factor, one can see the apparent nucleation of (oxide) patches on Ni(111). Further studies are still being pursued.

DISCUSSIONS

TDS studies show that with the presence of MnO_x or AlO_x on Ni surface, CO chemisorption is suppressed. The suppression varies linearly with the oxide coverage, implying a site-blocking effect [3]. For both model catalysts, the CO adsorption was completely suppressed at an average oxide coverage greater than 1. This indicates that three-dimensional oxide islands are formed on the Ni surface.

Results of the present study show that new chemisorption sites are created when a nickel surface is modified by reduced oxides. Although the exact configuration of these new sites is unknown, the fact that their presence requires the surface being partially covered by reduced oxides strongly suggests the involvement of the metal as well as the reduced oxide sites. At this moment, we favor a model similar to that proposed by Vannice, Sachtler, and their coworkers [14, 19]. In such a model, taking MnO_x/Ni as example, each new site consists of both MnO_x at the oxide island perimeter and Ni in the immediate vicinity, with the CO adsorbed on the Ni and the oxygen end of the CO activated by the MnO_x species. Our TDS and HREELS results reveal that such adsorbed CO has weakened C-Ni and C-O bonds, as indicated by the lower thermal desorption temperature and lower C-O stretching frequency.

CO chemisorption also indicates that at the optimum oxide coverage, the concentration of these perimeter sites is only about 10% of the available metal surface atoms. This does not contradict the proposed model that these perimeter sites are the active sites in CO hydrogenation because it is found that the number of active sites in a CO hydrogenation reaction is a very small fraction of the total metal surface sites [6, 13, 20]. It has also been found that, once the metal/reduced oxide ensembles are formed, they cannot be fully oxidized easily [21]. Therefore, the SMSI effects are quite durable under CO hydrogenation reaction conditions[22]. This indicates that the chemistry at these sites is not due to a simple superposition of the metal and the reduced oxide.

CONCLUSIONS

CO chemisorption on Ni(111) is suppressed almost linearly with increasing MnO_x or AlO_x coverage, implying a site-blocking effect. Complete suppression occurs at oxide coverage greater than one indicates three-dimensional island formation of these oxides on the Ni surface. The formation of these islands has been observed by STM.

From $MnO_x/Ni(111)$, CO chemisorption at metal/oxide perimeter sites is characterized by a thermal desorption peak at around 300 K, and a 1620-cm^{-1} HREELS peak. These metal/oxide ensembles are postulated to be the active sites in CO hydrogenation.

ACKNOWLEDGMENTS

Acknowledgment is made to the donors of the Petroleum Research Fund, administered by the American Chemical Society, for supporting this work.

REFERENCES

1. S.J.Tauster, S.C.Fung and R.L.Garten, *J.Amer.Chem.Soc.*100, 170 (1978).
2. S.Takatani and Y.W.Chung, *J.Catal.*90, 75 (1984).
3. D.J.Dwyer, S.D.Cameron and J.Gland, *Surf.Sci.*195, 430 (1985).
4. Y.B.Zhao and Y.W.Chung, *J.Catal.*106, 369 (1987).
5. M.A.Vannice and R.L.Garten, *J.Catal.*56, 29 (1979).
6. M.A.Vannice, S-Y.Wang and S.H.Moon, *J.Catal.* 71, 152 (1981).
7. R.A.Demmin, C.S.Ko and R.J.Gorte, *in* "Strong Metal-Support Interactions", ACS Symposium Series 298, p.48.Amer.Chem.Soc., Washington, DC.
8. Y.W.Chung, G.X.Xiong and C.C.Kao, *J.Catal.*85, 237 (1984).
9. X.Z.Jiang, T.F.Hayden and J.A.Dummesic, *J.Catal.*83, 168 (1983).
10. R.F.Hicks, Q.J.Yen and A.T.Bell, *J.Catal.*89, 498 (1984).
11. Y.-M.Sun, D.N.Belton and J.M.White, *J.Phys.Chem.*90, 5178 (1986).
12. Y.W.Chung and Y.B.Zhao, *in* "Strong Metal-Support Interactions", ACS Symposium Series 298, p.54.Amer.Chem.Soc., Washington, DC.
13. R.Burch, and A.R.Flambard, *J.Catal.*78, 389 (1982).
14. M.A.Vannice and C.Sudhakar, *J.Phys.Chem.*88, 2429 (1984).
15. W.Erley, H.Wagner and H.Ibach, *Surf.Sci.*80, 612 (1979).
16. J.C.Campuzano and R.G.Greenler, *Surf.Sci.*83, 301 (1979).
17. W.Erley, H.Ibach, S.Lehwald and H.Wagner, *Surf.Sci.*83, 585 (1979).
18. M.Ichikawa and T.Fukushima, *J.Phys.Chem.*89, 1564 (1985).
19. W.M.H.Sachtler, D.F.Shriver, W.B.Hollenberg and Lang, A.F., *J.Catal.*92, 429 (1985).
20. M.A.Vannice and C.C.Twu, *J.Catal.*82, 213 (1983).
21. C.M.Greenlief, J.M.White, C.S.Ko and R.J.Gorte, R.J., *J.Phys.Chem.*89, 5025 (1985).
22. D.J.Dwyer, J.L.Robbins, S.D.Cameron, N.Dudash and J.Hardenbegh, *in* "Strong Metal-Support Interactions", ACS Symposium Series 298, p.21.Amer.Chem.Soc., Washington, DC.

THE EFFECT OF SODIUM CATALYST DISPERSION ON THE
CARBON DIOXIDE GASIFICATION RATE

JIAN LI AND ADRIAAN R.P. van HEININGEN, Pulp and Paper Research Institute
of Canada and McGill University, Department of Chemical Engineering,
Montreal, Quebec, Canada, H3A 2A7

ABSTRACT

The CO_2 gasification rate of black liquor char (blc) is studied in a
thermogravimetric analysis set-up at temperatures between 600 to 800° C.
Blc is prepared via fast pyrolysis of the dry solids in spent liquor of
the kraft wood pulping process. Blc gasification by CO_2 is well described
by Langmuir-Hinshelwood type kinetics. The gasification rate of blc is
one order of magnitude larger than a high surface area activated carbon
impregnated with 12% Na_2CO_3. Also, the gasification rate of blc remains
high at sodium-carbon ratios where the rate of Na_2CO_3 impregnated chars
would be strongly reduced. With SEM-EDS mapping and line scans it is shown
that the unique gasification properties of blc are caused by a very fine
distribution of sodium in the carbon structure.

INTRODUCTION

The gasification of carbonaceous materials catalyzed by alkali metal
salts has been the subject of numerous studies. Comprehensive reviews have
been published by McKee [1], Wen [2], and Wood and Sancier [3]. Recent
research has provided more evidence that oxide groups on the carbon surface
are involved in the gasification [4,5]. In the latest mechanism proposed
by Sams and Shadman [5], alkali metal carbonate is first reduced

$$M_2CO_3 \rightleftharpoons (- CO_2M) + (- COM) \qquad (1)$$

followed by the redox sequence

$$(- CO_2M) + C \rightleftharpoons (- COM) + CO \qquad (2)$$

and

$$(- COM) + CO_2 \rightleftharpoons (- CO_2M) + CO \qquad (3)$$

where $(- CO_2M)$ and $(- COM)$ are believed to be carboxylic and phenolic
groups observed in IR studies [6,7]. It is also suggested that reaction
(2) is the rate determining step. From this mechanism it follows that the
number of alkali metal oxide surface groups, and thus the dispersion of
alkali on the carbon surface is an important factor determining the rate of
carbon gasification.

Black liquor is the spent liquid resulting from the digestion of wood
with a solution of NaOH and Na_2S. During pulping about 50% of the wood
substance, primarily hemicelluloses and lignin, is solubilized. Black
liquor char (blc) is the product found at the bottom of the furnace where
energy and chemicals are recovered from concentrated black liquor. The
char consists mainly of a mixture of carbon and Na_2CO_3 plus some Na_2SO_4 and
Na_2S. The porous char provides the conditions required for reduction of
sodium sulfate to sodium sulfide [8].

Blc is a unique carbonaceous material, since it is produced from a
liquid in which sodium is mixed and chemically bound with the char precursor on a molecular scale. This leads to a three dimensional and presumably
fine dispersion of the alkali metal catalyst. The gasification properties
of blc differ from other chars doped with alkali metal carbonate. These
chars are normally obtained by impregnation or by pyrolysis of a powdered
mixture of char and alkali metal carbonate. For example, the gasification
rate of "slow pyrolysis" blc is more than one order of magnitude higher
than coal char with an optimum alkali metal catalyst loading [9]. Also the
blc gasification rate remains high at sodium metal-carbon ratios far ex

Mat. Res. Soc. Symp. Proc. Vol. 111. ©1988 Materials Research Society

ceeding the optimum ratio for doped coal chars.

In this paper the gasification rate of "fast pyrolysis" blc is compared with alkali metal catalyzed gasification of activated carbon. The degree of sodium dispersion is quantified with the SEM-EDS technique with line scanning and used to explain the differences in gasification behavior.

EXPERIMENTAL

Black liquor, obtained by cooking black spruce wood chips to about 49% pulp yield, was oxidized under 200 psig oxygen at 130°C to convert all inorganic sulfur in Na_2SO_4. The liquor was subsequently dried below 120°C as a thin film on glass or teflon. This method was adopted to prevent enrichment of the inorganic salts in the mother liquor during drying. The chemical analysis of the black liquor solids is shown in Table I. It can be calculated from Table I that about 50% of the sodium is bound to SO_4^{2-} and CO_3^{2-}, indicating that the remaining sodium is chemically linked to lignin and hydroxy acids. After scraping from the drying surface, the solids were pyrolyzed for 20 minutes under nitrogen with 10% CO in a tube furnace preheated to 580°C. The volume of the pyrolyzed char is one to two orders of magnitude larger than the original volume of the solids. The solid to

TABLE I

Chemical Composition of Black Liquor Solid and Char

Black Liquor Sample	$[SO_4^{2-}]$ %	$[S^{2-}]$ %	$[SO_3^{2-}]$ %	$[CO_3^{2-}]$ %	$[Na^+]$ %	$[S_2O_3^{2-}]$ %
Solid	8.13	0.0	0	7.2	18.6	0
Char	< 0.1	5.7	0	33.0	33.2	0

char yield is about 70%. The char was ground and the fraction passing a 500 mesh (< 25 μm) was used. The sodium carbonate loading of blc just before gasification is 57.6%. A high surface area (2000 m^2/g) acid washed activated carbon from AC. Carbon Canada was also studied as a reference material. The coconut shell based active carbon was treated with boiling 4% HCl and washed with deionized water. The washed carbon was subsequently ground, and the fraction passing 500 mesh was impregnated with a Na_2CO_3 solution and dried at room temperature. The Na_2CO_3 loading of activated carbon is 12.8%.

A Cahn TGA 113 DC thermogravimetric analysis sytem was used for the gasification study. High purity carrier (He and N_2) and gasification (CO and CO_2) gases were used. In addition all gas lines contain an oxygen trap, while CO is further treated in a CO_2, H_2O and hydrocarbon trap because of its relatively low (99.9%) purity. A flat dish shaped pan of high purity alumina (99.8%) was used to eliminate chemical interaction with the sample and minimize external and internal mass transfer resistances. A GC with FPD was used for analysis of sulfurous gas components in the exhaust. CO and CO_2 were measured continuously with IR analyzers. Sulfate, sulfide, sulfite, thiosulfate, carbonate and sodium were measured with a Dionex IC with TCD and ECD. A JSM-T3000 was used for the SEM-EDS analysis.

After adding about 5 mg to the pan, the TGA system is evacuated three times and refilled with N_2 or He. Then the furnace temperature is raised from 20°C to 775°C at a rate of 25°C/min under a He or N_2 flow rate of 400 scc/min. In addition, 25 scc/min of CO is added above 500°C. After 2 minutes at 775°C, the temperature is reduced to the required reaction temperature. When the temperature and sample weight are stabilized, the CO concentration is adjusted and CO_2 is added to obtain the desired gasification atmosphere.

RESULTS AND DISCUSSION

A typical temperature and weight-loss curve for blc is shown schematically in Fig. 1. The small weight loss below 500°C is due to release of H_2O, CO_2 and CO. An apparent weight increase occurs at 500°C because of the increased drag when CO is introduced. CO is added to prevent

$$(- \text{CONa}) + C \rightleftharpoons (- \text{CNa}) + CO \tag{4}$$

and

$$(- \text{CNa}) \rightarrow \text{Na}_{vap} \tag{5}$$

which lead to vaporization of sodium and a weight-loss [6,10]. The weight-loss recorded between about 600°C and the final gasification temperature is caused by reduction of Na_2SO_4 to Na_2S with carbon [10]. A sharp weight increase is obtained when CO_2 is added as a result of the increased gas velocity and the fast reaction (3) [6]. The chemical composition of blc just before gasification is given in Table I.

Two reactions determine the weight change during gasification:

$$C + CO_2 \rightleftharpoons 2CO \tag{6}$$

and

$$Na_2S + 2CO_2 \rightleftharpoons Na_2CO_3 + COS \tag{7}$$

The carbon weight-loss is obtained after correction for the weight increase by reaction (7). The results for a typical run are shown in Fig. 2. The COS emission rate shown in the small graph in Fig. 2 is multiplied by 28/60 and then integrated to obtain the weight increase by reaction (7). The instantaneous rate of gasification is defined as:

$$- R_w = - \frac{1}{W} \frac{dW}{dt} \tag{8}$$

where W is the carbon weight at time t.

It is generally agreed that CO_2 gasification below 1300°K is governed by Langmuir-Hinshelwood (L-H) type kinetics [11], given as

$$- R_w = \frac{K \left[CO_2 \right]}{1 + K_{CO_2}[CO_2] + K_{CO}[CO]} \tag{9}$$

where K is the reaction constant, and K_{CO_2} and K_{CO} the adsorption constants Accordingly, the inverse of the carbon gasification rate of blc at conversions of 25%, 50% and 75% was plotted as a function of [CO] and $[CO_2]^{-1}$. A good linear relationship was obtained confirming the validity of L-H kinetics for black liquor char. The constants determined from these plots are listed in Table II. Also included in Table II are kinetic data from literature for some selected carbonaceous materials. It can be seen that the reactivity and CO adsorption of blc are much higher and lower, respectively, than the other materials at similar temperatures. The activation energy of blc gasification from 675 to 775°C was found to be 250 kJ/mol.

The gasification rate of Na_2CO_3 impregnated activated carbon and blc are plotted versus carbon conversion in Fig. 3. Generally the gasification rate of activated carbon is one order of magnitude lower than blc. To prove that the catalyst dispersion is responsible for the difference in

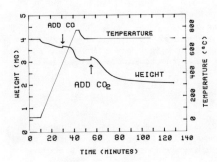

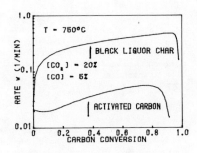

Fig. 1 Schematic picture of temperature and weight-loss curves
from TGA.

Fig. 3 Gasification rate for blc
and activated carbon.

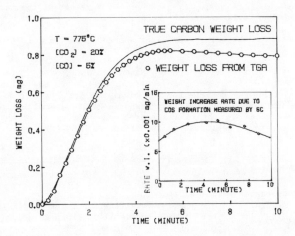

Fig. 2 Measured weight loss and carbon weight loss.

reactivity between blc and impregnated activated carbon, both samples were
examined by SEM-EDS.

Figs. 4a and 4b show a SEM-EDS picture of the sodium distribution on
the surface of blc and activated carbon, respectively. The sodium is
represented by the bright or colored spots. When comparing Fig. 4a with
4b, one can see that there are more sodium spots for blc and only bright
spots for activated carbon. The sodium represented by the dim spots could
be from sodium at varying depths below the carbon surface or from very
finely distributed sodium on the surface. A better indication of the sodium distribution is obtained when an EDS line scan is made over a horizontal area of the SEM video picture. Sodium and sulfur were measured for blc
(Fig. 5) while only sodium was scanned for activated carbon (Fig. 6). A
comparison of the figures shows a much finer sodium dispersion for blc than
activated carbon. The dispersion of sulfur in Fig. 5 is similar to that
of sodium. The intensities of sulfur are higher as a result of the higher
sensitivity of this technique for sulfur than sodium.

TABLE II

Reactivities of Carbonaceous Materials

Type of Carbon	Conv. (%)	Rate (min^{-1})	K (m^3/min. mol)	K_{CO_2} (m^3/mol)	K_{CO} (m^3/mol)	T (°K)	(CO$_2$) (atm)
Black	25	0.132	0.11	0.64	2.77	998	0.2
Liquor	50	0.185	0.217	0.864	3.26	998	0.2
Char	75	0.215	0.40	1.3	4.1	998	0.2
Electrode		1.6×10^{-5}	1.1×10^{-5}	0.34	4×10^3	1000	0.2
Carbon [12]		0.135	0.102	0.2	1.6	1400	0.2
Graphite 5%K$_2$CO$_3$ [13]		3×10^{-5}				1000	1
Coal 5% Na$_2$CO$_3$		4.4×10^{-4}				973	1
Char 10% Na$_2$CO$_3$		2.0×10^{-3}				973	1
[14] 20% Na$_2$CO$_3$		2.0×10^{-3}				973	1
Activated Carbon [15] 5% K$_2$CO$_3$		0.012	4.7×10^{-3}	0.47	3.1×10^2	1000	0.27

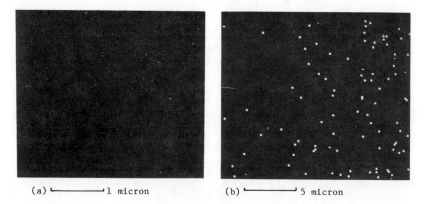

(a) ⊢————⊣ 1 micron   (b) ⊢————⊣ 5 micron

Fig. 4   Sodium distribution by EDS mapping for (a) blc, (b) activated carbon.

From these results it can be concluded that the high rates obtained
for blc are caused by the fine dispersion of sodium on the carbon surface.
When gasification proceeds, new small catalytic sites will be exposed because of the three dimensional nature of the catalyst dispersion in blc.
This could explain the sustained high rates at catalyst loadings where the
catalytic gasification of coal chars is substantially reduced due to pore
plugging and catalyst sintering [16]. Another confirmation of the importance of the dispersion of sodium is obtained when blc was prepared from
solids dried in a drying dish.   It was found that the reactivity of the

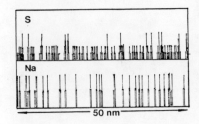
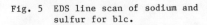
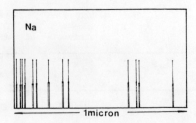

Fig. 5 EDS line scan of sodium and   Fig. 6 EDS line scan of sodium
sulfur for blc.   for activated carbon.

dish dried char was about 20 to 30% lower than the film dried char. On
examination of the dish dried char by SEM, it was found that some sodium
crystals were present, contrary to the film dried char where no crystals
were found. This shows that a loss in the sodium dispersion with the dish
drying method again results in a reduced gasification rate.

CONCLUSIONS

The very high gasification rate of black liquor char is explained by
the fine dispersion of sodium in the carbon structure. In addition, the
fine dispersion of sodium in the bulk of the carbon structure of blc rather
than just on the internal surface is probably responsible for the high
gasification rates at sodium-carbon ratios where the rate for Na_2CO_3 impregnated chars is normally strongly reduced due to pore plugging and sintering of the catalyst.

REFERENCES

1.   D.W. McKee, in Chemistry and Physics of Carbon, Vol. 16, (Marcel
     Dekker, Inc., New York, 19), p.1.
2.   W.Y. Wen, Catal. Rev.-Sci. Eng., 22(1), 1 (1980).
3.   B.J. Wood and K.M. Sancier, Final Report, SRI International, DOE
     Contract No. DE-AC21-80MC14953 (1984).
4.   C.A. Mims and J.K. Pabst, Fuel, 62(2), 176 (1983).
5.   D.A. Sams and F. Shadman, AIChE J., 32(7), 1132 (1986).
6.   I.L.C. Freriks, et al, Fuel, 60, 463 (1981).
7.   S.L. Yuh and E.E. Wolf, Fuel, 63, 1604 (1984).
8.   J. Li and A.R.P. van Heiningen, JPPS, 12(5), J146 (1986).
9.   J. Li and A.R.P. van Heiningen (submitted to J. Can. Chem. Eng.
     1987).
10. J. Li, Master Thesis, McGill University (1986).
11. P.L. Walker, Jr., et al, Advan. Catal., 11, 133 (1959).
12. L.G. Austin and P.L. Walker, Jr., AIChE J., 9(3),   (1963).
13. D.W. McKee, Carbon, 20(1), 59 (1982).
14. C.L. Spiro et al, Fuel, 62(3), 323 (1983).
15. T. Wigmans, Ph.D. Thesis, University of Amsterdam (1982).
16. D.A. Sams and F. Shadman, Fuel, 62(8), 880 (1983).

INFLUENCE OF POTASSIUM ON THE OXIDATION RATE OF CARBON

PETER SJÖVALL, BO HELLSING, KARL-ERIK KECK AND BENGT KASEMO
Chalmers University of Technology, Department of Physics, S-412 96 Göteborg, Sweden

ABSTRACT

The influence of K, deposited on a carbon surface, on the oxidation of carbon in O_2 was investigated. Reaction rate measurements, carried through in a UHV-system by use of AES, showed that potassium increases the reaction rate by up to ~ 10^4 times. A theoretical model, based on the assumption that O_2 dissociation is the rate limiting step, has been developed. The model shows that a charge transfer mechanism can explain the observed rate increase. Results from TPD/TPR-measurements indicate that the sticking probability for O_2 on a graphite surface with predeposited K is approximately independent on K coverage for coverages down to 0.5×10^{14} cm^{-2}, corresponding to an effective radius of K of ~ 7.3 Å.

1. Introduction

Reactions between solid carbon and different gaseous reactants (e.g. O_2, H_2, H_2O, CO_2, NO) play an important role in many technological areas, e.g. gasification of coal and reduction of soot particles from auto exhausts. These reactions can be accellerated if a catalyst (mainly alkali metal salts and transition metal oxides) is added to the carbon. Although this catalytic effect is well documented by reaction studies [1-7], there is still debate about the reaction mechanisms and there has been surprisingly few fundamental adsorption studies performed on well defined carbon surfaces [8-12]. In contrast, there are many studies of coadsorption of alkali metals and simple gas molecules on metal surfaces [13-21].

One common denominator between the alkali-on-metal and alkali-on-graphite studies is that they concern the function of promotors in catalysis. Because of the quite different electronic properties between carbon (graphite), being a semi-metal, and the metals one expects significant differences between the two systems, but there are also reasons to expect common features. For example, in both cases alkali metals may act as donors of electron charge. In this paper we report from an ongoing theoretical and experimental study of potassium promotion of oxidation of evaporated carbon films and graphite.

2. Experimental

The experiments were performed in a UHV-system with facilities for AES, work function measurements, mass spectrometry, TPD/TPR (temperature programmed desorption/reaction), ion sputtering, C-evaporation, K-evaporation and high pressure studies [22].

In the oxidation rate measurements (see ref. [23]) thin (~ 10 Å) evaporated carbon films were used as samples. Since the carbon film is gasified (forming CO and CO_2) during oxidation, the reaction rate can be determined by measuring the film-thickness by AES between successive periods of oxidation (i.e. sample heating during O_2-exposure). The amount of predeposited K was estimated (1) by measuring the evaporation rate with a quartz crystal microbalance and (2) by integrating the TPD-spectrum of K after oxidation. Absolute calibration of K-concentration was obtained using AES and work function measurements. On the horizontal scale in figures 1 and 2, 0.3 corresponds to 5×10^{14} K-atoms/cm^2. The oxidation rate measurements presented here were performed at T=726 K and P_{O_2}=1 x 10^{-5} torr for the K-doped carbon films and at T=773 K and P_{O_2}=5 x 10^{-4} torr for the pure carbon film. Work function measurements were accomplished by applying a negative voltage on the sample and measuring the low energy cut off energy in the electron energy spectrum recorded by the CMA-analyzer.

In the second type of study (the TPR/TPD-studies) highly oriented pyrolytic graphite (HOPG) was used. The sample was heated by passing a direct current through the sample. During K-evaporation and O_2-exposure the graphite sample was held at T=120 K in order to prevent intercalation. A temperature ramp (3.1 K/s) was then generated on the sample in front of the masspectrometer while the desorption spectra from mass numbers 28, 39 and 44 were recorded.

The TPD/TPR spectra were influenced by desorption from the Ta sample holder (some K always diffused from the graphite over to the Ta holder). However, by using the observation that desorbed K only reached the ion source by direct sight, and by positioning the sample at different distances from the ion source of the MS, it was possible to distinguish sample desorption from sample holder desorption.

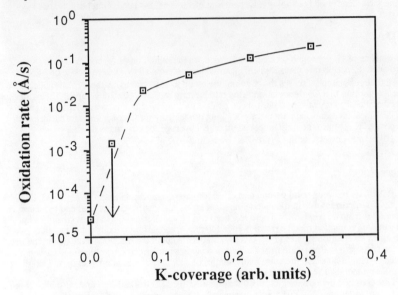

Fig. 1.   Oxidation rate of carbon films vs amount of predeposited K at T = 726 K and $PO_2 = 1 \times 10^{-5}$ torr (for $\theta_K = 0$, T = 773 K). The arrow indicates that the measured value at K-cov. = 0.03 is an upper limit for the oxidation rate at this coverage.

3. Results and discussion

The oxidation rate measured for carbon films with different amounts of predeposited potassium are summarized in fig. 1. There is a dramatic increase in the reaction rate when potassium is predeposited. The oxidation rate for the pure carbon film corresponds to a reaction probability of $\sim 7 \times 10^{-6}$ per O_2-collision with the carbon surface at 773 K while the maximum oxidation rate with deposited potassium corresponds to a reaction probability of $\sim 6 \times 10^{-2}$ at 726 K. Thus, potassium increases the oxidation rate by $\sim 10^4$ times.

We have recently developed a model for the promoting effect of potassium on the oxidation of carbon [23]. The model assumes that the dissociation of O_2 is the rate limiting step. This assumption is strongly supported by the observations that i) the sticking coefficient of O_2 on graphite is extremely low while the sticking of atomic oxygen is much higher [9] and ii) that atomic oxygen in contrast to O_2 easily reacts with soot films at RT [24]. We further assume that the O_2-dissociation on the surface is thermally activated, i.e. the dissociation probability is

$$P \sim \exp(-E_{diss}/kT)$$

The activation energy for O_2-dissociation at the surface is determined by the transient, fractional occupation (δn) of the antibonding $2\pi^*$ -orbital of O_2:

$$E_{diss} (\delta n) = E_{diss} (0) - (E_{diss}[O_2] - E_{diss}[O_2^-]) \delta n$$

where E_{diss} (O) is the dissociation energy of O_2 on pure graphite. E_{diss} [O_2] (= 5.1 eV) and E_{diss} [O_2^-] (= 4.1 eV) are the dissociation energies in gas phase for O_2 and O_2^- respectively. The

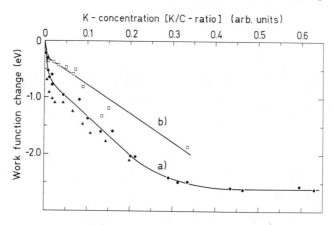

Fig. 2.   The work function vs amount of predeposited K on evaporated carbon at RT. (a) Before O_2-exposure (after K evaporation on a (•) 10 Å thick film and on a (▲) 100 Å thick film) (b) After saturating O_2-exposure.

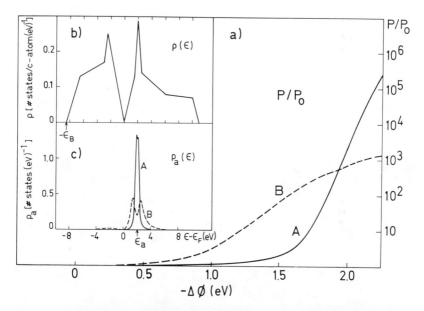

Fig. 3.   (a) The theoretically estimated increase in dissociation probability of O_2 on K-promoted graphite P(-ΔØ)/P(O) for two coupling strengths, corresponding to two different O_2 binding energies, A:ΔE = -0.18 eV [25] and B:ΔE = -1.1 eV. (b) Density of states with π symmetry for graphite. (c) Density of states projected in the $2\pi^*$ level of oxygen. A and B as in (a).

occupation of the $2\pi^*$ -orbital is determined by the shape and position in energy of this hybridized orbital relative to the position of the Fermi-energy of the sample; the orbital is occupied up to the energy corresponding to the Fermi-energy of the sample.

The important consequence of potassium adsorption is that the work function decreases, i.e.the Fermi-energy of the sample increases, which in turn increases the $2\pi^*$ occupation. Fig. 2 shows that the measured work function of the carbon film decreases by up to 2.6 eV upon K-deposition. Subsequent oxygen adsorption causes an increase in work function (fig. 2b) but the final value is still much lower than for the pure carbon film. The Newns-Anderson model was used to calculate the shape and position of the $2\pi^*$ -orbital of O_2 when it interacts with a graphite substrate (fig. 3c (for details see [23])). The occupation of this orbital was then calculated as a function of the potassium induced work function change and from eq (2) the decrease in dissociation energy of O_2 was calculated, and the rate enhancement was obtained from eq. (1). Fig. 3a) shows the theoretically estimated increase in dissociation probability for O_2 as a function of potassium induced work function change at T=726 K which is given by

$$p(-\Delta\varnothing)/p(o) = \exp (\delta n(-\Delta\varnothing)/kT)$$

The model approximately describes the large increase in reactivity by potassium promotion. The point we emphasize is that the simple model, assuming O_2-dissociation to be rate limiting and potassium induced charge transfer to $2\pi^*$ to be rate enhancing, can account for the observed rate increase by several orders of magnitude. (Note that the model does not include any fitting parameters.) We do not emphasize a detailed comparison between the experiment and theory, since the latter does not consider possible local effects such as a direct reaction between O_2-molecules and K-atoms on the surface. The model is thus most relevant in the low coverage limit, where non-local effects definitely seem to be important (see below).

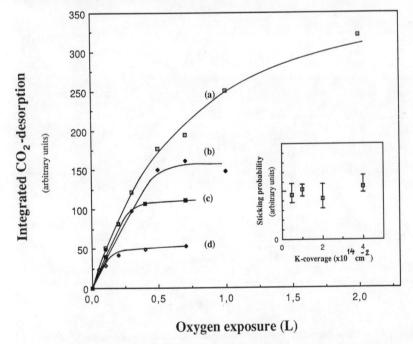

Fig. 4. The integrated CO_2-desorption in TPD/TPR-spectra vs oxygen exposure of a graphite surface with predeposited K. The K-coverages were (a) 4 x 10^{14} cm^{-2} (b) 2 x 10^{14} cm^{-2} (c) 1 x 10^{14} cm^{-2} (d) 0.5 x 10^{14} cm^{-2}. The inset shows the slope of these curves at O_2-exp. = O vs K-coverage.

The results obtained with evaporated films above motivated a continued study with graphite. We present here a progress report of TPD/TPR-studies (AES and $\Delta\emptyset$-results will be presented elsewhere). Since we believe that the dissociative adsorption of O_2 is the rate limiting step it is of interest to investigate how the sticking coefficient of O_2 changes with the amount of predeposited K on the surface. Fig. 4 shows the integrated CO_2-desorption signal vs O_2-exposure for different K-coverages. Since CO_2 is the main desorption channel for oxygen (see below) the integrated CO_2 signal gives information about the oxygen uptake kinetics. The initial sticking coefficient is proportional to the slope of the oxygen uptake vs O_2-exposure curve at zero coverage. The inset in fig. 4 shows that the sticking probability does not change appreciably with K-coverage for coverages down to 0.5×10^{14} cm^{-2}. This result can not be explained by a simple reactive impingement of O_2 on K-atoms on the surface, since the effective radius of K for O_2-sticking at the lowest coverage is $\geqslant 7.3$ Å, to be compared with the metallic radius of K, 2,3 Å. This observation supports a non-local promotion effect. (An alternative explanation which has not yet been explored is trapping of O_2 in a mobile precursor state followed by dissociation at K-atoms.)

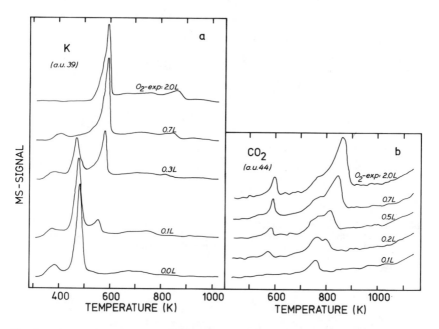

Fig. 5.   Desorption spectra of (a) M 39 (K) and (b) M 44 (CO_2) after constant depositions of K (3×10^{14} cm^{-2}) and varying oxygen exposures at T ~ 120 K.

Fig. 5 shows the desorption spectra of K and CO_2 after oxygen exposure of a graphite surface with a K-coverage of 3×10^{14} cm^{-2}. There is also a CO-desorption peak at T ~ 950 K but since this peak coincides with K-desorption from the sample-holder, we believe that the CO-desorption peak comes from the sample-holder. The K desorption peak at T = 390 K is also attributed to desorption from the sample holder. The remaining structures are attributed to desorption from the graphite sample. No signals were detected at M 32 (O_2) and M 94 (K_2O), thus CO_2 is the dominating reaction product. The coincidence of the K and CO_2-peaks around T ~ 600 K and T ~ 800 K suggests that they originate from thermal decomposition of K-C-O complexes on the surface. Further TPD/TPR and surface spectroscopic studies are underway to investigate the nature of these complexes.

452

Acknowledgement

This work is financed by the National Swedish board for Technical Development, grant no. 83-5405.

References

[1] For reviews see D.W. McKee, Chem. Phys. Carbon 16, 1 (1981); B. Wood and K.M. Sancier, Cat. Rev. Sci. Eng. 26, 233 (1984).
[2] Fuel vol. 62 (1983) and Fuel vol. 65 (1986) contain the papers presented at the "International Symposia on the Fundamentals of Catalytic Coal and Carbon Gasification", in the Netherlands, in 1982 and 1986.
[3] S.R. Kelemen and H. Freund, J. Catal. 102, 80 (1986).
[4] C.A. Mims, J.J. Chludzinski, Jr., J.K. Pabst and R.T.K. Baker, J. Catal. 88, 97 (1984).
[5] C.A. Mims and J.K. Pabst, J. Catal. 107, 209 (1987).
[6] M.B. Cerfontain, R. Meijer, F. Kapteijn and J.A. Moulijn, J. Catal. 107, 173 (1987).
[7] A.L. Cabrera, H. Heinemann and G.A. Somorjai, J. Catal. 75, 7 (1982).
[8] M.T. Johnson, H.I. Starnberg and H.P. Hughes, Surf, Sci. 178, 290 (1986).
[9] M. Barber, E.L. Evans and J.M. Thomas, Chem. Phys. Lett. 18, 423 (1973).
[10] S.R. Kelemen, H. Freund and C.A. Mims, J. Vac. Sci. Technol. A2, 987 (1984).
[11] S.R. Kelemen and C.A. Mims, Surf. Sci. 133, 71 (1983).
[12] S.R. Kelemen, H. Freund and C.A. Mims, J. Catal. 97, 228 (1986).
[13] H.P. Bonzel, Surf. Sci. Rept. 8, 43 (1987).
[14] L.Surnev, G. Rangelov and M. Kiskinova, Surf. Sci. 179, 283 (1987).
[15] G. Pirug, G. Brodén and H.P. Bonzel, Surf. Sci. 94, 323 (1980).
[16] L.J. Whitman, C.E. Bartosch and W. Ho, J. Chem. Phys. 85, 3688 (1986).
[17] L. Bugyi and F. Solymosi, Surf. Sci. 188, 475 (1987).
[18] F. Solymosi and A. Berkó, J. Catal. 101, 458 (1986).
[19] J.K. Nørskov, S. Holloway and N.D. Lang, Surf. Sci. 137, 65 (1984).
[20] N.D. Lang, S. Holloway and J.K. Nørskov, Surf. Sci. 150, 24 (1985).
[21] P.J. Feibelman and D.R. Hamann, Surf. Sci. 149, 48 (1985).
[22] K.-E. Keck, B. Kasemo and T. Högberg, Surf. Sci. 126, 469 (1983).
[23] P. Sjövall, B. Hellsing, K.-E. Keck and B. Kasemo, J. Vac. Sci. Technol. A5, 1065 (1987).
[24] B.G. Wicke, Chor Wong and K.A. Grady, Combust. Flame 66, 37 (1987).
[25] ΔE is estimated from the desorption temperature $T_d = 65$ K in a phase diagram of O_2 on graphite in P.W. Stephens, P.A. Heiney, R.J. Birgenean, P.M. Horn, J. Stottenberg and O.E. Vilches, Phys. Rev. Lett. 45, 1959 (1980). This gives $\Delta E = -0.18$ eV.

A SINGLE TURNOVER (STO) CHARACTERIZATION OF SUPPORTED PALLADIUM CATALYSTS

ROBERT L. AUGUSTINE AND DAVID R. BAUM
Department of Chemistry, Seton Hall University, South Orange, NJ 07079

ABSTRACT

While the STO catalyst characterization procedure has been applied to a variety of supported Pt catalysts, application of this technique to the characterization of supported Pd catalysts showed that there were several significant differences between the Pt and the Pd catalysts. Under STO reaction conditions each surface site on a Pt catalyst reacts only once so there is a 1:1 relationship between the product composition and the densities of the various types of active sites present. With Pd catalysts under these same conditions, alkene isomerization takes place so readily that the amount of isomerized product formed depends on the contact time of the reactant pulse with the catalyst so there is no direct relationship between the amount of isomerization and the number of isomerization sites present. On Pt there are some direct saturation sites present on which H_2 is rather weakly held. Such sites are not present on Pd catalysts. The reactive surface of supported Pt catalysts remains constant on long exposure to air. With Pd catalysts exposure to air results in a decrease in saturation site densities which can be reversed by re-reduction of the surface with H_2 under ambient conditions but not completely under what can be termed "reaction conditions" where the extent of surface re-reduction decreases with catalyst age.

INTRODUCTION

The relationship between the surface structure of a catalyst and its reaction characteristics is one of considerable importance in the development of selective catalysts for specific reactions. Extensive single crystal catalysis studies have shown that different types of surface atoms have differing reaction properties.[1,2] This has been accomplished by using high vacuum analytical techniques such as LEED to provide a definite description of the surfaces of these single crystal metal catalysts which were then transferred directly into a pressure reaction cell and used to promote a variety of catalytic reactions thus providing a relationship between the type of surface atoms present and the nature of the reaction taking place.

Unfortunately, such techniques cannot be used, at present, to determine the surface characteristics of the small metal particles present on supported metal catalysts. But, since these single crystal data indicate that different types of surface atoms can promote different catalytic reactions it would seem that one should be able to extrapolate from product composition to the nature of the surface of a dispersed metal catalyst. However, since it is unlikely that all of the sites react at the same rate and since the specific site reaction rates are unknown such an approach is not possible using standard steady state reaction product data. If, however, the reaction sequence occurs in such a way that each active site reacts only once then a product/site correlation can be made. This is the essence of the Single Turnover (STO) catalyst characterization procedure.[3,4] In this reaction sequence a sample of the catalyst in a small flow reactor is exposed to sufficient hydrogen to saturate the metal surface and the excess hydrogen is removed in the carrier gas stream. A pulse of 1-butene is then introduced to react with the adsorbed hydrogen in a stoichiometric manner so that each site reacts only once to provide the data for a direct product/site correlation.

Over Pt catalysts five different types of surface sites have been identified and measured by this STO procedure.[4] There are two types of sites which promote the direct one-step hydrogenation of the butene by reaction of the adsorbed hydrogen with the alkene. These sites differ in that on one the adsorbed hydrogen is rather weakly held while on the other the adsorption is stronger. A third type of site promotes a two-step hydrogenation of the butene while on another kind isomerization takes place. The fifth type of site is one which merely adsorbs hydrogen but does not take part in the alkene hydrogenation but may take part in other catalytic processes. A correlation of cyclohexane dehydrogenation data obtained over several STO characterized Pt catalysts[5] with comparable Pt single crystal results[2] indicated that all three of the hydrogenation sites are kink or corner atoms of one type or another. This finding supports the proposed reaction mechanism for these sites which is based on homogeneous catalytic analogies.[6] Further mechanistic considerations indicate that the isomerization sites may be edge or step atoms while the hydrogenation inactive sites could be terrace or face atoms.

EXPERIMENTAL

The apparatus used, the details of the gas purification procedure and product analysis have been described as has the STO reaction procedure and its application to the characterization of Pt/CPG catalysts.[3,4] The Pd/CPG catalysts used contain 1.8% Pd and were prepared by the ion exchange procedure already described[3] using the same CPG involved in the Pt/CPG preparation. The catalysts used were reduced at 150°C under a stream of H_2 using the isothermal technique previously discussed.[7]

For aging studies the catalyst, after reduction, was placed in a desiccator under air and samples taken out at appropriate intervals for STO analysis. The standard STO sequence involved the placing of a sample of the reduced catalyst between plugs of Pyrex glass wool in a reactor (6mm OD Pyrex tube) and purging the system with He for 30 min before the introduction of the reactant pulses. The STO pulse sequence (H_2-ene-H_2) was then run several times to affect the surface reduction of the catalyst under reaction conditions. During this time the amount of butane formed during successive STO reaction sequences increased somewhat and the surface reduction was considered complete when the product composition remained constant for subsequent turnovers.

Re-reduction: The surface reduction was also accomplished by placing the catalyst sample in the reactor in a stream of H_2 at room temperature for 30 min before the STO procedure was run. After purging the reactor with He, STO product composition was constant from the first turnover showing that surface reduction had taken place and, further, that no H_2 spillover had taken place during the re-reduction procedure.

in situ Reduction:[7] The obtaining of a catalyst which had not been exposed to air was accomplished by placing a sample of the unreduced supported salt into the reactor tube and reducing it in a stream of H_2 at 150°C. The system was purged with He and the excess H_2 present either interstitially or spillover was removed by titration with pulses of 1-butene until no butane formation was observed.[8] The reaction product composition from STO reaction sequences run after this titration remained constant from the initial turnover.

RESULTS AND DISCUSSION

The STO reaction procedure has been used to provide considerable information concerning the relationship between the surface characteristics of supported Pt catalysts and their reaction properties.[3,4,5,7,9,10] It was felt that an extension of this technique to other catalysts would be beneficial in developing a rational understanding of the observed differences between catalytically active species. Application of the STO procedure to the characterization of supported Pd catalysts showed some interesting differences between the Pd and Pt catalysts under these reaction conditions. In the first place no direct saturation sites with weakly adsorbed hydrogen were detected on Pd/CPG catalysts. It appears that there is a single type of direct, one-step, saturation site on Pd catalysts and that on it the hydrogen is chemisorbed rather strongly. These sites and the two-step saturation sites are assumed to be corner or kink atoms by analogy with homogeneous reaction mechanisms and the Pt results described above.

With Pt the product composition remained constant when the contact time of the 1-butene with the catalyst was varied by changing the size of the reactant pulse or the carrier gas flow rate.[3,4] Over Pd catalysts increasing the reactant pulse size or decreasing the carrier gas flow rate resulted in a consistent increase in the extent of isomeric products formed. Figure 1 shows the effect of the 1-butene pulse size on the composition of the product mixture obtained in STO reactions over a Pd/CPG catalyst. The first butane is that formed by reaction over the direct hydrogenation sites and the second butane is produced over the two-step saturation sites. The fact that butane formation, either direct or 2-step, remained constant, showed that reaction product formation was not controlled by diffusion parameters. Thus, the increase in the extent of isomerization must result from multiple isomerization turnovers taking place on the isomerization sites. Because the isomerization reaction is not stoichiometric a direct correlation between the extent of isomerization and the number of isomerization sites cannot be made. Changing the reactant contact time by varying the carrier gas flow rate has the same effect.

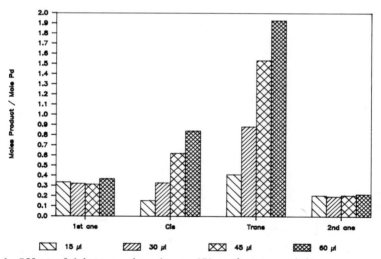

Fig.1   Effect of 1-butene pulse size on STO product composition.

Another difference between Pt and Pd catalysts was the relative stability shown by each on storage under air for prolonged periods of time. While both catalysts react readily with oxygen in air to form surface oxides, with Pt these oxides are relatively easily reduced and the reduction does not appear to change the surface morphology of the metal crystallites. Samples of the same Pt/CPG catalyst gave the same STO characterization results after one year storage in air as was observed after preparation.[7] This was not the case with Pd/CPG catalysts. The data presented in Fig.2

show that the STO reaction characteristics of Pd catalysts change
appreciably on standing in air. A freshly reduced Pd/CPG catalyst was
stored in a desiccator under air and samples taken out for STO characterization at various intervals. These samples were treated by the normal
STO procedure in which the catalyst was subjected to successive pulses

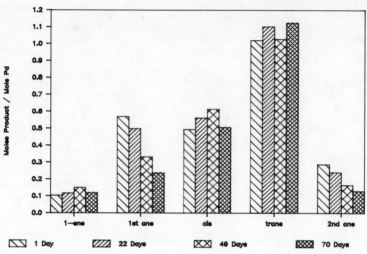

Fig.2   Effect of catalyst age on product composition after reduction under
STO reaction conditions.

of H_2-ene-H_2 in the STO sequence until the product composition remained
constant for successive turnovers. During this "surface reduction" procedure the product composition changed somewhat from turnover to turnover.

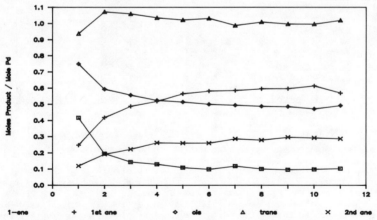

Fig.3   Product composition on successive turnovers run over a
Pd/CPG catalyst exposed to air for one day.

Figure 3 shows the product composition observed for successive turnovers run over the catalyst which had been exposed to air for only one day. It can be seen from these results that after 3-4 turnovers the product composition remained constant but that during the first few turnovers the amounts of butane and trans 2-butene increased and the amount of cis 2-butene decreased. Comparable data were obtained from STO sequences run over the same catalyst after it had been "aged" in air for times up to 70 days. The general trends in observed product composition were the same as seen in Fig. 3 but the amount of butane formed is progressively less. With the 70 day old catalyst the constancy of the data was not observed until 8-10 turnovers had been run. The data plotted in Fig. 2 are those obtained after the product composition had leveled.

The running of the STO reaction sequences in the way just described is considered to be similar to the exposure of the catalyst to standard reaction conditions; no catalyst pre-treatment which results in the surface reduction taking place with hydrocarbons also present. It was considered important to see if a similar decrease in saturation activity would be observed if the "aged" catalysts were first re-reduced in a stream of H_2 before being subjected to the STO reaction sequence. To accomplish this samples of a freshly reduced catalyst which was stored in air for varying lengths of time were placed in the STO reactor, re-reduced by exposure to a stream of H_2 for 30 min at room temperature and then characterized by the STO procedure. The results are summarized in Fig. 4. For catalysts treated in this way the product composition was essentially constant from the first turnover and there was no difference in STO characterization data between catalysts exposed to air for varying lengths of time. Surprisingly, butene titration[8] showed that there was no H_2 spillover or excess H_2 stored in the Pd interstices. For comparison a sample of the unreduced supported catalyst precursor was reduced in the reactor and used directly in the STO reaction sequence to establish the surface characteristics of a catalyst which had not been exposed to air. These results are also presented in Fig.4 as in situ reduction data. While in this instance butene titration showed the presence of spillover H_2, once this was removed the overall product

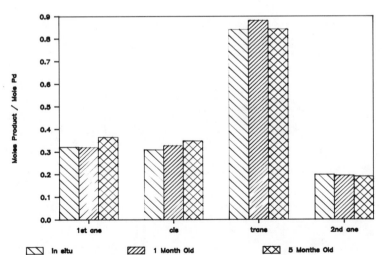

Fig.4   Effect of catalyst age on STO product composition after re-reduction by H_2.

composition was the same as that observed with the re-reduced catalysts.

CONCLUSIONS

The STO catalyst characterization procedure provides information concerning the nature of the metal surface of supported Pd catalysts. These catalysts have two types of alkene saturation sites, one involved in the direct, one-step addition of hydrogen to a double bond and the other promoting a two-step alkene saturation. The direct saturation site has the H_2 adsorbed on it rather strongly. The isomerization sites are so reactive that conversion of 1-butene into its isomers is non-stoichiometric under STO conditions with the amount of isomer formed dependant on the size of the reactant pulse and/or the carrier gas flow rate. The surface oxide produced by exposure of Pd catalysts to air becomes progressively more difficult to reduce under "reaction conditions" as the time of exposure increases. These catalysts have decreased saturation site densities. Re-reduction of this surface oxide by H_2 without the presence of hydrocarbon gives catalysts having the same surface characterization regardless of the time of "aging" in air so it appears that there must be some interaction of the hydrocarbon reactant with the partially reduced surface which hinders complete reduction. What this interaction could be is not apparent at this time. The marked similarity between the surface characterization data if the re-reduced catalysts and the in situ reduced species which had not been in contact with oxygen shows that this re-reduction can regenerate the original surface of the catalyst and that "aging" in air, at least for up to 5 months, does not change the overall surface characteristics of the Pd particles on the catalyst.

ACKNOWLEDGEMENT

This work was supported by the U.S.Department of Energy, Office of Basic Energy Science, through Grant No. DE-FG02-84ER45120.

REFERENCES

[1] G.A.Somorjai, Phil.Trans.R.Soc.Lond.A **318**, 81 (1986).
[2] R.K.Herz, W.D.Gillespie. E.E.Peterson, and G.A.Somorjai, J.Catal. **67**, 371 (1981).
[3] R.L.Augustine and R.W.Warner, J.Org.Chem. **43**, 2614 (1981).
[4] R.L.Augustine and R.W.Warner, J.Catal. **80**, 358 (1983).
[5] R.L.Augustine and M.M.Thompson, J.Org.Chem. **52**, 1911 (1987).
[6] S.Siegel, J.Outlaw, Jr., and N.Garti, J.Catal. **52**, 102 (1978).
[7] R.L.Augustine and K.P.Kelly, J.Chem.Soc.Faraday Trans.1 **82**, 3025 (1986).
[8] R.L.Augustine, K.P.Kelly, and R.W.Warner, J.Chem.Soc.Faraday Trans.1 **79**, 2639 (1983).
[9] R.L.Augustine, M.M.Thompson, and M.A.Doran, J.Chem.Soc.,Chem.Commun.,1173 (1987).
[10] R.L.Augustine, K.P.Kelly, and Y.-M.Lay, Appl.Catal. **19**, 87 (1985).

^{51}V NMR: A NEW PROBE OF STRUCTURE AND BONDING IN CATALYSTS

HELLMUT ECKERT* and ISRAEL E. WACHS**
*Department of Chemistry, University of California at Santa Barbara, Goleta, CA 93106
**Zettelmoyer Center for Surface Studies and Department of Chemical Engineering, Lehigh University, Bethlehem, PA 18015

ABSTRACT

Solid State ^{51}V NMR spectra are sensitive to changes in the surface structure of V_2O_5 dispersed on Al_2O_3 and TiO_2 supports. For V_2O_5 supported on alumina, wideline and magic-angle spinning experiments at 4.7 and 7.0 T reveal the presence of two vanadium species with different bonding environments. The relative proportions of these species change monotonically as a function of the vanadium concentration. In contrast, supported vanadium oxide on TiO_2 substrates show a much more V_2O_5-like environment. Structural inferences are discussed in connection with solid state ^{51}V NMR studies on model compounds with known crystal structures and site symmetries.

INTRODUCTION

Recent studies have shown that V_2O_5 supported on a metal oxide is a superior catalyst to unsupported crystalline V_2O_5 for the selective oxidation of many hydrocarbons. For example, the selective oxidation of o-xylene to phthalic anhydride only proceeds if V_2O_5 is supported on TiO_2 (anatase) [1]. Detailed studies of the reaction network and the kinetics of this reaction [2,3] have been complemented by structural investigations using laser Raman spectroscopy [1,3]. Those studies have shown that at low V_2O_5 contents the vanadium oxide species exist in the form of a monolayer of surface vanadia interacting with the TiO_2 support, whereas vanadia contents in excess of monolayer coverage are present as microcrystalline V_2O_5. The amount of V_2O_5 necessary to form a monolayer depends on the specific surface area of the system under consideration. Similar results have been obtained with other supports such as ZrO_2 and Al_2O_3 [4]. The complex vanadium oxide structural chemistry prevents a complete structural determination of the surface vanadium oxide phases solely from Raman spectroscopy [4]. A need for a complementary technique of structural characterization also arises from the inherent difficulty to derive quantitative information from Raman spectra.

Solid state ^{51}V NMR represents a promising approach to these systems. Owing to a large magnetic moment, high natural abundance (99.76 %) and favorable relaxation characteristics, this nucleus is very amenable to solid-state NMR investigation. A number of previous studies have shown that the ^{51}V chemical shielding tensor is very sensitive to changes in the coordination environments, and provides a facile means of discrimination between different

Mat. Res. Soc. Symp. Proc. Vol. 111. ©1988 Materials Research Society

vanadates in the solid state [5-11]. The scope of previous ^{51}V NMR applications to catalysts has been extremely limited [11-13], and, to our knowledge, no systematic studies have been carried out. We report here the first application of ^{51}V NMR wideline and magic-angle spinning (MAS) NMR spectroscopy to catalytic systems based on atomically dispersed surface vanadate. In an effort to correlate the spectroscopic information obtained with specific vanadium bonding environments in such systems, we have generated a database of ^{51}V chemical shift tensors of crystalline compounds with known structures. Using this information, we will discuss possible microstructures on catalysts containing pentavalent vanadia supported on alumina, titania, and zirconia substrates.

EXPERIMENTAL

The supported vanadium oxide on TiO_2 catalysts were prepared by the incipient wetness impregnation method with $VO(OC_2H_5)_3$ in ethanol on Degussa P-25 TiO_2 (surface area 50 m^2/g; anatase/rutile ca. 2). The impregnated samples were subsequently dried at room temperature for 16 hrs, dried at 110-120 °C for 16 hrs, and calcined at 450 °C for 2 hrs to form the supported vanadium oxide on TiO_2 catalysts. Vanadia contents were determined by ICP analysis. Raman characterization revealed that samples with up to 7 wt.% V_2O_5 contained only atomically dispersed surface vanadium oxide species, above this limit V_2O_5 crystallites were observed as an additional phase. Thus, a monolayer of surface vanadium oxide corresponds to ca. 7% V_2O_5/TiO_2 for this system.

The supported vanadium oxide on Al_2O_3 catalysts were prepared by the incipient wetness impregnation method with $VO(OC_3H_7)_3$ in methanol on Harshaw Al_2O_3 (surface area 180 m^2/g). The impregnated samples were subsequently dried at room temperature for 16 hrs , dried at 110-120 °C for 16 hrs, and calcined at 500 °C for 16 hrs to form the supported catalysts. Vanadia contents were determined by ICP analysis. Raman characterization of the V_2O_5/Al_2O_3 system revealed that the 1-20% V_2O_5/Al_2O_3 samples contained only atomically dispersed surface vanadium oxide species and that at higher vanadia contents V_2O_5 crystallites are observed in addition to the atomically dispersed species. Thus, a monolayer of surface vanadium oxide corresponds to ca. 20 % V_2O_5/Al_2O_3 for this system. Crystalline vanadates investigated to provide reference data were obtained from commercial sources (Alfa, Aldrich, Strem, Aesar) and their identity was verified by x-ray diffraction and Raman analysis. The sample of $Pb_5(VO_4)_3Cl$ (vanadinite) was taken from the mineral collection of the California Institute of Technology.

Room temperature wideline and MAS NMR studies were conducted at 4.7 T and 7.0 T. Measurements at the lower field (corresponding to a resonance frequency for ^{51}V of 52.65 MHz) utilized a homebuilt spectrometer system interfaced with a 293 B pulse programmer, Explorer fast digitizer, and Nicolet 1280 computer. ^{51}V NMR wideline experiments were carried out on ca. 100 mg of material, compressed into cylindrical sample tubes of 5 mm outer diameter in a 5 mm homebuilt probe. Magic-angle spinning NMR experiments at 4.7 T were carried out using a modified ^{13}C CPMAS probe from DOTY Scientific. The powdered materials were spun in sapphire rotors of 7.0mm outer diameter at typical rates of 3-5 kHz.

Measurements at 7.0 T (frequency for ^{51}V: 79.0 MHz) were carried out on a General Electric GN-300 spectrometer, equipped with Henry-Radio amplifiers and a Chem-Magnetics probe. Both wideline and MAS NMR experiments were conducted within KELEF rotors of 9.5mm o.d. which were spun at a number of different speeds up to 3.5 kHz.

All of the spectra at both fields were recorded with pulse lengths between 0.5 and 2 μs (corresponding to solid 90 degree pulses, depending on the pulse power available and the efficiency of the respective probe used). Typical recycle delays were 1-2 seconds, resulting in spectra free from relaxation effects. All ^{51}V NMR chemical shifts were determined with respect to the resonance location of solid vanadinite (which resonates at 514 ppm upfield of $VOCl_3$).

RESULTS AND DISCUSSION

Figures 1 and 2 show ^{51}V MAS NMR spectra of selected model compounds. For all of the samples under investigation, the isotropic shifts were found to be equal within experimental error at both external field strengths, thus confirming that second-order quadrupolar shifts [14] are minimal at the field strengths used, and that the observed spinning sideband patterns are dominated by the chemical shift anisotropy [15]. The isotropic shifts (determined from these spectra) as well the shielding tensor components S_1, S_2, and S_3 (estimated independently from singularities in the corresponding wideline NMR spectra) are shown in Table I. As discussed in more detail below, it is clear from this table that these chemical shift parameters are sensitive to the different vanadium coordination environments in these compounds.

In vanadium pentoxide, the V atoms are coordinated to five oxygen atoms at distances of 154, 188, 188, 202, and 177 pm, whereas the sixth oxygen ligand is further removed (281 pm). The shielding tensor in V_2O_5 is essentially axially symmetric [19], the dominant perpendicular component being located near -250 ppm (giving rise to a peak maximum near 300 ppm). A similar situation obtains in lead metavanadate, where two vanadium sites exist with V-O distances of 193, 192, 206, 167, 165, 273 pm, and 161, 172, 192, 192, 203, 257 pm, respectively. Both wideline and MAS NMR experiments detect only a single species, suggesting that the shielding tensors for both sites must be very similar. The chemical shift anisotropies of the tetrahedrally coordinated vanadates under study are much smaller, and depend on the number of nonbridging oxygen atoms, N_{no} attached to them. For example, N_{no} = 2 in the metavanadates $NaVO_3$ and NH_4VO_3, the structures of which are based on chains of $VO_2O_{2/2}$ tetrahedra[21,22]. Due to lack of cylindrical symmetry of such microstructures, the spectra of these compounds reveal a non-axially symmetric shielding tensor. In contrast, the compounds Na_3VO_4, $BiVO_4$, $Pb_5(VO_4)_3 Cl$, and $AlVO_4$ have isolated, more symmetric tetrahedra (N_{no} = 4), thus resulting in only small shift anisotropies [23,24]. In this context, the spectrum of $AlVO_4$ deserves specific comment. Although the structure of this compound remains to be determined, X-ray powder diffraction patterns and infrared spectra [25] suggest an isomorphous

462

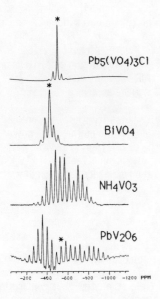

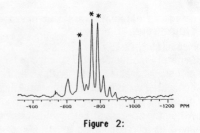

Figure 2:

52.65 MHz ^{51}V MAS–NMR spectrum of AlVO$_4$.

Figure 1: 79.00 MHz ^{51}V MAS–NMR spectra of model compounds. Central bands are indicated by asterisks.

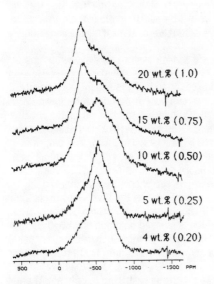

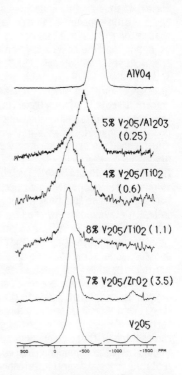

Figure 3: 52.65 MHz wideline ^{51}V NMR spectra of V$_2$O$_5$ supported on Al$_2$O$_3$ catalysts. Vanadia contents and surface coverages are indicated.

Figure 4: 52.65 MHz wideline ^{51}V NMR spectra of several catalysts and model compounds. Surface coverages in parentheses.

Table I: NMR chemical shift parameters for reference
compounds under study (in ppm vs. $VOCl_3$)

Compound	S_1	S_2	S_3	δ_{iso}	Structure ref.
	(± 10 ppm)			(±1 ppm)	
V_2O_5	-250	-250	-1300	-614	[16,17]
PbV_2O_6	-310	-320	-1000	-538	[18]
$NaVO_3$	-360	-530	-840	-584	[21]
NH_4VO_3	-370	-530	-830	-578	[22]
Na_3VO_4	*	*	*	-550	
$BiVO_4$	-365	-415	-520	-439	[23]
$Pb_5(VO_4)_3Cl$	*	*	*	-514	[24]
$AlVO_4$	*	*	*	-678	[25,26]
	*	*	*	-753	
	*	*	*	-786	

* not determined

relation with $FeVO_4$, which crystallizes in the space group $P\bar{1}$ and shows three
crystallographically inequivalent vanadium sites [26]. The MAS- NMR spectrum
(Figure 2) confirms this interpretation. Overall, Table I suggests that the
anisotropic shielding information is more valuable than the isotropic chemical
shift in discriminating between different vanadium environments.

Figure 3 shows the wideline NMR spectra of five V_2O_5/Al_2O_3 catalyst
samples, where the vanadium oxide is present as an atomically dispersed surface
phase. Up to concentrations of 5 wt.% V_2O_5, a broad resonance around -540 ppm
is observed. Linewidth, shape and chemical shift location of this resonance are
consistent with an assignment to a tetrahedral vanadate species, most likely
with $N_{no}=2$. However, magic-angle spinning experiments (5 kHz) do not result in
any line-narrowing, hence indicating a broad distribution of isotropic chemical
shifts. This result is suggestive of a distribution of vanadate tetrahedral
environments, possibly involving different numbers of non-bridging oxygen
atoms, within the surface layer. In addition, the NMR results give no evidence for
the formation of a bulk crystalline or amorphous $AlVO_4$ phase. Upon increasing
the vanadia content beyond 5 wt.%, a new downfield signal component around
-330 ppm emerges. The contribution of this signal increases monotonically with
increasing vanadia content, suggesting the appearance of a distinct species with
a different V(V) coordination. Both wideline and magic-angle spinning spectra
suggest that the short range order structure for this species is similar to that of
V_2O_5. Figure 4 illustrates, along with the wideline NMR spectra of some model
compounds, the influence of the support on the nature of the surface vanadia
species. In the TiO_2 supported catalysts studied, the V_2O_5 like species seems
more prominent in comparison to Al_2O_3 supported catalysts with comparable

464

coverages. The spectrum of V_2O_5 supported on ZrO_2 with a low surface area closely resembles that of crystalline V_2O_5 (Figure 4, bottom).

ACKNOWLEDGMENTS:

We would like to thank J. Johnson (Exxon) for synthesizing the $AlVO_4$ reference compound. Part of this research was conducted at the Southern California Regional NMR Facility at Caltech, supported by NSF grant number 84-40137. We thank Dr. James P. Yesinowski for helpful discussions.

REFERENCES:

[1] I.E. Wachs, R.Y. Saleh, S.S. Chan, and C.C. Chersich, Appl. Catal. 15, 339 (1985) and references therein.
[2] R.Y. Saleh and I.E. Wachs, Appl. Catal. 31, 87 (1987)
[3] R.Y. Saleh, I.E. Wachs, S.S. Chan, and C.C. Chersich, J. Catal. 98, 102 (1986).
[4] I.E. Wachs, S. D. Hardcastle, and S.S. Chan, this volume
[5] V.A. Gubanov, N.I. Lazukova, and R. N. Pletnev, Zh. Neorg. Khim. 23, 655 (1978) [Russ. J. Inorg. Chem. 23, 362 (1978)]
[6] R.N. Pletnev, N.I.Lazukova, and V.A. Gubanov, Zh. Fiz. Khim. 51, 2359 (1977) [Russ. J. Phys. Chem. 51, 1382 (1977)]
[7] V.A. Gubanov, R.N. Pletnev, V.N. Lisson, and A.K. Chirkov, Spectrosc. Lett. 10, 527 (1977).
[8] S. Ganapathy, S. Schramm, and E. Oldfield, J. Chem. Phys. 77, 4360 (1982).
[9] E. Oldfield, R.A. Kinsey, B. Montez, T. Ray, and K.A. Smith, J. Chem. Soc. Chem. Comm. 1982, 254
[10] S.L. Segel and R.B. Creel, Can. J. Phys. 48, 2673 (1970)
[11] V.M. Mastikhin, O.B. Lapina, V.N. Krasilnikov, and A.A. Ivakin, React. Kinet. Catal. Lett. 24, 119 (1984).
[12] V.M. Mastikin, O.B. Lapina, and L.G. Simonova, React. Kinet. Catal. Lett. 24, 127 (1984).
[13] T.P. Gorshkova, R.I. Maksimovskaya, D.V. Tarasova, N.N. Chumachenko, and T.A. Nikoro, React. Kinet. Catal. Lett. 24, 107 (1984).
[14] M.H. Cohen and F. Reif, in Solid State Physics, edited by F. Seitz and D. Turnbull, Vol. 5 (Academic Press, New York, 1957), p. 321.
[15] J.F. Baugher, P.C. Taylor, T. Oja, and P.J. Bray, J. Chem. Phys. 50, 4914 (1969).
[16] A. Bystrom, K.A. Wilhelmi, and A. Brotzen, Acta Chem. Scand. 4, 1119 (1950).
[17] H.G. Bachmann, F.R. Ahmed, and W.H. Barnes, Z. Krist. 115, 110 (1961)
[18] B.D. Jordan and C. Calvo, Can. J. Chem. 52, 2701 (1974).
[19] S.D. Gornostansky and C.V. Stager, J. Chem. Phys. 46, 4959 (1967).
[20] H. Eckert and I.E. Wachs, to be published.
[21] F. Maruma, M. Isobe, and S. Iwai, Acta Crystallogr. B30, 1628 (1974).
[22] H.T. Evans, Jr., Z. Krist. 114, 257 (1960).
[23] A.W. Sleight, H.Y. Chen, and A.Ferretti, Mat. Res. Bull 14, 1571 (1979).
[24] J. Trotter and H.W. Barnes, Can Miner. 6, 161 (1957-61).
[25] E.J. Baran and I.L.Botto, Monatsh. Chem. 108, 311 (1977).
[26] B.Robertson and E. Kostiner, J. Solid State Chem. 4, 29 (1972).

ATOMIC FORCE MICROSCOPY-FEASIBILITY FOR MATERIALS RESEARCH.

P.J. BRYANT, R. YANG and R. MILLER, University of Missouri-Kansas City; Physics Dept.; 1110 E. 48th; K. C., MO 64110.

ABSTRACT

The applicability of atomic force microscopy (AFM) to both conductive and dielectric materials is the subject of this study. A representative conductor, Cu, and two dielectrics, mica and selenite, were examined. Microstructure and single lamellar steps were resolved. Surface areas on Cu and mica generated reproducible images when scanned repeatedly. There was no evidence of damage to the probe or the sample as a result of the AFM investigations. Selenite did show evidence of change after repeated scans with an AFM lever of 12 N/m spring constant exerting a 10^{-6}N force.

ATOMIC FORCE MICROSCOPY

The atomic force microscope (AFM) was proposed in 1986 [1] based upon the technique of scanning tunneling microscopy (STM) [2]. Other versions of the force microscope have been developed recently [3, 4]. Atomic resolution has been demonstrated by the version utilizing the high sensitivity of a tunnel gap control mechanism [5].

The microscope design reported here is based on STM technology. It utilizes short range contact forces between a small stylus and a sample surface to produce high resolution images of defects and structural features of the surface. An adjacent tunnel gap control detects contact forces and provides a very sensitive feedback signal for a Z-direction piezoelectric drive to maintain a constant force value during raster scanning, see Fig. 1. This AFM adapter was attached directly to a NanoScope to convert it to AFM operation. Thus, with this new AFM attachment an STM may serve as an AFM. The same piezoelectric drive mechanism, electronics, and recording system was automatically employed to produce AFM images.

The feasibility of utilizing AFM for materials research is demonstrated here for both conductive and dielectric materials. Single lamellar steps on mica crystalline surfaces were well resolved by 2-D line scans earlier [6]. In this report 3-D scans of individual lamellae and structural features of surfaces are imaged.

Lamellar materials were useful for evaluation experiments, because they provided atomically smooth planes with single layer steps. The relatively large area planes were ideal surfaces for atomic resolution studies of crystalline structure and possible reconstruction. The known lamellar step dimensions were employed to evaluate the accuracy of the images produced by this new microscope facility.

Atomic Force Microscope Design

The version of an AFM reported here is of the stylus contact type utilizing a tunnel gap controller. The design of the AFM attachment is illustrated in Fig. 1. When operating as an AFM the small stylus type tip performs a raster scan of the sample surface while maintaining a constant contact force. The deflection of the lever multiplied by the spring constant of the lever determines the force value. The Z-direction motions of the assembly are controlled by a piezoelectric drive and an electronic feedback circuit. The current set-point regulates the tunnel gap and the contact force. The lever has high strength in the x and y directions

Mat. Res. Soc. Symp. Proc. Vol. 111. ©1988 Materials Research Society

Fig. 1. Diagram of AFM equipment, the stylus supported by a flexible beam utilizes a senstive tunnel gap (G) to image the surface profile of a material (M). The piezoelectric drivers (P) move the AFM unit in 3-D. The electronics package (E) contains the feedback circuit. Data is recorded by a storage oscilloscope (C).

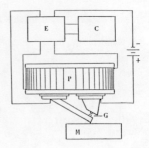

relative to the z-direction deflection motions so that the tunnel gap maintains a constant geometry and sensitivity.

This design does not utilize a modulated tunnel gap nor a vibrating beam and probe as noncontacting atomic force microscope designs. The use of contact forces between the probe and a sample is similar to the design of a stylus profilometer. However, the small tips, atomic scale contact areas, and the tunnel gap regulation of contact forces, provide this profilometer with exceptional resolution.

A standard STM unit was readily converted to AFM operation by adding a small AFM lever to the regular STM probe. Microsize force and energy recording cells were designed and tested, including: thin diaphrams, cantilevered fibers, evaporated films, and crossed wires. Contact force measurements, evaluation experiments, and single layer line scans of lamellar material surfaces, will be presented.

Two AFM lever designs were employed for the research described here. One was fabricated from platinum with the following dimensions: thickness (h) 14µm, width (w) 96µm, and length (L) 940µm. The spring constant (K) was calculated from the standard relations which depend directly on the beam thickness cubed and inversely on length cubed. The value 147 GPa for the modulus of Pt yields a spring constant of approximately 12 N/m for this lever. The moment of inertia of the lever in the bending mode was estimated to be 8.0×10^{-15} kgm^2. This parameter is relevant to the ability of the lever to follow the contours of the sample at high scan speeds. Fast scan rates are desirable to minimize the effects of thermal drift and vibrations. A high value of the natural resonant frequency for the lever is also important. The calculated value for this lever is 7.0 kHz. The second lever design consisted of platinum thin films enclosing a honeycomb core of low density carbon particles. The advantage of this composite lever design was the low value of the moment of inertia. The thickness of the honeycomb core maintained the spring constant level while reducing the inertia significantly. Thus the composite lever can follow the geometric contours of the sample accurately while moving at high scan rates.

Interpretation of AFM Patterns

The principle of operation of an AFM is more complex than that of an STM but the data reduction is simpler. For the stylus contacting design reported here, the data is a record of the electronic signal to the piezoelectric driver to maintain a constant strain in the lever. The strain of a lever in contact responds to geometric factors. Assuming that the stylus geometry remains constant, variations of the data signal should represent geometric features of the sample. To check for other variable factors it is imparative to perform multiple scans and record repetitive images of the same sample area.

The AFM images presented here are typical of repetitive scans. The lamellar materials studied were chosen because they are known to possess atomically smooth surfaces and uniform lamellar steps. These surfaces do not reconstruct. However, AFM is in the early development stages and the interpretations of data are subject to question.

The resolution of an STM or an AFM is tip dependent. The geometry of the tip is an important factor, especially when the tip, acting as a stylus, operates in contact with the sample. Several hypotheses have been proposed to describe the roll of the tip in the formation of STM images. The effect of multiple tips, sample additions to the tip, and foreign material in conjunction with the tip, have been described [7].

Another process, named the Self Imaging Distortion Effect (SIDE), is important for STM image interpretations and especially for a stylus type AFM operating at the level of atomic resolution. This effect consists of a distortion of the final image caused by the atomic contours of the probe on each of the sides as they interact with the asperities of the sample to produce a self image of the sides of the probe itself.

APPLICABILITY OF AFM TO MATERIALS RSEARCH

The STM operates only on samples which can support a tunnel current. The AFM design described here operates equally well on electrically conducting or nonconducting materials. The universal applicability of this AFM design results from the use of short range repulsive forces to image various atomic species. As a demonstration of applicability the following metallic and dielectric materials were examined by AFM: Cu, mica, and selenite. AFM proved to be applicable to each of these test materials. Individual characteristics of surface structure are shown for each material: Cu in Fig. 2, mica in Fig. 3, and selenite in Fig. 4.

(a)   (b)

Fig. 2. (a) Scanning tunneling microscope image of copper (100) etched surface, the scan dimensions are 6000 x 6000 Å, the largest step height seen is 100 Å. (b) AFM image of the same copper surface, the scan dimensions are 1500 x 1500 Å, the step height is 60 Å.

468

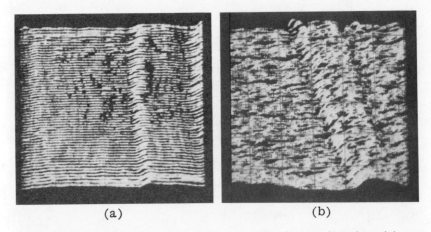

(a) (b)

Fig. 3. AFM images of lamellar planes and steps of muscovite mica: (a) scan dimensions are 400 x 400 Å, step height is 100 Å; (b) scan dimensions are 40 x 60 Å, step height is 10 Å.

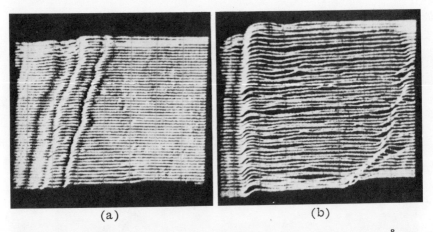

(a) (b)

Fig. 4. AFM images of selenite: (a) scan dimensions are 6000 x 6000 Å, the step height is 100 Å; (b) scan dimensions are 6000 x 6000 Å, the step height is 200 Å.

EXPERIMENTAL RESULTS

AFM was demonstrated first on two nonconductive materials: muscovite mica, $KAl_2(AlSi_3O_{10})(OH)_2$; and selenite, $(CaSO_4)(2H_2O)$. These two lamellar minerals have different interlayer binding forces: mica layers with ionic bonding from interlayer potassium ions; and selenite with hydrogen bonds between layers. Thus AFM with a contact stylus was tested on materials with a range of hardness from 110 for mica, to 32 on the Knoop Scale for selenite and a range of surface energy values, in air, from 700 ergs/cm^2 for mica [8] to 358 ergs/cm^2 for selenite [9]. The strength parameters of the test samples are important for evaluation of the stylus contacting type microscope being utilized here.

Lamellar solids were chosen because their easy cleavage planes should yield atomically smooth surfaces and known layer step dimensions. They are ideal samples for the evaluation of a new microscopy technology such as AFM.

Repetitive scans over the surface of selenite with an AFM, utilizing the relatively high spring constant lever, K = 12 N/m were successful, as shown in Fig. 4. After about ten or twenty repetitive images were formed the surface showed evidence of physical modification. Lamellar steps were eventually modified by the relatively high contact force of the stylus, estimated at about 10^{-6}N.

CONCLUSIONS

The applicability of atomic force microscopy has been tested with the following results:
1. The AFM design tested here is applicable to both conductive and non-conductive materials.
2. AFM data is representative of the geometric structure of a surface as evidenced by the correct profiles of layer step heights and features recorded from lamellar materials.
3. AFM is capable of imaging a sample at the rate of 20 lines per sec, with x and y image dimensions from 0.6 to 600nm.
4. Line scan images and grey scale AFM images of sample surfaces may be recorded, in real time, by a storage oscilloscope.

ACKNOWLEDGMENTS

This research was supported by the McDonnell Douglas Corporation Independent Research and Development Program.

REFERENCES

1. G. Binnig, C.F. Quate & Ch. Gerber, Phys. Rev. Lett. **56**, 930 (1986).
2. P.K. Hansma and J Tersoff, J. Appl. Phys., **61** (2) R1 (1987).
3. Y. Martin, et. al., J. Appl. Phys., **61** (10), 15 May 1987.
4. G.M. McClelland, R. Erlandsson and S. Chiang, Rev. Progr.in Quant. Non-Destrc. Eval., **6**, 1307 (1987).
5. G. Binnig, Ch. Gerber, E. Stoll, T.R. Albrecht and C.F. Quate, Europhysics Letters, **3** (12) 1281 (1987).
6. R. Yang, R. Miller and P.J. Bryant, J. Appl. Phys., Jan. 1988.
7. R.J. Colton, S.M. Baker, R.J. Driscoll, M.G. Youngquist and J.D. Baldeschwieler, J. Vac. Sci. and Tech., Mar./Apr. 1988.
8. P.L. Gutshall, P.J. Bryant and G.M. Cole, The Amer. Mineralogist, **55**, 1431, July/Aug., 1970.
9. M.L. Oglesby, P.L. Gutshall and J.M. Phillips, The Amer. Mineral., **61**, 295, Mar./Apr., 1976.

Author Index

Subject Index